STRUCTURE AND PROPERTIES OF ENGINEERING MATERIALS

FIFTH EDITION

MCGRAW-HILL SERIES IN MATERIALS SCIENCE

Barsoum
Fundamentals of Ceramics

Courtney
Mechanical Behavior of Materials

DeHoff
Thermodynamics in Materials Science

Dieter
Mechanical Metallurgy

Fontana
Corrosion Engineering

Gibson
Principles of Composite Material Mechanics

Henkel, Pense
Structure and Properties of Engineering Materials

Hyer
Stress Analysis of Fiber Reinforces Composite Materials

Kasap
Principles of Electrical Engineering Materials: Devices

Mahajan
Principles of Growth & Processing of Semiconductors

Schaffer, Saxena, Antolovich, Sanders, Warner
The Science and Design of Engineering Materials

Smith
Foundations of Materials Science and Engineering

Smith
Structure and Properties of Engineering Alloys

STRUCTURE AND PROPERTIES OF ENGINEERING MATERIALS

FIFTH EDITION

Daniel Henkel, PhD, PE

Alan W. Pense, PhD
Lehigh University

Boston Burr Ridge, IL Dubuque, IA Madison, WI New York San Francisco St. Louis
Bangkok Bogotá Caracas Kuala Lumpur Lisbon London Madrid Mexico City
Milan Montreal New Delhi Santiago Seoul Singapore Sydney Taipei Toronto

McGraw-Hill Higher Education

A Division of The **McGraw-Hill** *Companies*

STRUCTURE AND PROPERTIES OF ENGINEERING MATERIALS Published by McGraw-Hill, an imprint of The McGraw-Hill Companies, Inc. 1221 Avenue of the Americas, New York, NY, 10020.

This book is printed on acid-free paper.

1 2 3 4 5 6 7 8 9 0 FGR/FGR 0 9 8 7 6 5 4 3 2 1

ISBN 0-07-235072-5

President of McGraw-Hill Science/Engineering/Math: *Kevin Kane*
Publisher: *Thomas Casson*
Senior sponsoring editor: *Jonathan Plant*
Developmental editor: *Kristen Druffner*
Senior marketing manager: *John T. Wannemacher*
Project manager: *Laura Ward Majersky*
Senior production supervisor: *Michael R. McCormick*
Freelance design coordinator: *Mary L. Christianson*
Freelance cover designer: *Tin Box Studio*
Photo research coordinator: *Judy Kausal*
Senior producer/developer, Media technology: *Phil Meek*
Compositor: *York Graphic Services, Inc.*
Typeface: *10/12 Times Roman*
Printer: *Quebecor World Fairfield Inc.*

Library of Congress Cataloging-in-Publication Data

Henkel, Daniel P.
Structures and properties of engineering materials/Daniel P. Henkel, Alan W. Pense.–5th. ed.
p. cm.
Includes bibliographical references and index.
ISBN 0-07-235072-5
1. Materials. I. Pense, Alan W. II. Title.

TA403 .H397 2002
620.1'1–dc21

00-066435

www.mhhe.com

To my wife, Julie, and my children, Evan and Luke.
Dan Henkel

CONTENTS

SECTION 3

Metallic Materials Engineering 188

PREFACE

The previous fourth edition responded to the ever-expanding field of materials science with the addition of new chapters on engineering polymers and ceramics. At the time, it examined what were considered novel subjects such as directional solidification and splat cooling. However, twenty-three years have passed since the writing of that edition and substantial modernization was clearly overdue. The original premise of the book has not been changed in the fifth edition. It is still intended to be a textbook for intermediate level materials science courses, which would then follow the practicing engineer as an important reference on basic principles and practical information.

In general, the book gives an engineering perspective on physical metallurgy with some expansion into polymer science, ceramics and composites. An understanding of the contents will enable the reader to reason out many material selection issues and understand unusual responses to unique environments. The first two sections begin with discussions of atomic level structures and progress through the five fundamental strengthening mechanisms. The final section then provides what is considered the most useful information available on thirteen different classes of materials. It is an unusual combination of reference material in a language that a nonmetallurgical engineer should be able to comprehend.

Consideration was given to the placement of the important chapter on material deterioration, which was previously just an appendix. It does not belong in either the sections on strengthening or specific materials. Therefore, it was decided to place this review of the breakup of structure immediately after the introduction to structure. There was also some needed expansion on the topics of creep and fatigue. A more traditional arrangement of material classes is used than in previous editions with ferrous alloys, followed by nonferrous alloys, followed by nonmetallic materials. To maintain a similar length book, some subjects were reluctantly removed such as beryllium, steelmaking processes, and hardenability while new topics include coatings, composites, and aluminum-lithium alloys.

The book is intended to be a bridge between a theoretical text and an engineering handbook. There is an emphasis on phase transformations and the effects of processing variables that enable the prediction of both structure and properties. Property relationships are presented, usually without derivation, but with enough description to make them useful in actual practice. Many new references have been included on each subject to direct the reader to the theoretical sources. Updated problems are provided to stimulate some discussion and review of the text but they are not as extensive as those that may appear in an introductory materials science book. A final point of modernization involved replacement of most of the English units with the metric system.

We want to gratefully acknowledge the many contributors noted in previous editions that made this book a collection of more than fifty years of practical knowledge. Finally, Dr. Henkel wants to express his appreciation to Julie Klement Henkel for the motivation to pursue the tough projects and for her continuous inspiration.

Daniel P. Henkel Alan W. Pense

ABOUT THE AUTHORS

DANIEL P. HENKEL

Dr. Dan Henkel has twenty-five years of combined experience as a professional metallurgical engineer, materials research scientist, and college professor. He is presently the Senior Manager for Materials Development at Pall Corporation in Cortland, New York. Dr. Henkel received both his masters in metallurgy and material engineering and his doctorate in materials science from Lehigh University, and he received his bachelors in electrical engineering from the Pennsylvania State University. He was formerly president of Henkel Metallurgical Technologies, which performed surface science research and provided materials consulting to industry and government. Before HMT, Dr. Henkel was a Senior Research Scientist at NASA Langley Research Center. His specialty is the relationship between microstructure and the physical/mechanical properties of materials. While at NASA, he was also an adjunct professor in the Applied Science Program at the College of William and Mary, teaching courses in advanced microscopies and microstructural characterization. He holds two state licenses in metallurgical engineering, has three U.S. Patents and a list of archival publications, and is a member of ASM International. He was the recipient of an NSF research scholarship, a NASA research training grant, and a Department of Energy research grant.

ALAN W. PENSE

Professor and Provost Emeritus, Lehigh University

Dr. Alan W. Pense has been a member of the Lehigh University faculty since 1960, serving as professor, department chair, dean, provost, and vice president of academic affairs. In 1987, he co-authored the proposal that led to the Advanced Technology Center for Large Structural Systems (ATLSS), an NSF-sponsored engineering research center at Lehigh. A specialist in physical and mechanical metallurgy and welding, Dr. Pense has been teaching and conducting published research for forty years. In addition to several prestigious teaching awards, he was elected as a fellow of ASM International in 1989 and the National Academy of Engineering in 1993. He is also listed in *American Men and Women of Science, Who's Who in America, Who's Who in the World,* and *Who's Who in Technology Today.* In 1997, he retired and became professor and provost emeritus. Today, Dr. Pense continues to advance the field of welding as an ATLSS research associate and a consultant to government and private industry.

SECTION 1

INTRODUCTORY MATERIALS CONCEPTS

chapter 1

Structure and Properties

1.1 ATOMIC PACKING

Think of atoms in a solid as hard, round spheres that exert attractive forces in all directions. Given a number of these hard, spherical bodies, how can they be arranged to be as close-packed as possible, i.e., to occupy a minimum volume? It is easiest to consider the two-dimensional case first. When a given number of atoms are fitted together in a hexagonal pattern like the cells of a honeycomb, this criterion is achieved and they cover a minimum area.

Because they have relatively high densities, metals must consist of atoms that are packed very closely together as patterned in Fig. 1.1. It represents the closest possible packing of spheres on a flat sheet. To make a three-dimensional array of spheres occupying a minimum volume, it is necessary to place the centers of the atoms of the second sheet over the holes between the atoms of the first sheet. In Fig. 1.2, crosses show the locations of the centers of the atoms of the second sheet. Note that the atoms of the second sheet cover only half the holes in the first sheet. The other set of holes, the ones not marked by crosses, could have equally been used for the atom centers of the second sheet. The atomic structure of the two layers would look the same in either case.

Figure 1.1 Packing of equal-sized spheres on a single plane to occupy a minimum area.

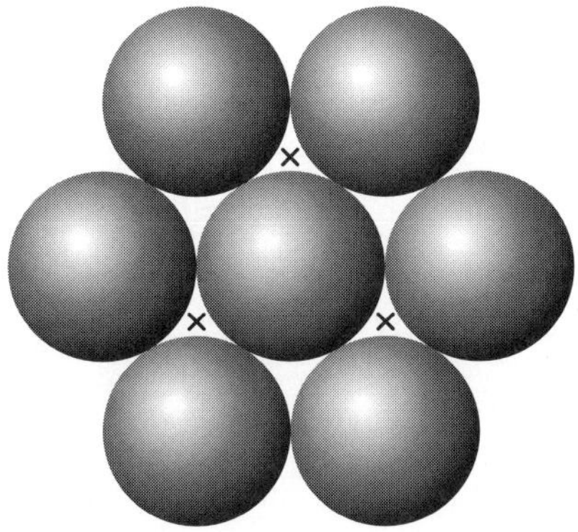

Figure 1.2 Location of the centers of the atoms in positions marked **x** of a second layer similar to Fig. 1.1.

The metal crystal that has been constructed by stacking up hard spheres is only two atomic layers thick. A third layer introduces a slight complication because there are two different ways in which it can be added. One way is to put the centers of the atoms of the third layer over the holes of the first layer that are not marked by crosses. In order to describe this structure, let the first layer be called *A*, the second layer, with its atom centers at the **x** positions, be called *B*, and the third layer, with its atom centers over the unmarked holes in *A*, be called *C*. The stacking sequence of close-packed layers is then *ABC* in this case. All the possible alternative positions for close-packed layers have been used up in this stack of three layers of atoms, but the fourth layer can be added in the *A* position again. In this way a sequence of layers *ABCABCA* can be built up to any desired thickness. The result is a three-dimensional array of spheres packed as closely as possible. Figure 1.3 shows a plane view of this structure.

There is a second way in which the close-packed layers of Fig. 1.1 can be stacked up to make a close-packed crystal. The first two layers are stacked as in Fig. 1.2, but the third layer is now added so that it is directly over the first. This structure is obviously close-packed, like the one derived above, but the stacking sequence is *ABABA*. Many metals are found to have either the *ABCABCA* or the *ABABA* type of close-packed structure. Some metals have more complicated structures; these will be described later.

For many purposes it is convenient to think of the close-packed structures as stacks of close-packed layers according to the description given above. Sometimes, however, it is convenient to single out a small group of atoms in the stack and then describe the atom arrangement in this group. The group of atoms chosen for this purpose is called a unit cell of the structure. One can choose a number of equally accurate ways of representing the unit cell, but experience has shown that, for each structure, there is one cell which is most easily visualized. This one best shows the symmetry of the atom arrangement.

For the close-packed structure of the *ABCABCA* type, the group of atoms forming the unit cell is shown in Fig. 1.4*b*. Note that the close-packed layers are in an

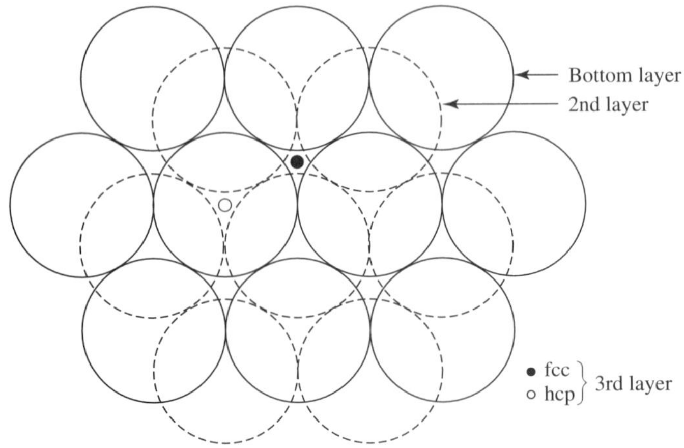

Figure 1.3 Solidly outlined circles represent a bottom plane of atoms, and dash-outlined circles represent a second plane nested on the first. In a close-packed structure, the third layer can be centered at the position of either the closed dot (*C* layer) or the open dot (another *A* layer).

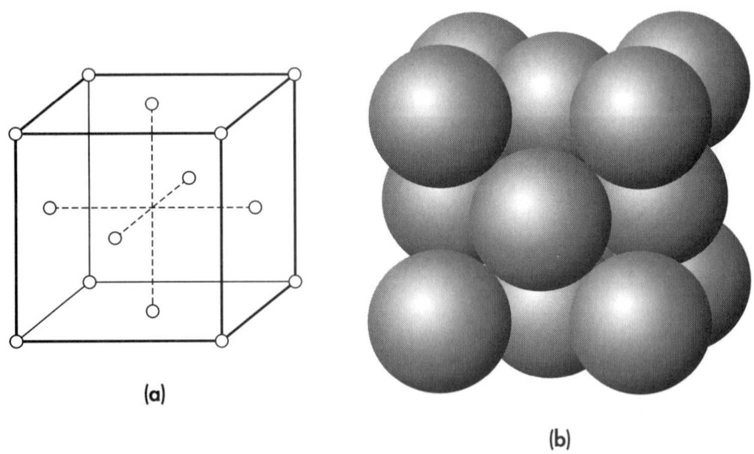

Figure 1.4 (*a*) The positions of the centers of the atoms in one unit cell of the fcc structure. (*b*) The packing of atoms in a unit cell of the fcc structure. Each atom on a face center touches each of the nearest corner atoms as well as the face-centered atoms in front and in back of it, making 12 nearest neighbors. There are four atoms in this unit cell (one-half of each face-centered atom plus one-eighth of each corner atom).

inclined orientation in this drawing. Figure 1.4*a* shows a schematic representation of the location of the atoms in this structure. In this representation, the cubic symmetry of the atomic arrangement is evident. Because of the arrangement of atoms in the unit cell, the *ABCABCA* type of close-packed structure is given the name face-centered cubic structure (usually abbreviated fcc). Typical fcc metals are copper, aluminum, and nickel.

The group of atoms forming the unit cell for the *ABABA* type of structure is shown in Fig. 1.5*b*. In this case the close-packed planes are horizontal and are easily recognized. The hexagonal symmetry of the atom arrangement can be seen from Fig. 1.5*a*. The structure is called hexagonal close-packed (hcp). Magnesium and zinc are examples of hcp metals.

If the hypothesis that the atoms of metals are hard, round spheres exerting attractive forces equally in all directions were true in all cases, then it would be expected that all metals would have either the fcc or the hcp structure. It is known from the study of chemistry, however, that many atoms do not attract other atoms equally in all directions but tend to form bonds preferentially in certain directions. Some metal atoms behave in this way, and the resulting crystal structures are not close-packed. One structure that deviates only slightly from being close-packed is the body-centered cubic (bcc) one shown in Fig. 1.6. Many of the important metals, including iron, chromium, and tungsten, have the bcc structure.

The structures of a number of metals deviate more markedly from close packing. Bismuth, antimony, and gallium have an atomic arrangement with the symmetry of a rhombohedron rather than that of a cube. These metals have so-called open structures, meaning that there is a substantial amount of empty space surrounding each atom, considered as a hard sphere. It is interesting in this connection that these

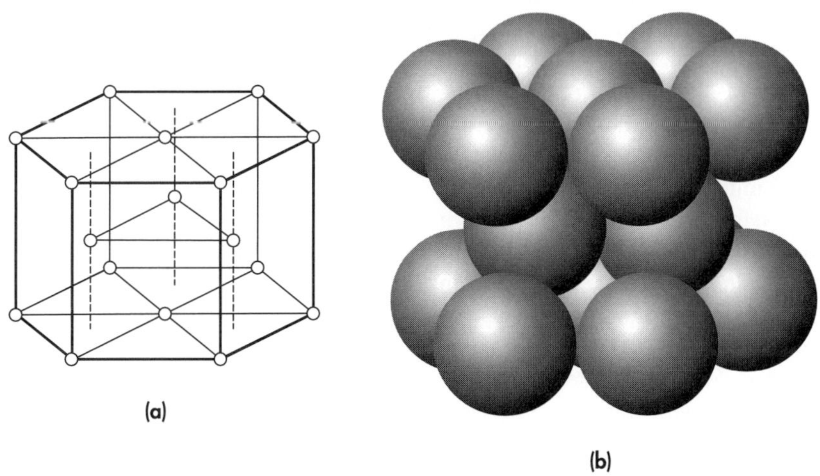

Figure 1.5 (*a*) The positions of the centers of the atoms in one unit of the hcp structure. (*b*) The packing of atoms in one unit cell of the hcp structure. As in the fcc structure, each atom has 12 nearest neighbors, with 6 on the same horizontal layer, 3 on the layer below, and 3 on the layer above.

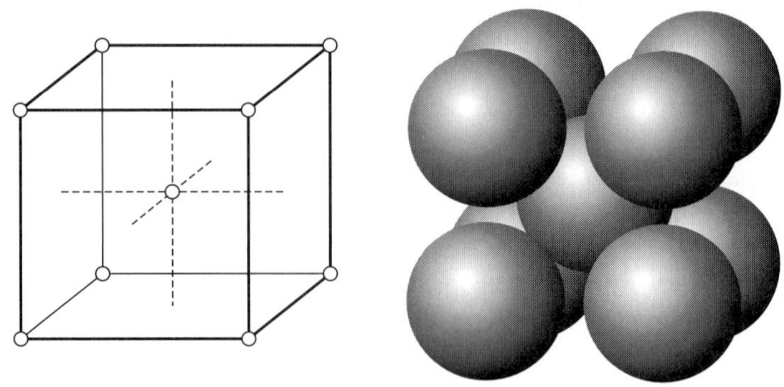

(a) (b)

Figure 1.6 (*a*) The positions of the centers of the atoms in one unit cell of the bcc structure. (*b*) The center atom touches each corner atom, but these do not touch each other. There are two atoms per unit cell (one centered atom plus one-eighth of each corner atom), and each atom has eight nearest neighbors.

metals decrease in volume when they melt, whereas most metals have a greater specific volume in the liquid state than in the solid state.

Atomic Size

To the extent that the atoms of a metal can be considered to be hard spheres, they can be assigned a definite diameter. The atomic diameter can be calculated when the crystal structure and lattice parameters of a metal are known. The atoms of a crystal structure are in contact along certain directions in the unit cell known as the close-packed directions. Inspection of Fig. 1.4 shows, for example, that the close-packed directions in the fcc structure are along the diagonals of the cube faces. If a is the lattice parameter, the length of a face diagonal is $a\sqrt{2}$ and the atomic diameter must be $\frac{1}{2}a\sqrt{2}$.

The hard-sphere model of the atoms of a metal is only a first approximation. It is found, in fact, that the atoms of a given metal may not always appear to have the same diameter. For example, iron atoms behave as if they had one diameter when in pure iron metal, another slightly different one when in an iron-nickel alloy, and still another in ferric chloride crystals. Nevertheless the concept of atomic diameter has proved to be a useful one in metallurgy and plays an important role in understanding the formation of alloys.

The atoms adjacent to a given atom along the close-packed directions of a crystal are called the nearest neighbors of the given atom. Each atom in the fcc structure has 12 nearest neighbors, as can be seen from Fig. 1.4. So does each atom in the hcp structure. But in the bcc structure each atom has only 8 nearest neighbors. The number of nearest neighbors found in a given crystal structure is said to be the coordination number of the structure. Thus, bcc iron has a coordination number of 8 and copper has a coordination number of 12.

1.2 CRYSTAL STRUCTURE

Lattice Parameters

The term lattice is frequently used in the description of crystal structures. A lattice is an array of atoms repeated regularly throughout space. If a point is placed at the center of each of the atoms in the fcc crystal structure, the array of points so generated is the fcc lattice. Lattices for the bcc and hcp structures can be generated in the same way.

An important characteristic for understanding alloying behavior of metals and ceramics is the lattice parameters. These are the lengths of any side of a unit cell of a given crystal structure. In the metals having cubic symmetry, the size of the lattice is fixed when the length of the edge of the cubic unit cell is given. Cubic metals have, therefore, only one lattice parameter. The lattice parameter of a metal can be measured by observing the diffraction of an x-ray beam passed through the metal. The results of such measurements are usually reported in angstroms, where $1\ \text{Å} = 10^{-8}$ cm. Typical values of the lattice parameters of cubic metals are Al, 4.04 Å, and Mo, 3.14 Å.

When the unit cell does not have cubic symmetry, more than one lattice parameter has to be specified. In hexagonal crystals, these are, first, the distance a between neighboring lattice points in the close-packed planes (or basal planes, as they are often called) and, second, the distance from the top to the bottom of the unit cell. If a hexagonal metal is truly close-packed, i.e., is made up of spherical atoms stacked *ABABA*, then the distances a and c must bear a fixed relation to each other. In fact the ratio $c/a = (8/3)^{1/2} = 1.633$ represents perfect close packing. The forces between the atoms in most metals that crystallize in the hexagonal structure are such that there are usually slight deviations from ideal close packing. In zinc, this deviation is unusually large, and the axial ratio c/a is 1.85. This affects the strain-hardening rate, which will be discussed later in this chapter.

Crystal Planes and Axes

The points of a crystal lattice define an array of crystal planes. Some of these planes are easily visualized. For example, the planes that mark out the unit cells of the fcc and bcc lattices shown in Figs. 1.4 and 1.6 constitute a set of crystal planes known as the cube planes. Each of these planes is repeated indefinitely throughout the lattice as an array of parallel planes separated by a distance equal to the lattice parameter in the simple cubic structure and to half the lattice parameter in the fcc and bcc structures. Some other important planes are shown in Fig. 1.7. Note that each set of crystal planes has its own characteristic interplanar distance.

Because a given type of crystal plane is repeated throughout the lattice, the location of a specific plane is not of much interest in crystallography. The orientation of a plane or a set of planes relative to the edges of the unit cell is important, however, in understanding deformation modes of the crystal. In order to specify this orientation, it

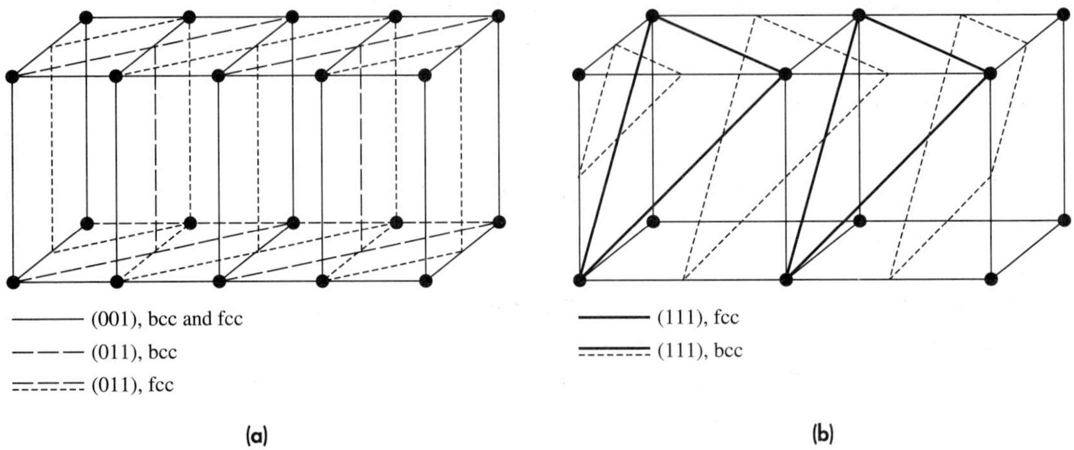

Figure 1.7 Two cells of a cubic lattice, showing only the corner atoms. (*a*) The cube-plane (001) spacing is the same for both the body-centered and face-centered structures. The dodecahedral-plane (011) spacing necessary to account for all atoms is twice as great for the body-centered as for the face-centered cube. (*b*) The reverse is true of the octahedral planes (111).

is convenient to establish, as a set of crystal axes, one set of the edges of one of the unit cells of the structure. For the fcc and bcc structures, then, the crystal axes are a set of rectangular cartesian coordinates. The intercepts of a crystal plane on these axes could be measured in units of centimeters or angstroms, but it proves to be more useful in crystallography to use the lattice parameters themselves as the units of measure on the crystallographic axes. Planes that are in the same orientation will then have the same intercepts in different crystals, even though the lattice parameters are different.

It has been found that six different kinds of crystal axes are needed to fit all the different possible kinds of crystal lattices. These six different sets of crystal axes are given in Table 1.1.

Table 1.1 The six crystal structures

	Axes		
Name	**Length of Unit Vectors**	**Axial Angles**	**Examples**
Cubic	$a = b = c$	$\alpha = \beta = \gamma = 90°$	
Tetragonal	$a = b \neq c$	$\alpha = \beta = \gamma = 90°$	Tin, indium
Hexagonal	$a = b \neq c$	$\alpha = \beta = 90°, \gamma = 120°$	Magnesium, zinc
Orthorhombic	$a \neq b \neq c$	$\alpha = \beta = \gamma = 90°$	Gallium, sulfur
Monoclinic	$a \neq b \neq c$	$\alpha = \gamma = 90° \neq \beta$	Phosphorus
Triclinic	$a \neq b \neq c$	$\alpha \neq \beta \neq \gamma$	Antimony, mercury

Allotropic Forms

Some metals exist in more than one crystalline form. Iron is bcc at temperatures ranging up to 910°C, but at this temperature it undergoes an allotropic transformation and becomes fcc. The fcc phase is stable to 1400°C, at which temperature it transforms back to the bcc structure, which it then retains up to the melting point. The different crystalline forms of a metal are said to be allotropic forms. Many metals such as aluminum and copper have only one crystal structure, but many others, particularly among the elements in the transition series of the periodic table, have two or more allotropic forms.

Allotropic transformations occur because in certain temperature ranges one structure is more stable than another is. It is found that the energy differences between the different structures of a metal that undergoes allotropic transformation are very small. Thus small changes in atomic forces can change the structure of a metal from one type of crystal to another. It is also found that in many cases allotropic transformations are due to magnetic interactions between the atoms, but in most cases theorists have not yet attained a full understanding of these forces.

Structure of Nonmetallic Solids

The crystal structures of the common metals, discussed above, are exceedingly simple in comparison with most of the structures found among the species of the mineral kingdom. The structures of some typical nonmetallic crystals are described below.

A number of different schemes can be used to classify substances when their structures are to be described. In mineralogy, for example, a classification based on chemical composition is generally used, and crystals are arranged according to whether they are elements, oxides, sulfides, silicates, etc. A classification based on the type of bonding forces acting between the atoms of the crystal will be used here. On this basis four main classes of crystals may be recognized. They are:

1. Metallic, already described above
2. Ionic, in which the interatomic forces are between electrically charged ions such as Na^+ and Cl^-
3. Covalent, in which the atoms are bound together with chemical covalent bonds
4. Molecular, which includes crystals made up of chemically saturated molecules held together by van der Waals forces

A few representative examples of the last three classes are given below.

Ionic Crystal Structures

One of the most common of the ionic crystal structures is that of sodium chloride (Fig. 1.8). This structure is based on the fcc lattice, but instead of having one atom

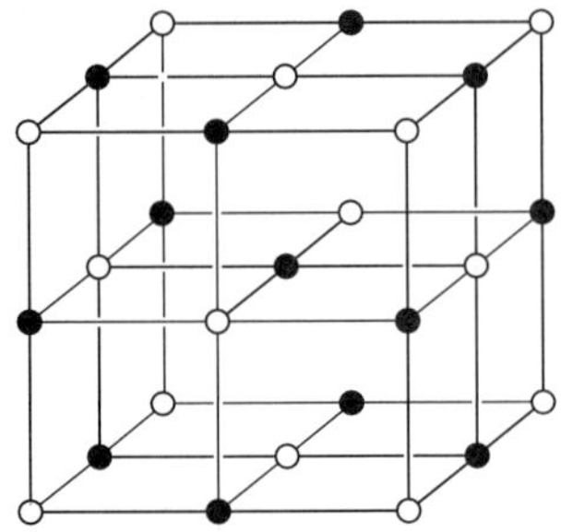

Figure 1.8 The crystal structure of rock salt, solid NaCl. Note that the close-packed directions here are cube edges rather than face diagonals.

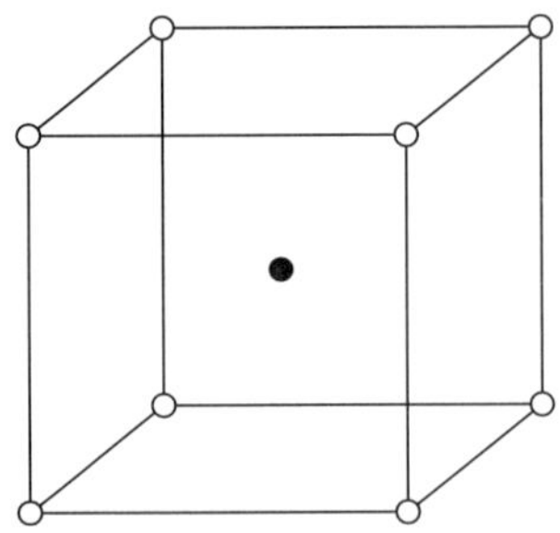

Figure 1.9 The crystal structure of cesium chloride, CsCl. The open circles are Cs ions and the closed circle is a Cl ion, or vice versa.

associated with each lattice point as in most metals, there are two atoms per lattice point, one centered on the point and another a distance ½a away along the edge of the fcc unit cell. The "atoms" in this case are actually ions, and of the pair associated with a lattice point, one will be Na^+ and the other Cl^-. Note that six negatively charged nearest neighbors surround each positive ion, and, conversely, each negative ion has six positive ions as nearest neighbors, as would be expected, since unlike charges attract.

Another simple ionic crystal structure is that of cesium chloride (Fig. 1.9). It may be derived from a simple cubic lattice by placing one ion on a lattice point and another, of opposite sign, halfway along the body diagonal of the unit cell. Thus an ion of one sign is surrounded by eight ions of the opposite sign.

In the NaCl and CsCl structures it is seen that there are no individual salt molecules present, that the entire crystal can be regarded as one gigantic molecule. In some ionic crystals there is more of a tendency to form molecules. In pyrite (Fig. 1.10) there are effectively FeS_2 molecules at each of the lattice points of an fcc lattice.

Covalent Crystal Structures

Diamond (Fig. 1.11) is a typical covalent (shared electrons) crystal that, as can be seen from the drawing, is based on an fcc lattice. The lines drawn between the atoms in the figure represent the covalent bonds, four for each carbon atom as in the molecule CH_4. In addition to diamond, the element silicon, germanium, and one

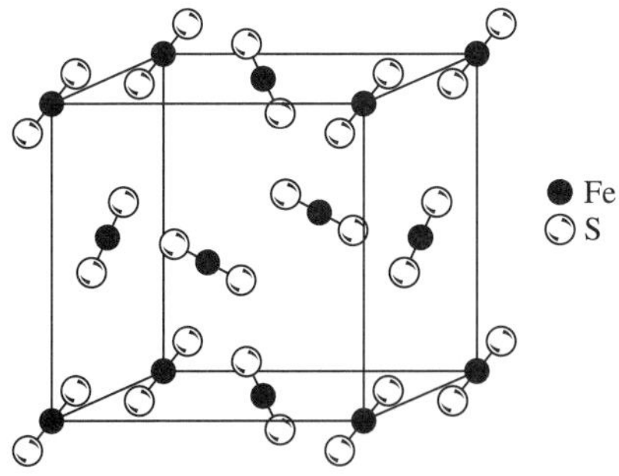

Figure 1.10 The structure of the mineral pyrite, FeS_2.

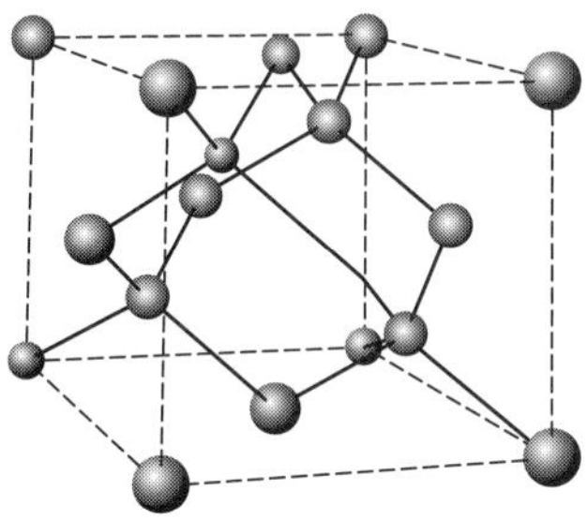

Figure 1.11 The crystal structure of diamonds, one of the allotropic forms of pure carbon.

allotropic form of tin have this diamond cubic structure. The network of covalent bonds extending throughout the structure is responsible for the great hardness of these elements.

The silicates represent a very important and very large class of covalent crystals. The basic unit from which these structures are built is the tetrahedral array of one Si and four O atoms shown in Fig. 1.12. Because these SiO_4^{4-} tetrahedra can be linked up in many different ways with each other and with various positive ions, a great variety of different silicate structures are possible. These are often rather complex, as is seen in Fig. 1.13. Silicate minerals constitute the great part of the lithosphere of the earth and so are of considerable interest to the geologist and mineralogist. They are also essential constituents of many of the refractories used in metallurgical furnaces.

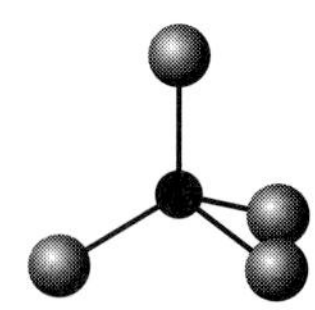

Figure 1.12 A tetrahedral array of four oxygen atoms and one silicon atom forms the group SiO_4^{4-}.

Amorphous and Polymer Structures

Substances such as glass and polymers are not crystalline. They have no regular, periodic arrangement of atoms. Most glasses consist primarily of silicate ions, SiO_2, to which an appreciable number of large-sized atoms such as Na and Ca have been added. The added atoms, since they do not fit into the silicate structure very well, make it more difficult for crystallization to occur when the melt is cooled. Glass is then a supercooled liquid having a very high viscosity. Glasses are discussed in more detail in Chap. 19.

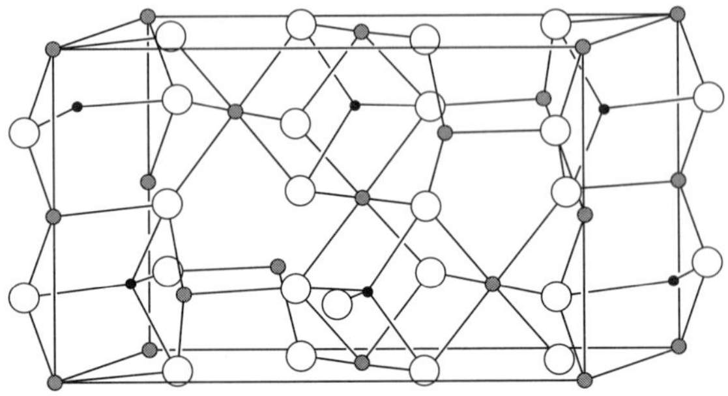

Figure 1.13 Crystal structure of olivene, Mg_2SiO_4.

Polymers, such as rubber or lucite, consist of very long organic molecules twisted up and intertwined with each other. Some polymers do, however, show a degree of crystallinity: in some regions of the polymer the long molecules are all lined up parallel to themselves so that, within this region, there is a fairly regular pattern of molecular arrangement. Such polymers are sometimes said to be crystalline, although they do not have a regular crystal structure in the usual sense of the term. The structure and properties of polymers are covered in Chap. 18.

1.3 GRAIN STRUCTURE

Metals are essentially always crystalline. Although there are amorphous metals, known as metglasses, there is no such thing as "molecules" of solid metals. The unit cell of a metal crystal repeats itself in three dimensions (with occasional imperfections) forming 10^3 to 10^8 atom diameters. For example, a rod of metal 2.5×20 cm, containing 10^{25} atoms, may be produced as a single crystal. However, most metals are polycrystalline, i.e., made up of many crystals, which typically will be from 10 to 100 micrometers (μm) in diameter. The individual crystals of a pure metal are identical except for the orientation of their crystal axes with respect to an external reference system. They are called grains, and their average diameter is termed the grain size. As will be seen in Chap. 4, the grain structure is one of the primary characteristics in determining a material's mechanical properties. Each individual grain is in atomic contact with all of its neighbors; there normally are no voids; see Micro. 1.1.

When two crystalline grains of different orientation with respect to an external reference meet in contact, it is evident that at the interface, atoms cannot match up perfectly with both crystal lattices. Between the two grains, there must be a transition layer where the structure is that of neither one grain nor the other. Because the atoms in this transition layer are not in their proper places with respect to either grain, the degree of disorder is higher and, therefore, the increased entropy at the grain boundary results in higher energy than the material in the neighboring grains.

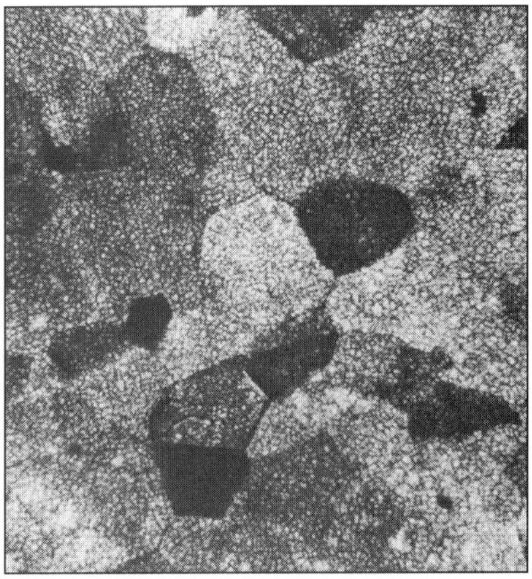

Micrograph 1.1 Transmission electron micrograph at ×26,000 magnification of vacuum-deposited aluminum film 0.5 μm (microns) thick. Grains appear different shades because of the difference in the number of electrons diffracted according to different orientations of crystals. The cellular networks within grains are dislocation subboundaries.

Exposed to a chemical etching solution, then, the grain boundary will be dissolved away more rapidly than the surrounding material and, in an etched microstructure, will appear as a groove in an otherwise flat surface. Even when the etching solution used does not differently color the differently oriented grains, the grain boundaries of a microstructure will be revealed as black lines.

Evidence available today indicates that the transition layer of irregular atomic structure in a grain boundary is only a few atom layers. It is quite permissible, then, to think of the boundary as a plane interface characterized by certain energy per unit area. Experimental measurements of grain-boundary energies show that, as long as the angular misorientation between a pair of grains is greater than about 10°(ten degrees), the grain-boundary energy is very nearly constant, independent of both the misorientation between the grains and the direction of the boundary. Furthermore, it can be shown statistically that the occurrence of boundaries of less than 10° misorientation is a very rare event in a polycrystalline material. To a rather good approximation, then, grain boundaries can be regarded as having a constant energy per unit area. This will usually be expressed in units of ergs per square centimeter or the equivalent units of dynes per centimeter. Thus the grain boundary, like a liquid-vapor interface, can be thought of as having a surface tension. It is this surface tension which is the principal factor determining the equilibrium (fully annealed) grain shape in polycrystalline materials.

If three grain boundaries meet, the surface tensions of the three grains must form a system of forces in equilibrium. Since the three boundaries have the same energy

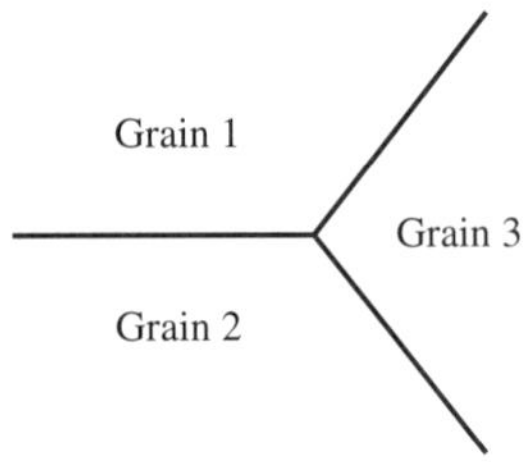

Figure 1.14 Intersection of three grain boundaries at angles of 120° to each other.

per unit area, and therefore the same surface tension, they will meet, as in Fig. 1.14, at equal angles of 120° to each other. If only two boundaries were to meet, it is clear that they could do so only in a straight line. The intersection of four boundaries is a relatively rare event. The shape of the polyhedron that has faces meeting at angles of 120° and edges at angles of 109°28′ (the angle between a pair of four lines meeting symmetrically at a point) is a tetrakaidecahedron. These can be stacked with other similar polyhedra so as completely to fill space as shown in Fig. 1.15. It is expected that the grains of polycrystalline metal will take up a similar shape under ideal conditions. "Ideal conditions" means, in this case, that the grain boundaries have had a chance to migrate to stable positions. The grain structure of pure metals often approximates the ideal of Fig. 1.15. One of the best ways to observe this is by the following experiment. A drop of liquid gallium is placed on the surface of a piece of coarse-grained aluminum, and the aluminum is then scratched with a sharp-pointed instrument under the gallium. After some minutes, the liquid gallium will penetrate the aluminum grain boundaries and the individual grains can be pulled out of the metal. They will be polygons approximating the shapes of those in Fig. 1.15.

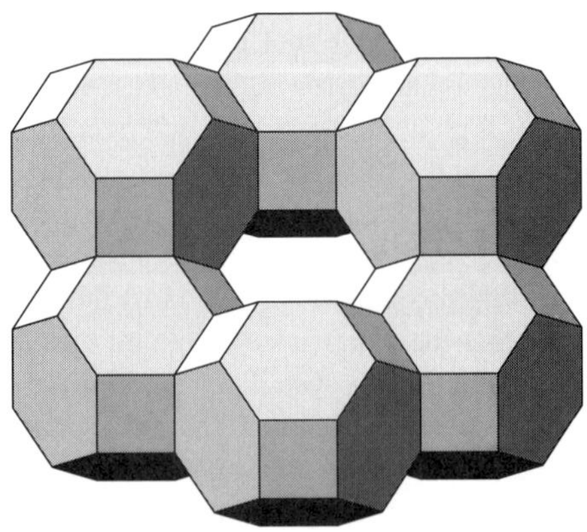

Figure 1.15 A stack of tetrakaidecahedra, showing how metal grains having these polyhedral shapes can completely fill space. It minimizes the high-energy line boundaries, i.e., junctions of three grain boundaries at or approaching 120° to each other, and also minimizes point junctions so that only four line boundaries meet at a point with angles to each other approaching the minimum energy condition of 109°.

A microsection of a metal having ideal grain shape does not reveal quite as much regularity as might be expected from the above discussion. This is because the plane of the section passes through the stack of polyhedra in a random orientation, cutting through large parts of some grains and small parts of others. Micrograph 1.1 does show clearly, however, that almost all grain-boundary intersections involve just three boundaries.

In addition to grain shape, which has a significant influence on strength, the grain size in a sample is often important. Grain size can range from large enough to be easily seen by the naked eye to so small as to be barely resolvable with the most powerful optical microscope.

Grain size can be measured from a micrograph taken at known magnification by finding the number of grains per unit area. The ASTM grain-size number N is defined to be such that 2^{N-1} equals the number of grains per square inch at linear magnification of 100 times. Table 1.2 compares the ASTM grain-size numbers with other measures of grain size that are sometimes used. The ASTM has published a number of standard grain-size comparison charts to facilitate the rapid estimation of grain sizes from direct observation through the microscope.

Table 1.2 Comparative systems of reporting grain size

ASTM	Grains/in^2	Average Grain	
No.	**At ×100***	**Grains/mm^2**	**Diameter, mm**
−3	0.06	1	1.00
−2	0.12	2	0.75
−1	0.25	4	0.50
0	0.5	8	0.35
1	1	16	0.25
2	2	32	0.18
3	4	64	0.125
4	8	128	0.091
5	16	256	0.062
6	32	512	0.044
7	64	1,024	0.032
8	128	2,048	0.022
9	256	4,096	0.016
10	512	8,200	0.011
11	1024	16,400	0.008
12	2048	32,800	0.006

* For ×50, report two ASTM numbers lower; for ×200, report two ASTM numbers higher.

1.4 MECHANICAL PROPERTIES AND TESTING

Theory of Mechanical Properties

For purposes of discussion, it is convenient to classify mechanical properties as elastic or plastic and as structure-sensitive or structure-insensitive. Elastic properties are the appropriate elastic constants, while plastic properties include the strength properties and the creep, fatigue, and fracture characteristics (since fracture, except in ideally brittle materials, always involves some plastic flow). Structure-sensitive properties are those that depend in a sensitive way on the extent and distribution of atomic-scale imperfections in the material and so can vary widely from sample to sample of the same material. The yield strength of a pure metal depends, for example, on the past thermal and mechanical history of the particular sample tested and can vary by 100 percent or more from sample to sample. Structure-insensitive properties such as the elastic and optical constants of a metal are pretty much the same from sample to sample.

Elastic Properties The elastic constants of a crystal are essentially a measure of the force required to displace the atoms of the crystal relative to one another and so are directly related to the binding forces between the atoms. Figure 1.16 shows how the energy per atom varies as a function of the interatomic distance. The equilibrium interatomic distance is that corresponding to the minimum in the curve and so would be at point a. Applying a tensile stress to the crystal along the x direction causes the interatomic distance to increase and the energy of the crystal to rise. It can be shown that the elastic modulus for strain in the x direction is directly proportional to the curvature of the E-x curve at its minimum.

Because the elastic constants are so closely related to the interatomic forces, very little can be done to alter them in any given material. They can be moderately changed by alloying but are relatively insensitive to cold work, irradiation, or other treatments that can be applied to a material.

Amorphous materials are elastically isotropic: the Young's modulus and shear modulus are the same, regardless of the direction in which the material is stressed. Even cubic crystals, however, are elastically anisotropic. Young's modulus in copper,

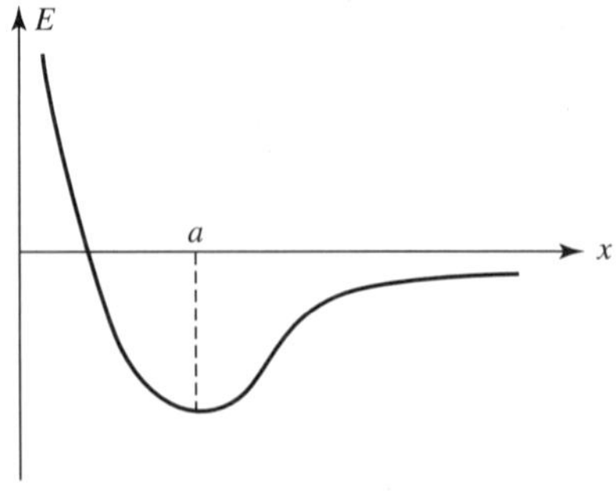

Figure 1.16 An energy-vs.-displacement curve for expansion or contraction of a crystal along a direction x. Point a corresponds to the equilibrium interatomic distance along the x-axis.

for example, varies from 19,400 kg/mm^2 in the [111] direction to 6800 kg/mm^2 in the [100] direction. While an isotropic material has two independent elastic constants, a cubic crystal has three, and a calculation of the Young's modulus in an arbitrary crystallographic direction requires use of all three constants. A polycrystalline metal, when the grains are at random orientations with respect to each other, behaves as if it were an isotropic elastic material with elastic constants that are averages of those of the individual grains.

Plastic Properties The plastic properties of crystals are structure-sensitive, i.e., dependent on the imperfections present in a particular sample of material as well as on the characteristics of the perfect crystal. They are discussed in detail in Chap. 4 but are considered here as they relate to mechanical testing.

Mechanical Property Tests The mechanical properties of a material determine its response to the application of stress. Stresses may be in the tensile or shear mode and may be applied to a material singly or in combination. The description of the response of a material to a system of stresses can be an exceedingly complex problem in engineering mechanics. In order to study the properties of materials, the materials engineer relies on tests in which the stress pattern applied to a specimen is purposely made simple, as in the tensile test. There is, of course, a danger here in that the results obtained under the idealized conditions of the laboratory may not be applicable in a real situation where the stress pattern is not simple. Many spectacular failures of engineering structures have resulted from inadequate attention to just this problem. Nevertheless, idealized laboratory tests remain the starting point in any study of mechanical properties.

The Tensile Test For over 100 years, the breaking of tensile test specimens has been a primary source of information on the mechanical properties of materials, and it is not likely that this method of testing will be supplanted for a long time to come. Tensile tests are considered to have an immediate practical value, but the interpretation of tensile test data is not so straightforward. The engineer must make knowledgeable decisions on the application of tensile data, acquired under laboratory conditions, to real service conditions where additional factors such as temperature and corrosion may enter the system.

Testing Machines and Tensile Specimens The principal problems that must be solved to perform a successful tensile test are the following:

1. The specimen must be held firmly in the grips of the testing machine, without slippage as the load is applied. During loading, deformation must be confined to a known part of the specimen and, in particular, must not occur in the grips. This can be accomplished by machining a gage length on the specimen such that the cross-sectional area in the gage length is substantially less than that in the grips. It is important in designing and making tensile specimens to avoid sharp corners or deep

scratches on or near the gage length, as these can give rise to stress concentrations that may result in the premature failure of the specimen.

2. The state of stress in the specimen must be determined. Usually this is done by measuring the load on the specimen and assuming that this load gives rise to a uniform tensile stress in the specimen whose magnitude is the load divided by the cross-sectional area in the gage length.

3. The measurement of how much the specimen has been strained is the most difficult part of the tensile test. If the specimen deforms homogeneously in its gage length, does not slip in the grips, and does not deform in any places other than the gage length, then the strain can be figured as the total elongation (measured as the crosshead displacement) divided by the gage length. Because of spring and slack in the grip and crosshead assemblies of most tensile machines, the above-mentioned conditions are not often realized, and accurate strain measurements are hardly ever achieved by the above methods except when the strain is very large. Accurate determinations of strain can be made using strain gages cemented directly on the specimen. The strain gage is essentially a piece of wire for which the change of resistance for a given elongation is accurately known. The amount of strain in the specimen is observed by following the increase in resistance of the wire during the tensile test. Frequently, a strain-gage bridge can be attached to the specimen instead of the cemented gage.

Some tensile-test machines are arranged to automatically record on a chart the elongation observed from a strain gage and the load as obtained from a load cell. These machines, then, automatically plot a load-elongation curve as the test proceeds.

Test Results A typical load-elongation curve for a pure metal is shown in Fig. 1.17. The initial straight-line part of the curve running up to point *A* results from the elastic elongation of the specimen. If, at any point up to *A*, the load is taken off the specimen, it will return elastically to its original length. After point *A* is passed, however, it does not return to its original dimensions. Continuing the loading of the specimen

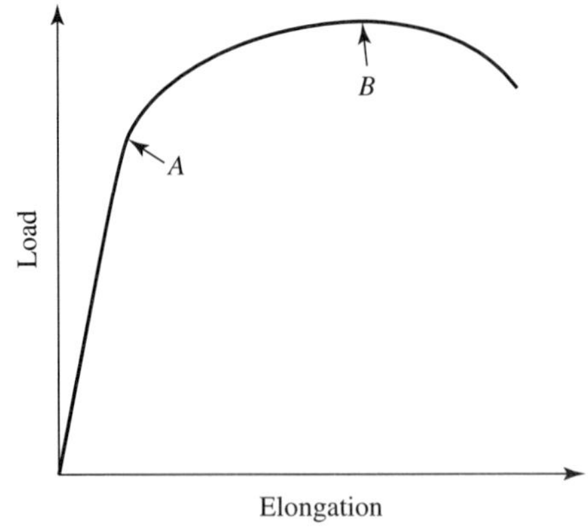

Figure 1.17 Load-elongation curve obtained in a tensile test. Plastic deformation begins at point *A*, necking at point *B*.

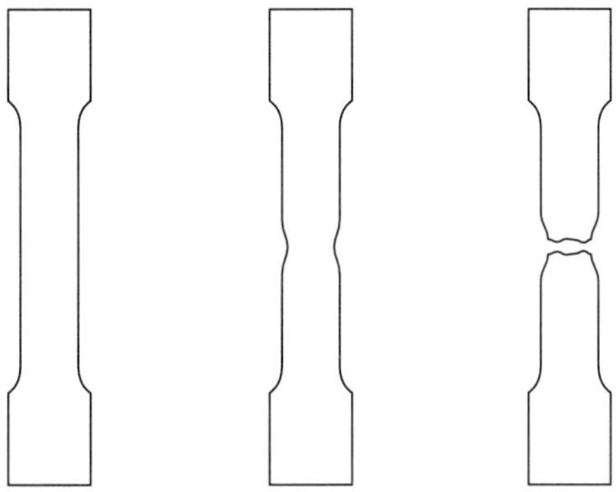
(a) (b) (c)

Figure 1.18 (*a*) Tensile specimen before deformation, (*b*) after onset of necking, and (*c*) after fracture.

results in further deformation throughout the gage length. When point *B* is reached, a condition of plastic instability sets in. The specimen begins to neck down, with the plastic flow now occurring within a small part of the total gage length (Fig. 1.18). The result is a very rapid decrease of the cross-sectional area with increasing strain. Because this area decreases so fast when necking sets in, the load on the specimen decreases even though the stress in the necked region is still increasing; the load-elongation curve bends over and continues downward until the specimen breaks.

Strain is defined as the increase in length per unit length of material $\Delta l/l$, while stress is defined as load per unit area. Because the length l and cross-sectional area of the tensile specimen are continually changing as a tensile test proceeds, the calculation of the true stress and true strain in the sample is not always an easy matter. For some types of work this problem is avoided by utilizing the engineering stress, load divided by original cross-sectional area, and engineering strain l/l_o, where l_o is the original gage length of the specimen. It will be recognized that these "engineering" quantities bear little relationship to the true state of stress and strain in the specimen in the later parts of the tensile test. In fact, when necking sets in, it is almost impossible to calculate the true stresses and strains. As long as the deformation is uniform, however, the true strain ε is given by

$$\varepsilon = \int_{l_o}^{2} dl/l = \ln l/l_o$$

and the true stress by the load divided by the actual cross-sectional area. In the elastic region the "engineering" and true quantities can be taken as equal for all practical purposes.

A number of important quantities can be calculated from the load-elongation or stress-strain curve of a material. Some of these, such as the elastic modulus, are true physical properties of the material, while others, such as the tensile strength, are arbitrarily defined parameters useful in design work.

Young's Modulus This is defined as the stress divided by the strain for elastic deformation, which is also the slope of the linear part of the stress-strain curve.

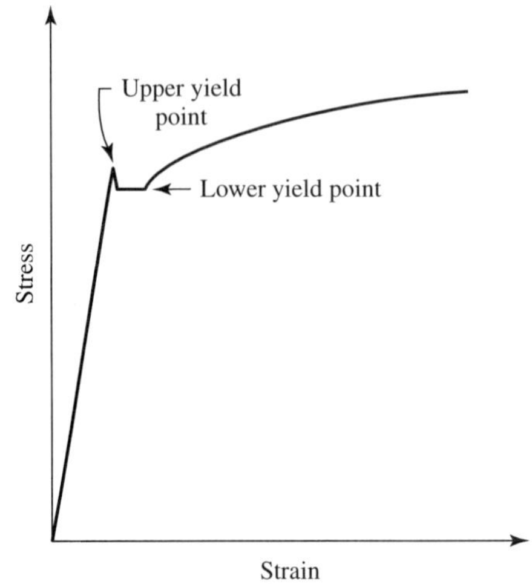

Figure 1.19 Tensile stress-strain curve showing a sharp yield point.

In engineering practice, the Young's modulus of a material is primarily of interest as a measure of the amount of deflection that will occur in a structure under a given load. The lower the modulus, the greater the elastic deflection for a given stress.

Yield Strength When a material under tension reaches the limit of its elastic strain and begins to flow plastically, it is said to have yielded. The yield strength is then the stress at which plastic flow starts. Some materials under certain well-defined test conditions show a sharp yield point on their stress-strain curves, as illustrated in Fig. 1.19. In most cases there is a gradual transformation from elastic to plastic behavior. In fact, if you work with sufficiently sensitive techniques, you can find evidence of permanent deformation in the material even in the early part of the elastic region. It is clear, then, that the yield strength is, at best, an arbitrarily defined point on the stress-strain curve. The most commonly used method of defining the yield point is as follows. As in Fig. 1.20, a line is constructed parallel to the elastic part of the stress-strain curve but displaced to the right an amount equivalent to a strain of 0.20%. The stress at which this line intersects the stress-strain curve is the 0.20% offset yield strength.

Two other quantities are sometimes used as measures of the yield point. The proportional limit is defined as the stress at which the stress-strain curve first shows a measurable deviation from linearity. The elastic limit is the maximum stress to which a metal may be subjected without permanent plastic deformation. The difficulty with these quantities is that they depend entirely on the sensitivity of the test equipment used and so cannot be expected to be reproducible by different observers. Also, the elastic limit can be found only by repeated loading and unloading of the specimen, a tedious procedure at best. The above-defined yield point is a much more useful property, and use of the terms elastic limit and proportional limit should be discouraged.

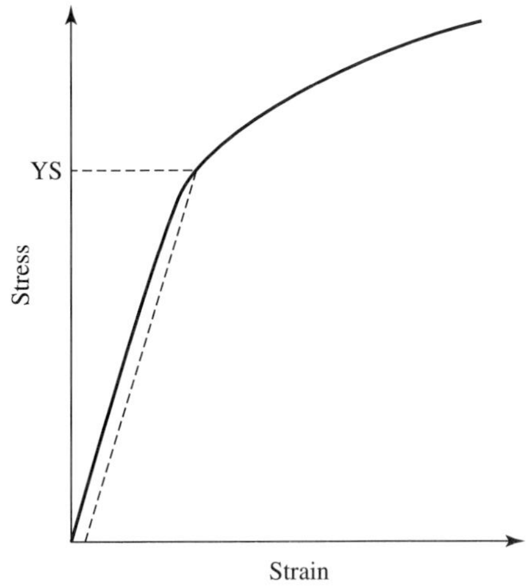

Figure 1.20 Offset method of determining the yield stress. The dashed line is drawn parallel to the elastic part of the stress-strain curve but is offset by an amount corresponding to a strain of 0.20%.

Tensile Strength This is a mechanical property defined as the maximum load sustained by the specimen during the tensile test, divided by its original cross-sectional area. It is sometimes called the ultimate strength of the material. It should be realized that it is an entirely arbitrary quantity having no real physical significance, although it is useful as a quality assurance indicator.

True Breaking Strength This is the load on the specimen at the time of fracture divided by the cross-sectional area at the fracture. It is therefore the actual stress that the material sustained when it finally failed and is a true physical quantity.

Tensile Elongation This is frequently taken as an indicator of the ductility of the material under the tensile test. It is determined, not from the load-elongation curve, but from examination of the failed specimen. To determine the elongation, the increase in distance between two reference marks, scribed on the specimen before the test, is measured with the two halves of the broken specimen held together. The percentage elongation is 100 times the quotient of the increase in length and the initial distance between the scribe marks. Because of restraints to transverse contraction by the unstrained and lower-stressed sections at the ends of the reduced gage section, elongation is not necessarily uniform within this volume of metal between the scribe marks. Therefore, the "percent elongation" is a function of the initial gage, which should be specified as a percent within a given gage length. Furthermore, localized elongation is very high in the necked section adjacent to the fracture, and this contributes even more to the effect of gage length on reported percentage elongation. Because of this factor and others as well, it is quite possible to fabricate cold-rolled steel that has a tensile elongation in 2 in of only 1% and obtain localized deformations of 20% without fracture.

Reduction of Area This is the quotient of the decrease in cross-sectional area at the plane of fracture and the original area at that plane (times 100, to express as a percentage). Similarly to percent elongation, this number, while related to ductility, does not differentiate between the localized reduction in area associated with necking and the general reduction in area throughout the gage length caused by uniform elongation.

Rate of Strain Hardening The slope of the stress-strain curve beyond the yield strength (see Figs. 1.19 and 1.20) shows the rate of increase of stress with increase in plastic strain. The steeper this part of the curve, the greater is the hardening effect of the plastic deformation. While this section of the stress-strain graph is curved, it ordinarily becomes a straight line in a loglog plot and thus conforms to the relation $\sigma = \kappa \varepsilon^n$, where n defines the slope of the early plastic deformation part of the stress-strain plot and is called the strain-hardening exponent. This useful dimensionless quantity can be shown to be numerically equal to the strain at maximum load, namely, the elongation at B in Fig. 1.17. Materials with a low strain-hardening rate have little uniform elongation and start to neck locally early in the tensile test. Materials with a very high rate of strain hardening will tend to have a large amount of uniform elongation and no localized necking under tension.

Uses and Limitations of the Tensile Test Tensile testing is the principal source of information on the strength properties of materials, providing the information most directly useful to the designer. In using and interpreting tensile-test results, however, certain complications and limitations should be kept in mind. First, the test results for practically all materials are sensitive to test conditions. The importance of the design of the specimen has already been discussed. Whether the machine is "hard" or "soft," i.e., applies load at a constant rate, will influence the shape of the load-elongation curve. The rate at which the specimen is strained is also often of considerable importance, the properties of some materials at high strain rates being quite different from those observed at low strain rates. Particularly with materials of limited ductility, the precision with which the specimen is lined up with the tension axis of the testing machine is important, as poor alignment can cause premature failure. Repeated tests under such conditions result in a scatter of results from which reliable information can be obtained only by statistical methods. Of course, if only one test had been made, its unreliability probably would not have been appreciated. The tensile test is destructive, so a fresh specimen has to be prepared for each test. This often discourages experimenters from running a sufficient number of repeat tests to gain confidence in their results.

Hardness The hardness of a material can perhaps be best defined as that property that is measured by a hardness tester. Actually, hardness is not a true physical property but depends on a combination of different physical properties. There are a number of different methods of measuring hardness. Each method measures a different combination of the true properties of the material. Consequently, hardness as measured by one test is not necessarily comparable with hardness as measured by another test. The three principal types of hardness test are the scratch, rebound, and penetration tests. In the scratch test, the resistance of the surface of the material to scratching is determined by comparison with a set of standard materials ranging from talc (softest) to diamond (hardest). The test is used primarily in testing minerals and refractories. The rebound (scleroscope) test involves observing the height of rebound of a hardened steel ball dropped from a given height onto the test piece; the harder the material, the greater the rebound. What is being measured in this case is the amount of energy absorbed in the material during impact. The rebound test is not extensively used in modern practice.

Nearly all hardness testing of materials today is done by penetration testing. While there are a variety of these tests in use, they all have the common feature that the amount of penetration of a standard indenter into the metal under a known load is observed. The most widely used penetration tests are the Brinell and the Rockwell. In the Brinell test, a hardened steel ball is forced by a known load into the metal being tested, and the diameter of the impression is measured and, by use of a formula or tables, converted into an empirical number called the Brinell hardness number (BHN). The Rockwell test is similar except that steel balls of various diameters or a diamond cone (Brale) may be used. The diameter of the impression made by the indenter is not measured in the Rockwell test. Instead, its depth is automatically indicated on a dial. In both these indentation tests, the results are affected not only by the original resistance of the metal to deformation but also by the rate at which this resistance changes in the vicinity of the indenter during the tests. The results are also affected by elastic properties of the metal, since the diameter or depth of the indentation is measured after the load has been released, with an accompanying elastic recovery or slight reversed dimensional change. Another variable that should be recognized is the relative volumes of material displaced by varying depths of indentation of a sphere and a cone. In general, Brinell hardness cannot be converted to Rockwell, nor can hardness readings on one Rockwell scale be converted to another scale. However, for a particular material fairly reliable conversion tables can be made up empirically; such tables should not be used to convert hardness readings taken on other materials.

Hardness measurements are extremely useful as a quick and rough indication of the mechanical properties of a metal. The Brinell hardness number, for example, may be multiplied by 500 to obtain an approximation of the tensile strength in pounds per square inch of most carbon steels. The Rockwell test has the advantage of ease and rapidity of measurement and a small size of indentation that does not noticeably mar the surface or affect the usefulness of the part after testing. Most of the hardness data in this book and in the literature are expressed in terms of Brinell values or Rockwell numbers (R with another letter designating the load and indenter used; for example, B = 100-kg load and a $\frac{1}{16}$-in ball; C = 150-kg load and the Brale indenter).

There are a number of variants of the Rockwell test useful in special applications. The Rockwell superficial-hardness tester, for example, is designed to test relatively thin surface layers or sections of metal by keeping the depth of penetration small. Various microhardness indentation tests utilize a diamond indenter and very light loads, the indentation being so small that it must be measured under the metallurgical microscope. The hardness of individual grains in the metal or of particles within grains can be measured by this test. However, the results of microhardness tests depend greatly on the actual loads employed; 1-g-load test data and 50-g-load test data will not agree in a quantitative sense.

Impact, Creep, and Fatigue When the mechanical properties of a material are tested under conditions of very rapid application of load, the test becomes an impact test. Similarly, when a very low strain rate is obtained, one is dealing with a creep test. Fatigue testing involves the repeated application of a load, usually in the vicinity of the yield strength, until failure of the test piece results. All these different types of tests reproduce conditions met in the practical utilization of materials under service conditions, and their use is largely an empirical method of finding better materials to meet certain applications. Impact tests in particular are difficult to interpret in scientific terms.

Impact tests are made by applying a sudden load, or impulse, to the test piece. The impulse is derived from a falling weight or, most commonly, a swinging pendulum. What is usually measured is the energy absorbed when the impact is great enough to break the specimen. The "tougher" the material, the greater the energy absorption. A brittle material, such as glass or fully hardened steel, on the other hand, absorbs virtually no energy upon breaking.

The commonly used Izod or Charpy notched impact test is supposed to reveal the toughness of metals under impact-load conditions. Actually, the test indicates the ability of metals to absorb energy by local deformation under the biaxial stress conditions of a notch; i.e., it is a notch test, not an impact test. For example, the energy required to break an unnotched bar in tension under impact loading (in the tensile impact test) correlates well with the area under a true-stress–true-strain curve in a static tensile test. In both cases, the strong but brittle metal and the weak but plastic metal show lower energy absorbed to fracture than a moderately strong, moderately plastic specimen. Thus energy absorption, in the absence of a notch, is considered a good evaluation of toughness.

Creep is the flow of a material over a period of time when under a load too small to produce any measurable plastic deformation at the time of application. The simplest type of creep test is made by just hanging a weight on the test specimen and observing its elongation as a function of time, by use of a measuring microscope or other sensitive detector of strain. In more refined creep apparatus, a system of levers and cams may be employed to apply the load in such a way that the stress on the specimen remains constant, the load being decreased as the cross-sectional area of the specimen decreases during its elongation. The rate at which a material creeps is sensitive to the temperature of the test relative to the melting temperature T_m. Below a temperature of about $0.4T_m$, creep is not appreciable, but above this point it increases rapidly with increasing temperature. A furnace in which the temperature can be accurately controlled is therefore a necessary part of any apparatus for creep experiments above room temperature.

Since many engineering structures made from metals would be useless if strained more than 0.01 to 0.10%, creep is often of great practical importance, particularly in service at elevated temperatures as in steam or gas turbines. On the other hand, high-temperature strength may be needed for only a few seconds or minutes, as in rocket-propulsion units. In this case the property desired is not creep strength but short-time, high-temperature strength.

Fatigue testing determines the ability of a material to withstand repeated applications of a stress which in itself is too small to produce appreciable plastic deformation. Situations involving fatigue frequently occur in engineering practice. A common example is a rotating beam loaded transversely to the axis of rotation, e.g., an axle. The material at the surface is alternately stressed in tension and compression during each cycle of rotation. If the strains are completely elastic, nothing happens. If, however, the stresses are high enough for some small region of the metal to be even minutely deformed, permanent damage begins to build up in that region. The high stresses may be from overloading, from excessive vibration, loud noise (sonic fatigue occurring in jet aircraft), or from local stress concentrations at notches (keyways, etc.), when it was thought that the stresses were in a safe range. In any case, the accumulation of permanent damage in a local region eventually results in the formation of a crack. Once this occurs, the stress concentration at the crack root is greatly increased. This leads to the propagation of a crack through the bar and finally the occurrence of a "notch" type of failure showing no evidence of ductility. In a corrosive environment, the susceptibility of a metal to fatigue failure may be significantly increased, a phenomenon known as corrosion fatigue.

In a rotating-beam type of fatigue-test machine, a shaft of the material is loaded transverse to the axis of rotation at one end (Fig. 1.21), and the number of revolutions before failure occurs is counted. In a push-pull type of test, alternate tension and compression are applied to the test piece. Again, the number of cycles of stress sustained before failure is observed. The results of fatigue tests on a material are usually presented on a stress-cycle, or S-N, diagram (Fig. 1.22), where the number of cycles to failure is plotted against the applied stress on the specimen. If the S-N curve becomes horizontal, a level of alternating stress has been reached that can be sustained by the material indefinitely. This stress is called the *endurance limit*.

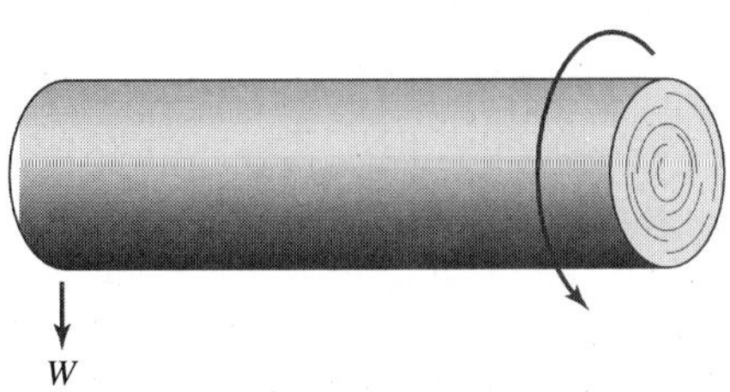

Figure 1.21 Method of loading specimen in rotating-beam fatigue-testing machine.

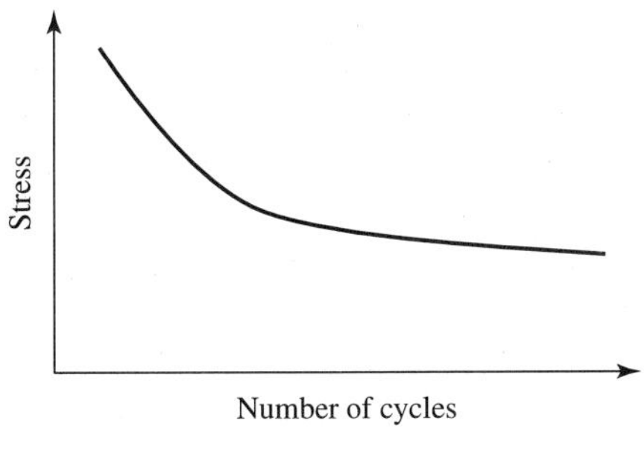

Figure 1.22 An S-N curve obtained from fatigue tests. The number of cycles to failure is plotted against the applied stress.

1.5 PHYSICAL PROPERTIES

Density If the lattice parameters of a metal and the mass of its individual atoms are known (from measurements with a mass spectrometer or calculated as the atomic weight over Avogadro's number), it should be an easy matter to calculate its density. The first step in carrying out such a calculation is to find the number of atoms contained in a volume equal to that of the unit cell. In simple structures this number can be found by inspection: it is necessary only to imagine the unit cell displaced slightly so that it contains whole lattice points and then count the number of points in it. An alternative method that does not tax the observer's three-dimensional visualization so much is illustrated in the following examples. Consider a bcc unit cell. Each of the corner atoms is shared between eight cells and so the total number of atoms in the cell is

$$1 \text{ center atom} + (8 \times \tfrac{1}{8}) \text{ corner atoms} = 2 \text{ atoms per unit cell}$$

For the fcc unit cell one finds in the same way that

$$(6 \times \tfrac{1}{2}) \text{ face atoms} + (8 \times \tfrac{1}{8}) \text{ corner atoms} = 4 \text{ atoms per unit cell}$$

In order to find the density of a crystal, all that is necessary is to find the mass of all the atoms in the unit cell and then divide by the volume of the cell. For example, the density of iron, which is bcc, would be calculated in the following way:

$$2(55.85)/6.025 \times 10^{23} \times (2.8610 \times 10^{-8})^3 = \text{density of iron, g/cm}^3$$

since the atomic weight of iron is 55.85 and the lattice parameter is 2.8610 Å. The density so calculated is the density of a perfect iron crystal. The presence of voids in the crystal will make the measured density of the crystal less than that calculated, and the presence of an impurity will increase it. Thus density measurements can be used as a measure of crystal perfection.

Thermal Properties The response of a material to the application of heat is determined by (1) its specific heat, which determines the amount of heat energy that must be supplied to produce a temperature rise of 1°C, (2) its thermal conductivity, which specifies the rate at which heat can penetrate the material, and (3) its thermal expansion. In metals and other solids having simple crystal structures, these properties are related to the vibration of the atoms about their normal sites in the crystal.

Specific Heat According to the kinetic molecular theory the thermal energy of an ideal gas resides in the kinetic energy of translation of its molecules, $\frac{1}{2} Nmv^2$, where m is the molecular mass, v the velocity, and N the number of molecules. When heat is supplied to a crystal, it would be expected, then, that the thermal energy taken up would reside in the motion of its atoms. In crystals, the atoms are located at definite sites in the lattice, held there by the cohesive forces existing between atoms. To a useful degree of approximation, these cohesive forces can be thought of as springs connecting the atoms together. In a simple cubic structure, each atom would be attached to its six nearest neighbors by six springs. If energy is supplied to this system, the atoms will vibrate back and forth, always in motion but with average positions that are their true lattice sites. This is just the situation in real crystals. In their vibratory motion, the atoms of a simple crystal have six degrees of freedom. Three degrees are associated with the velocity of each atom as it vibrates just like the gas molecule and the other three with its potential energy, since this is determined by its position (which must be specified by three coordinates) relative to its atom site. The thermal energy of a mole of crystal is then $\frac{6}{2}RT$, and its specific heat

$$C_v = \frac{\delta}{\delta T}\frac{6}{2}RT = 3R \approx 6 \text{ cal/mol}\cdot°\text{C}$$

The specific heat of most metals (at constant volume) is, in fact, quite close to 6 cal/mol·°C. For conductive metals, free electrons can also store heat and one can get to 8 cal/mol·°C.

According to the above theory, C_v should be independent of temperature. This is nearly true for most metals except when the temperature becomes very low. Then, as shown in Fig. 1.23, C_v decreases very rapidly.

Thermal Conductivity Metals are characteristically good conductors of both heat and electricity. Heat is conducted through metals by two distinct mechanisms. One involves the transfer of thermal energy by the free electrons in the metal, the same electrons that conduct electric current. This electronic contribution will be discussed in the next section. The other, called lattice conductivity, results from the coupling together of the vibrations of the atoms in the crystal. In metals and alloys, both the electronic and lattice contributions to the conductivity are important. In insulators, the electronic contribution is completely absent.

The origin of the lattice contribution to the thermal conductivity can be seen in the following way. When one end of a bar of metal is heated, the atoms at that end are excited to larger amplitudes of vibration than those at the cool end. Since an atom is, in effect, connected to neighboring atoms by the equivalent of springs (bonding forces), its increased vibration is communicated to its nearest neighbors, from them to their neighbors, and so on through the crystal. A quantitative theory of thermal conductivity must be based on the quantum mechanics of lattice vibrations, but

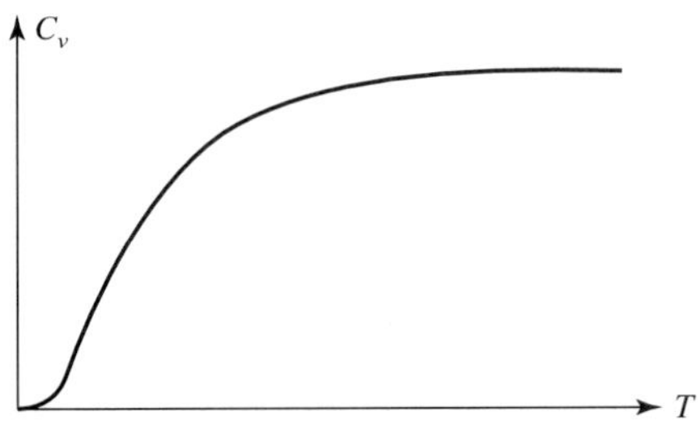

Figure 1.23 The specific heat at constant volume of a metal as a function of temperature. As the temperature falls to 0 K, the specific heat approaches zero.

there are no simple rules governing thermal conductivity as there are for specific heats. Theory does show that impurities and imperfections impede the propagation of lattice vibrations through a crystal, which means, however, that the lattice part of the conductivity can be altered by alloying.

Thermal Expansion Associated with the thermal vibration of the atoms of a crystal is a gradual shift in their average position—a change in the dimensions of the crystal lattice—with temperature. Generally this is an expansion, although there are some materials that show thermal contraction in certain crystal directions. Thermal expansion is usually linear with temperature except at very low temperatures and in the vicinity of temperatures where allotropic or magnetic transformations occur. A dilatometer, an instrument that makes accurate measurements of the length of a specimen as a function of temperature, can often be used to detect allotropic transformations because of differences in the volumes and expansivities of the different phases.

It has been found that the thermal-expansion coefficient α is related to the compressibility κ, specific volume V, and specific heat C_v of a material by

$$\alpha = \gamma \, \kappa C_v / 3V$$

where γ is a constant known as Grüneisen's constant and having a value of about 1.8 for most metals. The above relation, while not always accurately obeyed, is often useful in estimating the value of one quantity when data on the others are at hand. Alloying can substantially alter the magnitude of the thermal expansion of materials, particularly metals having strong magnetic properties.

Electrical and Magnetic Properties

On the basis of their electrical properties, materials may be classified as being conductors, semiconductors, or insulators. In fact, electrical conductivity is the property that varies between the widest limits among different classes of materials. In a good conductor, the resistivity may be as low as 10^{-6} $\Omega\cdot$cm, while in a good insulator it may be 10^{18} $\Omega\cdot$cm. Semiconductors, with resistivities of the order of tens of ohm-centimeters, are intermediate between conductors and insulators. Besides conductivity, the dielectric constant is an important electrical property of semiconductors and insulators. Also, the optical properties of all three classes of materials can be regarded as a special case of the electrical properties, since the exposure of a material to light is equivalent to applying a high-frequency electric field.

Electrical Conductivity Metals are characterized by a high electrical conductivity that decreases with increasing temperature. The electric current in metals is carried by electrons, which are free to migrate through the crystal. To a first approximation, the conduction electrons can be regarded as a sort of "electron gas" confined in a box whose walls are the surfaces of the metal. It is the valence electrons of the metal atoms that become delocalized in forming the electron gas; the closed-shell electrons remain attached to their individual atoms. The electron gas plays a fundamental role in determining the properties of a metal, for it is the interaction between the metal atoms and the electron gas which gives rise to the bonding forces between the atoms of a metal.

Although an accurate theory of the conduction electrons in a metal must be made in terms of quantum mechanics, a useful description of the electrical properties of metals may be formulated by considering the electron gas to have the properties of an ordinary molecular gas. One imagines the electrons to be moving about at random with a high velocity, colliding from time to time with the atoms of the crystal. When an electric field is applied, the electrons take up a drift velocity in the opposite direction to the field, owing to their negative charge; this gives rise to an electric current. The number of collisions between conduction electrons and atoms limits the magnitude of the current for any given field. It turns out in the quantum theory of conduction that these collisions are more likely to occur as an atom when it is farther from its exact lattice site. Thus, at higher temperatures, where the atoms are vibrating with greater amplitude, the electrical resistivity of metals is greater.

The electron gas, just like an ordinary molecular gas, can conduct heat. The electron gas is responsible for the electronic part of the thermal conductivity of metals. If, as is the case in pure metals, the electronic part of the thermal conductivity is large compared with the lattice part, then it would be expected that those metals which are good conductors of electricity should also be good conductors of heat, and conversely. Such a relationship does exist and is known as the law of Wiedemann and Franz. According to this law:

$$\text{Thermal conductivity}/\text{electrical conductivity} \times \text{temperature} = \text{constant}$$

In using this relationship, the constant, which is known as the Lorenz number, and the conductivities must be expressed in a consistent set of units. In the cgs system of units the Lorenz number is 2.48×10^{-13}.

Not all materials are equally good conductors of electricity. Two conditions that favor high conductivity are: one valence electron per atom in the electron gas and high purity. The reason for the first condition is not apparent from the simple electron-gas picture. It is a result of a quantum-mechanical effect whereby the ability of electrons to carry current is restricted when the concentration of the electron gas is too high. Electrical resistivity also results from scattering of the conduction electrons by atoms that are off their lattice sites because of thermal vibration. High purity is necessary because impurity atoms dissolved in a metal also act as scattering centers for conduction electrons. Of the common metals, copper best fulfills the first of the above conditions. Aluminum has the advantage of its greater valence-electron concentration but is not quite so good a conductor as copper. However, when light weight is required, aluminum has a distinct advantage. Copper conductors, on the other hand, are more easily joined by soldering.

Except at very low temperatures the electrical resistivity of metals is quite accurately proportional to temperature. This is the basis of the resistance thermometer. As the temperature approaches absolute zero, the resistivity of pure metals

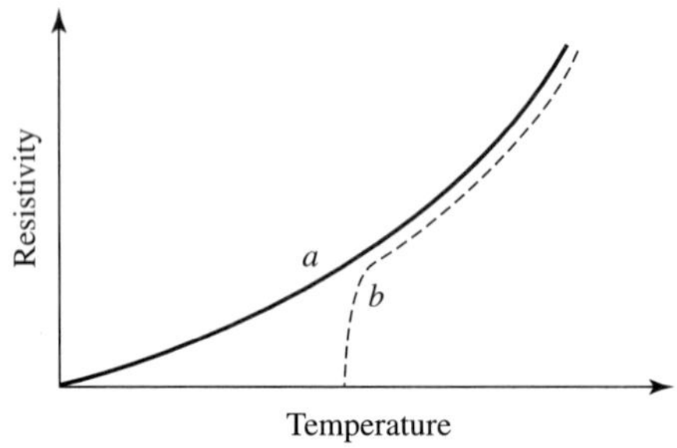

Figure 1.24 The decrease of the resistivity (*a*) of a pure, normal metal and (*b*) of a superconductor as the temperature approaches absolute zero.

becomes vanishingly small, as shown in Fig. 1.24, curve *a*. Some metals at low temperatures transform to the superconducting state, where they have no resistivity, as shown by curve *b* in Fig. 1.24. Mercury, tin, and zinc are among the superconducting metals. Unfortunately, there is a practical limitation on the use of superconductors for carrying current: When the current becomes too large, the metal reverts to the normal nonsuperconducting state because of the influence of the magnetic field that accompanies the current.

Magnetic Properties All materials show some response to an applied magnetic field, but it is only in the ferromagnetic and ferrimagnetic cases, where there is a spontaneous magnetization of the material, that these effects are large enough to be readily observable. The magnetic properties of materials are important industrially in the generation and transmission of electric power, where losses in the magnetization of rotors, stators, and transformer cores are a principal limitation on the efficiency of electrical equipment, and in electronic circuits of all kinds. Three physical quantities enter into any discussion of the magnetic properties of materials:

Magnetization. J is defined as the magnetic moment per unit volume of material and is a measure of the extent to which a material is magnetized.

Magnetic field strength. H is the intensity of a magnetic field expressed in oersteds. The greater H, the greater is the force on a unit magnetic pole placed in the field.

Induction. B is the actual magnetic flux density existing within a material due to an externally applied field and the magnetization of the material. It is expressed in gausses.

In materials that are not spontaneously magnetized (not ferromagnetic or ferrimagnetic), it is found that the magnetization J is always proportional to the applied field H, that is,

$$J = \chi H$$

where the constant of proportionality χ is called the magnetic susceptibility of the material. Because J is proportional to H, it turns out that B is also proportional to H,

$$B = \mu H$$

and μ is known as the *permeability*.

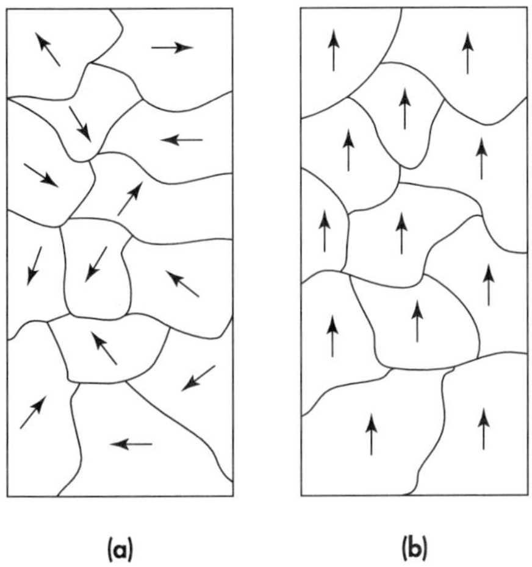

Figure 1.25 Magnetic domains in (*a*) an unmagnetized bar and (*b*) the same bar when fully magnetized.

In some metals, notably iron, nickel, and cobalt, there exists a spontaneous magnetization; that is, J has a definite value even when $H = 0$. These metals are said to be ferromagnetic. The magnetization existing within a piece of, say, iron is not ordinarily observed because there are, within the iron bar, domains, or small regions, within which the magnetization is all in the same direction. The directions of J in the various domains, however, will all be different so that, in any piece of metal larger than a few micrometers in size, the magnetizations of the different domains all cancel out, leaving no external evidence of the existing magnetization (Fig. 1.25). "Magnetizing" a bar of iron involves forcing J to be in the same direction in all the domains (Fig. 1.25*b*). This can be accomplished by applying an external magnetic field H to the bar with a resultant flux density of B.

As the temperature of a bar of iron is increased, the strength of the magnetization within its domains decreases until, when the Curie temperature is reached, the spontaneous magnetization disappears altogether. Ferrimagnetism is a variant of ferromagnetism occurring in certain oxides, such as magnetite, in which spontaneous magnetization is not so complete as in the ferromagnetic case.

The magnetic properties of a ferromagnetic material are determined by measuring the total magnetization of a sample B as a function of the applied field H. The total magnetization is a measure of the extent to which the magnetizations within the individual domains are lined up together. The result of such a test is a B-vs.-H curve, as shown in Fig. 1.26. Starting at the origin with $H = 0$ and $B = 0$, B increases with increasing H, rapidly at first and then more slowly until a maximum, or saturation, value of B_{max} is attained. Beyond this point where the magnetizations in all the domains are lined up, further increase in H produces no greater magnetization in the sample. If H is now decreased, the magnetization decreases, but not along the original magnetizing curve. Instead, it follows the upper branch of the curve in Fig. 1.26, so that when $H = 0$ the sample remains partially magnetized. The amount of magnetization left at $H = 0$ is called the remanence. To reduce the B to zero, a

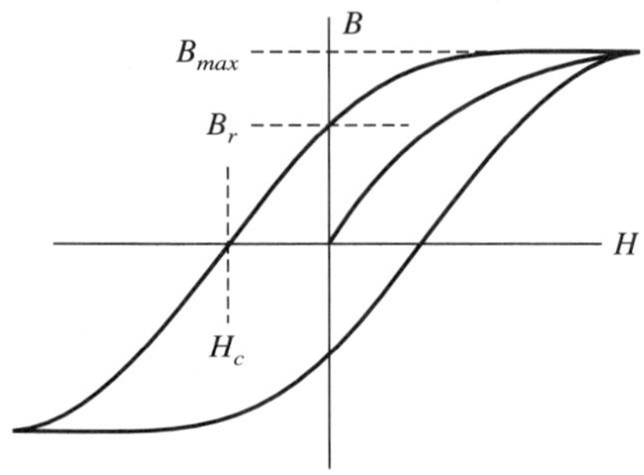

Figure 1.26 Hysteresis loop for the magnetization of iron. H is the applied magnetic field and B the resulting magnetization of the sample. B_{max} is the maximum magnetic flux density attainable, B_r is the residual magnetism, or remanence, left after removing the applied field, and H_c is the amount of reversed magnetic field (coercive force) required to bring the magnetism to zero.

field must be applied in the reverse direction, the magnitude of which is called the coercive force. Continuing to cycle H results in tracing out the complete hysteresis loop shown in Fig. 1.26. The amount of energy dissipated in the specimen during one complete cycle of the magnetizing field is equal to the area enclosed by the hysteresis loop. This energy appears as heat, and in an iron-transformer core, where the hysteresis loop is traced out 60 times each second, the heat so generated can be a major cause of inefficiency in the transformer. The metallurgical problem with iron for electric-power applications is, then, to try to make the hysteresis loop as small as possible. On the other hand, in making a material for permanent magnets, where it is desired that the remanence and coercive force be as large as possible, the solution of this problem calls for entirely different metallurgical conditions.

1.6 CHARACTERISTICS OF UNALLOYED SOLIDS

Whether a particular element is a metal or not depends on certain of its properties. There is, however, no universal agreement as to what property or combination of properties determines a metal. To the chemist, an element is a metal if its oxide dissolved in water produces an alkaline solution; to the physicist, an element is a metal if it displays good electrical conductivity that decreases with increasing temperature. The materials engineer, being primarily concerned with the mechanical, electrical, and magnetic properties of materials, generally regards an element as a metal if it meets the physicist's requirement and also can be plastically deformed to some extent. In fact, the engineer's principal business is finding out how to modify and change these metallic properties through control of the composition and structure of metals. An engineer who understands the degree to which metallic properties can be altered can utilize metals and alloys for structural and other uses more intelligently than by relying solely on handbooks.

The term unalloyed material rather than pure material is chosen here, since purity is never absolute and varies from 99.999% in some electro- and zone-refined elements to 98.0% or less for others. In this chapter, we will be concerned with commercially pure metals having no alloying elements deliberately added. Reference is made also to certain nonmetallic materials that, over the past several decades, have become increasingly important as replacements for metals in many services.

In Table 1.3, physical-property data are listed for some of the more important metals. For the most part, only structure-insensitive properties are considered. It is manifestly impossible to tabulate values for the structure-sensitive properties, since these vary widely from sample to sample. In classifying materials, it is convenient to use the framework of the periodic table, as in the discussion below.

The IA, or Alkali, Metals

The alkali metals include lithium, sodium, potassium, rubidium, and cesium. They are characterized by high chemical reactivity, low densities, and low melting points. All have the bcc structure and are open metals in the sense that, if their atoms are regarded as hard spheres, these spheres are not in contact in the crystal structure. Consequently, these metals have large compressibilities. Because they are soft and have low strengths, they are not used for structural purposes. As liquids, low-melting-point alkalies are useful as heat-transfer media.

The IB, or Noble, Metals

Copper, silver, and gold, like the alkalies, have a valence of 1 and, like the alkalies, are good conductors of electricity. However, the noble metals have close packed structures (fcc) and relatively high melting points. They show good strength and ductility, the latter being particularly important since large quantities of copper are drawn into wire for use in the transmission of electric power. Gold has the unique property among the metals that its oxide is unstable; its surface will remain bright indefinitely. Silver, which has the highest conductivity of all the common metals, finds use in electrical contacts, where its corrosion resistance is also important.

Although most metals find their major industrial use in the form of alloys, this is not true of copper. This metal is used most extensively as an electric conductor and in radiators. Both applications require high conductivity, electrical in one case and thermal in the other, which in turn requires high purity. Electrolytic refining reduces the content of metallic impurities to tolerably low levels, leaving oxygen (unavoidably introduced during melting of the cathode copper) as the principal impurity in the metal. Adding phosphorus to the liquid metal can produce oxygen-free copper. Ordinarily the residual phosphorus content reduces the conductivity of the metal below that required for electrical applications. Oxygen-free high-

text is continued on p. 38

Table 1.3 Physical properties of the metals

Element	Symbol	Density 20°C, g/cm^3	Coefficient of Thermal Expansion, 0 to 100°C $\times 10^{-6}$/°C	Electrical Resistivity at 18°C, $\Omega \cdot cm$	Electrical Conductivity (Compared with Copper), %	Lattice Type at 20°C (or Indicated Temperature)		Lattice Parameter (Length of Base of Unit Cell) $a \times 10^{-8}$ cm
Aluminum	Al	2.70	24.0	2.72	61.8		fcc	4.0490
Antimony	Sb	6.62	9.8	39.8	4.23		rh	4.5064
Beryllium	Be	1.82	12.4	6.3	26.7		cph	2.2854
Bismuth	Bi	9.80	11.8	118.0	1.41		rh	4.7356
Cadmium	Cd	8.65	29.8	7.25	23.2		cph	2.9787
Calcium	Ca	1.55	....	4.5	37.3	α	fcc	5.57
						$\beta 300 < T < 450$°C		
						$\gamma T > 450$°C	cph	3.99
Carbon	C	2.22	1.2	3500	....		dia cubic	3.568
						Graphite	hex	2.4614
Chromium	Cr	7.19	6.1	13.0	....		bcc	2.8845
Cobalt	Co	8.92	12.8	6.8	24.0	α	hcp	2.507
						β	fcc	3.552
Columbium	Cb	8.57	7.2	13.1	8.4		bcc	3.3007
Copper	Cu	8.96	16.7	1.68	100		fcc	3.6153
Gold	Au	19.3	14.3	2.21	76.1		fcc	4.0783
Hafnium	Hf	13.3	....	32.0	5.26		cph	3.206
Iron	Fe	7.87	12.3	8.7	19.3	α	bcc	2.8664
						$\gamma 908 < T < 1405$°C	fcc	3.656
						$\delta T > 1403$°C	bcc	2.94
Lead	Pb	11.34	28.3	20.7	8.13		fcc	4.9495
Magnesium	Mg	1.74	26.0	4.3	39.1		cph	3.2092
Manganese	Mn	7.44	19.7	185.0	0.91	α	cubic	8.912
						$\beta 727 < T < 1095$	cubic	6.313
						$\gamma 1095 < T < 1173$	fcc	3.782
						$\delta T > 1173$		
Mercury	Hg	13.55	42.2(86 K)	95.4	1.76		rh	2.006
Molybdenum	Mo	10.2	4.9	4.72	35.6		bcc	3.1466
Nickel	Ni	8.9	13.3	7.35	22.9		fcc	3.5238
Palladium	Pd	12.0	11.7	10.75	15.6		fcc	3.8902

Lattice Parameter (Height of Unit Cell) $c \times 10^{-8}$ cm or Axial Angle	Distance of Closest Approach $\times 10^{-8}$ cm	Specific Heat (at Room Temp.) cal/g·°C	Heat of Formation of Lowest (Metalous) Oxide, kcal/g mol	Heat of Fusion ΔH_{fus}, kcal/mol	Melting Point, °C	Ultimate Tensile Strength of Annealed Metal, MPa	Elongation of Annealed Metal, %	Brinell Hardness of Annealed Metal	Young's Modulus of Elasticity, GPa
....	2.862	0.217	389.5	2.57	659.7	47	49	16	55
57°6.5′	2.903	0.051	165.4	4.74	630.5	11	....	30	78
3.584	2.225	0.425	146	2.36	1350	186	....	97	290
57°14.2′	3.111	0.030	135.5	....	271.3	....	....	....	....
5.617	2.979	0.059	65.2	1.45	320.9	71	50	21	69
....	3.94	0.145	151.7	....	810	59	53	17	21
6.53	3.95								
....	1.544	0.165	26.4	....	3550	....	....	....	5
6.7014	1.42								
....	2.498	0.110	267.4	4.20	1615	483	....	110	248
4.069	2.506	0.104	57.5	3.64	1495	255	....	48	....
....	2.511								
....	2.859	....	....	....	2500	345	30	....	....
....	2.556	0.093	34.9	3.12	1084	220	42	....	103
....	2.884	0.031	−12	2.95	1063	131	45	25	80
5.087	3.15	....	....	....	1700	....	....	....	....
....	2.481	0.107	64.04	3.7	1535	289	....	77	207
....	2.585								
....	2.54								
....	3.499	0.030	52.47	1.225	327.4	12	30	4.2	22
5.2103	3.196	0.245	145.76	2.16	651	194	....	29.4	41
....	2.24	0.1211	90.8	3.5	1260	Brittle	....	....	159
....	2.373	....	....	....	....	Brittle			
3.533	2.587	....	....	....	....	496	40	....	159
70°31.7′	3.006	0.033	21.7	....	−38.87	....	....	....	....
....	2.725	0.065	131.4	6.71	2620	683	....	144	290
....	2.491	0.105	57.83	4.21	1455	407	40	100	724
....	2.750	0.058	21.5	3.84	1553	145	24	49	117

(Continued)

Table 1.3 *(Continued)*

Element	Symbol	Density 20°C, g/cm^3	Coefficient of Thermal Expansion, 0 to 100°C $\times 10^{-6}$/°C	Electrical Resistivity at 18°C, Ω·cm	Electrical Conductivity (Compared with Copper), %	Lattice Type at 20°C (or Indicated Temperature)		Lattice Parameter (Length of Base of Unit Cell) $a \times 10^{-8}$ cm
Platinum	Pt	21.45	8.9	10.5	16.0		fcc	3.9237
Potassium	K	0.86	83.0	6.9	24.3		bcc	5.344
Rhenium	Re	20.53	...	...	...		cph	2.7609
Rhodium	Rh	12.44	9.6	5.0	33.6		fcc	3.8034
Selenium	Se	4.81	37	...	...		hex	4.3640
Silicon	Si	2.33	3.1	...	...		dia cubic	5.4282
Silver	Ag	10.5	18.8	1.58	106.3		fcc	4.0856
Sodium	Na	0.97	62.2	...	...		bcc	4.2906
Tantalum	Ta	16.6	6.7	14.7	11.4		bcc	3.3206
Tellurium	Te	6.25	16.75	...	...		hex	4.4559
Tin	Sn	7.28	26.92	11.3	14.9	α, gray $T < 13.2$°C	dia cubic	6.47
						β, white	tetrag	5.8311
Titanium	Ti	4.5	8.5	89	1.89	α	cph	2.9504
						$\beta T > 900$°C	bcc	3.33
Tungsten	W	19.3	4.3	5.32	31.6		bcc	3.1648
Uranium	U	18.7	...	...	...	$\alpha T < 665$°C	orthorh	2.858
						$\beta 665 < T < 775$°C	low symmetry	
						$\gamma 775 < T < 1130$°C	bcc	3.49
Vanadium	V	5.96	7.7	25.9	6.5		bcc	3.039
Zinc	Zn	7.14	26.28	5.95	28.3		cph	2.664
Zirconium	Zr	6.4	5.2	45	3.74	α	cph	3.230
						$\beta T > 867$°C	bcc	3.62

Lattice Parameter (Height of Unit Cell) $c \times 10^{-8}$ cm or Axial Angle	Distance of Closest Approach $\times 10^{-8}$ cm	Specific Heat (at Room Temp.) cal/g·°C	Heat of Formation of Lowest (Metalous) Oxide, kcal/g mol	Heat of Fusion ΔH_{fus}, kcal/mol	Melting Point, °C	Ultimate Tensile Strength of Annealed Metal, MPa	Elongation of Annealed Metal, %	Brinell Hardness of Annealed Metal	Young's Modulus of Elasticity, GPa
...	2.775	0.032	17.0	5.27	1773.5	117	30	...	166
...	4.627	0.192	86.26	0.57	62.3	...	...	0.037	...
4.4583	2.740	0.035	...	...	3000	...	...	...	...
...	2.689	0.058	...	...	1985	503	...	139	294
4.9594	2.32	0.077	56.42	1.3	220	...	...	...	...
...	2.351	0.181	198.3	9.48	1420	93	...	...	113
...	2.888	0.0558	6.95	2.855	960.8	159	48	28	76
...	3.715	0.295	99.16	0.63	97.5	...	...	0.07	...
...	2.860	0.036	500.12	...	3027	345	40	46	186
5.9268	2.87	0.048	78.3	9.28	452	...	...	...	...
...	2.81	0.054	69.8	1.72	231.9	14	96	5	41
3.1817	3.022								
4.6833	2.89	0.1125	217.4	...	1800	538	27	...	116
...	2.89								
...	2.739	0.034	126.2	8.07	3370	1000	310	...	352
4.955	2.77	0.028	256.6	...	Ca. 1133	...	...	...	...
...	3.02								
...	2.632	0.1153	209	...	1710	...	...	...	...
4.945	2.644	0.0925	84.4	1.765	419.47	110	...	...	69
5.133	3.17	0.068	178	...	1900	240	...	...	76
...	3.13								

conductivity (OFHC) copper is melted and cast in a carbon monoxide atmosphere and is used in electrical applications.

The presence of oxygen as an impurity in copper leads to two difficulties from the standpoint of the mechanical properties of the metal: it somewhat decreases the ductility and it can lead to embrittlement if the metal is heated in an atmosphere containing hydrogen.

The IIB Metals

Zinc and cadmium have the hcp structure, while solid mercury is rhombohedral. Actually, zinc and cadmium have axial ratios that considerably exceed the ideal value of 1.633 (Zn 1.856, Cd 1.886) and so are not truly close-packed. With a valence of 2, they are not particularly good conductors of electricity. Zinc owes one of its chief uses to the fact that it can be coated readily on the surface of iron by immersing the iron in liquid zinc or by electrodeposition; it will thereafter protect the iron from rusting or corrosion in mildly corrosive media. The protection is electrochemical, since zinc is anodic to iron and will go into solution while iron acts as a cathode and is unaffected as long as zinc is present and electrically connected to the iron.

Zinc has a relatively low melting point and thus, in the alloyed form, is particularly suitable for making pressure die castings. However, the die castings first made from this metal proved very unsatisfactory, particularly in warm, humid climates. They would swell enough to jam mechanisms in which they were used, and the intergranular corrosion, which caused the swelling, greatly reduced the strength properties. It was found that, by keeping the total content of lead and cadmium impurities below 0.01%, the die castings would indefinitely resist intergranular corrosion. Thus, zinc for die-casting alloys must have a purity of 99.99%, whereas that used for alloying with copper (to make brass) or for galvanizing iron has considerably higher permissible impurity limits.

The Light Metals

This category includes the metals having low density and reasonably good mechanical properties—aluminum, magnesium, and beryllium. Aluminum is fcc in structure and shows good ductility. It is slightly stronger and less ductile in the commercially pure form where iron, silicon, and copper impurities are present (as in alloy 1100), but in most alloy applications these are relatively unimportant. They have some influence on alloy casting properties and heat-treatment temperatures, and the amounts present should be controlled for reproducible optimum properties.

Aluminum has an electrical conductivity about two-thirds that of copper on a volume basis; it is a better conductor on a weight basis, since it weighs only about one-third as much as copper. Since the iron and silicon impurities form constituents that are insoluble in solid aluminum and do not materially reduce conductivity, commercially pure aluminum is becoming widely used for long-distance high-voltage power-transmission lines. For such cables, aluminum wires surround a steel wire, present to increase strength, and the assembly is sufficiently light to increase spans between supporting towers and materially reduce line-installation costs.

Magnesium is hcp in structure and at ordinary temperatures has only limited ductility. Above about 250°C, however, it can be extensively worked. Magnesium, like aluminum, is generally used in alloyed form when its mechanical properties are important. Beryllium is also hcp and ordinarily displays very limited ductility. If the problems of limited ductility and limited availability could be solved, the high elastic modulus, high melting point, and strength-weight properties of beryllium would ensure it important uses in aircraft and space vehicles.

The Transition Metals

This group includes all the transition elements of the periodic table, scandium to nickel, yttrium to palladium, and lanthanum to platinum. Some transition metals are little more than laboratory curiosities while others are our most important industrial materials. All are characterized by the fact that they have incompletely filled d shells in their electronic structure. A consequence of this is that they have high cohesive strengths and high melting points; they have good strength properties and retain these up to high temperatures. Some—iron, cobalt, and nickel—are ferromagnetic, but all of them are poor electrical conductors as compared with copper or aluminum.

Most iron is used in the alloyed form, steel. In the pure state, iron has much better corrosion resistance than in the relatively impure form of steel, even low-carbon grades. Ingot iron finds its most important applications in enameled ware and in fields where better corrosion resistance than that of steel, but not particularly high strength, is required.

The commonest undesirable impurity element in all iron and steels is sulfur, which, with iron, forms the low-melting-point constituent iron sulfide and thus causes hot shortness. This is a condition of deficient properties at hot working temperatures. The presence of manganese in amounts of about five times the sulfur content (in excess of the stoichiometric requirement) converts the sulfur to innocuous (high-melting-point) manganese sulfide. When present in comparatively large amounts, the globular manganese sulfide, by interrupting somewhat the continuity of the plastic ferrite matrix, permits the steel to be machined faster, with less power, and with a better surface finish.

Nickel, in commercially pure form, may contain a slight amount of sulfur from the fuel used in melting furnaces. This may form a continuous envelope of brittle nickel sulfide at the grain boundaries, thus embrittling the entire structure. The amount of sulfide can be so small as to be undetectable by ordinary micrographic techniques. The addition of about 0.05% Mg causes sulfide to form in an innocuous dispersion of particles like MnS in steel and permits the metal to display its inherent plasticity or malleability. Similarly, lead may be present as an impurity in gold in amounts small enough to escape detection by the microscope and yet form a thin, weak envelope at grain boundaries, which, being continuous or nearly so, embrittles the entire structure.

Nickel is a moderately expensive metal, and its relatively high cost has limited its uses to some extent. Its very good resistance to corrosion is most often utilized by electroplating a thin layer on the base metal or on an intermediate copper plate. The nickel plate is most frequently covered with a very thin layer of chromium electroplate, the chromium being harder, brighter, and therefore more pleasing to the eye. However, the corrosion protection depends on the nickel, since the chromium deposits are always somewhat porous. When the so-called chrome plate on an automobile begins to rust, it is generally because too thin a layer of nickel was deposited underneath the chromium.

The electrical and electronic industries depend on nickel for various components because of its unique electron-emission and expansion (for sealing in glass) characteristics. Nickel is also important as a catalyst in certain chemical industries. However, the major uses of nickel are as an alloying element, particularly in steels.

Metals of High Valence

Included here are gallium, indium, thallium, tin, lead, bismuth, antimony, and arsenic. With the exception of lead, which is fcc, these metals all have relatively complex crystal structures; they have low melting points and low strength. Some of the higher-valence metals, such as arsenic, are hardly recognizable as metals. In fact, they may be regarded as materials, which are not clearly metallic or nonmetallic. Tin is extensively used in plating steel to make tin cans, and lead is used for storage batteries.

Semiconductors

In metals, there are plenty of conduction electrons available to carry the current and resistivity arises from thermal scattering. In semiconductors, such as silicon and germanium, the number of electrons available to carry current is very much smaller, so small as to result in a much greater resistivity. At 0 K, a semiconductor has no delocalized electrons: all electrons are bound to individual atoms. As the temperature is raised, some electrons are freed from their atoms, i.e., become delocalized, and are available to carry electric current. At higher temperatures, the number of delocalized electrons is increased and the resistivity correspondingly decreased. This, then, is the characteristic difference between metals and semiconductors. In a sense, semiconductors are more closely related to insulators than to metals, for semiconductors have a continuous gradation in the ease with which electrons can be released from their atoms by heating. In insulators the energy barrier to this release (the band gap) is large, in semiconductors it is small, and in metals the conduction electrons are delocalized even at 0 K. Some typical values of band gaps are listed in Table 1.4.

Table 1.4 Band gaps of semiconductors

Material	Band Gap, eV
Tin (gray)*	0.08
Germanium	0.6
Silicon	1
Diamond	10

* Tin has two allotropic forms. White tin, stable above 18°C, is a metal. Gray tin, stable at lower temperatures, is a semiconductor.

The electrical properties of semiconductors are strongly influenced by the presence of even minute concentrations of impurities because the energy barrier to the release of electrons from most dissolved impurities is very much less than that to the release of atoms of the host material. Small amounts of impurities therefore have a very marked effect on the conductivity of semiconductors, and for this reason control of purity is a matter of major importance in their preparation. A large part of the technology in the manufacture of semiconductor devices such as the transistor is concerned with the addition of the correct amount of the desired impurity element (doping) to a purified base material. The semiconductor industry has been primarily responsible for the development of modern methods of purifying the base materials.

PROBLEMS

1. Show that the c/a ratio for an ideal hcp structure is $(8/3)^{1/2} = 1.633$. Compare this value with those observed for various hcp metals.
2. Suppose that silver atoms can be regarded as spheres having a radius of 1.44 Å. Silver metal has an fcc structure. On the assumption that silver atoms are just in contact along the direction of closest packing in the crystal, what is the lattice parameter of silver?
3. Using the lattice parameter obtained above and the atomic weight of silver, calculate the density of silver metal.
4. Two samples of copper wire of the same purity are observed to have values of Young's modulus that differ by almost 50%. What is the probable cause, and how could your hypothesis be checked?
5. Prove by differentiation of the equation relating stress α to strain ε beyond the yield strength in tensile tests, namely, $\alpha = \kappa\alpha^n$, that $n = \varepsilon$ (at maximum load). (Note that it is necessary to substitute for stress and strain the equivalent load, area, and elongation expressions.)
6. Using the value of the resistivity of copper given in Table 1.3, calculate the resistance of a copper wire 1 m long and 3 mm in diameter.
7. Which is the least desirable type of impurity to have present in aluminum for electrical use, soluble or insoluble?
8. Would you say that there was any reasonable probability of making a transparent metal?
9. Considering the energy-vs.-displacement curve in Fig. 1.16, what do you conclude would be the effect of increasing temperature on the Young's modulus of a metal?
10. Considering just the thermal part of the electrical resistivity, what would be the effect of the application of a large hydrostatic pressure to a metal on its resistance?
11. A salesman claims he has a new alloy with a specific heat of 12 cal/mol·°C. What is your reaction to his claim?

REFERENCES

Flinn, R. A., and P. K. Trojan: *Engineering Materials and Their Applications,* 4th ed., Wiley, New York, 1981.

Reed-Hill, R. E., and R. Abbaschian: *Physical Metallurgy Principles,* 3d ed., Van Nostrand, New York, 1994.

Askeland, D. R.: *The Science and Engineering of Materials,* 3d ed., PWS Pub. Co., Boston, MA, 1994.

Barrett, C. S., and T. B. Massalski: *Structure of Metals,* 3d ed., Pergamon Press, Oxford, UK, 1980.

Callister, W. D., Jr.: *Materials Science and Engineering—An Introduction,* 5th ed., Wiley, New York, 2000.

Budinski, M. M., and M. R. Budinski: *Engineering Materials, Properties and Selection,* 6th ed., Prentice-Hall, New York, 1999.

VanVlack, L. H.: *Elements of Materials Science and Engineering,* 6th ed., Prentice-Hall, New York, 1989.

Mangonon, P. L.: *The Principles of Materials Science for Engineering Design,* Prentice-Hall, New York, 1999.

Cullity, B. D.: *Elements of X-ray Diffraction,* 2d ed., Addison-Wesley, Reading, MA, 1978.

Dieter, G. E.: *Mechanical Metallurgy,* 3d ed., McGraw-Hill, New York, 1986.

ASM Handbook: Vol. 8, *Mechanical Testing,* ASM International, Materials Park, OH, 1995.

Cullity, B. D.: *Introduction to Magnetic Materials,* 2d ed., Addison-Wesley, Reading, MA, 1972.

Pollock, T. C.: *Properties of Matter,* 5th ed., McGraw-Hill, New York, 1995.

chapter

2

Deterioration of Material Properties

Of the various ways in which a material may deteriorate and fail in service, fracture or breaking into separate pieces is the most conclusive and dramatic. A primary aim in the selection of a material and the design of the parts for a particular application is the suppression of fracture. It may occur as the sudden breaking of an apparently sound structure such as the hull of a ship, as a result of fatigue in a part subject to repeated loading, or as the end result of extensive plastic deformation.

A less dramatic change in a material than its complete fracture may result in the inability of a part to perform its proper function; changes in dimensions through creep or even through elastic deflection may lead to the failure in a system where close dimensional tolerance must be held. However, these lesser failures are easier to anticipate and eliminate in advance, whereas the sudden and complete fracturing of a load-carrying member in a structure is often difficult to anticipate and frequently leads to dramatic consequences. For this reason, the development of an understanding of how material properties deteriorate is as important as having the knowledge to select the initial properties.

The control of this deterioration is only partly a materials science problem. Proper part design and use of the techniques of flaw detection and quality control are also important. Through care in the selection of material for a given application and in manufacture and inspection, a great deal can be done by way of preventing unexpected failures. These, however, can never be completely eliminated, and in many cases it is as important to design for adequate reliability in a structure as for adequate load-carrying capacity. Design for reliability implies that, in addition to the strength of the material under consideration, such factors as the statistical scatter in mechanical-test data, the possible presence of corrosive environments, conditions of dynamic loading, and design of multiple-load paths in cases of partial failure must be considered.

The fracturing of a material is the end result of a series of complex deterioration processes occurring within the material. Because the events leading to fracture are on a submicroscopic scale, occur rapidly, and usually are not visible at the surface, direct experimental study of the fracture process is exceedingly difficult. Because of the variety of mechanisms at work during the fracture process, it is often difficult to know to which cases a particular theory applies. The materials engineer must therefore be cautious about making generalizations about fracture for particular alloys.

Two basic types of fracture can be recognized in materials: brittle fracture and ductile fracture. A fracture in a specific material may be partly brittle and partly ductile, but frequently fractures that are almost entirely brittle or entirely ductile are encountered.

A brittle fracture in a metal is one in which the fracture crack propagates throughout the material at or before the moment when plastic flow first begins. Perfectly brittle fracture occurs in noncrystalline materials, such as glasses, when the temperature is so low that no atomic mobility exists. Under these conditions, fracture can occur without any trace of plastic deformation. In metals, however, it appears that brittle fracture is initiated by the onset of plastic flow within the material. A good example of brittle fracture can be observed by breaking a bar of gray cast iron with a hammer. The fracture surface will show many bright, flat facets, and there will be no visible evidence of any plastic deformation (although some local microscopic deformation will have taken place).

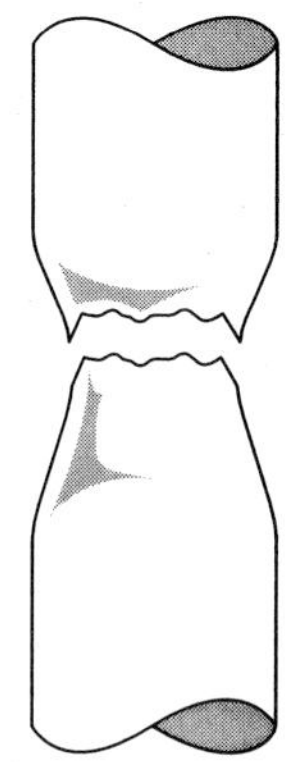

Figure 2.1
Cup and cone ductile fracture of a tensile specimen.

Ductile fracture occurs at the end of extensive plastic deformation. It may be observed by pulling a bar of soft copper, for example, to failure in a tensile testing machine. The fracture surfaces of the test piece will have the characteristic "cup and cone" appearance illustrated in Fig. 2.1. Ductile fracture begins after the test bar has begun to neck down. Stopping the tensile test short of failure and sectioning the specimen has showed this. As illustrated in Fig. 2.2, an internal crack can be observed inside the neck.

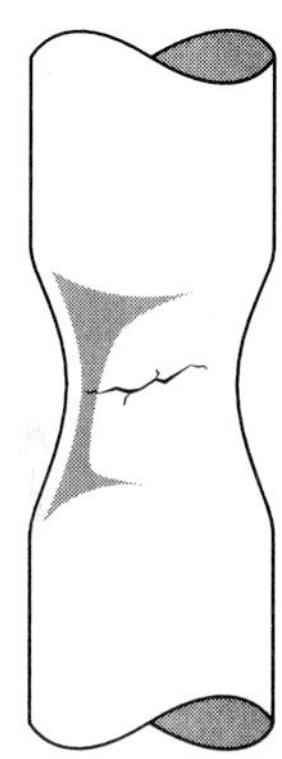

Figure 2.2
Longitudinal sections of a tensile-strained specimen just prior to fracture, showing the development of an internal ductile fracture crack starting at the center of the necked-down portion of the tensile specimen. Just prior to formation of the crack, substantial triaxial tensile stresses were present.

The foregoing experiment illustrates one of the most characteristic differences between brittle and ductile fracture: ductile fracture proceeds only so long as the material is being strained. Stop the deformation, and propagation of the ductile fracture crack stops. In brittle fracture the crack, once initiated, propagates through the material with a velocity comparable with that of the speed of sound, and there is virtually no possibility of arresting it in transit. Brittle fracture in structures is particularly dangerous because of this characteristic: there is no external warning of the imminence of fracture.

In minimizing deterioration and failure of materials through fracture, the perfection of the material itself is often an important factor. Internal defects such as cracks originating during the fabrication of the material and inclusions can, like notches, act as stress concentrators and so initiate fatigue or brittle failures. Also, parts containing slow-growing fatigue cracks can often be removed from service before they reach a critical length, if suitable testing methods can locate the cracks. Tests used in any attack on these problems must be nondestructive, for obviously all the material in question must be tested in such a way as not to impair its future usefulness. A number of techniques for flaw detection are available. A common problem is distinguishing between the signals that originate from defects and those that come from the true structural features of the material. Large defects, i.e., discontinuities that are large compared with the features of the microstructure, are easy to detect, while small defects, i.e., those comparable with the features of the microstructure, are very difficult to find.

There are three general methods of nondestructive testing. In radiography, a photograph of the piece to be tested is made by using either x-rays or gamma rays. Magnetic testing finds irregularities in the magnetic field surrounding the material as it carries a heavy electric current. In ultrasonic testing, use is made of the ability of defects to reflect sound waves. Acoustic methods are also being used on materials suspected of having internal damage. Frequently, they can pick up the minute acoustic signals emitted by such a phenomenon as dislocation motion, an intermittent step of crack growth.

2.1 FRACTURE IN A DUCTILE MATERIAL

It has been known for several years that ductile-fracture cracks are nucleated at impurity particles, such as oxide inclusions in metals. If a specimen of metal truly free of inclusions could be made, it is expected that it could show a 100% reduction in area in a tensile test. Figure 2.3 shows the development of ductile-fracture cracks in a tensile specimen of a ductile metal such as copper. When a tensile bar begins to neck down and is about to fracture, it appears that many such cracks are nucleated and that they grow together. Continuation of the plastic deformation produces a cavity that grows outward in much the same manner as the neck grows inward.

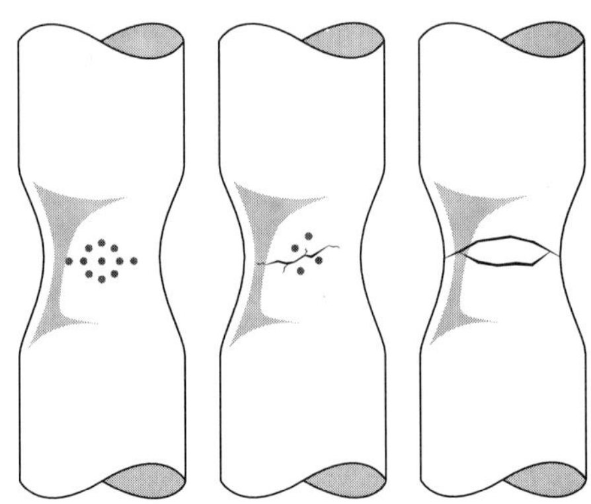

Figure 2.3 During deformation of a tensile specimen of a ductile material, small internal cracks are nucleated at impurity particles, grow, and coalesce until final failure occurs by shear in the outer ring.

2.2 FRACTURE IN A BRITTLE MATERIAL

Brittle fracture can be regarded as occurring in two distinct steps, viz., the initiation or nucleation of a crack, and its subsequent propagation. This division is reasonable, since it is possible, in some circumstances, for cracks to initiate but not propagate (Micro. 2.1).

To initiate an intergranular crack, it is necessary to develop, in a local region, a normal stress across a pair of crystal planes that is greater than the cohesive strength of the crystal on these planes. Theory shows that the order of magnitude of the stress required is about 10^6 lb/in^2. In the absence of a stress concentration of this magnitude, no crack can be formed. Under some conditions, however, stress concentrations of this magnitude can be built up in the crystal during the initiation of plastic deformation.

The two conditions for the nucleation of a crack by plastic deformation are as follows:

1. Dislocations set in motion must interact with barriers or other dislocations to establish large local stress concentrations.
2. The stresses so built up must not be relieved by plastic deformation of the surrounding material.

Definite evidence of crack nucleation by plastic flow has been found in the bcc transition metals as shown in Fig. 2.4. One mechanism by which large, local normal stresses can develop is illustrated in Fig. 2.5. As plastic flow begins in one grain, dislocations moving on a pair of intersecting {110}-type planes come together. It

Micrograph 2.1 Microstructure of iron slightly deformed in tension at −140°C. Cracks were nucleated in the structure but did not propagate.

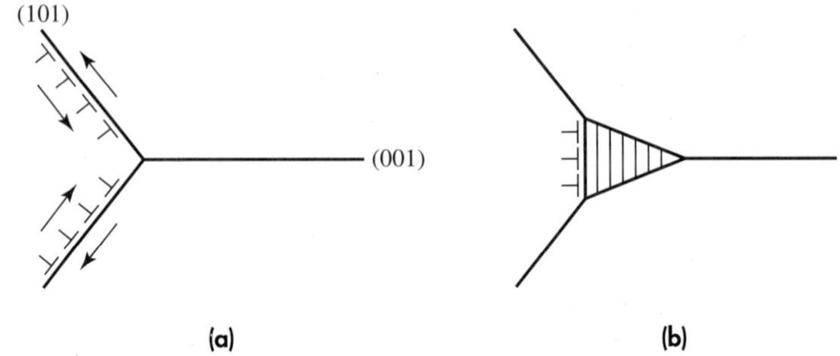

Figure 2.4 Motion of dislocations on intersecting {110} planes in the bcc structure, resulting in the nucleation of a void, which forms a crack on a {100} plane.

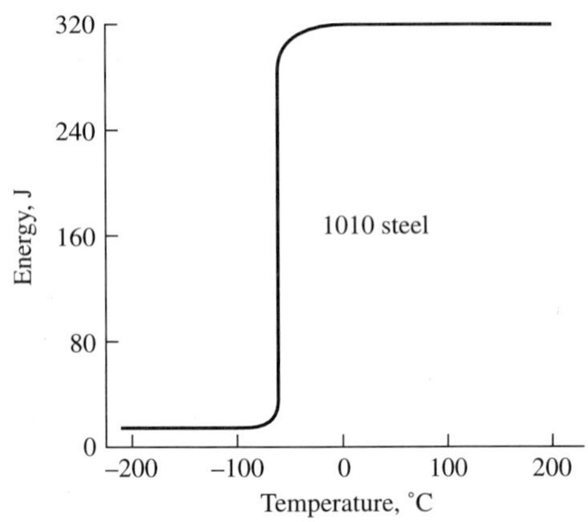

Figure 2.5 Ductile-brittle transition-temperature curve for low-carbon steel as revealed by V-notch impact tests over a range of temperatures.

can be shown that the energy of this array can be lowered if pairs of dislocations on the {110}-type planes combine to form new dislocations with extra half planes parallel to the {100} crystal planes. Since {100} planes are not slip planes in bcc metals, these new dislocations are not mobile, and, as shown in Fig. 2.4, they act as a wedge tending to split the crystal apart on the {100} planes. If the dislocations in the surrounding grains cannot move because of the presence of interstitial solute atoms, the stress concentration developed by the dislocation interaction just described will not be relieved and a crack will be nucleated. In agreement with the above mechanism are the experimental facts that the bcc metals do cleave along {100} planes in brittle fracture and that iron free of all interstitial solutes (carbon, nitrogen, boron, and oxygen) is ductile even at temperatures as low as 4 K. In the case of the fcc metals, it can be shown that the dislocation interactions of the type shown in Fig. 2.4 are energetically unfavorable; in fact, brittle fracture never occurs in the fcc metals unless a second phase is present or the metal is in an environment leading to stress-corrosion cracking.

If the stress required to nucleate cracks by a mechanism such as that discussed above is greater than the stress required to propagate a crack, the material will be brittle. If it is less, it may or may not be brittle, depending on whether or not the crack can propagate. For a given state of stress in a material, the factors resisting crack propagation are the increase in energy resulting from the formation of two new surfaces and the work done in plastically deforming the material ahead of the running crack. Thus, the plastic properties of the material are important in determining whether or not a crack can propagate. Large resistance to plastic flow means that less plastic work will be done when a crack runs and that a crack of a given size is more likely to propagate. Decreasing the plasticity of a material, i.e., increasing its yield strength, can generally be expected, then, to increase the susceptibility to failure by brittle fracture.

2.3 SUPPRESSION OF BRITTLE FRACTURE

Several considerations are important in the design of steel structures if brittle fracture is to be avoided: temperature of operation, composition and grain size of the steel, service environment, and the stress distribution must all be considered.

Temperature

At subzero temperatures, steel exhibits a greater resistance to plastic flow and hence is more susceptible to failure by brittle fracture. As measured by a particular test, such as the energy absorbed in breaking a specimen of standard size and shape, the transition from ductile to brittle behavior occurs at a fairly sharply defined temperature known as the ductile-to-brittle transition temperature (DBTT). As illustrated in Fig. 2.5, a sharp drop in the energy required to break the specimen occurs when brittleness sets in. It should be remembered that the transition temperature is not a uniquely defined temperature; it depends on the testing method used, the shape, the alloy composition, and even the size of the piece of material in question. A material that is ductile as a thin plate at a given temperature may be brittle when in the form of a thicker plate at the same temperature.

Composition and Grain Size

In ordinary plain-carbon steel, increasing the carbon content increases the brittleness of the material; i.e., it raises the transition temperature and lowers the energy absorbed in breaking above the transition temperature. Of the other elements commonly present in plain-carbon steel, excess phosphorus is particularly undesirable,

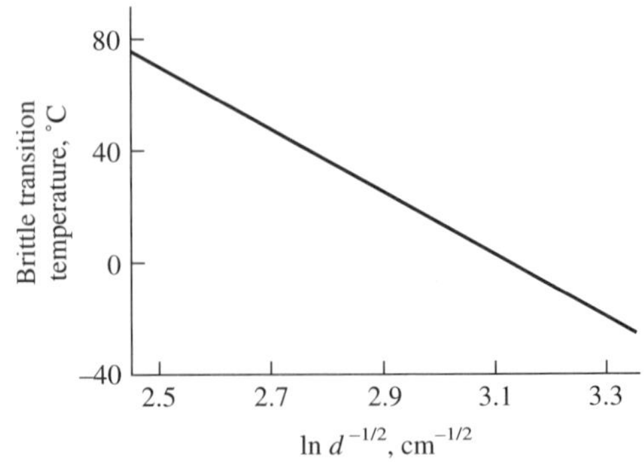

Figure 2.6 Dependence of brittle transition temperature on square root of ferrite grain size in a 1010 steel (*d* represents average grain diameter).

as the presence of this element greatly raises the transition temperature. Of the various elements used in alloy steels, only manganese and nickel seem to have any really beneficial effect on the transition temperature.

Ferrite grain size has an important effect on the brittleness of steel. The smaller the ferrite grain size, the lower the transition temperature. As shown in Fig. 2.6, there is a linear relation between the transition temperature and the logarithm of the square root of the ferrite grain size. The smaller the grain size, the shorter are the slip planes in any one grain, which means that the number of dislocations interacting is less and there is less chance of building up a stress concentration sufficient to nucleate a crack.

Environment

Certain environments can greatly increase the susceptibility of materials to brittle fracture. For example, any environment that leads to hydrogen being dissolved in steel is particularly dangerous, as the steel may be embrittled even at room temperature. A high-strength steel, for example, may be rendered brittle by certain types of electrolytic plating. Hydrogen is dissolved by the steel while in the electrolyte, with results as discussed in the next section.

Stress Distribution

Materials that are completely ductile under uniaxial tensile loading may be completely brittle when subjected to the complex combined stresses found in a complicated structural member. Two considerations are important in dealing with the influence of the state of stress on the occurrence of brittle fracture: stress-concentration effects and tensile-stress components in directions other than loading. These may be illustrated by considering the effect of a notch on the fracture properties of a plate loaded in tension (Fig. 2.7). As shown in Fig. 2.8, there is a large concentration of tensile stress in the direction of stress application (vertical)

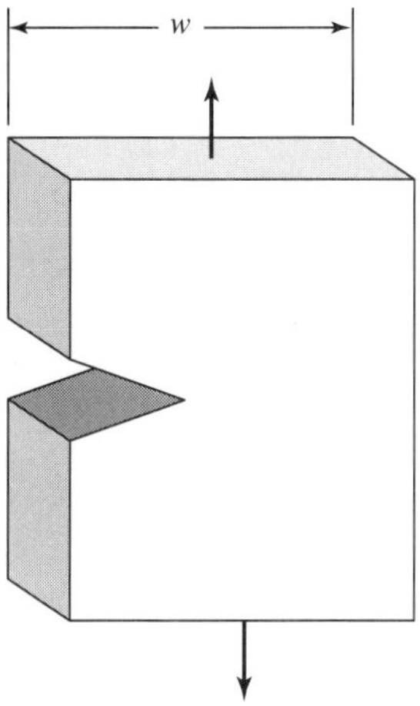

Figure 2.7 Notched plate loaded in tension.

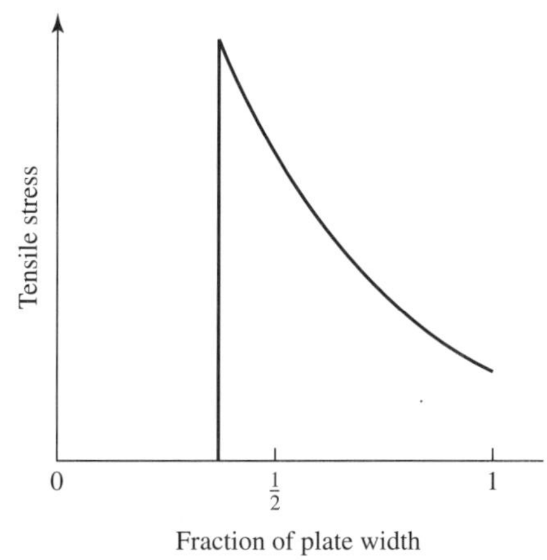

Figure 2.8 Tensile stress across the plane of the notch in the specimen of Fig. 2.7.

at the root of the notch. Fracture, if it occurs, is expected to be initiated in this region of stress concentration. In addition to the concentration of the vertical tensile stress, tensile components develop in the horizontal plane passing through the center of the notch. These stress components, which arise through the Poisson expansion of the material, become greater as the plate's thickness increases. The material at the root of the notch enters a state of triaxial tension and increases its susceptibility to fracture. Because of this effect, notched thick plates may be brittle while notched thin plates of the same material may be ductile at a given temperature (grain size in this case is usually an additional factor).

The engineer, through elimination of stress concentrators such as notches, can greatly minimize the possibility of brittle fracture, thereby allowing the use of higher-strength material in any given application. Notches leading to brittle fracture in highly stressed parts can also arise through faulty workmanship. Machining marks and poor welding are two common causes of trouble. In the microstructure of the material itself, the graphite flakes in gray cast iron act like notches and deprive this material of almost all its ductility.

2.4 LINEAR ELASTIC FRACTURE MECHANICS

The significance of notches or cracks in terms of brittle fracture resistance, particularly when heavy sections of material are involved, has led to the development of an approach to fracture called linear elastic fracture mechanics. The analysis assumes that all real structures contain one or more sharp cracks, as a result of either fabrication defects or material flaws. Thus the problem becomes one of defining exactly the level of stress that may be applied to this flaw before it becomes a moving brittle crack.

The elastic stress field in the vicinity of a crack tip can be described by a single parameter known as the stress intensity factor. It is a function of the flaw geometry and the nominal stress acting in the region in which the flaw resides. Therefore, if the relationship between the stress intensity factor and the important external variables (applied stress and flaw size) is known for a given structural geometry containing a particular defect, the stress intensity in the region of the crack tip can be established from knowledge of the applied stress and flaw size alone.

A critical value of the stress intensity factor, conventionally designated K_C, can be used to define the critical crack-tip stress condition for failure. For the opening mode of loading (tension stresses perpendicular to the major plane of the flaw) under plane-strain conditions (strain under constraint that prevents any strain in the direction parallel to the crack front), the critical stress intensity factor for fracture instability is designated as K_{IC}.

For materials that can develop a region of plastic deformation at the crack tip—for example, most metals—rapid crack extension is basically related to a plastic strain limit (ductility) of the metal crystals located in the plastic zone when stored energy is present. Unstable crack movement occurs when the plastic zone reaches a critical size; the larger the plastic zone ahead of the fracture, the more energy is consumed in propagating the fracture and the tougher the material. If the structure dimensions are such that the plastic zone size is very small compared to the flaw size, the fracture toughness is the lowest possible value K_{IC} that will be exhibited by the metal. It is on this basis that K_{IC} is considered a fundamental material parameter.

Analysis by linear elastic fracture mechanics is accurate provided that the plastic zone at the tip of the crack is small compared with the general specimen dimensions. As the ratio of plastic zone size to specimen size increases, linear elastic fracture mechanics methods become inapplicable. For constructional metals in heavy section thickness, plastic constraint due to thickness may produce plane-strain conditions even in normally high plasticity materials. Measurements of K_{IC} in these materials would then be a potentially useful tool in fracture control, since it is a

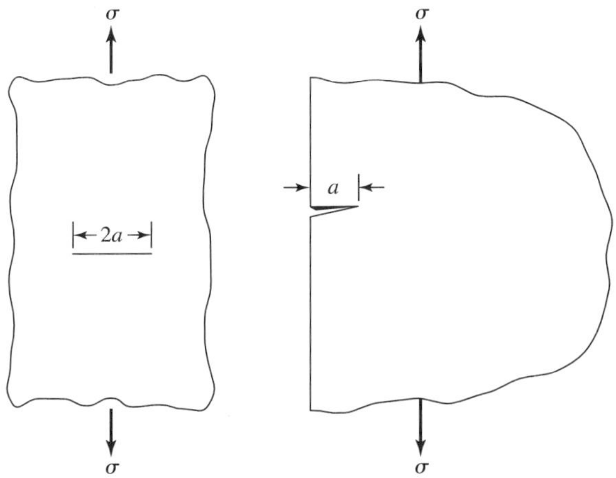

$K = \sigma\sqrt{\pi a}$ $\qquad$ $K \cong 1.1\sigma\sqrt{\pi a}$

Figure 2.9 Some stress intensity factors for simple configurations. Cracks are through the thickness of the plate. When K reaches K_{IC} of the material, unstable growth of the crack occurs in the case of thick sections, which by definition meet plane-strain criteria.

minimum level of toughness that could be expected under the most adverse stress states.

In many cases, the mathematical calculation of K values for service conditions (for flaws known or suspected to exist) is simple. Examples of two common cases are seen in Fig. 2.9. Using these formulations and the known K_{IC} for a given material, the largest flaw that can exist without producing brittle fracture can be calculated. This permits the evaluation of material known to contain flaws for continued service if already in service (for example, a bridge that was found to contain cracks) or setting conditions for use of materials in terms of stress–flaw size combinations that are permissible for a material with a given K_{IC}.

2.5 PROPERTY DETERIORATION AT HIGH TEMPERATURE

The endpoint of the creep process is failure by creep rupture, but in most situations where creep is important, "failure" has occurred by the time the part in question has changed its dimensions outside of design tolerances; usually this is long before rupture occurs.

Under stresses less than their yield stresses, metals display a slow plastic flow—creep—resulting from the action of two processes that were both in operation: grain-boundary sliding, and the movement of dislocations past obstacles by climb. Both processes depend on thermally activated atomic mobility. The magnitude of this mobility can be illustrated in the following way: The time τ that elapses between successive atomic jumps can be calculated from the self-diffusion coefficient D by the relation

$$D = \frac{\alpha a^2}{\tau}$$

where α is a constant that depends on the crystal structure but is always of the order of magnitude of $\frac{1}{10}$ and α is the lattice parameter. We know that D depends on temperature through the relation

$$D = D_O e^{-Q/RT}$$

and it is found that for self-diffusion the ratio Q/T_m, where T_m is the melting temperature in Kelvins, is nearly constant for all metals, equal to about 37 kcal/mol·K. By using the above expression, then, τ can be calculated as a function of the ratio T/T_m, and the calculated values of τ so obtained will apply to all metals. Table 2.1 shows the results. At very low temperatures the atoms can be regarded as permanently fixed in their places in the crystal lattice, while when the melting temperature is approached, the atoms change places with one another very rapidly indeed.

The data given in Table 2.1 are for atomic mobility within the grains themselves; at the grain boundaries atomic mobility is always much greater. In fact, experiments at moderately high temperatures show the grain boundary acting as if it were a viscous fluid. Under stress at moderate or high temperature, a metal deforms by grain-boundary sliding owing to viscous flow at the grain boundaries. This is one mechanism of creep strain.

If the applied stress on a metal is less than the yield stress, then its dislocations will presumably not be set in motion. This will be true at absolute zero, but at higher temperatures thermally activated atomic mobility can, over a period of time, aid dislocations in moving past barriers which they otherwise would not be able to overcome. One mechanism by which this can occur is illustrated in Fig. 2.10. Suppose that dislocations are piled up at a barrier on the slip plane, the applied stress not being sufficient to allow them to break through. Any dislocation that can move to a parallel slip plane above the obstructed one could move again. This process, known as dislocation climb, can occur in a finite time if the atomic mobility is great enough. All that is necessary is for vacancies to diffuse to the extra half plane of the dislocation until one row of atoms is removed. The dislocation then terminates on the next slip plane above and is free to move on. Table 2.1 shows that the contri-

Table 2.1 Atomic mobility in metals

T/T_m	0.1	0.2	0.3	0.4	0.5	0.6	0.7	0.8	0.9
τ, s	10^{66}	10^{25}	10^{11}	10^{4}	10	10^{-2}	10^{-4}	10^{-6}	10^{-7}

Figure 2.10 Climb of dislocation over a barrier during creep of a metal.

bution of dislocation climb to creep will increase rapidly with temperature. In fact, as T approaches T_m, it becomes more and more difficult for a metal to show any true rigidity.

Both grain-boundary sliding and dislocation climb contribute to the creep strain of a stressed metal. When the stresses are low and the creep rates small, it is believed that grain-boundary sliding is the more important process. At high stresses and temperatures, dislocation climb becomes more important, presumably because the motion at the grain boundaries is restricted, not by the viscous properties of the boundary itself but by restraints offered by neighboring grains.

The continuation of creep in a metal eventually results in its failure by rupture. Generally the failure that results follows along the grain boundaries, resulting in what is known as an intercrystalline fracture. Two types of intercrystalline failure are observed to occur. In the first, fracture is nucleated at points of particularly high stress concentration along the boundaries. The lines where three or more boundaries intersect and where, consequently, it is difficult to accommodate the slip occurring on all three boundaries are most likely to be the places where cracks nucleate. The other type of failure involves the growth of voids along grain boundaries, a type of failure most likely to occur when small stresses are applied for long periods of time. The voids are observed to form primarily on boundaries that lie normal to the direction of tensile stress. With the passage of time, they grow and coalesce, developing into cracks. It is believed that the voids grow by the agglomeration of vacancies generated within the grains during their deformation.

As a practical matter, suppression of creep involves, first of all, the use of a material with a melting point as high as possible relative to the temperature of service. Elements in solid solution and a large grain size also tend to reduce creep rates. A technique known as dispersion strengthening is particularly effective, since the second-phase particles of a few nanometers in size are themselves hard at the temperature of operation. The most effective dispersant for high-temperature service is Al_2O_3, a finding that could have been predicted on the basis of the high melting point of this oxide. These dispersed particles must be neither too large nor too small if they are to act as effective barriers to the dislocation motion responsible for creep. If the particles are too small, then the dislocations can easily push through them in the matrix. If they are too large, the dislocations can pass through the relatively large amount of space between particles (for a given amount of precipitate, the smaller the particle size, the smaller the distance between particles). A spacing of about 10^3 atom distances between particles is found to be most effective.

An important problem in producing a creep-resistant alloy by use of second-phase precipitates is to keep them from coarsening at the service temperature. Because the surface-to-volume ratio for fine particles is large, the interfacial energy of the particles is an effective driving force tending to cause coarsening. Since the temperature is high and the interparticle distance small, there will be a tendency for matter transport from small to large particles by means of matrix diffusion. To suppress this tendency, the composition of the particles must be such that there is but limited solubility in the matrix or the particle-matrix interfacial energy must be very low, thus reducing the driving force for agglomeration. In the highly creep-resistant material known as SAP (sintered aluminum powder), Al_2O_3 particles are dispersed in aluminum and, because of limited solubility of the oxide in the matrix, remain dispersed even at high temperatures. In this material, the ratio of the maximum service temperature to the melting temperature T_s/T_m is about 0.72. Service at much higher temperatures is attained with iron- or nickel-base alloys, although with these the T_s/T_m ratio is usually not so favorable as with SAP.

2.6 PROPERTY DETERIORATION FROM CYCLIC LOADING

It is well established that the phenomenon known as fatigue starts at the tensile-strained surface of the material, subjected to cyclic loading, and grows at a very slow rate. In contrast to creep, it does not occur through thermally activated atom movements. Fatigue failures can be readily produced at temperatures as low as 4 K. The initial stage of the fatigue process is found to be the formation of a small opening at the place where a slip band reaches the surface.

This can be shown to be a result of the back-and-forth motion of small groups of dislocations moving near the surface. Once an opening in the surface has been formed, it is only a question of time before the fatigue crack grows to a critical size. However, if the tensile component of the cyclic applied stress remains below the endurance limit, either no localized dislocation motion at the surface of the type described above occurs, or, if it does, work hardening occurs fast enough to arrest it before permanent damage is done.

Increasing the yield strength of a material by alloying will usually make it more resistant to fatigue damage. However, the avoidance of fatigue failure is as dependent on proper design as it is on material properties. Stress concentrations play an important role in initiating fatigue cracks. Sharp corners in a stressed member or even surface roughness left by machining are often the causes of fatigue failures. Small changes in the design of a part or structural member may be sufficient to reduce high stress concentrations. Sometimes it is possible to incorporate residual compressive stresses, as by shot peening, into a part that is to be subjected to alternating stresses in service. These compressive stresses greatly reduce the probability of a fatigue crack being initiated. The condition of the surface is of prime importance in the control of fatigue failure. Not only must it be smooth and free of stress concentrations, it should be as strong as or stronger than the underlying structure. A small amount of decarburization at the surface of steel may drastically reduce the endurance limit, while nitriding may have a beneficial effect.

With the increasing interest in characterizing materials by fracture mechanics concepts, emphasis in fatigue has shifted to crack growth behavior. The reasoning behind this shift is twofold. First, since most engineering components contain some kind of flaw or discontinuity, it is the crack growth, not the crack-initiation portion of the fatigue life, that is of interest. Endurance limit tests are a summation of both. Second, a fracture mechanics analysis of a component requires a knowledge of flaw sizes at all points in the service life of the material. Since initial flaws grow by fatigue during this life, the rate of flaw growth or crack propagation rate must be known to assess the flaw size at the end of life and ensure that it does not exceed the critical fast-fracture flaw size for the material.

Fatigue-crack growth-rate tests usually are made using the fracture mechanics parameter K as the means of characterizing growth-rate behavior. The variations in K, or ΔK, caused by variations in stress acting upon the flaw size, are correlated to the rate of crack growth by an expression of the type

$$\frac{da}{dN} = A(\Delta K)^m$$

where a is crack length, N is the number of cycles, ΔK is the stress intensity factor range, and A and m are constants reflecting such variables as mean stress, material properties, and environment. Test specimens used in studies of this type are center-notched or center-cracked panels, wide enough for a fatigue crack to grow through the specimen thickness and then laterally toward the specimen edges. The total life of the component is determined by knowing or assuming an initial flaw size and using the measured growth rate, integrating to find the total life.

A corrosive environment will greatly accelerate fatigue, a phenomenon known as corrosion fatigue. So sensitive is fatigue to environment that even the presence of moist air can reduce the fatigue life of a metal as compared with its life in a test under vacuum. Exposure of a steel gear to seawater, for example, may result in its premature failure by fatigue upon use after removal from the water, even though it was carefully cleaned after exposure. The remedy for corrosion fatigue is the selection of a material resistant to corrosion in the service environment.

PROBLEM

1. A material possessing a plane-strain fracture toughness of 50 MPa-m$^{1/2}$ and yield strength of 1000 MPa is made into a large panel.

(*a*) If the panel is to be stressed to a level of 250 MPa, what is the maximum size flaw that can be tolerated before catastrophic failure occurs? (Assume a center notch configuration.)

(*b*) At the point of fracture, what is the size of the plastic zone at the middle of the panel along the crack front?

(*c*) If the panel was 2.5 cm, would this constitute a valid plane-strain condition?

(*d*) If the thickness was increased to 10 cm, would there be a change in critical size of the flaw calculated in part (*a*)?

REFERENCES

Hertzberg, R. W.: *Deformation and Fracture Mechanics of Engineering Materials,* 4th ed. Wiley, New York, 1996.

Garrett, G. G., and D. L. Marriot (Ed.): *Engineering Applications of Fracture Analysis,* Pergamon Press, New York, 1979.

Campbell, J. E., W. W. Gerberich, and J. H. Underwood (Ed.): *Application of Fracture Mechanics of Metallic Structural Materials,* ASM Int., Materials Park, OH, 1982.

Conway, J. B.: Creep-Fatigue Interaction, in *Mechanical Testing,* vol. 8, 9th ed., *Metals Handbook,* ASM Int., Materials Park, OH, 1985.

Broek, D.: *The Practical Use of Fracture Mechanics,* Kluwer Academic Publishers, Dordrecht, Netherlands, 1989.

Bannantine, J. A., J. J. Comer, and J. L. Handrock: *Fundamentals of Metal Fatigue Analysis,* Prentice-Hall, New York, 1990.

Fuchs, H. O., and R. I. Stephens: *Metal Fatigue in Engineering,* Wiley, New York, 1980.

SECTION

2

STRENGTHENING MECHANISMS

chapter

3

Solid-Solution Strengthening

Metals are alloyed by the addition of other elements, generally combined in the liquid state, in order to achieve an increase in the relatively low strength of pure metals and to improve other properties. The simplest and most general alloy-strengthening mechanism is that achieved by adding another metal that is so related to the base metal that after solidification of the liquid a single-phase solid solution results. In this chapter, the simple Cu-Ni system will be used as an example of solid solutions and associated strengthening.

3.1 FORMATION OF SOLID SOLUTIONS

A solid solution is a solid containing two or more elements atomically dispersed at random in a single crystalline structure. In the substitutional type of solid solution, atoms of the solute element are substituted for solvent atoms at random points in its lattice. In the interstitial type of solution, the solute atoms are present at interstices of the solvent structure. Interstitial solutes are limited to the small atoms, generally B, C, N, and O.

In a binary alloy system, four principles define the degree of substitutional solid solubility that a given solvent may have for various solute atoms.

Crystallography Complete solid solubility, i.e., from 0 to 100% of the added metal, is possible only if both elements have the same crystal structure. Many systems meet this requirement, but in only a few, for example, Cu-Ni and W-Mo, does this complete solid solubility exist at all temperatures. In many other cases, for example, Fe with Cr, Ni, or V, allotropic transformations of one or both metals limit complete solid solubility to certain temperature ranges. In still other cases, for example, Cu-Au, the random distribution of solute atoms changes to a preferred, or "ordered," distribution at certain atom ratios, for example, 1:3 or 1:1 for the four-atom lattice.

Size Factor A necessary condition for the formation of a solid solution when two metals are alloyed is that their atomic sizes or effective radii be within 15% of one another. There are other factors that may prevent the formation of a solution when the size factor is favorable, but if the size difference is greater than 15%, solubility will be very limited.

A large size difference between solute and solvent means that there is a large elastic strain set up around each solute atom in the solid solution. With increasing solute content, the strain energy in the solid increases, making the solution unstable.

Valence Factor As the electron valences of the solute and solvent become more unlike, solid solubility becomes more restricted. The valence difference of the solute and solvent determines the electron-to-atom ratio e/a of the alloy, i.e., the number of valence electrons present per atom. For example, $e/a = 1.5$ for an alloy of 50 at. % Cu (valence 1) with 50 at. % Zn (valence 2). Alloy structures can generally tolerate only a certain change, increase or decrease, in their e/a value before they become unstable or of too high energy and transform to another lower-energy structure. This is illustrated in alloys of copper with metals of higher valence. As the valence of the solute increases, its maximum solid solubility decreases. The data in Table 3.1 show that the maximum solubilities of Zn, Ga, and Ge, which are in the same row of the periodic table as Cu, occur at an approximately constant value of e/a.

Table 3.1 Maximum solubilities of succeeding atomic numbered metals in copper

Atomic No.	Solute	Valence	Maximum Solubility, At.%	e/a at Max. Solubility
29	Cu	1		
30	Zn	2	38.3	1.39
31	Ga	3	19.9	1.40
32	Ge	4	11.8	1.35

Electronegativity Elements such as F and Cl are strongly electronegative, meaning that they have a strong affinity for electrons so that chemical bonds are formed. In metals such as Na and Mg, electronegativity is very low, while elements such as Pb and Sn near the center of the periodic table have intermediate electronegativities. When the electronegativity difference between two elements is large, the elements tend to form compounds of definite composition rather than solutions. Thus a large electronegativity difference between two elements means that the solubility of one in the other will be limited. Mg and Sn, for example, differ appreciably in electronegativity and form the compound Mg_2Sn. The solubility of Sn in Mg is only 3.4 at. %, even though the size factor is favorable.

3.2 MECHANISM OF SOLIDIFICATION

It is free energy that determines whether or not one phase is stable relative to another phase, and Gibbs free energy is defined by the equation

$$G = U + PV - TS$$

where P is pressure, V is volume, and U is the internal energy, the amount of work that must be done to completely separate its atoms to infinity. The greater the cohesive forces between the atoms, the greater this work and the more negative[1] the internal energy. At constant pressure, the sum of $U + PV$ occurs often and is referred to as enthalpy. Therefore, the Gibbs-Helmholtz equation is the form used for metallurgical systems

$$G = H - TS$$

In the second term, T is the absolute temperature and S the entropy, a measure of the amount of disorder in the arrangement of atoms in the phase. In a solid crystal, entropy exists primarily because of the thermal vibration of the atoms. In a liquid, where there is no regular atomic arrangement, the entropy is due both to vibration and to structural disorder. The zero point from which internal energy is measured is taken as that state where the atoms of the phase are all separated at an infinite distance from one another. Work must be done on the phase to make this separation, and internal energy is therefore always a negative quantity.

Figure 3.1 shows schematically the variation of the free energies of the solid and liquid phases of a metal with temperature. At $T = 0$ K, $F = E$, and the solid, because its interatomic cohesion is greater, has the lowest internal energy and hence the lowest free energy. Increasing the temperature increases the relative importance of the TS term and, since $S > 0$, causes the free energy to decrease. But the entropy of the liquid phase is greater than that of the solid, and so the free energy of the liquid phase decreases more rapidly than that of the solid as the temperature is raised. Consequently, the two free-energy curves must cross eventually, with the liquid having the lower free energy at higher temperatures. At low temperatures, the solid is stable and at high temperatures the liquid is more stable. The point where the two curves cross corresponds to the melting temperature.

Ideally, as a liquid metal is cooled, it should transform to solid as soon as it reaches the freezing point. Actually, this cannot happen unless, at essentially one instant of time, every atom in the melt transports itself to its proper position in the solid crystal. Since the probability of all the atoms present making such coordinated moves is negligible, solidification must start first in some localized region of the melt with the formation of a small particle of crystalline solid. As soon as such a piece of solid forms, a liquid-solid interface is created. This interface has a definite energy per unit area and increases the total free energy of the solid. If, as is the case

[1] The zero point from which internal energy is measured is taken as that state where the atoms of the phase are all separated at an infinite distance from one another. Work must be done on the phase to make this separation, and internal energy is therefore always a negative quantity.

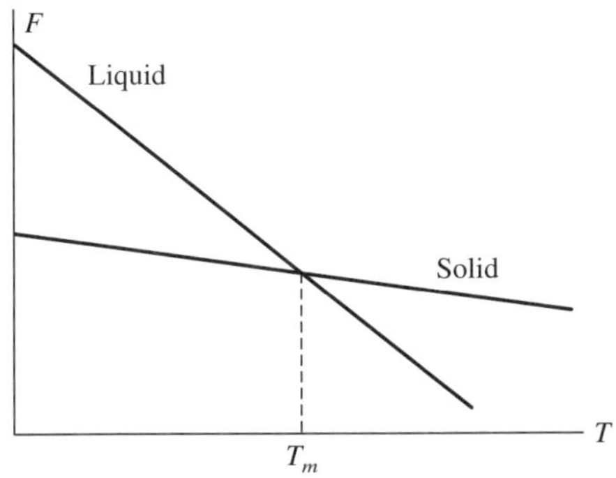

Figure 3.1 Schematic curves showing the variation of the free energy *F* of the liquid and solid phases with temperature *T*. Because of the greater entropy of the liquid, its curve has the greater slope.

at the freezing point, the free energies of the bulk liquid and solid are equal, the additional surface energy of the small particle of solid makes its total free energy greater than that of a corresponding mass of liquid, and it is therefore unstable. If such a particle of solid were to form in the liquid at the freezing temperature, it would immediately redissolve. To get the solidification process started, the liquid phase must be undercooled, cooled to a temperature below the freezing point. Then, because the bulk of solid has a lower free energy than the corresponding amount of liquid, the surface energy of a small particle of solid will not raise its total free energy above that of the liquid; it is stable and can grow. The formation of particles of solid that can grow in the liquid is called nucleation, and the particles capable of growth are called nuclei (Fig. 3.2).

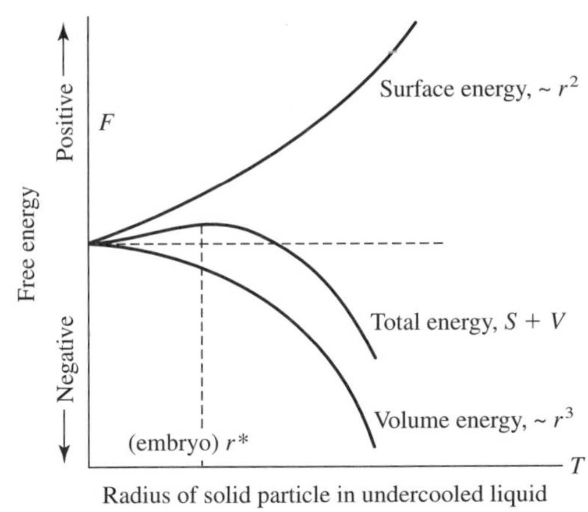

Figure 3.2 Schematic surface-vs.-volume-energy relationships for a solid particle in an undercooled liquid. The critical particle radius r^*, which determines whether the particle will tend to disappear or to grow, is a function of the degree of undercooling.

There is an important relation between the size of nuclei and the amount of undercooling of the melt. A small particle has a relatively large amount of surface compared with its volume. This is easy to see for spherical particles where the surface-to-volume ratio is

$$S/V = 4\pi r^2/\tfrac{4}{3}\pi r^3 = 3/r$$

In a small particle, the surface energy is an important part of the total energy, but in a large particle the surface area, and hence the surface energy, is relatively unimportant. With a small amount of undercooling, only large particles can become nuclei, whereas, with large undercooling, the particles can be much smaller and still be stable or grow. Because solid particles are formed by chance fluctuations in the liquid, the rate of nucleation is expected to be greater, the greater the undercooling. With small undercooling one may have to wait some time for a particle large enough to be capable of growth to form by nucleation. With more undercooling the chance of formation of a smaller nucleus is much greater.

Theory indicates that the amount of undercooling required to start the solidification of metals is quite large. In laboratory experiments, liquid metals have been undercooled by amounts as great as a third of their melting temperatures. In practice, however, only a small amount of undercooling is found necessary to initiate solidification. This is because of the phenomenon of heterogeneous nucleation. The surfaces of foreign solid particles in the melt and of the container walls are sites where nuclei of the solid phase can form with a smaller increase in surface energy than in the bulk of liquid. The energy barrier to nucleation is therefore lower at these interfaces, and the probability of a nucleus forming at a given degree of undercooling is correspondingly greater. Homogeneous nucleation, nucleation in the bulk of pure melt, is not ordinarily observed unless special precautions are taken to suppress heterogeneous nucleation.

Once a nucleus is formed, it can proceed to grow as fast as the latent heat of solidification can be carried away. Thus the thermal conductivities, relative masses, and shapes of the melt, the solid, and the mold all influence the growth of the solid phase. The importance of the relative temperatures of the solid and liquid is illustrated by the phenomenon of dendritic growth. If there is a plane interface between solid and liquid, with the solid at a lower temperature than the liquid, then the flow of the latent heat released at the liquid-solid interface will be into the solid. If, on the other hand, a greatly undercooled liquid is cooler than the solid, a situation of instability obtains; if one part of the interface advances ahead of the rest, it grows into cool liquid and can more readily dissipate the latent heat of solidification. This allows it to grow even further, with the result that a small irregularity on the solid surface can rapidly grow out into a projecting spike. Such a growing spike may send out side branches, creating a network of solid fibers growing into the liquid. This is dendritic growth, and the individual spikes are the dendrites. Figure 3.3 presents a schematic drawing of an idealized dendrite growing in a liquid. Usually, the dendrite would be attached to a cooler surface, such as a mold wall, enabling removal of the heat of solidification. It is observed that in metals dendritic growth occurs in certain preferred directions, the [001] direction for fcc and bcc and [0001] for hcp metals; these are the crystallographic directions along the length of the dendritic arms.

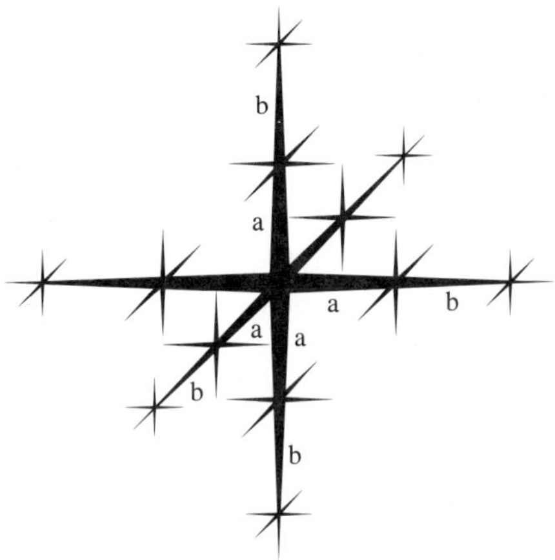

Figure 3.3 Schematic view of a solid cubic crystal dendrite growing in a liquid. The rapid-growth directions are the three cubic axes. The appearance of a two-dimensional view, i.e., a slice cut through this dendrite, would depend on the angular relationship between the plane of view and the *X*, *Y*, *Z* cubic axes.

In addition to the influence of the thermal gradient on dendritic growth, the compositional gradient at the freezing interface may also be important. Small amounts of an alloying element generally decrease the freezing temperature of the metal. As freezing proceeds, the atoms of the alloy element tend to concentrate in the liquid phase rather than the solid, and thus a thin layer of solute-rich liquid builds up just ahead of the solidifying metal. Since the freezing temperature of the liquid is roughly proportional to its composition, the liquid next to the solid-liquid interface will have a lower freezing temperature than the liquid farther removed. The tip of the dendrite spike extending through the solute-rich layer into this region will find liquid that has a greater degree of undercooling by virtue of the fact that although its actual temperature may be similar to that of the solute-rich layer, its freezing temperature is much higher. This effect, called constitutional supercooling, accounts for the stability of the dendritic growth mode in many alloys. This same phenomenon may also lead to the formation of equiaxed grains in the center of castings, as described in the next section. In this case, the center of the casting becomes undercooled before the solute-rich layer allows dendritic solidification to be complete. New grains nucleate in the undercooled liquid, and this region consists of randomly oriented equiaxed grains (grains with generally equal dimensions). Another possible reason for the formation of this region may be the fact that growing dendrite spikes extending through the solute-rich layer are broken or melted off due to thermal fluctuations or convective currents occurring during solidification. These broken dendrites float into the central undercooled region and act as nuclei for the equiaxed zone.

3.3 SOLIDIFICATION OF PURE METALS

Upon very slow cooling of a melt, with all the metal maintained at a uniform temperature, solidification will start just below the freezing point. A few nuclei, distributed throughout the liquid, will form, and each will grow in all its preferred directions (as dendrites) to form a coarse, equiaxed grain structure. If the entire liquid cools uniformly but rapidly, undercooling will be greater and many more nuclei will originate in the melt, producing a finer-grained equiaxed structure. If one part of the liquid cools rapidly and another slowly (as in industrial casting processes where a hot liquid is in contact with an originally cool mold), nuclei will form only where undercooling is first attained, i.e., at the mold wall, and these nuclei will grow in the direction of the themal gradient, giving elongated or columnar crystals. Later, the center, or hotter part, of the casting may reach the freezing temperature before columnar crystals have grown into this section. Equiaxed grains may be found here.

In castings, it is possible to have combinations of coarse and fine, columnar, and equiaxed crystals. By controlling liquid and mold temperatures, thermal conductivities, and relative masses, it is possible to exercise considerable control of cast structures. Liquid-metal temperatures just above the freezing temperature mean that removal of a small amount of heat will be sufficient for heterogeneous nucleation of the solid crystals to start. Nuclei will form and grow first at the mold wall but seldom can grow extensively before heat flow has brought the next layer of liquid to the temperature for nucleation of the solid. Thus new nuclei will tend to form before the first crystals have grown to this zone. Therefore, the tendency is for all grains to be equiaxed, with their size determined by the rate of heat removal, i.e., the temperature, mass, and thermal conductivity of the mold.

Considerable superheat of the liquid, i.e., a temperature well above its freezing point, inevitably means that steeper thermal gradients in the liquid metal are present. Thus, after the first nuclei form at the mold wall, they may be able to grow as fast as heat flows in the opposite direction. The hot liquid may not cool to the freezing temperature appreciably in advance of the growing crystals, and therefore no new nuclei can form. In this case, columnar grains are certain to form, but again their size will be determined by the rate of heat removal. It should be noted, though, that the long axis of the columnar grains would always be normal to the mold wall. In addition, since a crystal does not grow in all crystallographic directions at the same rate, favorably oriented nuclei grow faster, shutting off less favorably oriented crystals. Thus, there is a tendency for the axis of the columnar grain to be in the specific crystallographic directions of preferred growth. In the case of fcc crystals, for example, [100] directions will be perpendicular to the mold wall.

Typical examples of some possible ingot macrostructures are presented in Fig. 3.4. They happen to represent an aluminum solid solution but are typical of any metal solidifying quietly without gas evolution. Figure 3.4*a* shows a very fine equiaxed grain structure, Fig. 3.4*b* a coarser equiaxed grain structure, and Fig. 3.4*c* a completely columnar structure.

Sectioning, grinding smooth, and deep etching of a cast metal will not only reveal grain size and shape but will also often disclose voids.

Unfortunately for illustrative purposes, but fortunately for their subsequent use, the small ingots of Fig. 3.4 are free of these voids. They may originate from gas in solution in the liquid metal that is concentrated in the liquid during solidification and is evolved only when the concentration reaches a critical value. If the metal contained little gas, that point might be reached only in the final stages of solidification and might result in a few voids in the top center part or the part that is last

(a) (b) (c)

Figure 3.4 Macrographs of aluminum-base solid solution. (*a*) Fine-grained ingot obtained by pouring liquid metal at only a few degrees above its melting point into a cold iron mold; (*b*) coarse-grained ingot obtained by pouring liquid metal at only a few degrees above its melting point into a hot iron mold; (*c*) columnar-grained ingot obtained by pouring liquid metal superheated well above its melting point into a cold iron mold.

to freeze. If the original gas content is high, the gas bubbles probably will be distributed as voids throughout the casting.

A second source of voids appearing in the macrostructure or microstructure is the shrinkage of most metals during solidification, which results from the greater density of atom packing in the solid than in the liquid state. If freezing starts at all surfaces and fixes the external dimensions at nearly the same as those of the liquid, then there inevitably will be 4 to 8% voids in the structure, these figures representing the range of solidification shrinkage encountered. Only having an external reservoir of liquid metal to supply the deficiency due to shrinkage can eliminate shrinkage voids. In ingots, this liquid may be supplied by insulated "hot tops" or an electric arc that keeps the top liquid or by continually pouring additional liquid in the top and freezing from the bottom. In sand castings, the liquid is supplied by a reservoir called a riser, larger in section than the castings and connected if possible directly to the thickest section of the casting.

Both shrinkage voids and gas pores are necessarily interdendritic. Gas voids are spherical and seldom continuous, but under conditions where gas evolution is unusually large, e.g., from surface contamination of the mold or excessive gas in the liquid, long semicontinuous voids called wormholes can form. If physical conditions make nucleation and growth of a bubble of molecular gas possible, the concentration of gas in the adjacent liquid is sufficiently diminished for normal solidification to proceed for at least some specific time interval before a critical solute concentration of gas is attained again. On the other hand, shrinkage voids are necessarily continuous throughout any area in which they occur. Shrinkage cavities are therefore more likely to cause "leaky" castings, i.e., cast metals that permit water or other liquids under pressure to leak through the supposedly solid metal.

A fine equiaxed grain structure is usually desired for its greater strength and hardness. If impurities are present at grain boundaries, they will be more finely dispersed in a fine-grained structure and therefore are less troublesome. This is particularly true when the casting is an ingot that is subsequently to be rolled; a coarse-grained structure is far more likely to crack in the early stages of working, and the brittleness is related to impurity concentrations at grain boundaries.

3.4 SOLIDIFICATION OF METAL ALLOYS

Charts showing the phases present in an alloy system as functions of the temperature and composition are called phase, constitution, or equilibrium diagrams. All the diagrams given in this textbook are equilibrium diagrams, meaning that the alloy phase condition indicated for a given temperature and composition will show no change or tendency to change with time. The stable, or equilibrium, condition is dynamic; atoms are not stationary, but the gross summation of all movements is zero. The following generalizations will be helpful in analyzing charts of specific binary solid-solution alloy systems, e.g., the Cu-Ni system of Fig. 3.5:

1. Single-phase fields, e.g., those marked "liquid" or "α" must be separated by a two-phase field containing some of each single phase, e.g., the field of liquid plus α. The upper line defining the liquid plus α region is called the liquidus and the lower line the solidus.

2. When an alloy of fixed composition is heated or cooled past the temperature indicated on a diagram by a line, there is a partial (for sloping lines) or complete (for certain points on horizontal lines) change of phase and a concomitant absorption or release of energy in the form of heat. Thus, on cooling a 70% Cu–30% Ni alloy (composition a of Fig. 3.5) past the liquidus, some solid crystals of the α phase start forming, and the release of their heat of formation causes a change in the slope of the cooling curve. Observation of this effect is utilized to determine the temperature at which solidification begins or the minimum temperature to which an alloy must be heated to ensure complete melting.

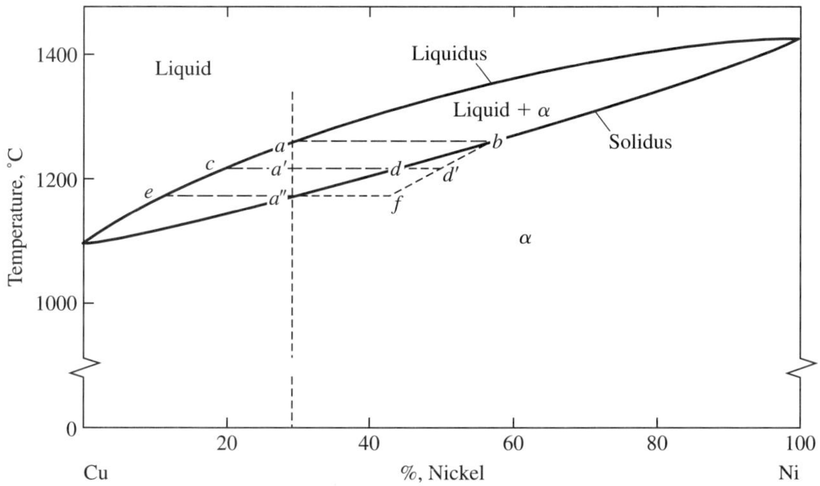

Figure 3.5 Phase diagram of the Cu-Ni alloy system.

3. In a two-phase field, the composition of each phase at a specific temperature is given by the intersections of a horizontal line, drawn at this temperature, with the phase-field boundary lines. Thus, at the temperature indicated by the horizontal line *ab*, the liquid has the composition of *a* (30% Ni) and the solid the composition of *b* (50% Ni); at the line *cd*, the liquid- and solid-phase compositions are, respectively, 22 and 40% Ni; etc.

4. The proportions of each phase in a mixture of two phases, with the temperature and composition of the alloy known, are given by the lever rule. This states that for an alloy in a two-phase field the proportionate amount of each phase is given by the ratio of the difference between the gross alloy composition and that of the other phase to the difference in composition of the two phases. Thus, in the diagram, the 70% Cu–30% Ni alloy at the temperature of the horizontal *cd* contains α of composition *d* and liquid of composition *c*. The proportionate amounts of each would be as follows:

$$[(d - a')/(d - c)]\,(100) = [(40 - 30)/(40 - 22\,)](100) = 55\%\ \text{liquid}$$

$$[(a' - c)/(d - c)]\,(100) = [(30 - 22)/(40 - 22)]\,(100) = 45\%\ \alpha$$

Generalization 3 requires that the composition of the solid phase must change during the interval of solidification over a decreasing temperature. Under equilibrium conditions, this adjustment of composition would occur throughout the solid phase, which would be forming dendrites. However, equilibrium is practically never achieved in commercial casting processes, and the first dendritic nuclei are richer in the higher-melting-point element than the successive layers formed at lower temperatures. The average composition of the total solid phase, in this case, will not be that shown by the phase diagram for a given temperature, e.g., point *d* at temperature *cd*; it will contain more of the higher-melting-point element, perhaps that given by point d'. This difference in composition from center to edge of a dendrite may remain as a result of the slowness of atomic interchange or diffusion. (Chinese bronzes over 3000 years old show this dendritic-composition difference.) The 70-30 alloy should be completely solid when it reaches the temperature of the lower phase-field boundary line (at point a''), but under nonequilibrium conditions the average solid-phase composition will be at some point near *f*, and some liquid will remain; specifically,

$$(f - a'')/(f - e) = [(37 - 30)/(37 - 15)](100) = 32\% \text{ liquid, } 68\% \text{ solid}$$

The presence of some liquid in a solid-solution alloy, cooled from the liquid state to the solidus line, results from the failure of diffusion between the two types of atoms to maintain the composition of the growing dendrites of the solid phase at the equilibrium concentration. The amount of liquid present here depends on the time available for diffusion during solidification. The more rapid the freezing, the further the departure from equilibrium and the greater the amount of liquid present. Even under relatively slow cooling conditions, such as might be employed in thermal analyses for the determination of this phase diagram, the solidus temperature is never well marked on a cooling curve. After the solidification of the last liquid is completed, cooling may speed up slightly, but seldom is there a well-defined change in slope at a specific temperature.

Liquidus temperatures can be depressed by undercooling because of slowness in the formation of the first solid nuclei, but this effect can be minimized by agitating or stirring the melt or, in some cases, by artificial nucleation. The mushy condition of the nearly solidified alloy prevents stirring, and although the required diffusion would be accelerated by deformation, that is difficult to accomplish with the alloy in a crucible or mold and still partly liquid. Solidus temperatures, however, can readily be determined by heating a homogeneous solid solution to successively higher temperatures. If the alloy is simultaneously subjected to a slight stress, it deforms plastically while entirely solid, but as soon as the solidus temperature is reached, liquid (enriched in the lower-melting-point element) forms at the grain boundaries and the alloy breaks with an intercrystalline failure (hot shortness). If the alloy is quenched from just above the solidus, evidence of the existence of the liquid phase at that temperature is preserved and can be identified micrographically by reason of its different composition. This constitutes a second method of determining solidus lines.

A powerful guide in the construction of phase diagrams from experimental data is the phase rule, developed through thermodynamic reasoning by Willard Gibbs. This rule, considering pressure to be constant,[2] is as follows:

$$F = C - P + 1$$

[2] The usual variables are temperature, pressure, and composition of each phase including gas. Metals and alloys are almost always studied at atmospheric pressure, and the gas phase is ignored; otherwise the general form of the phase rule would apply: $F = C - P + 2$.

where F = number of independent variables

C = number of components in system; in alloys, number of elements

P = number of coexistent phases

Upon applying the Gibbs phase rule to the Cu-Ni system, for each pure metal at its melting point, there are one component and two phases, liquid and solid metal, coexisting during freezing. Therefore, $F = 1 - 2 + 1 = 0$, the system is invariant, and the metal must freeze at constant temperature. If the rate of heat removal is constant near the freezing point, this means that a "cooling" curve of temperature vs. time would show a horizontal break for the duration of freezing, on the assumption that equilibrium exists.

The freezing of any alloy in the Cu-Ni system also involves only two phases, the α-solid and the liquid solution. However, with two components, $F = 2 - 2 + 1 = 1$, there is one independent variable. Thus the temperature of the system may be varied while two phases coexist, but the composition of each phase is fixed for any selected temperature. If the composition of one phase, e.g., the solid α is specified, the composition of the other phase and the temperature are thereby fixed. Another way of explaining the significance of $F = 1$ is to say that the alloy freezes over a range of temperatures, with compositions of both liquid and solid phases dependent on the temperature. The result is that solidification of a solid solution is not manifested by a horizontal break on the cooling curve. The heat of crystallization of the solid α, released upon its formation in the liquid, slows the rate of temperature drop and results in a well-defined point of inflection at the start of freezing.

3.5 DIFFUSION

General Laws

Solidification of a Cu-Ni alloy under equilibrium conditions requires that the solid phase continually change composition as solidification proceeds. Although it may seem at first glance that once a solid solution forms, the atoms should be locked in place, the simple observation above on homogenization shows that this cannot be true—atoms can migrate through a solid. Since we know that the atoms of a crystal vibrate with larger amplitude the higher the temperature, it would be expected that this thermal vibration might cause an intermixing of atoms. In a close-packed structure, however, there is not room for the atoms to squeeze past each other unless a very high-energy barrier is overcome. Calculations show that the amount of diffusion resulting from such direct interchange of atoms is negligibly small, even at temperatures near the melting point of the metal. If, however, there were some vacant lattice sites in the crystal, as illustrated in Fig. 3.6, then a shifting around of the atoms becomes relatively easy, because the energy barrier for the motion of an atom into a neighboring vacancy is relatively small. Any desired interchange of atoms can be worked out by moving the vacancy around as on a puzzle board.

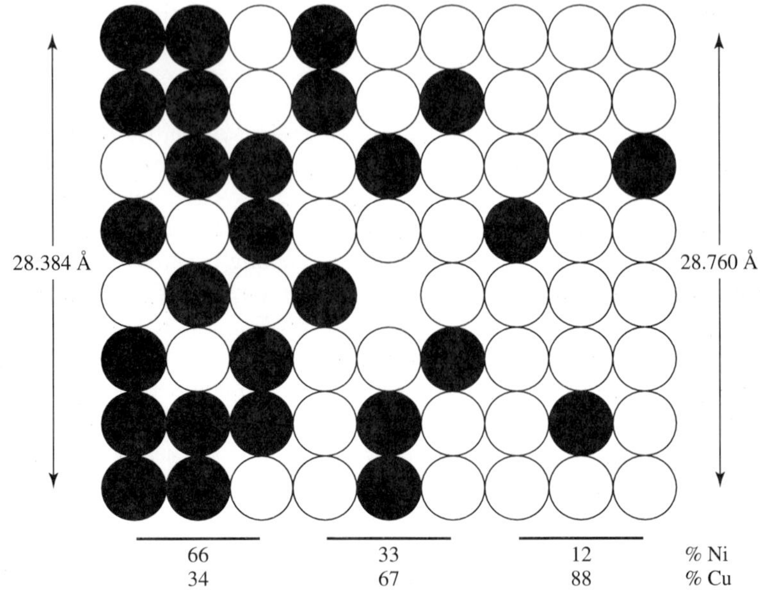

Figure 3.6 Atomistic picture of an exaggerated, sharp compositional gradient in a solid solution, e.g., strong coring in a chill-cast cupronickel. Accompanying the compositional gradient will be a gradient in the size of the unit cube, with extremes suggested on the sketch in angstroms (10^{-8} cm). A vacant atom site in the center is postulated, and this permits atom movements that would make the redistribution of nickel atoms (black) and copper atoms (white) achieve a uniform or homogeneous solid solution, if time and temperature permit.

Although the presence of vacancies in crystals is predicted theoretically, there is also direct experimental evidence for their presence. This is obtained by comparing precision measurements of the lattice parameter and length of a crystal as the temperature is raised. When vacancies form at a given temperature, the average distance between atoms remains the same but the crystal becomes larger, since there must be an extra atom on the outside for every vacancy inside. Careful experiments show that the volume of metal crystals increases faster than the lattice expands as the temperature is raised, indicating, then, that the vacancy concentration increases with temperature. Such measurements on aluminum show that close to the melting point the fraction of vacancies present is 9.4×10^4. The statistical mechanics of crystals predicts that the vacancy concentration C should depend on temperature through an expression of the form

$$C = C_0 e^{-E_f/kT}$$

where C_0 = const

E_f = energy required to form vacancy in crystal

k = Boltzmann's constant

T = absolute temperature

Experimental observation of a phenomenon known as the Kirkendall effect provides conclusive evidence of the vacancy mechanism of diffusion in close-packed alloys: it is observed that the rates of diffusion of the two species of atoms in a binary alloy are generally different; in brass, for example, the zinc atoms diffuse faster than the copper. If diffusion occurred by direct interchange of atoms, both species would have to diffuse at the same rate. The difference in diffusion rates is easily explained by the vacancy mechanism.

According to the above model of diffusion, even the atoms of a pure metal should be continuously interchanging positions with one another. This phenomenon is called self-diffusion. That it exists can be demonstrated experimentally by making some of the atoms on one surface of a pure metal specimen radioactive. With the passage of time the radioactive atoms are found to make their way from one surface of the metal specimen to the other.

Laws of Diffusion

Consider an alloy in which there is a concentration gradient. The atoms are continually changing position, and there is no preferred direction of atomic jump: atoms on any side of a vacancy have an equal chance of jumping into it. Simply because there are more atoms of one species in a given region, there will be a net flow of these atoms out of the region of high concentration. If there is a concentration gradient in the X direction and J is the flux of atoms (the number of atoms crossing a plane of unit area normal to the X direction in unit time), it is found that

$$J = -D\frac{dc}{dx}$$

where D is a constant at a given temperature and is known as the diffusion coefficient. The above equation is called Fick's law. The solution of a diffusion problem involves the integration of this equation to find c as a function of position and time, given certain boundary conditions. Mathematically the problem is similar to that of finding the temperature distribution in a piece of metal after heat has been flowing under an initially given temperature gradient. For estimating the distance over which an appreciable amount of diffusion can occur in a given time, the following approximate rule, derived from Fick's law, is very useful. Let the initial condition be, as in Fig. 3.7*a*, one with an alloy of concentration C_0 in contact with pure metal. The concentration as a function of x, the distance along the bar, is as shown in Fig. 3.7*b* at time zero; it drops abruptly from C_0 to zero at the interface between the alloy and the pure metal. After diffusion has proceeded for some time, the concentration distance, or penetration, curve will be as in Fig. 3.7*c*. The distance x' as defined in the figure is the mean penetration distance and is a measure of the distance over which an appreciable amount of diffusion has occurred. By solving Fick's law under these conditions, it is found that

$$x^2 = Dt$$

where t is the elapsed time.

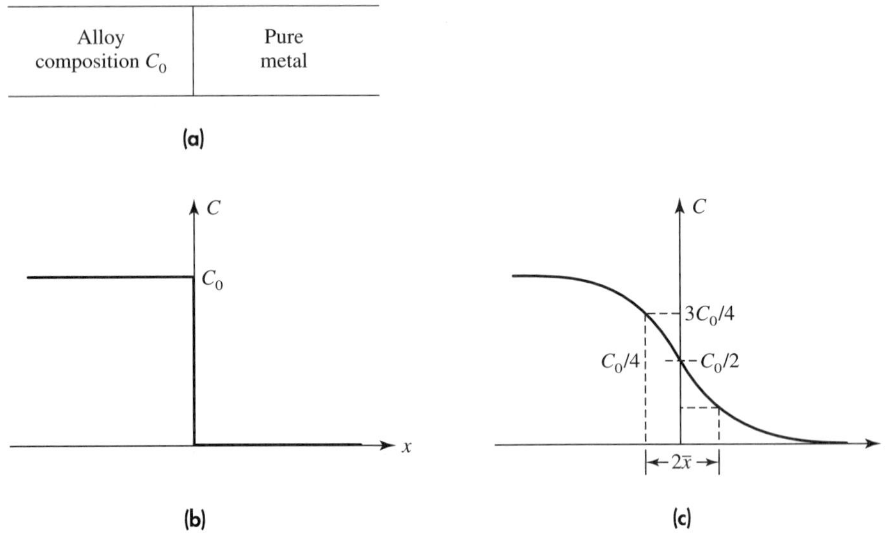

Figure 3.7 A simple diffusion experiment. (*a*) A long bar of alloy having concentration C_O of solute in a given solvent metal is bonded to a long bar of pure solvent metal; (*b*) composition as a function of distance along the bars at a time zero when no diffusion has occurred; (*c*) after holding at an elevated temperature, some solute atoms have diffused out of the alloy and into the pure metal with solvent atoms moving in the reverse direction. x' is called the mean penetration distance.

As an example of the use of this relation, consider the problem of homogenizing a cast alloy by holding it near its solidus temperature. Most substitutional alloys have D about equal to 10^{-8} cm^2/s near the solidus. Since 1 day is about 10^5 s, $x' = 0.3$ cm, and it is clear that homogenization by diffusion will be a slow process.

Temperature Dependence of the Diffusion Coefficient

The rate at which diffusion proceeds in a close-packed metal is proportional to the concentration of vacancies and to the rate at which atoms change places with adjacent vacant sites. The vacancy concentration is

$$C = C_O e^{-E_f/KT}$$

while the rate of jump into vacant sites is proportional to $e^{-E_j/kT}$, where E_j is the energy required for an atom to squeeze past its neighbors into an adjacent vacancy. The diffusion coefficient is therefore given by

$$D = D_O e^{-E_f/kT} e^{-E_j/kT} = D_O e^{-Q/kT}$$

where D_O is a constant of proportionality and $Q = E_f + E_j$ is the activation energy for diffusion. According to the above equation, a plot of ln D vs. $1/T$ should be a

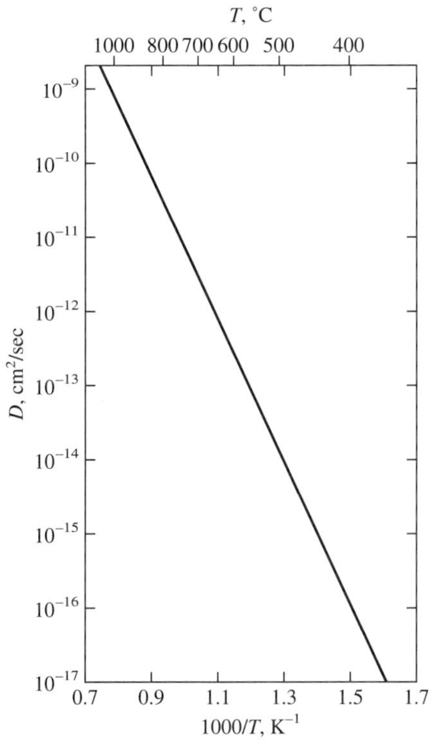

Figure 3.8 Diffusion coefficient for diffusion of gold into copper as a function of temperature.

straight line. Figure 3.8 shows such a plot for diffusion in an Au-Cu alloy, from which it can be seen that the magnitude of D decreases very rapidly with temperature. It is clear from this graph and the relation

$$x^2 = Dt$$

(x = distance) that when it is desired to achieve a maximum amount of diffusion, as in homogenization, the highest possible temperature must be used; at lower temperatures the time required to reach equilibrium is so exceedingly long that the alloy is, for all practical purposes, "frozen" into the nonequilibrium state.

In the case of the diffusion of interstitial solutes, such as carbon in iron, the interstitial solute atoms simply move from one interstitial site to another. Because the solubility of interstitial atoms is low, there are always neighboring empty sites available to the interstitial atoms. Hence, the diffusion coefficient in this case is simply

$$D = D_0 e^{-E_j/kT}$$

where E_j is the energy barrier that must be overcome when a solute atom moves from one interstitial site into another. The activation energy for diffusion is relatively low in this case, and the diffusion rates of interstitial solutes are much faster than those of substitutional solutes.

3.6 SEGREGATION IN METAL ALLOYS

The solidification rate of solid-solution alloys is rarely slow enough for equilibrium to be maintained through the liquidus-solidus interval. Therefore, it is characteristic for compositional gradients to exist from the center of dendritic arms to the center of interdendritic spaces. Since the rate of chemical attack varies with composition, the proper etching of polished surfaces usually reveals the dendritic structure. The degree of dendritic segregation or coring depends on the diffusivity of the two unlike atoms in the solid solution and the time available for diffusion. The latter in turn depends on the solidification rate, being short for chill casting and relatively long for casting in sand, with its lower heat-transfer rates.

Dendritic segregation on a microscopic scale is called coring; it can be explained by the use of a phase diagram in the manner already discussed. On a macrographic or full-size scale, a similar effect may be noticed in that the first parts of a casting to freeze are enriched in the higher-melting phase, while the parts last to solidify (generally, top center sections) are enriched in the lower-melting-point constituents. The effect is statistical in nature, since both sections will exhibit coring. The resulting nonuniformity of chemical composition is known as normal segregation and differs only in dimensions from coring. A third type of segregation is the reverse of this; i.e., the macroscopic parts of the casting first to freeze are enriched in low-melting-point constituents. The effect is called inverse segregation. It is primarily caused by the contraction of solidifying dendrites, which tends to enlarge the interdendritic channels. As these open up, a resultant suction effect draws residual liquid metal, enriched in solute atoms (or low-melting-point constituents), through the channel to the surface. An internal pressure from the release of dissolved gases may aid the action considerably. Thus "tin sweat," exudations rich in tin, may form on the surface of tin bronze castings when the liquid metal contains appreciable amounts of dissolved hydrogen that is released as gas in a late state of solidification and forces the tin-rich liquid at the center of the casting through the interdendritic channels to the surface.

Coring can be completely eliminated by diffusion (homogenization) treatments at high temperatures, as shown by Micros. 3.1 to 3.4. Normal and inverse segregation are little affected by such treatments because of the tremendous distances (on an atomic scale) involved.

Zone Refining

The segregation of constituents, which occurs during the nonequilibrium solidification of alloys, can be utilized as a technique for refining metals. What is done is to produce the segregation along the length of a bar with the end containing the higher

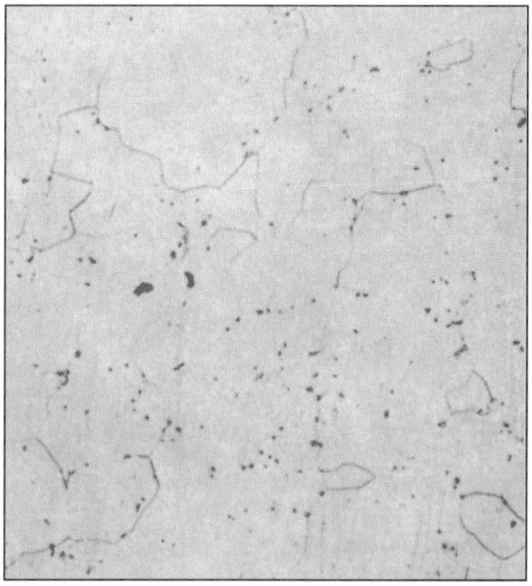

Micrograph 3.1 85% Cu–15% Ni, as chill-cast and heated 9 h at 950°C; ×50. This lengthy, high-temperature treatment has completely homogenized the cast structure, i.e., equalized the composition at all points. Grain boundaries are clearly evident, and their irregular shape is frequently encountered in cast and homogenized solid solutions, the irregularity being related to interpenetration of dendrites growing in the liquid alloy. Black particles are copper oxide or nickel oxide inclusions. The grain size is no larger than in the original casting, since grain growth does not usually occur in castings, except when they have been previously strained by some stress (externally applied or originating from contraction during cooling).

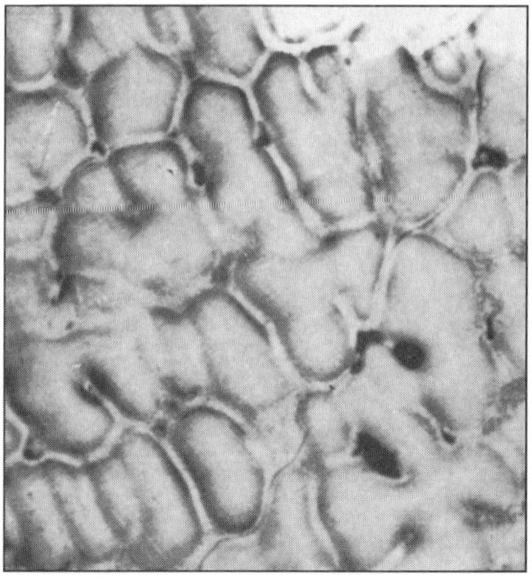

Micrograph 3.2 85% Cu–15% Ni, as cast in a hot mold and slowly solidified; ×50. The cells represent nickel-rich areas (low hills) and the narrow, approximately parallel lines outline the interdendritic copper-rich valley areas. The black, vaguely outlined areas are shrinkage cavities, which, it should be noted, occur in the parts last to freeze, i.e., the copper-rich areas. A single grain boundary diagonally traversing the field along the interdendritic spaces is also visible.

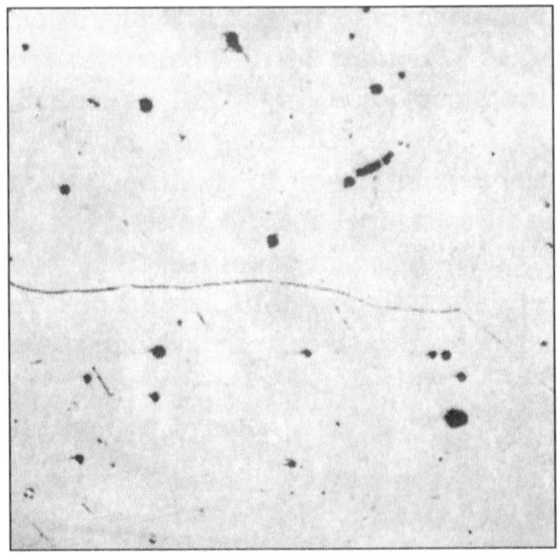

Micrograph 3.3 85% Cu–15% Ni, as cast in a hot mold, slowly solidified, and then reheated 15 h at 950°C; ×50. In this homogenized structure the grain size is considerably coarser than that of Micro. 3.2, more so than is evident in this photograph, which was taken at the intersection of three grains. Again, copper or nickel oxides are visible, and also some shrinkage cavities. The time required for homogenization of the coarse dendrites was greater than that for the fine dendrites, in spite of the smaller initial composition differences across the coarse dendrites. The reason is that the distance through which copper and nickel atoms must diffuse is so much greater in the coarse structure.

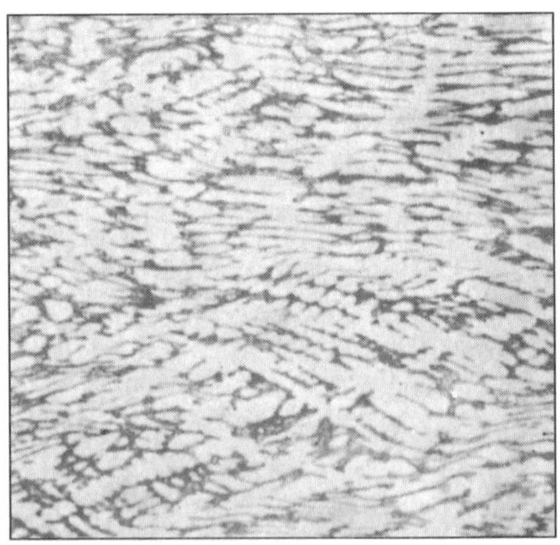

Micrograph 3.4 64% Cu–18% Ni–18% Zn; ×100. This is a longitudinal section of a wrought alloy, called nickel silver because its color resembles that of silver. It is frequently used as a base for plated silverware and other applications for which its color, corrosion resistance, and strength are adapted. This micrograph of commercial metal, in the hot-worked condition, shows that the cored dendritic structure has not been homogenized in spite of the fact that deformation accelerates homogenization. The increased softness of the homogeneous metal in these alloys is seldom worth the costs of the prolonged, high-temperature anneals necessary to obtain complete homogeneity.

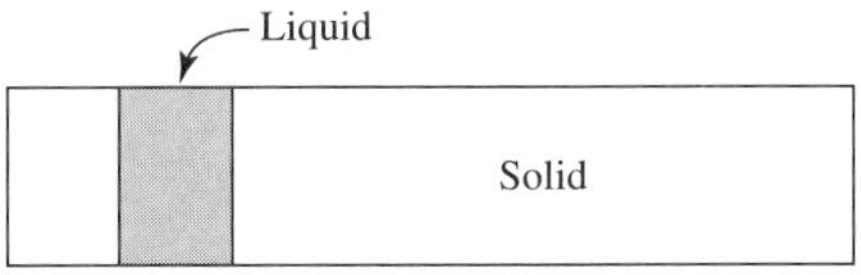

Figure 3.9 In zone refining the liquid zone is passed along the bar, e.g., from left to right, concentrating impurities into the right end.

concentration of solute cut off and discarded. Starting with a long bar of the material to be refined, a molten zone is formed at one end (Fig. 3.9), usually by induction melting. The induction-heating coil is then passed along the length of the bar from left to right. Since the solubilities of most impurities are greater in the liquid than in the solid, the liquid zone carries along with it a relatively high concentration of solute, leaving partially purified metal behind. By repeatedly passing molten zones along the bar, the impurity concentration can be built up on one end, leaving relatively pure material on the other. By using 10 or more passes, very high-purity material can be obtained over most of the length of the bar.

3.7 REAL SOLID SOLUTIONS

The Gibbs free-energy model provides a useful picture of randomness in mixing for ideal systems based on configurational entropy. However, in practical systems, it does not take into consideration the dependence of lowest-energy state on composition and temperature. A random arrangement of atoms may not be the equilibrium or most stable arrangement and may not give the minimum calculation of free energy. In practical solutions, there is a compromise between the lowest internal energy, related to atomic bond energy, and the highest entropy, or randomness, to achieve a minimum free energy.

In a binary solution consisting of *A* and *B* atoms, the state closest to equilibrium may be either an ordered substitutional, clustered, or random interstitial arrangement. If the internal energy of the system can be reduced by increasing the number of *A-B* atomic bonds, then an ordered substitutional solid solution will form. Clustering will occur when it is energetically favorable to increase the number of *A-A* and *B-B* bonds. When the atomic size difference between *A* and *B* is very large, then a random interstitial solid solution can be expected.

A crystal of a compound, such as NaCl, can be thought of as a solid solution (since the Na and Cl atoms are intimately mixed) in which the distribution of atoms on lattice sites is not random but fits a definite pattern. Some alloys have the property that they form disordered solid solutions at high temperatures, but when the temperature is lowered to a certain point, they transform to an ordered structure. An example is an alloy of Cu plus 50 at. % Zn. In the disordered state, above about 460°C, the Cu and Zn atoms are arranged at random on the points of a bcc lattice; below 460°C this alloy is ordered as shown in Fig. 3.10. The ordering results because of the force of interaction between Cu and Zn atoms: this must be such that a Zn atom prefers to be surrounded by Cu atoms as nearest neighbors, just as Cs prefers Cl as nearest neighbors in CsCl crystals.

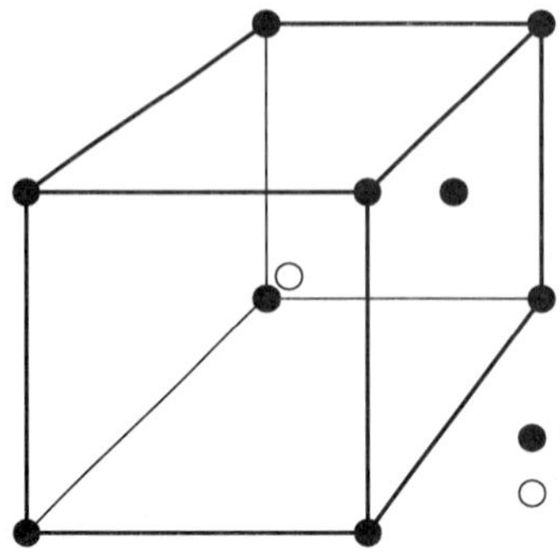

Figure 3.10 The ordered structure of α brass (or of CsCl). The disordered structure is bcc.

The electrical resistivity shows a very marked drop when ordering occurs in a disordered or random solid solution; it may reach a value nearly as low as that of the pure parent metals. Ordering usually results in an increase in strength of the pure parent metals and a decrease in ductility of an alloy.

3.8 GENERAL PROPERTIES OF SOLID SOLUTIONS

Solute Strengthening

In general, the influence of solute atoms is to introduce disorder into the crystal lattice because the solute atoms are different in size from those of the solvent. If the solute is substitutional in the solvent lattice, the replacement of the solvent with an atom of different size will create localized lattice strains as the surrounding atoms shift to accommodate the size disparity. If the solute is interstitial, much the same effect will occur since the interstitial atom tends to expand the crystal lattice in the region of the interstitial space it occupies. These lattice strains tend to increase as the concentration of solute increases. In systems showing complete solid solubility, the lattice strains are maximum around the 50 at. % composition.

The significance of the localized strains in the vicinity of solute atoms arises from their interaction with lattice defects called dislocations. These defects are discussed in more detail in Chap. 4, where their role in mechanical deformation is described. In connection with the mechanical behavior of solid solutions, it is sufficient to note that the strength of the solid is directly related to the ease with which dislocations can move through the crystal lattice. The strain fields surrounding solute atoms may also interact with dislocations, since the dislocations themselves have an associated local strain field. The interaction of these strain fields produces stable configurations of solute atoms such that the solutes and the dislocations are linked to each other. As a result, the motion of the dislocations is hindered, and the alloy exhibits higher strength and hardness than a pure metal. The association of solute atoms with dislocations is called a Cottrell atmosphere and, in the case of interstitial solutes, frequently accounts for the sharp yield-point phenomenon in some metals.

Grain Strengthening

Grain boundaries can hinder the motion of dislocations, and thus fine-grained alloys have higher yield and tensile strengths than coarse-grained ones. The interaction of grain boundaries with dislocations has been both theoretically and empirically related to the mechanical properties of solid solutions through the Hall-Petch relationship, which states that

$$\sigma_{ys} = \sigma_o + Kd^{-1/2}$$

where σ_{ys} = yield strength of alloy

σ_o, K = constants based in solid-solution properties

d = average grain diameter

Thus, there is an effective summation of contributions from solid-solution strengthening and grain strengthening. The fracture stress in metals usually exhibits a dependency on grain size similar to the Hall-Petch relationship. Fine grain size has thus been employed in some commercial solid-solution alloys to improve mechanical properties.

At a constant grain size and homogeneity, the properties of solid solutions show a gradual and continuous change as the concentration of solute is increased: the strength increases, ductility usually decreases, and the electrical resistance is increased. Typical data for Cu-Ni alloys are shown in the graphs of Fig. 3.11.

The properties of annealed solid solutions are so affected by the grain size of test specimens and by soluble impurity elements that it is difficult to obtain comparable data for specimens representing concentrations across an alloy diagram. Recrystallization temperatures and grain-growth characteristics are affected by both solute concentration and impurities. Consequently, the data given in Fig. 3.11 are not necessarily the same as would be obtained with industrial alloys and should be taken only as qualitatively indicative of the effects of dissolved elements. While progress has been made in understanding the mechanism of solution-hardening alloy crystals, there are few generalizations which allow the prediction of the property changes which will result from the addition of a particular solute to a polycrystalline metal; empirical data must be relied on for such information in most cases.

Solid-solution alloys are not customarily used where strength or hardness is of paramount importance. Their basic properties are ductility, approaching or in some cases surpassing that of the pure solvent metals, moderately increased strength properties, and other special properties derived from one or more of the components. The α brasses are cheaper than copper (since the addition agent, zinc, is cheaper) and stronger and yet show excellent ductility and good corrosion resistance. Cupronickels have much improved corrosion resistance as compared with both copper and brasses but are more expensive. In all other cases, comparable compromises of special properties can be obtained in solid-solution alloys.

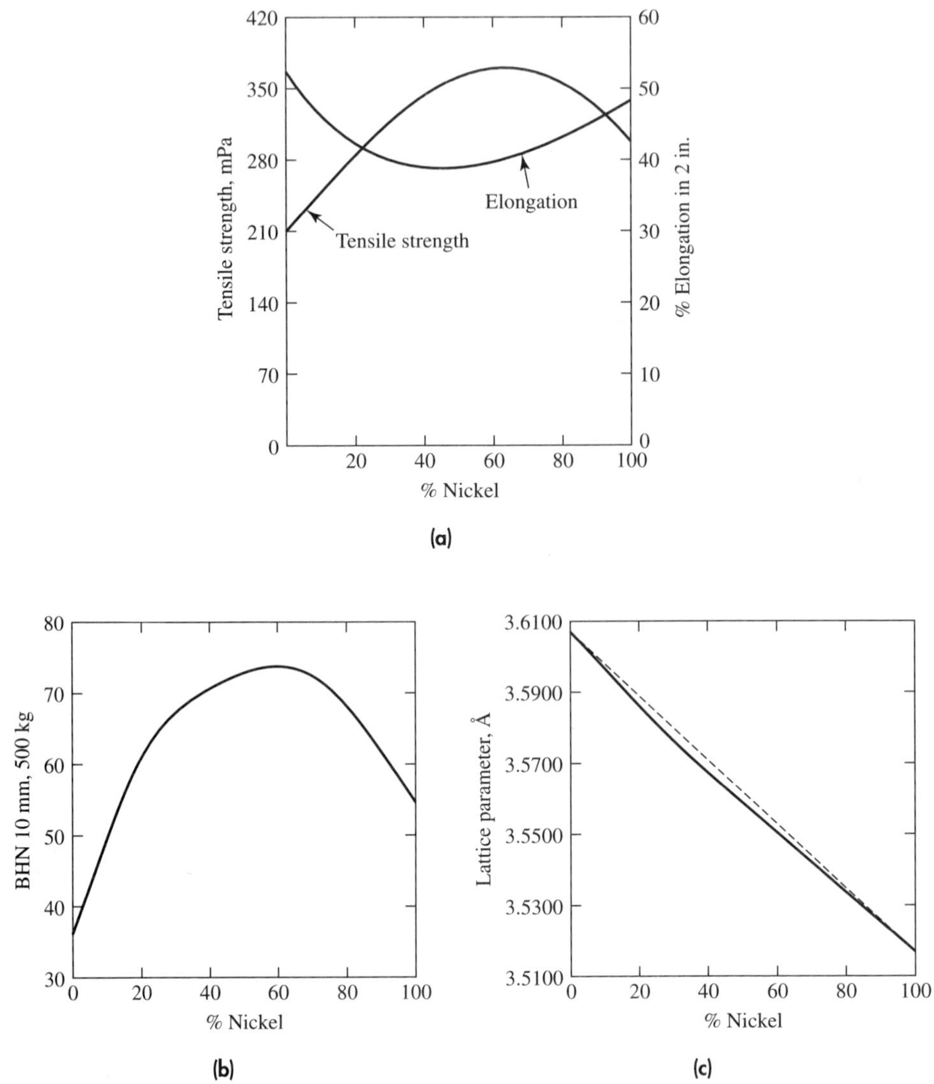

Figure 3.11 Properties of annealed Cu-Ni solid-solution alloys: (*a*) strength and ductility, (*b*) hardness, (*c*) lattice parameter.

It is interesting to list alloy systems that show complete solid solubility at all temperatures. The more important ones, together with their crystal structures, include:

Ni-Cu, Ni-Co, Ni-Pt, Ni-Pd (fcc)

Ag-Pd, Ag-Au, Au-Pd, Pt-Rh, Pt-Ir (fcc)

W-Mo (bcc), Bi-Sb (rhombohedral hexagonal, rh hex)

Electrical Properties of Solid Solutions

A solute atom perturbs the regularity of the crystal structure of a metal and acts as a scattering center for conduction electrons; the presence of solute atoms invariably leads to an increase in the resistivity of a metal. In dilute solid solutions the effect of solutes on the resistivity is described by the two following rules.

Matthiessen's rule states that the increase in resistance of a metal due to a small concentration of another metal in solid solution is independent of the temperature. Another way of saying this is that the rate of increase of resistance with temperature is independent of concentration. This is illustrated by the data in Fig. 3.12. At 0 K the only contribution to the resistivity of an alloy is that due to solute atoms in solution; this is called the residual resistivity and can often be used as a measure of the purity of a metal. A perfectly pure metal would have zero residual resistivity.

Linde's rule states that the increase in the resistivity of a metal due to a given atomic percent of solute atoms is proportional to the square of the difference of atomic number between solute and solvent. The data in Fig. 3.13 illustrate this for the case of Cu solid solutions.

It can be shown theoretically that the dependence of resistivity on concentration of solute in a solid solution is that

$$\text{Resistivity} = kc(1 - c)$$

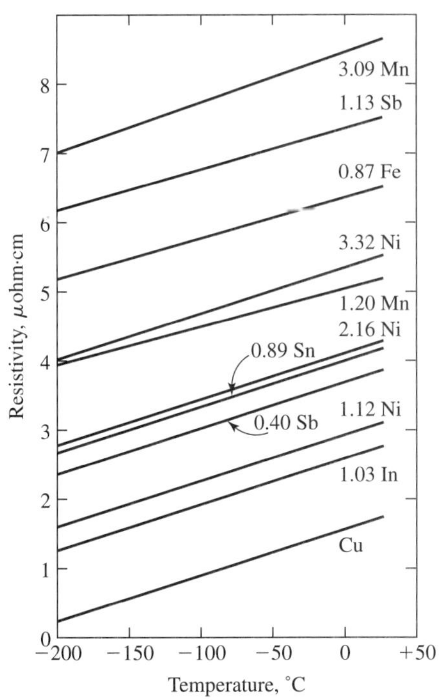

Figure 3.12 Temperature dependence of the electrical resistivity of dilute copper alloys. Compositions are shown in atomic percent. According to Matthiessen's rule, the curves should all be parallel.

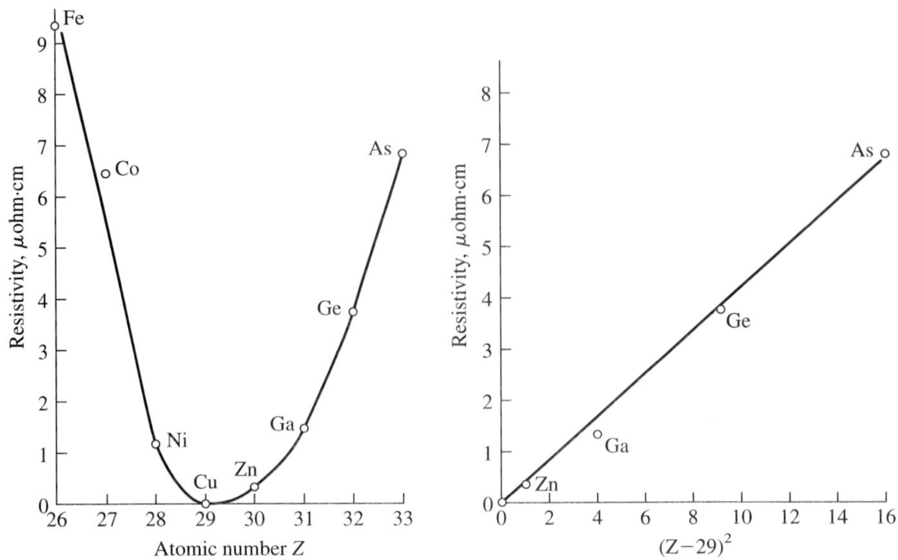

Figure 3.13 Resistivity increase due to the presence of 1 at. % of various metals in solid solution in copper.

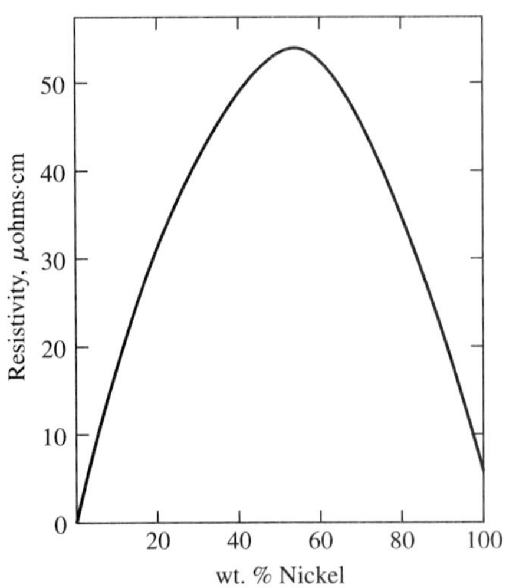

Figure 3.14 Resistivity measured at 0°C as a function of composition for Cu-Ni alloys.

where k is a constant and c is the atomic percent of solute. Figure 3.14 shows resistivity data for Cu-Ni alloys; they are seen to follow the above rule quite well. In very dilute solutions the addition of a small amount of solute causes a large increase in resistivity, while in concentrated solutions the resistivity changes much more slowly. This is to be expected, since in dilute solutions one is making a perturbation in an almost perfect structure, while in a concentrated alloy the perfection is already considerably impaired and further addition of solute does not make it much worse.

PROBLEMS

1. What would be the differences, if any, in microstructure between the following alloys when chill-cast and then homogenized: (*a*) 85% Cu–15% Zn, (*b*) 70% Cu–30% Zn, and (*c*) 50% Cu–50% Ni?
2. In comparison with their base metal or solvent, are solid solutions always (*a*) stronger, (*b*) less ductile, (*c*) more resistant to corrosion, (*d*) poorer electrical conductors, and (*e*) larger in atomic spacing?
3. If insoluble impurities are present in a solid-solution alloy, where in the microstructure will they be found if the alloy is in (*a*) the cast condition and (*b*) the wrought condition?
4. If a specific casting has a grain size that is too coarse, what can be done about this particular casting?
5. How can ordering of atoms in a solid-solution alloy be detected? Why is ordering in a bcc structure most likely to be found at a 50-50 at. % composition, while in an fcc cubic structure it is equally likely at the two 75-25 at. % compositions?
6. Specify three general effects of rapid solidification on the microstructure and macrostructure of a solid solution.

REFERENCES

Flemings, M. C.: *Solidification Processing,* McGraw-Hill, New York, 1974.

Kurz, W., and D. J. Fisher: *Fundamentals of Solidifications,* Trans Tech Publications, Rockport, MA, 1984.

Porter, D. A., and K. E. Easterling: *Phase Transformations in Metals and Alloys,* Van Nostrand Reinhold, Berkshire, England, 1983.

Smallman, R. E.: *Modern Physical Metallurgy,* 3d ed., Butterworths, Washington, 1970.

Hume-Rothery, W., R. E. Smallman, and C. W. Haworth: *The Structure of Metals and Alloys,* Institute of Metals and Institution of Metallurgists, London, 1969.

Swalin, R. A.: *Thermodynamics of Solids,* 2d ed., Wiley, New York, 1972.

Shewmon, P. G.: *Diffusion in Solids,* McGraw-Hill, New York, 1963.

chapter

4

Deformation Hardening and Annealing

The ability of materials to sustain extensive plastic deformation, i.e., permanent changes of shape without fracture, is a property of major importance in our civilization. Plasticity not only permits the economical forming and fabrication of desired shapes but also enables them to carry structural loads without sudden catastrophic failure.

Accompanying plastic deformation of a material are marked changes in its structure and properties. Strength properties are considerably increased with associated reductions in ductility. In contrast with solid-solution hardening, however, the increased strength and structure changes from deformation can be eliminated by subsequent annealing. Control of material and process variables is necessary for efficient utilization of deformation hardening and annealing characteristics.

4.1 PLASTICITY OF METALS

Plasticity of Single Crystals

In order to understand the process of deformation in polycrystalline materials, it is first necessary to know how the individual grains deform under stress. Thus, experiments on single crystals are the starting point for the study of plastic deformation. In examining a single-crystal specimen of a metal, the following facts are quickly observed:

1. Single crystals are remarkably soft. A single crystal of high-purity magnesium, for example, can be deformed by the application of only a few pounds per square inch of load.

2. Plastic flow takes place by shear only on certain crystallographic planes and in certain crystallographic directions. It is generally found that the planes on which slip occurs are the close-packed (most widely spaced) planes of the crystal, the (111) planes in fcc crystals, and the {0001} planes in hcp crystals. Similarly, the direction in which slip occurs on these planes is that along which the atoms are most closely spaced, the (110) directions in fcc and the $(11\bar{2}0)$ in hcp crystals.
3. In crystals of comparable purity and perfection of a given material, slip always begins when the shear stress across the slip planes reaches a certain definite value known as the critical resolved shear stress. The actual stress required to start deformation depends on the orientation of the slip planes relative to the applied stress. If the crystal in the form of a cylinder of cross-sectional area A is pulled in tension by a force F as in Fig. 4.1, the resolved stress on the slip plane in the slip direction is

$$\sigma_r = (F/A) \cos \phi \cos \lambda$$

If the slip planes are nearly perpendicular to the crystal axis, ϕ approaches zero and a very large force is required to initiate slip. This situation can arise in crystals of hcp metals, such as zinc and magnesium, where there is only one set of slip planes; in the fcc case there are four differently oriented sets of {111} planes so that, regardless of the orientation of the crystal axis, there will be at least one set of slip planes more or less favorably oriented for slip.

When deformation occurs on a single set of parallel slip planes, it can continue until the crystal is literally sheared apart, with very little increase of stress above that needed to start deformation. If the crystal, as in the fcc case, has several sets of planes that can slip, then while deformation may begin on only one set, it will eventually start on others. The occurrence of slip on two or more sets of (111) planes means that active slip planes will interact with each other. This hinders the glide on both sets, with the result that the stress must be continually increased to keep the deformation

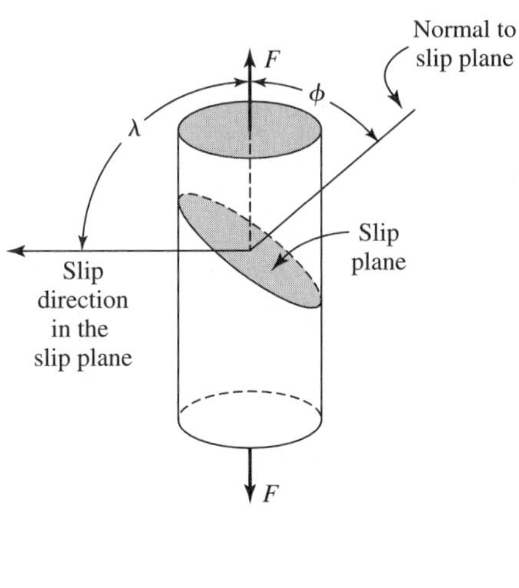

Figure 4.1 Application of the tensile force F to the single crystal shown causes slip to occur on a definite crystallographic plane in a definite direction. The angle ϕ is measured between the normal to the slip plane and the axis of tension.

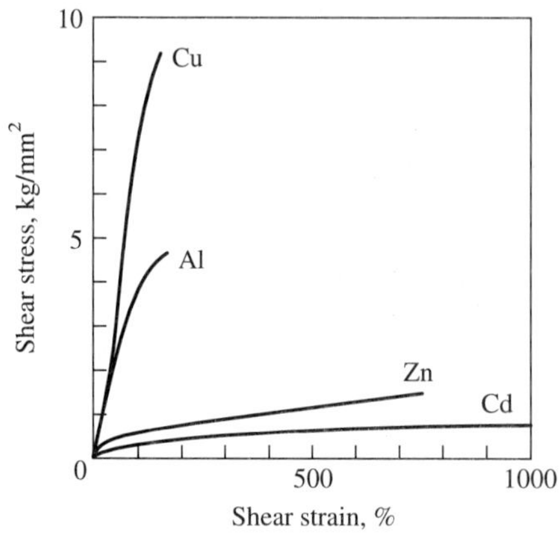

Figure 4.2 Stress-strain curves for single crystals of various pure metals. Strain hardening is very much less in the hcp metals because of the absence of intersecting slip planes.

going. This is illustrated by the stress-strain data in Fig. 4.2. In single crystals of the hcp metals zinc and cadmium, the strain hardening—the rate of increase of stress with strain—is very small, whereas, in the fcc metals copper, silver, and gold, it is very much greater because of the occurrence of intersecting slip systems.

Dislocation Theory of Plastic Flow

The cohesive strength of metals is very large in the sense that it takes stresses of the order of millions of pounds per square inch to separate a pair of atom planes by applying a force normal to those planes. Theoretical calculations for perfect crystals indicate that they should be comparably strong in shear; i.e., a shear stress of the order of 7 GPa should be required to start slip. The actual critical resolved shear stresses for single crystals are about 100,000 times smaller than theory indicates, and since there is no reason to believe the theory to be wrong, it must be that there are imperfections present in metal crystals that make deformation by slip an easy process.

One form of defect that makes slip at low applied stress possible is the edge dislocation[1] shown in Fig. 4.3 for the case of a simple cubic structure. It is a plane of atoms (the line *OC* in the two-dimensional view) that terminates in the structure. The edge of this plane of atoms is a line running normal to the plane of the paper through the entire crystal; it is called the dislocation line. The slip plane is the plane containing the line *OA* and the dislocation line. Under the influence of the shear stress σ_S, shown by the arrow, slip occurs through the motion of the dislocation line along the slip plane in the direction *OA*. The motion of the dislocation line to the end of the crystal produces a slip step on the surface as in Fig. 4.4; continued motion of many dislocations across the crystal can result in the occurrence of any desired amount of slip.

The critical resolved shear stress in metals is low because the shear stress required to set a dislocation in motion along its slip plane is small. To see this, consider

[1]A somewhat similar lattice defect, of particular importance to the deformation of bcc metals, is the screw dislocation, which is described in the reference books on dislocation theory.

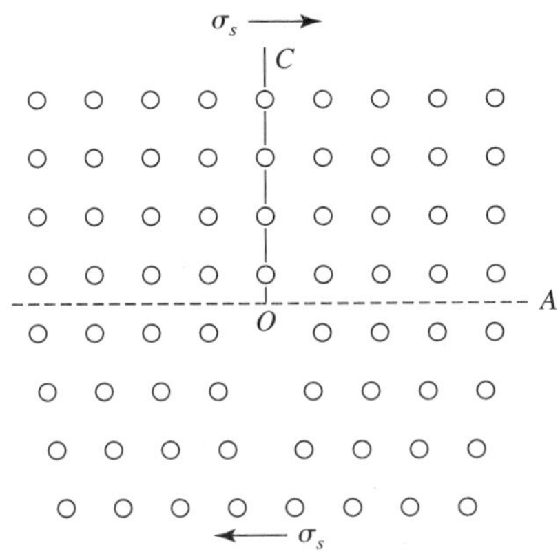

Figure 4.3 Two-dimensional view of a dislocation in a simple cubic lattice. The extra half plane of atoms *OC* extends throughout the crystal normal to the plane of the paper. The slip plane is *OA*.

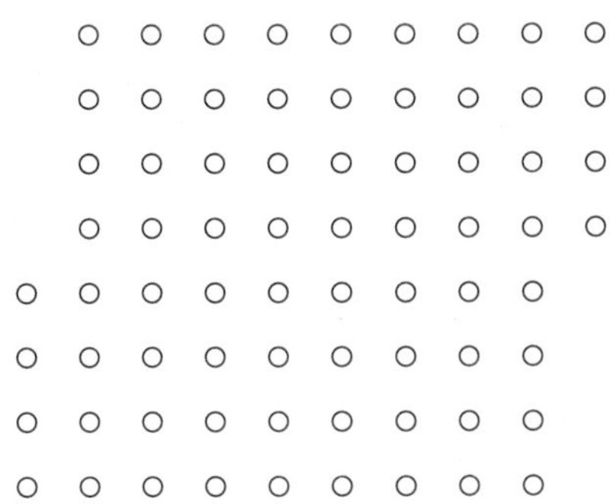

Figure 4.4 Two-dimensional view of the atomic structure of the crystal shown in Fig. 4.3 after the passage of the dislocation entirely across the slip plane.

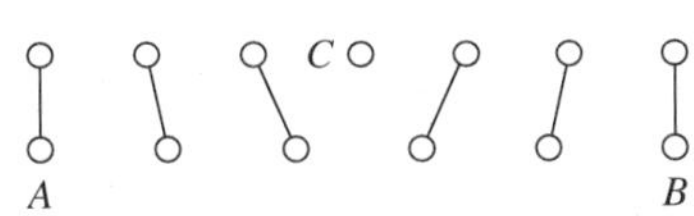

Figure 4.5 Schematic representation of the positions of the atoms along the slip plane of a dislocation. The force resisting horizontal displacement of the atom marked *C* is very small because of the symmetry of the atomic displacements to the right and left of *C*.

a group of atoms about the dislocation line shown in Fig. 4.5. *C* is the atom on the dislocation line. The displacement of the atoms on the right and left sides of a vertical plane running through *C* is perfectly symmetric, and the horizontal forces on *C* are just in balance. The horizontal force needed to start *C* moving along the slip plane is small. Once *C* starts to move, a slight asymmetry is set up in the horizontal forces and a very small applied stress is needed to keep the dislocation moving until it reaches its next position of symmetry. Theoretical calculations show that this force is smallest when the dislocation is moving on a close-packed plane of the crystal in a close-packed direction.

Crystals of materials other than metals, e.g., germanium and rock salt, contain dislocations but still do not show significant plasticity. This is because the ease with which a dislocation moves is determined by its width, the distance *AB* in Fig. 4.5, where the atoms of the slip plane are out of their proper places. The wider the dislocation, the more nearly balanced the horizontal forces on *C* as it moves along the slip plane and the lower the stress required to keep it moving. Dislocations in metals, because the bonding forces between the atoms are nondirectional, are very wide and consequently easy to move. In covalent crystals such as germanium, there are directed bonds between the atoms. The dislocations are narrow and difficult to move.

It might be supposed that, as deformation continues, all the dislocations in a crystal might be used up (by moving to a free surface) so that slip would have to stop. Experiments have shown that fresh dislocations are continuously generated during plastic flow so that slip on any given slip plane can continue indefinitely. The mechanism by which new dislocations can be generated is described in all modern reference books on dislocation theory, the most common model being the Frank-Read source.

Twinning

There is another mechanism of deformation, important in some materials, which does not involve slip. This is the formation of twins in a crystal under applied stress. A twin orientation is illustrated in Fig. 4.6, where the upper half of the

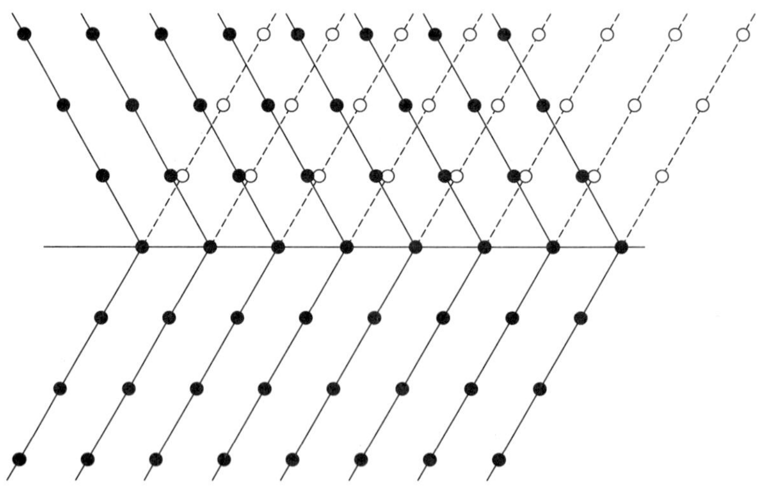

Figure 4.6 Schematic representation of the shift in atom positions which occurs upon twinning. The open circles and dashed lines represent the structure before twinning occurred.

crystal is a mirror image of the lower half. The lateral shear motions required to form a twin are not the gross movements shown, i.e., the displacements between the solid and dashed lines in the upper half, but the minimum movements from open-circle positions in each layer. Mechanical twinning is an important mode of deformation for hcp metals when crystal orientation is not favorable for slip. It is encountered in bcc metals at low temperatures. In all cases, for twinning to occur, it must lead to an extension of the crystal in the direction of the tensile stress. Twins form instantaneously, often with the emission of sound, e.g., the well-known "cry" of tin.

Deformation of Polycrystalline Materials

Extensive slip is particularly easy in single crystals of hcp metals because there is nothing, such as intersecting slip planes, to impede motion along the one set of slip planes that is active, the {0001} planes. In a polycrystalline hcp metal with random grain orientations, however, slip will be blocked at every grain boundary. As shown in Fig. 4.7, much greater stresses are required to deform a polycrystalline specimen, and the amount of deformation that can be achieved before failure is much less. It will be readily appreciated that deformation of polycrystalline metal is an exceedingly complex process, one that has so far withstood accurate scientific analysis.

In commercial working operations such as rolling, drawing, and spinning, a complex system of stresses is applied to the worked metal. Deformation will start first in those grains most favorably oriented for slip, and hardening induced by strain will occur in these grains before it does in the others. Stresses developed at the boundaries of neighboring grains will initiate slip in these until soon all grains are undergoing deformation. Because the grain boundaries hold together and deformation is occurring in different directions in different grains, all the slip systems in the different grains will come into operation. As deformation proceeds, bending, rotation, and other distortion of the slip planes will develop. During

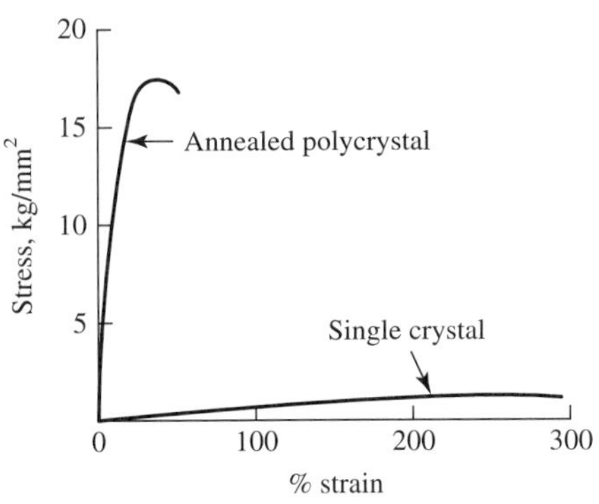

Figure 4.7 Stress-strain curves for pure zinc in polycrystalline form and as a single crystal.

deformation, the concentration of dislocations in the metal will be greatly increased. Undeformed metals usually have dislocation concentrations of the order of 10^6 cm of dislocation line per cubic centimeter of metal; in a heavily worked metal this number may be as high as 10^{12}. Micrographs 4.1 to 4.6 show the development of slip and, eventually, severe distortion and bending with increasing degrees of deformation.

In any micrographic examination of a cold-worked metal, the relation of the plane of polish to the direction of cold working is important. For example, in cold-rolled metals, the crystals on the rolling plane appear elongated, with the width unchanged; on a transverse cross section, they appear flattened, with the width unchanged; on the longitudinal cross section, they appear both elongated and flattened. Thus the deformation is most obvious when examined on the longitudinal cross section.

A fine-grained metal containing crystals with a random distribution of orientations will, in most respects, behave as an isotropic substance despite the anisotropic behavior of its individual component grains. After considerable plastic deformation, however, with crystallite fragments in each grain rotating toward a common position with respect to the flow direction, the metal ceases to be isotropic, both crystallographically and mechanically. It will show somewhat different properties depending on the direction of measurement, i.e., the position of the axis of the test specimen with respect to the flow direction or direction of rolling.

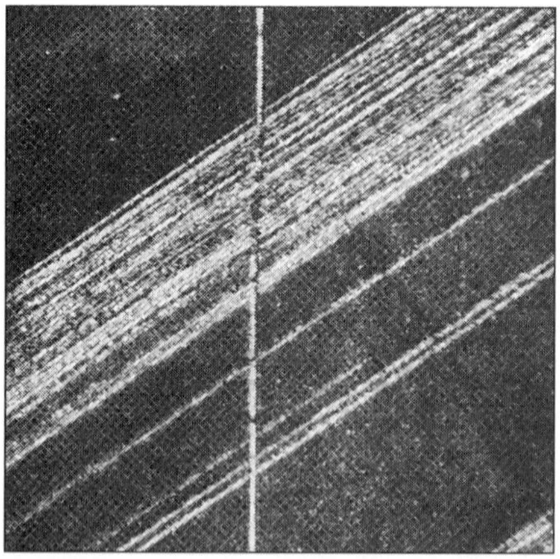

Micrograph 4.1 Single crystal of α brass strained 0.2% in tension, ×200. This specimen was polished, etched with ammonia peroxide, strained, and then photographed with the light at an oblique angle. The undisturbed surface of the crystal is dark, but light has reflected from the sides of steplike discontinuities or lines indicating the planes of slip. The single vertical line was scratched on the polished surface before straining. The slight offset from the original straightness indicates that the amount of plastic displacement at each visible slip line amounts to about 700 to 800 atom distances.

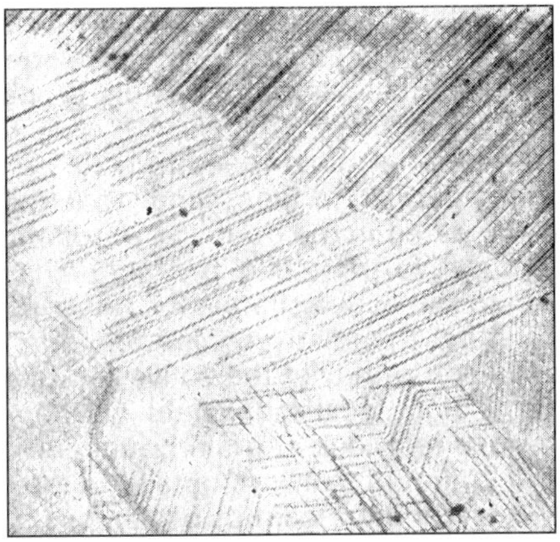

Micrograph 4.2 Polycrystalline annealed brass, polished, etched with ammonia peroxide, and then squeezed slightly in a vise, ×100. With normal illumination, the color variations of this grain structure are related to orientation differences (page 13, Micro. 1.1). The parallel dark lines are steplike discontinuities resulting from slip. Note that when these reach grain boundaries, they stop or change direction. In some cases, they are parallel to annealing twins (the planes of slip in brass are also potential twinning planes). Where the active slip plane is not parallel to the twin but intersects it, note the change of direction at the twin and resumption of the original direction on the other side. This is further evidence of the change of orientation in twins and yet the atomic conformity between the twin and original crystal. Note also that the bottom crystal shows two sets of intersecting lines within some of the twin bands, indicating that more than one set of the potential slip planes functioned during this slight deformation. All these markings appear only by reason of a difference in surface level on either side of the slip plane; repolishing the specimen would remove the line markings.

Micrograph 4.3 Armco iron polished, etched with Nital, and then squeezed in vise, ×100. Body-centered cubic iron has no single set of well-defined slip planes. That, rather than any lesser perfection of the crystal structure, is the probable explanation of these characteristically forked and wavy slip lines.

Micrograph 4.4 Alpha brass cold-rolled to a 30% reduction; NH_4OH-H_2O_2 etch, ×200. This structure of a surface parallel to the rolling plane (with the rolling direction vertical) shows grains somewhat elongated in the direction of rolling. The formerly straight twin bands now show curvature indicative of lattice bending. The numerous curved dark lines in the crystals are parallel to the active slip planes but appear for an entirely different reason than those of Micro. 4.2; after the relatively high reduction of 30% (Micro. 4.2 was deformed less than 1%), atomic nonconformity at the active slip planes, i.e., dense tangles of dislocations, has created a zone of high localized instability where etching attack proceeds more rapidly, in a manner analogous to attack at grain boundaries. Three sets of markings in one crystal indicate that three different sets of octahedral planes were active in the deformation process. Since these markings cannot be removed by repolishing, they are called noneffaceable deformation lines, etch markings, or simply strain markings.

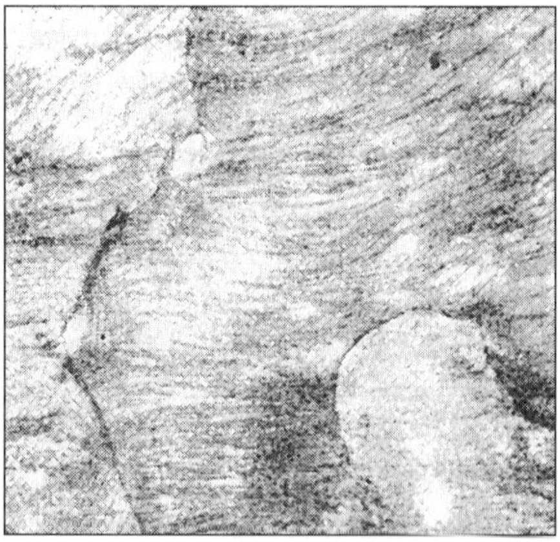

Micrograph 4.5 Alpha brass cold-rolled to a 60% reduction; NH_4OH-H_2O_2 etch; ×200. After a greater reduction, this structure of the rolling plane (rolling direction again vertical) shows strain markings that are less well defined and more wavy, curved, or branched, indicating greater lattice distortion and fragmentation. It is also evident that the strain markings (on the active slip lines), instead of being randomly disposed, grain-to-grain, as a result of random crystal orientation, are now tending to assume a common general position on the sheet surface, approaching perpendicularity to the rolling direction. Thus the rotation of individual crystals or banded sections of crystals into symmetry positions with respect to the direction of flow is generating a preferred orientation.

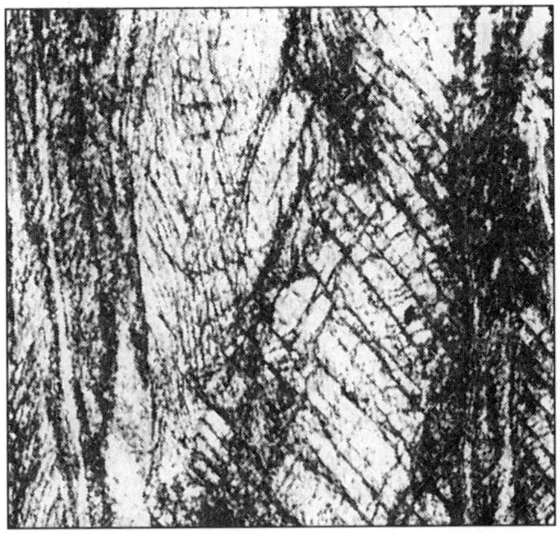

Micrograph 4.6 Same as Micro. 4.5 but photographed at ×200 on a plane parallel to the rolling direction and normal to the rolling plane (longitudinal section). The markings that tended to be perpendicular to the rolling direction in Micro. 4.5 are now seen to represent twisted and warped planes tilted at approximately 45° to the surface of the rolled sheet.

4.2 PROPERTY CHANGES IN DEFORMATION-HARDENED METALS

Strain Hardening

Although the grains of a material become anisotropic as it is cold-worked more than about 30%, the grain size remains constant. The requirement that grain boundaries maintain cohesion during cold working leads to severe bending of crystal planes (Micro. 4.4) related to atomic misfits. The increase in dislocation density from 10^6 to 10^{12} cm/cm^3 with most edge (line) dislocations intersecting causes dislocation tangles to form. Newly generated dislocations can move through these tangles only with difficulty and the application of greater forces. This is the model of strain hardening that was briefly mentioned in the previous sections.

Strain softening is possible under conditions of reversed strain. If a low-carbon steel is cold-rolled, its yield strength will increase from 240 to 860 MPa. Several cycles of subsequent reversed flexure where the metal surface layers are alternately stretched and compressed just beyond the yield point can reduce the yield strength to 690 MPa with an increase of ductility from 1% tensile elongation to 8% in 50-mm gage lengths. Accompanying this strain softening is a reduction in the density of dislocation tangles.

Variables Affecting Strain Hardening

1. The type of metal or solid solution alloy being plastically deformed is an important variable determining the rate of strain hardening possible.
2. The degree of deformation and the type of cold work, i.e., whether rolling, extrusion, drawing, etc., are, respectively, major and minor strain-hardening variables.
3. The temperature and rate of deformation determine whether it is cold working or hot working or a mixture of them. At temperatures less than one-third of the melting point (in kelvins), it will be cold work. At temperatures above two-thirds it will be hot work with no strain hardening. At temperatures in between, very rapid deformation will be cold work, i.e., lead to strain hardening, whereas slow deformation will not be cold work.

Property Changes from Strain Hardening

The extent of the various property changes is of importance in evaluating potential engineering applications of cold-worked metals. Although both hardness and strength increase, they do not follow parallel courses when plotted against the reduction by cold rolling (see Fig. 4.8). Strength generally increases more or less linearly, whereas hardness increases very rapidly in the first 10% reduction and then more slowly at successively higher reductions. In Fig. 4.8, for example, the increase of Rockwell B hardness and the tensile strength of 70-30 cartridge brass are shown for increments of reduction up to about 70% total deformation. The very steep initial increase of the hardness values and later leveling off may be partly explained in terms of two factors: (1) the nonlinearity of the Rockwell scale and (2) the depth of indentation. All Rockwell scales are very sensitive in the low numerical ranges and relatively insensitive as they approach 100; cold rolling of most metals involves a greater elongation (and lateral spread) in the surface layers than in the center, as is evidenced by the commonly observed concavity of ends and edges. Since the Rockwell impression is made on the surface, this would also magnify initial increases.

Ductility, as shown by the elongation in a tensile test, follows a course opposite to that of hardness, a large initial decrease in the first 20% reduction and then a gradually slower rate, asymptotically approaching a relatively low value. The data of Fig. 4.8 show the effect of increased zinc content in increasing both the solid-solution tensile strength (initial ordinate values) and the strain-hardening response. The relative effect of zinc in solid solution on plasticity and ductility is shown in two ways, by the elongation values and the yield strength (at a 0.2% offset) for tough-pitch electrolytic copper and 70-30 cartridge brass. The elongation data show only a slight beneficial effect from zinc at intermediate reductions. The yield-strength data in comparison with the corresponding tensile-strength data, the ratio of yield strength to tensile strength, are more significant in discussing ductility. The yield strength of copper, although very low in the annealed state, more closely approaches the tensile strength of copper, above reductions of 40%, than is the case with cartridge brass.

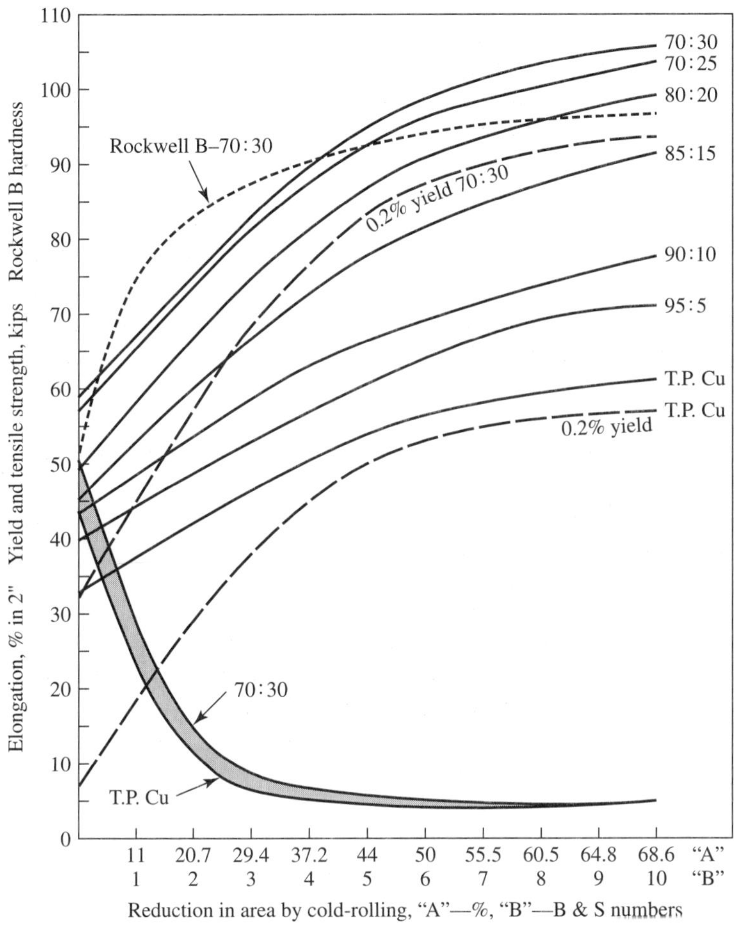

Figure 4.8 Effect of cold rolling on the tensile properties of Cu and various Cu-Zn alloys single-phase solid solution (first figure at upper right is % Cu) known as brasses. All tensile-test elongation values fall within the shaded area.

The greater spread between yield strength and tensile strength for cartridge brass than for copper means that cold drawing of brass can be done more readily without danger of cracking. Cold rolling is not so affected because of the absence of tensile stresses, but in cold working where tensile stresses are present, cartridge brass is notably superior to copper and, in fact, to almost all other metals.

In addition to mechanical properties, cold working changes physical properties. It has a deleterious effect on the magnetic properties of soft iron and decreases the electrical conductivity of pure metals slightly (about 2 to 5%) and of some alloys, such as α brass, strongly (about 20%). It also decreases the density of the metal by up to 0.1% through the generation of dislocations and vacancies.

4.3 ANNEALING

During plastic deformation of a metal, work is done on the material. The bulk of this work is dissipated as heat (increased temperature of the workpiece) but up to 10% of it is retained in the metal as stored energy. This stored energy is associated with the large numbers of dislocations and other defects that are generated in the metal during deformation. Cold-worked metal is unstable in the sense that, given the opportunity, it will lower its energy by returning to the undeformed condition. There is an energy barrier that stands in the way of this return which can be overcome by heating the metal to a suitable temperature for a certain time, a process called annealing. Many complicated arrays of different kinds of defects are generated in a metal during deformation; during annealing these defect configurations rearrange and eliminate themselves by a variety of different processes, some occurring at different temperature levels and others simultaneously at the same temperature. In these terms, annealing of cold-worked metal is a highly complex process. In terms of changes in microstructure and mechanical properties, three more or less distinct stages of annealing can be recognized, namely, recovery, recrystallization, and grain growth.

Recovery

This is the first stage of annealing, and it occurs at relatively low temperatures. In fact, examination with sufficiently sensitive techniques reveals recovery after deformation occurring even at room temperature (and below) in many materials. Rearrangement of dislocations to more stable configurations, apparently without much change in the total number of dislocations present, is the principal structural change during recovery. From a practical point of view, the principal change in the metal during recovery is the reduction of macro- and microstresses resulting from rearrangement of the dislocations introduced during deformation. Macro- and micro- have the same significance here as when applied to photographs of structures. The word macrostresses refers to elastic stresses existing, in a balanced state, over large areas of the metal. When the balance is upset by machining away part of the metal, the unbalanced stresses will redistribute themselves by a distortion of the metal; e.g., a slit cold-drawn tube may open up at the cut, increasing the diameter of the tubing. On the other hand, microstresses, while also of necessity in a balanced state, are on so small a scale or so localized in extent that they cannot cause a change of dimensions upon machining of the metal and are detectable only by x-ray diffraction or comparable methods.

Without the benefits of annealing, residual stresses in a cold-worked metal may approach the strength of the material and exceed it if localized surface notches are created by the attack of certain specific corrosive agents, e.g., ammonia or mercurous solutions in the case of brasses. When localized tensile stresses exceed the strength of the material or reduce its strength by the action of a corrodent, cracks start forming. Since the corrosive attack usually creates notches at grain boundaries, the cracks start and propagate along grain boundaries, resulting in intergranular failure of the type known as season cracking or, more accurately, stress-corrosion cracking.

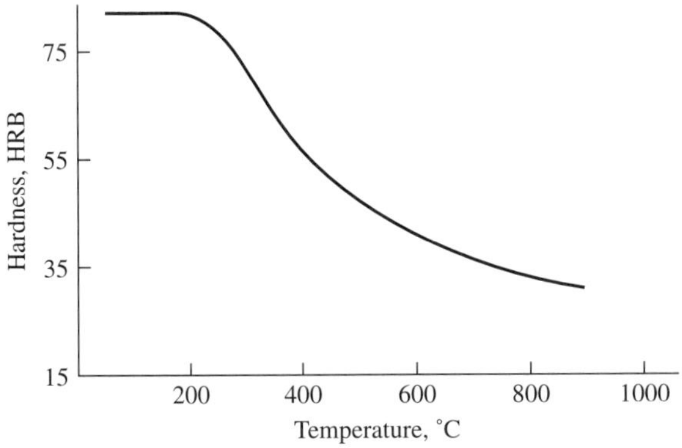

Figure 4.9 Hardness of samples of cartridge brass (30% Zn) cold-rolled to an 80% reduction in area and then annealed at various temperatures for 30 min.

A heat treatment of cold-worked metal to reduce the possibility of stress-corrosion cracking is called a stress-relief anneal and is done at a temperature in the recovery range. Since lattice distortions that impede the motion of dislocations through the structure are not affected, hardness and strength may not be noticeably decreased; in fact, in some solid-solution alloys, such as the α brasses, the hardness and strength may increase slightly. (Placing a test specimen of cold-rolled brass on a hot radiator for a couple of hours may bring its strength up to specifications if it originally was a little below the minimum.) This slight increase in hardness and strength is believed by some to be caused by precipitation of a phase soluble in the stable structure and insoluble in the distorted structure. There may also be some very minor structural changes, but at most these will result only in a decreased tendency to show etch or strain markings in microstructures of cold-worked and stress-relief annealed metal. This effect illustrates the decrease in localized stresses that had given the original etching effect.

Recrystallization

At temperatures just above the recovery range, hardness starts to decrease rapidly (Fig. 4.9). Simultaneously, minute new strain-free crystals, identical in composition and lattice structure to the original undeformed grains, make their appearance in the microstructure. These crystals are not elongated as are the fragmented, deformed grains. They appear to be approximately equiaxed, with their diameters about the same in any measured direction.

The crystals appear first in the most severely distorted part of the worked structure, usually at former grain boundaries. The process of formation of recrystallized grains is one of nucleation. For a grain of recrystallized metal to form, it is necessary that the energy decrease associated with the transformation of distorted to undistorted material be great enough to overcome the energy increase associated with the formation of an interface between the new grain and the original distorted structure.[2] As the temperature is raised, the probability of a thermal fluctuation being

[2]If a material has been lightly deformed so that its stored energy is low, it may be impossible to make it recrystallize. Abnormal grain growth may occur if the material is highly pure. In fact, this is one way to grow single crystals from a polycrystalline high-purity metal.

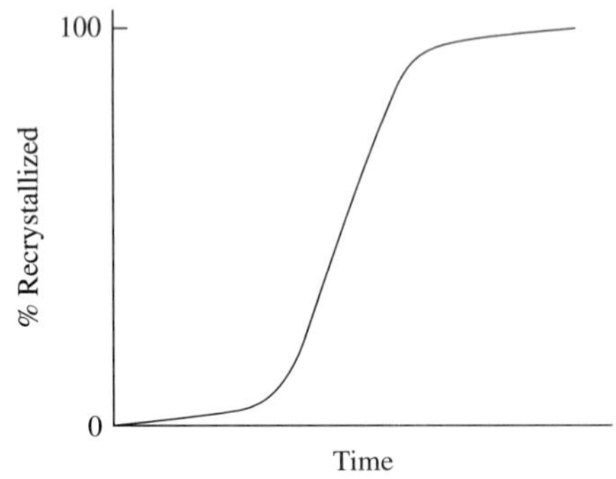

Figure 4.10 Extent of recrystallization in a cold-worked metal held for increasing times at a constant temperature within the recrystallization temperature range.

large enough to start a particular distorted region recrystallizing increases: the higher the temperature, the greater the rate of recrystallization. For a given annealing temperature, the progress of recrystallization with time is along a curve of the form shown in Fig. 4.10. Since it is difficult to say from this curve when complete recrystallization has occurred, it is convenient to define it on the basis of both time and change in some criterion such as decrease in hardness.

As shown by the data in Fig. 4.12, recrystallization will be completed at a lower temperature if the time at temperature is increased. With a recrystallization time of 1 h, a recrystallization temperature can be defined as the temperature required for recrystallization to become 95% complete. As shown in Table 4.1, there is a close relation between the minimum recrystallization temperatures of metals of ordinary purity and their melting temperatures.

Grain Growth

During recrystallization, grains of undistorted material nucleate and grow into the surrounding deformed material. The growth of a grain at the expense of surrounding recrystallized grains is called grain growth. There is usually no sharp distinction between the recrystallization and grain-growth stages of annealing, since grain growth may be going on in the parts of the structure that recrystallized first while other regions are still recrystallizing. The driving force for grain growth is the surface energy of the grain boundaries of the recrystallized grains: the larger the grain size, the smaller the amount of grain boundary in the material and the lower its energy.

The shape of individual grains that are in intimate contact with all adjacent grains to pack or fill space completely is also determined by interfacial energies. In an annealed polycrystalline metal, there will be two-dimensional surfaces between two contacting grain faces, one-dimensional lines at the intersections of three grains, and zero-dimensional points at junctions of four grains. The planar surfaces

Table 4.1 Relation between recrystallization and melting temperatures

Metal	Minimum Recrystallization Temp., °C	Melting Temp., °C	Ratio of Recrystallization Temp. to Melting Temp., K
Tin	< Room temp.	232	<0.60
Cadmium	~ Room temp.	321	~0.51
Lead	< Room temp.	327	<0.50
Zinc	Room temp.	420	0.43
Aluminum	150	660	0.45
Magnesium	200	659	0.51
Silver	200	960	0.38
Gold	200	1063	0.41
Copper	200	1083	0.35
Iron	450	1530	0.40
Platinum	450	1760	0.35
Nickel	600	1452	0.51
Molybdenum	900	3560	0.31
Tantalum	1000	3000	0.39
Tungsten	1200	3370	0.40

should be flat; otherwise, these will tend to migrate in the concave direction to achieve minimum energy by flatness. The three planar surfaces meeting at a line should be at 120° to each other to balance relative energies and become stable. Finally, the four lines that meet at a point juncture of four grains should be at an angle of 109°28′ to balance the interface line forces or energies. Kelvin's tetrakaidecahedra (page 14, Fig. 1.15) meet all these requirements except for the 109°28′ angle of the four line junctions.

Experimental studies show that during grain growth these theoretical considerations appear to be operative, since curved boundary surfaces do move to become flat (or straight lines in the two-dimensional microsections). The intersections of three boundaries also move to develop angles of 120° between three intersecting boundaries (in planar section normal to the boundaries). In general, as boundary migrations in directions so defined consume certain grains and cause surviving larger grains to more closely approach the ideal grain-packing topology, the structure becomes more stable so that further grain growth takes place more slowly.

Micrographs 4.7 to 4.12 illustrate the successive changes in microstructure during annealing. The numerous twins that appear in many of the fcc metals and solid solutions after annealing are called *annealing twins*. As can be seen from the micrographs, prolonged annealing at high temperatures can result in very large grain sizes.

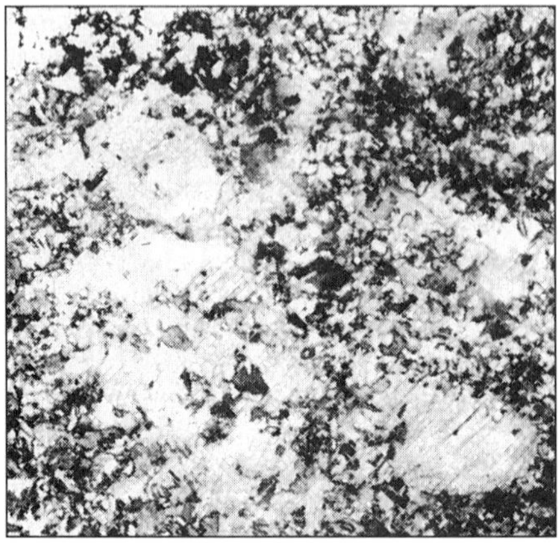

Micrograph 4.7 Alpha brass, cold-rolled 60% and heated to a temperature in the recrystallization range (30 min at 300°C); ×75. This structure shows masses of tiny new crystals, not very well resolved here, and some areas of the old deformed structure containing strain markings.

Micrograph 4.8 Same as Micro. 4.7 at ×500. The structure of the new crystals and of the strain markings is somewhat better resolved at this magnification. The average diameter of the new crystals is about 0.002 mm.

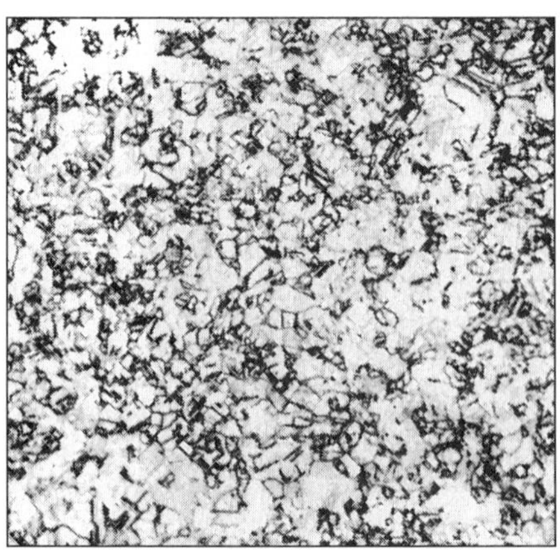

Micrograph 4.9 Same specimen as Micro. 4.8 reheated 30 min at 400°C; ×75. After recrystallization has been completed and after some crystal growth, the average grain diameter is about 0.020 mm.

Micrograph 4.10 Same specimen, reheated 30 min at 500°C; ×75. Additional growth has increased the average grain diameter to about 0.045 mm.

Micrograph 4.11 Same specimen, reheated 30 min at 650°C; ×75. The average grain diameter is now about 0.15 mm.

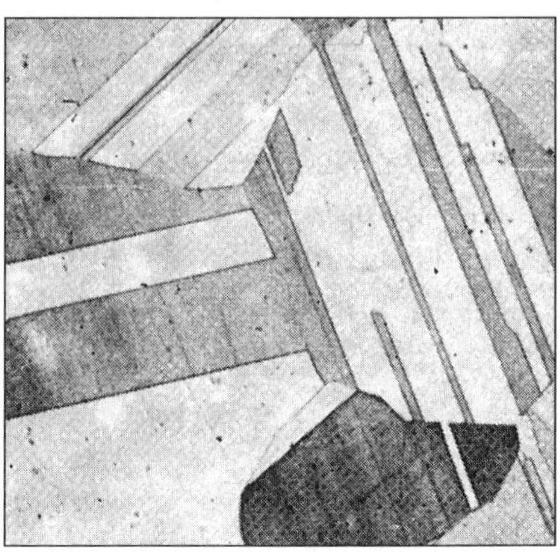

Micrograph 4.12 Same specimen, reheated 30 min at 800°C; ×75. Inspection of a larger area at a lower magnification is required to determine that the average grain diameter is now about 0.25 min.

Factors Influencing Annealing

The major factors influencing the time and temperature required to anneal a specific cold-worked metal are the following:

1. Recrystallization starts at a lower temperature and is completed within a narrower temperature range:
   a. The greater the prior deformation
   b. The finer the prior grain size
   c. The purer the metal
   d. The longer the time of annealing
2. The recrystallized grain size will be smaller
   a. The lower the temperature (above the minimum required for recrystallization)
   b. The shorter the time at temperature
   c. The shorter the time of heating to temperature (increased nucleation)
   d. The heavier the prior reduction
   e. The more insoluble particles present or the more finely they are dispersed (Fig. 4.11)

It is evident from statements 1*c* and 2*e* that soluble impurities or alloying constituents such as zinc and copper raise the recrystallization temperature, while insoluble constituents such as Cu_2O in copper do not noticeably affect the temperature of recrystallization but decrease the recrystallized grain size. The latter effect is widely used commercially to obtain fine-grained structures in annealed metals. How it arises is illustrated in Fig. 4.11. Grain boundaries have a certain energy per unit area, as do the interfaces between inclusion particles and the matrix; by entrapment of an impurity particle, the grain boundary in Fig. 4.11*b* reduces the total interfacial "boundary and particle" area and so represents a lower energy configuration. Thus grain boundaries in recrystallized metals containing inclusion particles will tend to stabilize when they encounter particles and not move freely as required for extensive grain growth.

The grain size obtained after holding for a specific time at a given annealing temperature will be increased if the metal is subsequently reheated to a higher temperature but will be stable, unaffected by all lower temperatures, unless the

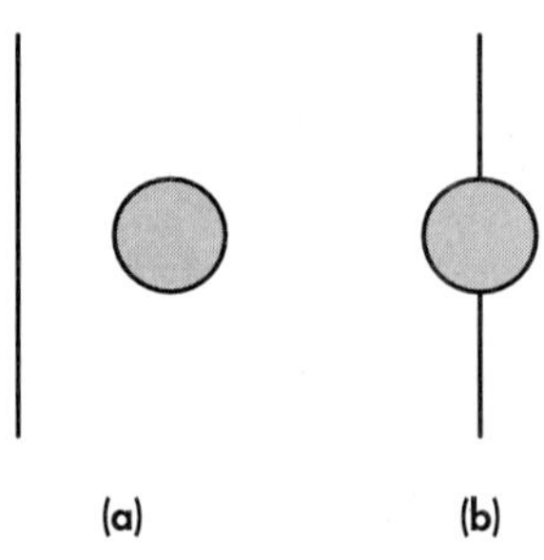

Figure 4.11 (*a*) A grain boundary near a nonmetallic inclusion; (*b*) the configuration corresponding to a minimum energy of this system.

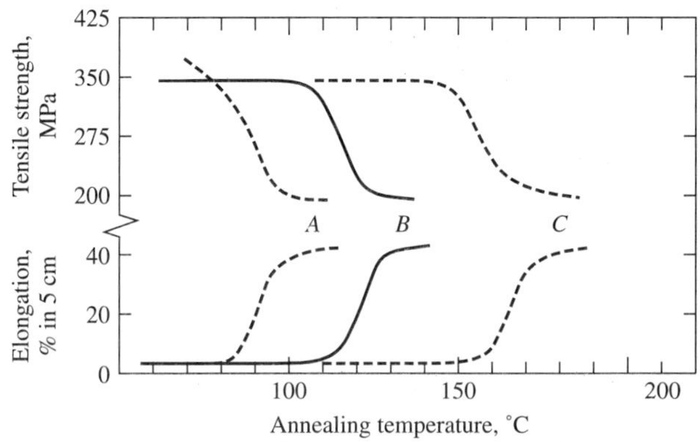

Figure 4.12 Change in the tensile strength and elongation of cold-worked, high-purity copper upon annealing, showing the effect of prior reduction and annealing time on the recrystallization temperature. *A*: 88% reduction, 24-h anneals; *B*: 50% reduction, 24-h anneals; *C*: 50% reduction, 1-h anneals.

time is increased very considerably, e.g., multiplied by about 1000 for 100°C or lower.

The effects of some of these variables are also shown in the annealing curves of high-purity copper (Fig. 4.12). Changing the annealing time from 1 to 24 h lowers the softening or recrystallization temperature range of 50% cold-rolled metal by 40°C. Changing the reduction from 50 to 88% reduces the softening temperature for 24-h anneals by 30°C.

4.4 PROPERTY CHANGES IN ANNEALED METALS

In the recovery range of annealing, hardness and strength are little affected, but stresses are at least partially removed, and thus susceptibility to stress-corrosion cracking is diminished or eliminated. The recrystallization range, in which the deformed structure is replaced by new undistorted crystals, is a range of rapid transition of properties, from those of a cold-worked or strained to those of a strain-free structure. Thus hardness and strength diminish, and ductility, as shown by elongation values in the tensile test, increases (Fig. 4.13). Higher annealing temperatures, which increase the grain size, correspondingly decrease the amount of grain boundary area per unit volume. Since grain boundaries offer discontinuities to slip or the movement of dislocations, grain coarsening is accompanied by further decreases in strength and hardness and by increases in plasticity.

Hot Working

Hot working is plastic deformation at temperatures and rates such that recrystallization occurs and the final structure is substantially free of strain hardening.

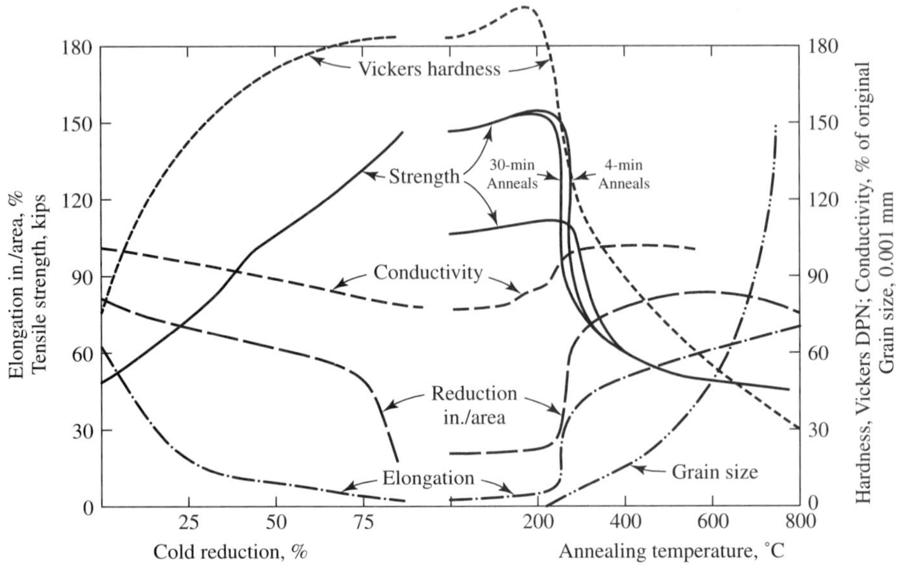

Figure 4.13 The effect of cold working and annealing on some properties of common high brass (35% Zn). Annealing times were 30 min at temperature, except for the indicated section of 4 min. The lower initial "strength" annealing curve is for metal cold-rolled only 50%, whereas the upper curve is for metal cold-rolled 87%.

Although it is usually stated that hot working is the equivalent of cold working and annealing, it must be appreciated that time and rate factors may cause very appreciable differences in structures. For example, very pure copper when severely cold-worked may recrystallize and soften within 24 h at 100°C, as shown by Fig. 4.12. However, the same copper probably could never be considered as hot-worked if rolled at any temperature close to 100°C, because of time and rate limitations. Laboratory tests have shown that cast tough-pitch copper hot-rolled at 800°C has the same hardness and strength as similar copper cold-rolled the same degree and annealed for 1 h at 350°C. However, the Cu_2O inclusions in the two samples are distributed somewhat differently, and the ductility is less in the cold-rolled and annealed structure.

It has already been pointed out that, upon cold rolling, deformation is greater in the surface zones in contact with the rolls, particularly in the early stages of deformation. This results in concave ends and edges. The opposite is true of hot working. Since the rolls or other working surfaces are usually cooler than the metal, the surface layers of the metal become cooler and harder. Therefore, deformation is greater in the center, and consequently the ends and edges of sections are likely to be convex or bulged.

In all deformational processes, a large part of the energy employed to change the shape of the metal is transformed into heat. In hot working with no strain hardening, practically all the energy becomes heat. If the surface-volume ratio of the metal being worked is large, heat is lost to the surroundings faster than it is generated in the metal and cooling will occur during working. However, large sections with a relatively small surface-volume ratio may become appreciably hotter and may even partially melt if deformed too rapidly. This factor limits the maximum deformation temperature. The minimum may be determined by strain hardening to the degree that applied forces can no longer deform the metal or by cracking in the case of metals and alloys that have limited ductility at ordinary temperatures.

Some metals that show only limited ductility in cold working may be extensively hot-worked. Magnesium (hcp), for example, has limited ductility at room temperatures where only the basal plane functions in the slip process. However, at elevated temperatures other crystallographic planes function, and the metal becomes very plastic, with the final structure substantially free of strain hardening.

4.5 PREFERRED ORIENTATION AND DIRECTIONAL PROPERTIES

The process of cold working results in rotation of grains with respect to the direction of flow. This rotation in polycrystalline alloys occurs, not in the form of entire crystals rotating uniformly but as sections, fragments, or bands within grains. The slip planes, however, do not necessarily tend to become parallel to the flow direction. In fcc metals, at least, a second set of slip planes begins to function when the first set reaches an angle of about 55° from the flow direction, and the position of stability is reached where the actual position of crystal blocks is as indicated in Fig. 4.14. Not all grains are in exactly this position (or its symmetrical mirror image),

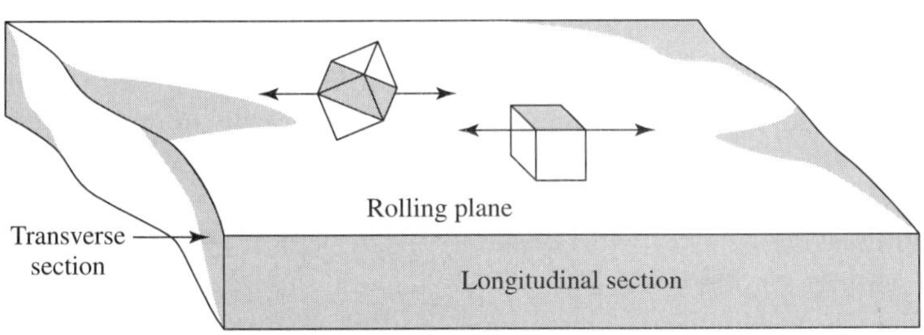

Figure 4.14 The usual preferred orientation in cold-rolled copper is shown by the left-hand cube, which has a body diagonal plane parallel to the rolled surface, oriented in the direction shown or the mirror image thereof. After annealing, this copper sheet would show the preferred orientation indicated by the right-hand cube, having a cube face parallel to the rolled surface and a cube edge in the rolling direction. Somewhat different preferred orientations are shown by brasses, aluminum, and other fcc metals, while bcc metals show very different preferred orientations.

but there is a tendency to approach the orientation shown. Body-centered cubic and close-packed hexagonal metals, with different slip mechanisms, have different end positions of the grains or grain fragments.

The development of a cold-rolled or cold-drawn preferred orientation is very greatly dependent on the state of the metal (random or preferred orientation) before cold working. In any event, a new cold-worked orientation does not usually become pronounced until the cold reduction exceeds 50%. Consequently, there is no direct relationship between strain hardening and preferred orientation.

Since most metal crystals are individually anisotropic, if the crystalline blocks all tend to align themselves in the same position with respect to the direction of the flow, it is not surprising that mechanical and other properties vary depending on the direction of testing with respect to the direction of the preceding flow. If the axis of bending is parallel with the rolling direction, the tendency to crack along the bend is far greater than when the axis of bending is at right angles to the rolling direction.

The preferred orientations found in deformed metals after relatively high reductions are frequently changed but never obliterated by the recrystallization and grain growth accompanying subsequent anneals. In fact, the directionality of properties may be greater. It is certain to be more troublesome, since disks blanked from rolled and annealed sheet are frequently drawn into cups or tubes. If the crystals are oriented in preferred directions, the flowability of metal will be greater in certain directions. Correspondingly, the drawn cups will not be uniform at the top edge but will have high sections (ears) of less wall thickness.

The directional properties of different metals vary upon cupping. Some show four ears at 0 and 90° to the rolling direction; others four ears at the 45° position; while hexagonal and, occasionally, cubic metals may show six ears. The directional properties generally increase, not only with increased prior reduction but with increased temperature of annealing. It would therefore seem that nuclei tend to show a preferred orientation not only during recrystallization but more importantly during grain growth; crystals oriented close to certain positions are favored and absorb their less fortunately situated neighbors.

The considerable amount of research in the field of preferred orientation in annealed metals has indicated that generally the preferred orientation develops from a preferred growth of nuclei having a specific crystallographic relationship to the deformed matrix lattice. In a more practical sense, it has been found that the conditions that tend to increase directionality are heavy penultimate (next to last) reductions, low penultimate annealing temperatures, and high temperatures of final annealing.

The change of properties with direction of testing of wrought metals is also known as fiber, or texture. The fiber of these rolled and annealed metals, based on preferential crystallographic orientations, should not be confused with the fiber of wrought iron or similar metals, where the effect is mechanical, caused by the presence of slag stringers all distributed in the rolling direction. Even the relatively small inclusions in ordinary rolled steels are sufficient to impair somewhat the transverse properties of annealed steels and even more affect the properties perpendicular to the rolled surfaces.

An important application of controlled preferred orientation is in the manufacture of iron sheet for transformer cores. The number of transformers in the electric-power distribution system of a country is very large, large enough for the total cost of the electrical energy dissipated as heat in the iron cores of the transformers to be a significant factor in the cost of distributing electricity. Thus, money invested in research that improves the magnetic properties of transformer iron has been found to be many times regained in savings in total power costs. The most commonly used transformer iron contains about 3% Si as an alloying addition. The silicon, in addition to narrowing the hysteresis loop, also increases the electrical resistivity of the iron so that the energy lost through eddy currents is reduced. Iron is most easily magnetized in the <100> crystallographic directions. A further improvement in the properties of the sheets used to make transformer cores is obtained by rolling and annealing the sheet to develop a strong preferred orientation. The <100> axes of the grains are all aligned in the same direction and in the direction in which the sheet is magnetized when in service. This product is known as grain-oriented sheet. In a typical sheet, the core loss for an induction of 10 kG is 0.16 W/N when the power frequency is 60 Hz.

PROBLEMS

1. Consider a cylindrical single crystal under a tensile load F applied along its axis. What are the magnitude and direction of the maximum shear stress in the crystal? What will happen to the crystal as F is increased if all the slip planes are normal to the tension axis?
2. How can measurements of mechanical.properties be used to distinguish between single crystals of an fcc and an hcp metal?
3. How can you distinguish between (*a*) slip lines, (*b*) lines of deformation (strain markings), (*c*) annealing twins, and (*d*) mechanical twins?
4. If a 70-30 brass strip were to be annealed and blanked into disks to be drawn to shape and electroplated as searchlight reflectors, what would be the most desirable annealing temperature? Why?
5. Suppose that a cast-brass part had numbers stamped on the surface and then filed off. How would you identify the original stamped numbers? If the numbers were cast in the mold, would it be possible to identify them once they were filed off?
6. Suppose that aluminum sheet, after cold rolling presumably to final thickness and annealing, was found to be 2% oversize in thickness. What would be the effect of cold rolling to size and annealing to remove the hardening effects of this slight reduction?

7. Assume that silicon-bronze parts, stressed in service while wrapped with friction tape, are found to break with an intercrystalline failure. Diagnose the probable cause of the trouble and suggest a means of checking the diagnosis and eliminating the difficulty.

REFERENCES

Courtney, T. H.: *Mechanical Behavior of Materials,* 2d ed., McGraw-Hill, New York, 2000.

Dieter, G.: *Mechanical Metallurgy,* 3d ed., McGraw-Hill, New York, 1986.

Roylance, D.: *Mechanics of Engineering Materials,* Wiley, New York, 1996.

Weertman. J., and J. R. Weertman: *Elementary Dislocation Theory,* Oxford University Press, New York, 1992.

Barrett, C. S., and T. B. Massalski: *Structure of Metals,* 3d ed., McGraw-Hill, New York, 1966.

Davis, H. E.: *Testing of Engineering Materials,* McGraw-Hill, New York, 1982.

chapter

5

Multiphase Strengthening

Alloys previously considered here have been solid solutions and, by definition, have consisted structurally of a single crystalline phase; impurities such as oxides or sulfides have been ignored. In this chapter, attention will be directed to alloys that, in the solid state, consist structurally of two or more different crystalline phases.

The presence of crystals of a different structure introduces new interfaces in addition to normal single-phase grain boundaries. These new interfaces together with the shape, size, and properties of the new phases introduce additional barriers to the free movement of dislocations. Therefore, two-phase alloys tend to be stronger than single-phase materials.

The simplest alloy systems giving two or more solid phases are eutectics. Such eutectiferous systems are very common in metallic alloys and in nonmetallic systems of mixed chlorides, oxides, etc.

5.1 BINARY EUTECTICS

Since the primary guide to multiphase reactions is the phase diagram, study of the phase diagram is the first step in understanding the structures and properties of multiphase alloys. The phase diagram of the Pb-Sb system (Fig. 5.1) is a typical eutectic diagram. The α phase, at the left end of the horizontal line of Fig. 5.1, represents solid lead with up to about 3.5% Sb in solid solution. This solubility decreases with temperature, as shown by the course of the left-hand line (solid solubility or solvus, line), which veers to the left with decrease of temperature, indicating a solubility limit at room temperature of 0.3% Sb. Under equilibrium conditions, an alloy of lead with 2% Sb will solidify as a solid solution that is stable until the temperature falls to about 220°C. Upon further cooling, the alloy crosses the solvus line and enters a two-phase field, with a resultant formation of α crystals dispersed throughout the solid β phase. The change from a one-phase solid to a two-phase

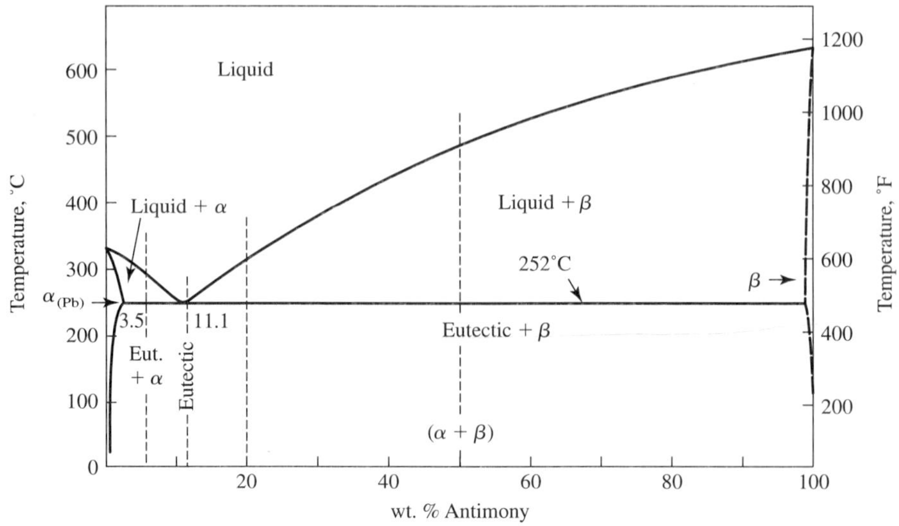

Figure 5.1 Phase diagram of the Pb-Sb alloy system, a simple eutectic.

structure on cooling, as related to this type of solvus line, is essential for age hardening (to be discussed in Chap. 6). Under equilibrium conditions, this 2% Sb alloy would show a two-phase structure at room temperatures but no eutectic.

The β phase, at the right end of the horizontal line, represents solid antimony with some lead in solution. The actual degree of this solubility is not definitely known and is of no commercial importance, since antimony and thus the α phase are relatively weak and brittle.

Along the horizontal line at 252°C, three phases may exist in equilibrium. In accordance with Gibbs' phase rule for fixed pressure, three phases can exist in a binary system under equilibrium conditions only at a constant temperature and with a fixed composition of each phase. This situation is best represented by the reversible reaction

$$\text{Liquid}_{\text{Pb}+11.1\%\text{Sb}} \rightleftharpoons \alpha_{3.5\%\text{Sb}} + \beta_{3\%\text{Pb}} + \text{heat}$$

The reaction proceeds to the right upon cooling, releasing the heat of crystallization of the α and β phases, and under equilibrium conditions solidification takes place at a constant temperature, 252°C. (For many years, eutectics were thought to be definite chemical compounds since they froze at a constant temperature and exhibited fixed concentrations, i.e., in this case, 88.9% Pb + 11.1% Sb.) On heating, the reaction goes to the left, absorbing the heat of crystallization of α and β phases and forming the liquid phase containing 11.1% Sb. Since the eutectic alloy (11.1% Sb) melts completely at a constant temperature, the word eutectic, from the Greek meaning "well melting," has come into use to describe this type of reaction or phase change. The prefix hypo-, from the Greek meaning "less than," is applied to alloys having less than the eutectic concentration of an alloying element and more than the solid-solution limit (here, 3.5 to 11.1% Sb); the prefix hyper-, meaning "more than," is applied to alloys to the right of the eutectic (here, 11.1 to about 97% Sb).

Under equilibrium conditions, hypoeutectic alloys solidify from the liquid state as follows: (1) On reaching the liquidus, nuclei of primary crystals of β form, with the composition given by the intersection of horizontal line with the solidus curve. (2) Cooling through the α + liquid field results in growth of the primary dendrites of β while their composition changes with temperature, as shown by the solidus line; at the same time, the formation of a lead-rich phase causes the residual liquid to become enriched in antimony so that its composition changes along the liquidus line. (3) At 252°C, primary α crystals and liquid exist in the ratio given by the lever rule; the eutectic liquid freezes, forming α and β phases in a fine mechanical dispersion. The dispersion is called mechanical since there is no continuous atomic conformity or necessarily any close bonding relationship between atoms of the two phases at their interfaces or planes of contact.

Under nonequilibrium conditions, freezing begins, not at the liquidus line, but at a temperature a few degrees under it (supercooling). The average composition of the primary crystals lies not on the equilibrium solidus but on a "metastable" solidus, and consequently the percentage of liquid at 252°C will be somewhat greater than that shown by the diagram. The low concentration of alloying element in the primary dendrites requires a metastable, leftward prolongation of the eutectic horizontal to intersect the metastable solidus. As a result, an alloy containing only 1.5% Sb, which should show no eutectic structure, usually will do so when solidified at rates encountered in normal castings. The eutectic freezing may be delayed by undercooling as well as the crystallization of the primary crystals; if one phase of the eutectic undercools more than the other, there can be a displacement of the eutectic concentration as well as temperature.

Undercooling of the eutectic liquid has a dual effect. Thermally, it causes the eutectic reaction to occur at a temperature of several degrees (perhaps 5 to 30°C) under that shown by the equilibrium diagram. Structurally, it causes a refinement of the particle size of the phases participating in the reaction, in the same way as solid-solution dendrites are refined by chill casting. When α containing 3.5% Sb and α–Sb containing about 97% Sb form from a homogeneous liquid, there must be a counterdiffusion of the two kinds of atoms in the liquid to each nucleus of the two phases. The rate of solidification when the liquid is suddenly chilled does not permit much time for even relatively rapid liquid diffusion, and simultaneously there is a very great increase in nucleation points for the reaction. Both factors operate to change the size of crystallites in the eutectic.

The Al-Si phase diagram (Fig. 5.2) shows this to be a simple eutectiferous system between aluminum containing a maximum of 1.65% Si in solid solution and nearly pure silicon. Supercooling, achieved partially by chill casting or more completely by small additions of metallic sodium which suppresses the nucleation of silicon crystals, lowers the eutectic temperature from 577°C to 550 to 560°C and increases the silicon in the eutectic from 11.6 to 13 to 14%.

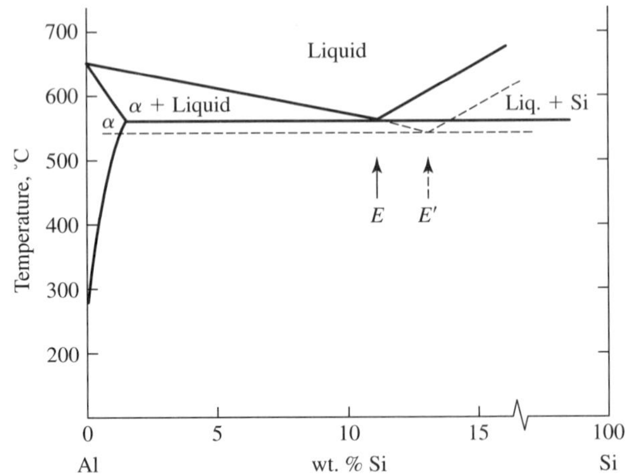

Figure 5.2 Aluminum-silicon phase diagram under equilibrium conditions (solid lines) and as chill-cast or sodium-modified (dashed lines), showing eutectic temperature and composition displacement E to E' resulting from preferential undercooling of the silicon phase.

Inspection of this phase diagram leads to certain useful qualitative conclusions. Since the aluminum solid solution makes up 85 to 90% of the eutectic, this phase should be and is continuous in the microstructure and therefore the eutectic structure shows some ductility. However, since silicon is hard and brittle, hypereutectic alloys containing relatively large primary crystals of silicon would not be expected to be as useful as eutectic or hypoeutectic alloys.

5.2 INTERMETALLIC COMPOUNDS

In the previous eutectic phase diagrams, there are two terminal solid solutions separated by a two-phase field. Often, however, intermediate phases occur in the phase diagram. These are crystallographically different structures whose composition can vary over certain limits. Sometimes intermediate phases exist over only a very small range of compositions. In this case they are called intermetallic compounds—compounds because they have a sharp melting point (or decomposition point, namely, peritectics) and other characteristics of distinct chemical species and because they often have compositions that are simple ratios when expressed as atom fractions. These compounds can be one constituent in eutectics. For example, the aluminum end of the Al-Mg system (Fig. 5.3) shows a eutectic at 450°C between aluminum containing 15.35% Mg in solid solution and an α phase which in composition approaches the stoichiometric ratio Al_3Mg_2. This intermetallic phase is quite hard and brittle. By use of the lever rule, one can see that the eutectic contains

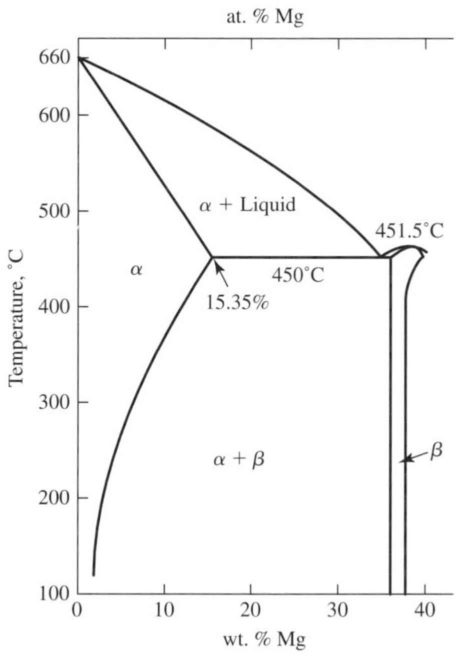

Figure 5.3 Phase diagram of the Al-Mg alloy system, from 0 to 40 wt % Mg, a eutectic between solid-solution alpha and the intermetallic compound beta.

so much α phase that the eutectic structure could be expected to be brittle—and is. Therefore, useful alloys must contain less magnesium than the maximum soluble in the solid solution, i.e., less than 15.35% Mg.

5.3 MULTICOMPONENT EUTECTICS

The temperature-phase relations for a three-component alloy system cannot be represented on a two-dimensional diagram, since there are three independent variables (two compositions and temperature) that require a three-dimensional construction. Ternary phase diagrams are usually constructed by plotting the amount of each element present along the sides of an equilateral triangle, so that each point in the triangle corresponds to one particular composition, and the temperature along a vertical axis. Figure 5.4 illustrates a ternary-alloy system of the elements *A*, *B*, and *C*, where the binary systems *A*-*B*, *A*-*C*, and *B*-*C* are all simple eutectics. The binary liquidus lines in three dimensions form three liquidus surfaces intersecting in three valleys that meet at a low point somewhere within the area of the triangle. Thus the addition of a third element depresses the freezing point of each binary eutectic and may also alter the relative proportions of the constituent metals. The ternary eutectic is at the minimum temperature where a liquid phase containing all three components can exist.

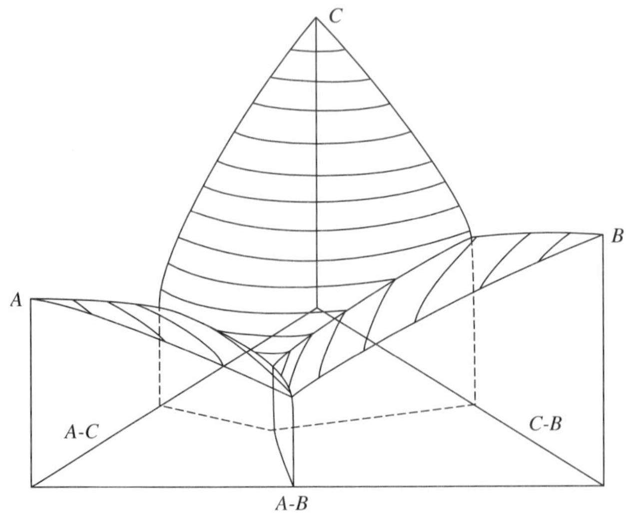

Figure 5.4 Isobaric sketch of three-dimensional phase diagram, showing a ternary eutectic between metals *A*, *B*, and *C*. Temperature is plotted vertically; composition is plotted in the equilateral triangular base. The liquidus surface has curved isothermal contour lines meeting at binary eutectics that converge to a low point, the ternary eutectic. Projections of the binary eutectic valleys are dashed in on the base.

The phase relations for alloys of four or more components cannot be represented in a single diagram, since a space of more than three dimensions would be required. Sections of multicomponent diagrams can be constructed, but they are cumbersome to use. As a practical matter, one often uses quasi-binary phase diagrams to describe complex alloy systems, such a diagram showing the phases as the concentration of one species is varied, the others being held constant. Several of these diagrams appear in later chapters of this book.

While complete phase diagrams have been worked out for a number of the simpler ternary systems, few if any complete diagrams have been determined for systems of four or more components.

5.4 MICROSTRUCTURE OF MULTIPHASE MATERIALS

Cooling-Rate Variables

The second phase in a eutectic will form in much smaller and correspondingly more numerous crystallites if cooling is rapid. A relatively slow cooling rate will not only have the expected opposite effect but in hypoeutectic alloys it also causes *eutectic divorcement.* In this case, when α and β must form eutectically in the presence of primary α crystals, the eutectic α will simply grow on the primary β. The structural result is that rather coarse β crystals are observed between α primary dendrites with no characteristic α-β eutectic structure observable.

The most striking case of eutectic divorcement ever observed was in northern Michigan, where in silver-bearing native copper a mass of copper containing 8% Ag was found with one crystal of silver containing 8% Cu attached. Silver and copper form a eutectic, solid solution of these compositions freezing from a liquid of 72% Ag + 28% Cu. The silver-bearing copper had solidified within a mass of initially molten magma over a period of geologic time, perhaps hundreds of years.

The opposite extreme of cooling rate is *splat cooling.* A shock wave is used to eject small droplets of liquid metal or alloy which impact or splatter at high velocity as thin films against a cold copper surface. Cooling rates are from 10^6 to 10^9 °C/s. Among other new structures so attainable (see Appendix 3) are highly supersaturated solid solutions. For example, instead of forming the normal eutectic structures, the Ag-Cu system shows highly supersaturated solid solutions across the entire system.

Shape Variables of Eutectic Structures

The shapes of the second phase in eutectics can vary from particles to rods, platelets, or lamellae. The crystallographic factor is strong; if the second phase is a brittle noncubic material, for example, Sb or a compound like Mg_2Sn, the eutectiferous second phase will show some idiomorphic crystallinity (Micro. 5.5) whereas cubic crystalline phases are less likely to show sharp crystalline faces and angles.

The relative proportion of the two phases is an important factor, as illustrated by Fig. 5.5. The phase that is dominant in volume will always tend to be continuous, the other eutectic phase occurring as discrete particles. If both are present in roughly the same proportion, lamellar structures are observed in which neither phase is continuous.

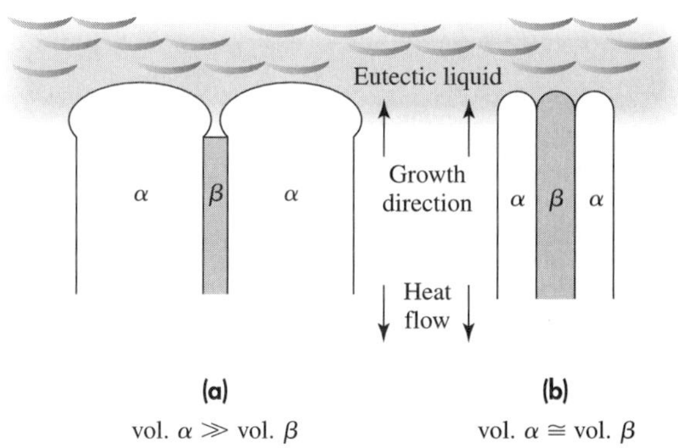

Figure 5.5 Schematic section of a eutectic reaction, solids α and β growing from a eutectic liquid. (*a*) The large volume of α causes it to grow and cut off the growth of the β crystal, resulting in short particles or platelets of β. (*b*) Approximately equal volumes of α and β can result in growth of a laminar or rod structure.

Finally, the development of commercial methods for directionally solidifying eutectics has permitted the production of eutectic structures, particularly those with rod-type second phases, where the rods are all aligned in one direction.

Eutectic Microstructures

Micrographs 5.1 to 5.5 show typical structures resulting from solidification of Pb-Sb alloys. Micrograph 5.6 is of a ternary hypereutectic alloy. Microstructures of Al-Si eutectic-type alloys are reproduced in Micros. 5.7 to 5.10.

An example of how eutectic microstructures can be changed by control of solidification is seen in Micros. 5.11 and 5.12, which are of the Al-Cu system eutectic. The eutectic in this system occurs at 33.2% Cu and is between a terminal solid solution of composition Al + 5.6% Cu and the intermetallic compound, $CuAl_2$, of composition Al + 52.5% Cu. With approximately equal volumes of both phases, this eutectic normally consists of colonies of lamellae, randomly oriented. If the eutectic structure is solidified directionally, i.e., a thermal gradient is imposed on the eutectic composition liquid such that freezing occurs progressively from one end of a small ingot to the other, orderly regions of eutectic structure over small colony areas are apparent. This colony structure is seen in Micro. 5.11. If the thermal gradient and solidification rates are adjusted to appropriate values (normally slow solidification rates, between 1.3 and 5.0 cm/h), the eutectic structure can become nearly completely aligned, as seen in Micro. 5.12, with a concomitant development of directional properties.

The structures seen in Micros. 5.11 and 5.12 are lamellar in nature, but the rodlike structures mentioned previously are generally more attractive in terms of properties. An example of such a rod structure is seen in Micro. 5.13. This eutectic structure consists of MnBi intermetallic rods in a nearly pure Bi matrix and has strong ferromagnetic anisotropy due to the mechanical and crystallographic alignment of the MnBi rods.

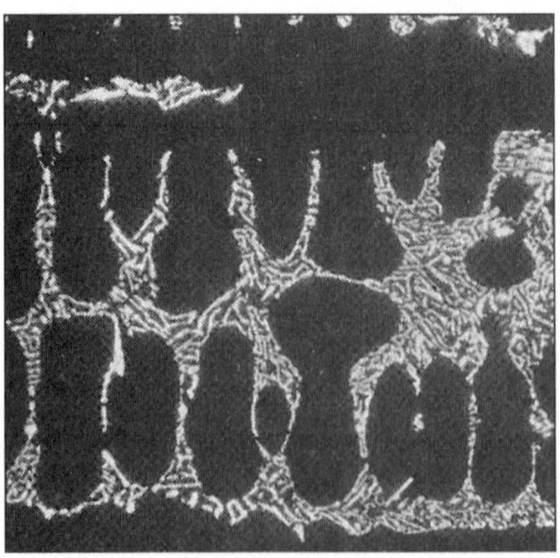

Micrograph 5.1 Pb + 6% Sb; ×50. This is a typical hypoeutectic structure consisting of primary α dendrites (black), plus an interdendritic filling of eutectic. The white particles are β crystals, which, together with the black α in which they are embedded, comprise the two-phase structure. There is a difference in appearance of primary α and the eutectic α, but both form a continuous plastic structure.

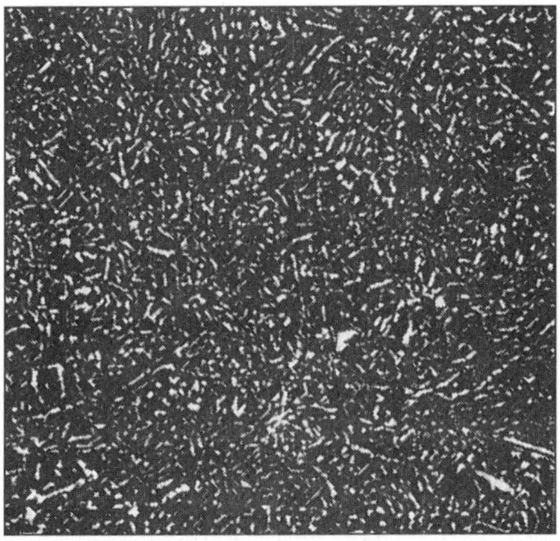

Micrograph 5.2 Pb + 11.1% Sb; ×50. At this concentration, the structure is completely eutectiferous, with white brittle crystallites of α dispersed in a continuous matrix of plastic α. Differently oriented "colonies" of β particles indicate different starting points for the eutectic reaction.

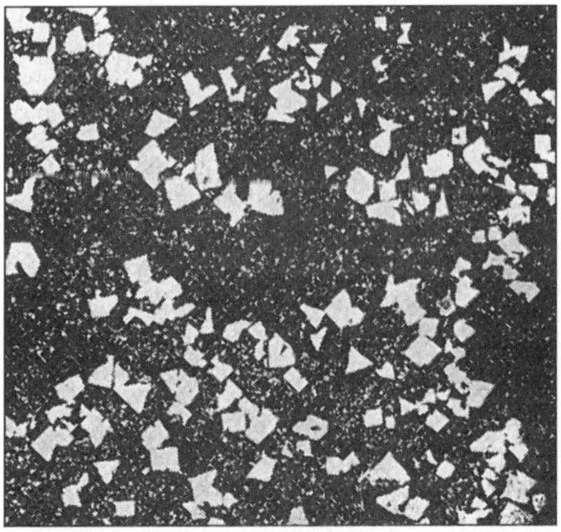

Micrograph 5.3 Pb + 20% Sb; ×50. The hypereutectic structure shows primary β crystals in a eutectic matrix. The primary β crystals are angular rather than rounded like primary α, presumably because of lower surface-tension forces. Note the clear-cut distinction in appearance, not in structure or composition, between the primary and the eutectiferous crystallites of antimony. There is a striking difference in the size of the eutectiferous β particles of this structure and those in Micro. 5.1. This is caused by a difference in solidification rate; the 20% Sb alloy was chill-cast, whereas the 6% Sb alloy solidified slowly in the melting crucible. If the 20% alloy were slowly cooled in the same manner, it would be found to show a marked gravity segregation; the lower half would be a completely eutectiferous structure, with 11.1% Sb, while the upper part would be crowded with the lighter primary β crystallites and might contain 50 to 60% Sb, a case of gravity segregation.

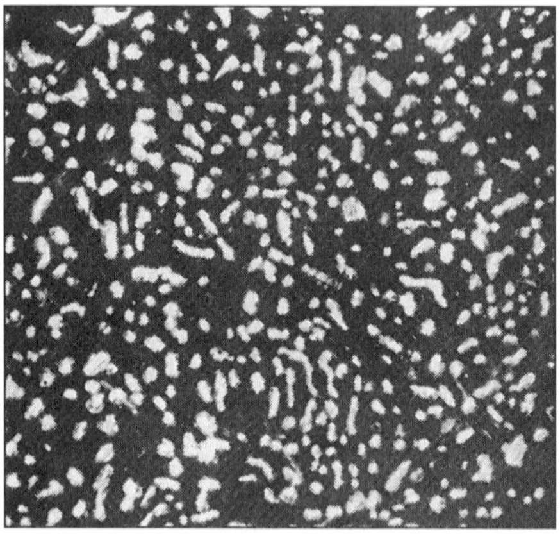

Micrograph 5.4 Pb + 20% Sb; ×500. This microstructure, of the same alloy as Micro. 5.3, is deceptive in that it shows only the eutectic part of the structure. Highly significant is the contrast in the shape as well as size of the eutectiferous α particles here and those in Micro. 5.1. Rapid solidification not only has caused the eutectic particles to be much smaller but also has prevented them from becoming platelike. Here they are much more rounded in shape.

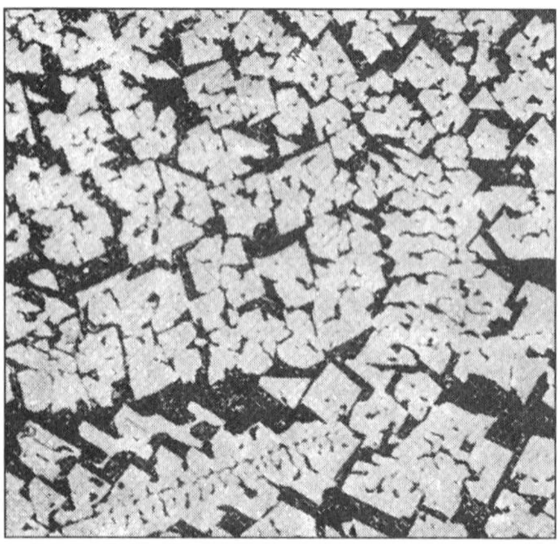

Micrograph 5.5 Pb + 50% Sb; ×50. This hypereutectic alloy contains more of the primary α crystals in a eutectic matrix. Note that the primary crystals, although characteristically sharply angular, now show a distinctly dendritic pattern or arrangement.

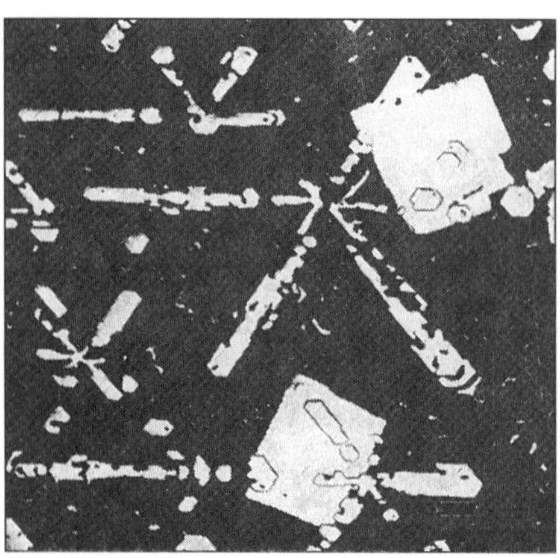

Micrograph 5.6 Hard babbitt of 84% Sn, 7% Cu, 9% Sb; ×50. This hypereutectic ternary (three-component) alloy shows primary clusters of CuSn (an intermetallic compound) crystals, arranged in a star-shaped dendritic pattern, and large rectangular crystals of primary SnSb compound in a ductile ternary eutectic consisting of these compounds and a tin-rich solid solution. The CuSn compound has a higher freezing point than the SnSb, since it seems to exist inside the latter phase, i.e., during solidification, the SnSb formed on some of the CuSn crystals already present.

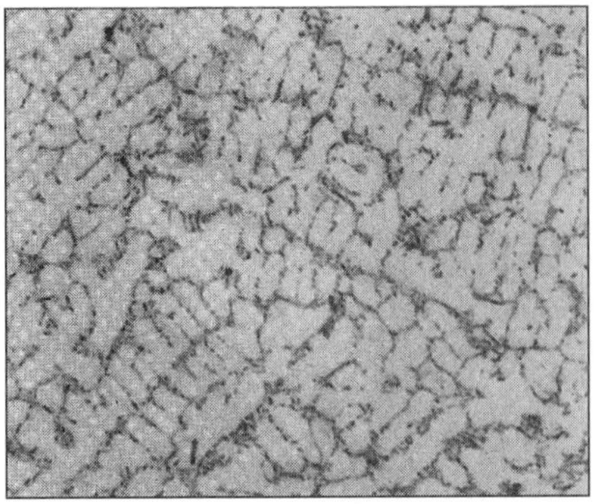

Micrograph 5.7 Al + 6% Si, chill-cast; ×100; 0.5% HF etch. Chill casting has resulted in small primary dendrites of Al and very fine particles of Si in the eutectic.

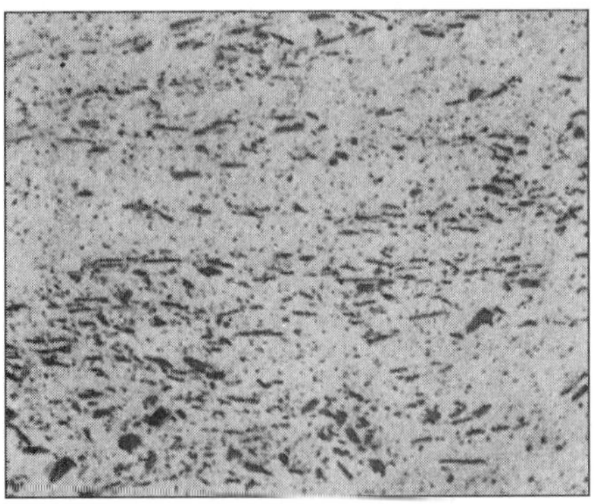

Micrograph 5.8 Al + 6% Si, hot-forged at about 500°C; ×100; 0.5% HF etch. Forging has broken up the eutectic structure and aligned silicon crystallites in the flow direction. The temperature of hot work has also induced growth of the eutectiferous crystallites. Slow cooling from the forging temperature has caused precipitation of silicon from solid solution, which gives the "dirty" background.

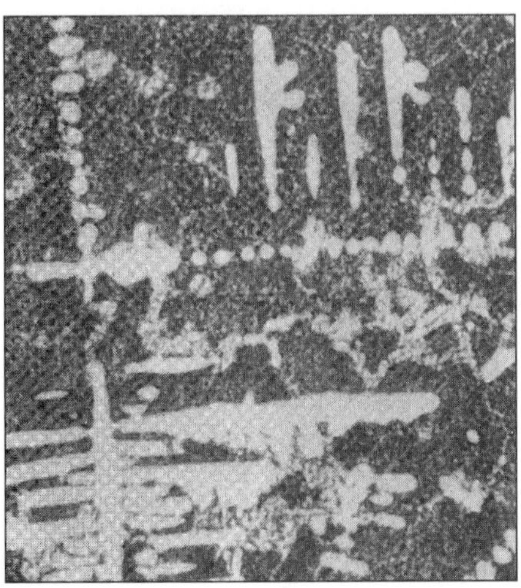

Micrograph 5.9 Al + 13% Si + about 0.8% Cu and Fe as impurities ×75; 0.5% HF etch. This is a hypoeutectic structure showing a few primary dendrites of αAl in a fine eutectic structure. Since the equilibrium eutectic composition is at 11.6% Si, this alloy would normally be hypereutectic, but the addition of 0.25% Na 15 min before casting suppresses both the formation of primary silicon crystals and the eutectic reaction. The liquid cools until it reaches the temperature of the metastable prolongation of the hypoeutectic liquidus to form a few primary αAl dendrites, after which the eutectic reaction starts. The undercooled eutectic liquid forms a very finely dispersed two-phase structure (at about 564°C) rather than the equilibrium temperature of 578°C.

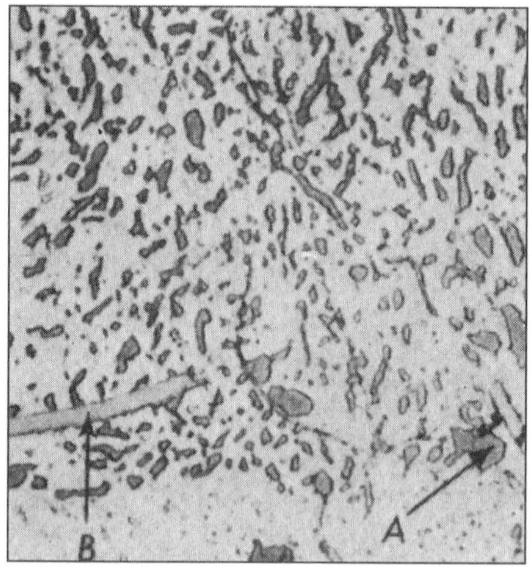

Micrograph 5.10 Same alloy (13% Si) ×1000; 0.5% HF etch. Dark gray particles, such as that marked *A*, are eutectiferous silicon crystallites; lighter gray needles, marked *B*, are an Al-Fe-Si compound originating from the iron impurity. Note that the αAl phase is continuous, as predictable from the approximate relative proportions of αAl and silicon in the eutectic.

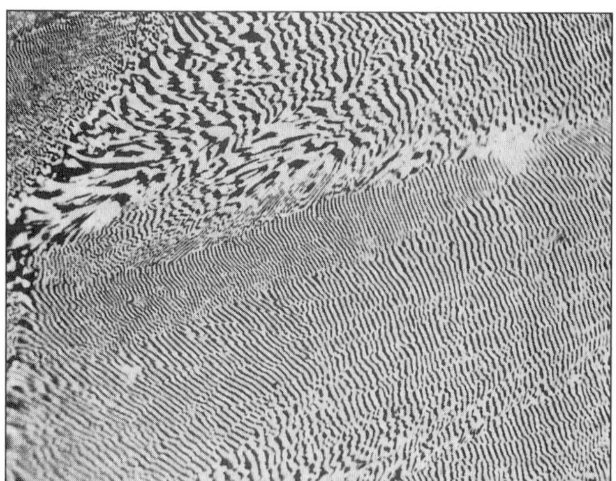

Micrograph 5.11 Al + 33.2% Cu, ×250. Etched with 20% HNO_3 and 80% H_2O. The white phase is alpha Al and the dark platelets are $CuAl_2$ compound. Normally this eutectic structure shows random colonies of lamellae such as in the top third of the photograph. This, however, is a transverse section of a directionally solidified rod, so the colonies are more regular in appearance.

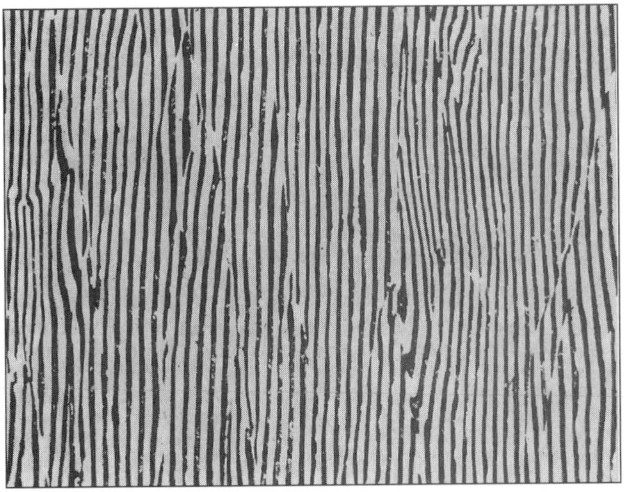

Micrograph 5.12 Al + 33.2% Cu, ×500. Etched with 20% HNO_3 and 80% H_2O. Aligned microstructure with same constituents as Micro. 5.11. Minor local variations in growth rate produced faulted structure rather than perfect alignment of the lamellae. Here the section is parallel to the growth direction.

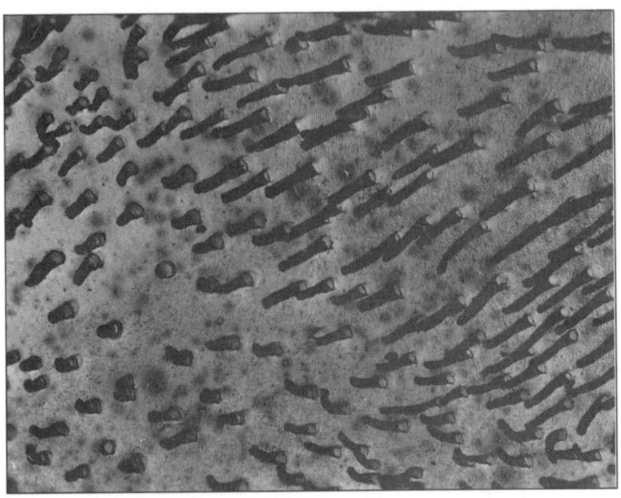

Micrograph 5.13 Bi + 0.6% Mn, ×5000. Electrolytic etch saturated KI and 2% HCl, 0.2 A/mm^2 for 20 s. Electron micrograph from shadowed plastic replica of a directionally solidified Bi-BiMn compound eutectic. This micro shows an overetched structure in which the Bi matrix has been preferentially attacked, allowing the rods of BiMn to stand out in relief. Note the nearly perfect alignment of the rods transverse to the plane of polish.

5.5 GENERALIZED PROPERTIES OF MULTIPHASE MATERIALS

The properties of a series of alloys across a eutectic horizontal will naturally be a function of the two solid phases present. In the Pb-Sb system and in a majority of commercially important eutectiferous alloys, one phase is relatively weak and plastic, the other relatively hard and brittle. As the antimony content is increased from 3 to about 97%, the proportionate amount or fraction of α crystallites increases linearly, but the strength does not increase in the same way because of the difference in dispersion or size of the α solid-solution crystals, depending on whether they are primary or eutectiferous. There is a rapid rate of increase in strength from 3 to 11% and a lesser rate from 11 to 97% Sb. Hardness, shown in Fig. 5.6, behaves in much the same way. In the hypoeutectic range, fine eutectiferous crystallites of βSb are increasing in amount; in the latter range, the number of small particles (or the amount of eutectic) is decreasing, with a corresponding increase in the number and size of large primary crystallites of antimony. The result is an inflection at the eutectic point in the plot of any mechanical property against alloy concentration across a eutectic series; in fact, it may be not only an inflection but also a maximum, particularly of strength (Fig. 5.7).

Examination of the microstructures shows that in all hypo- and hypereutectic alloys, the eutectic structure is continuous, as would be expected, since, during freezing, eutectic liquid surrounds the primary dendrites. If in the eutectic structure the plastic phase is continuous as in the Pb + 11.1% Sb alloy, then the entire series of alloys must have some plasticity. If, on the other hand, the brittle phase is continuous as in Al-Mg alloys, the entire series will be brittle.

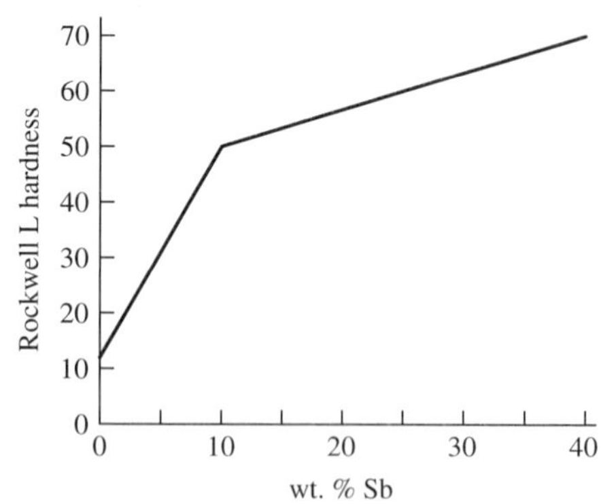

Figure 5.6 Rockwell L hardness of cast Pb-Sb alloys. The sharp break in the curve appears at the eutectic concentration of Sb.

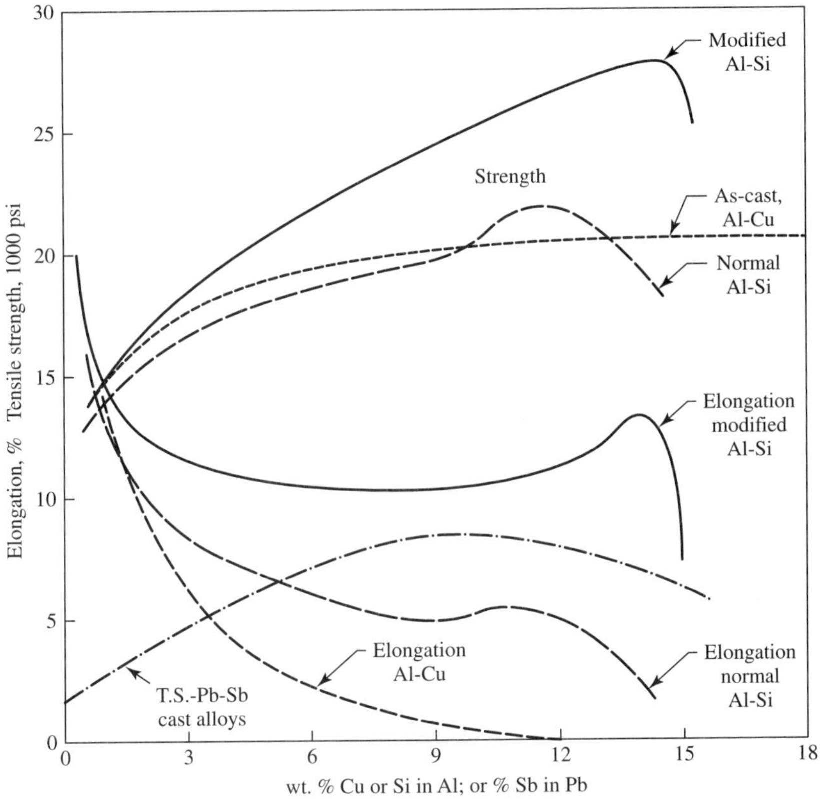

Figure 5.7 Strength properties across the hypoeutectic sections of the Al-Cu, Al-Si, and Pb-Sb systems. The diagram also shows the effect of modification (see Micro. 5.9) on the strength and elongation of the Al-Si alloys.

All the Al-Si alloys show some ductility, since αAl is the continuous phase in the eutectic. The superior ductility of the modified alloys is related to the shape as well as the size of the eutectiferous silicon crystallites. In the normal alloys, they tend to be angular plates whose sharp edges act as internal notches in the structure, whereas this effect is nearly absent with the rounded silicon particles of the modified alloy (Micro. 5.10).

The Al-Cu system has a eutectic composition nearer the brittle phase ($CuAl_2$) than the plastic βAl phase, and thus the αAl + eutectic structure is inherently brittle. The effect of fairly rapid cooling in casting and the related position of the metastable solidus result in some eutectics in cast alloys with as little as 2% Cu present. Figure 5.7 shows that there is a moderate increase in strength and strong decrease in ductility (as shown by test elongation values) as the amount of copper in aluminum is increased. When the eutectic structure becomes continuous, the strength is almost at a maximum and ductility almost nil.

PROBLEMS

1. Draw cooling curves for alloys of lead with 1, 6, 11, 20, and 50% Sb. The significant temperatures should be correlated with the phase diagram. Assume that equilibrium is attained.
2. Calculate the percentages of structural constituents, i.e., primary β or α and eutectic present in each of the alloys of Prob. 1. Do this on both a weight and a volume basis.
3. In the discussion of the microstructure of a Pb + 20% Sb alloy, it was pointed out that chill casting was required to prevent a gravity-segregation effect. Why is this true for the 20% Sb alloy and not for the 6% Sb alloy?
4. Many soldered joints are "wiped" by employing a solder that in the "mushy" state can be pushed around with a gloved hand. Why would the Pb + 62% Sn alloy be specifically suited or unsuited to this usage?
5. If the microstructure of a Pb-Sb alloy showed only traces of fine eutectiferous antimony crystallites in an interdendritic dispersion, what would be the probable composition of the alloy? If the few eutectiferous α particles were coarse, would the composition be any different? Why?
6. Explain the presence and type of coring in a Pb + 6% Sb alloy.

REFERENCES

Wark, K., and D. Richards: *Thermodynamics,* 6th ed., McGraw-Hill, New York, 1999.

Gaskell, D. R.: *Introduction to the Thermodynamics of Materials,* 3d ed., Taylor and Francis, Washington, D.C., 1995.

Okamoto, H.: "Reevaluation of Thermodynamic Models for Phase Diagram Evaluation," *J. Phase Equilibria,* vol. 12 (6), 1991.

Chadwick, G. A.: "Eutectic Alloy Solidification," *Prog. Mater. Sci.,* vol. 12 (2), 1963.

ASM Handbook: vol. 9, *Metallography and Microstructures,* 10th ed., ASM Int., Materials Park, OH, 1995.

ASM Handbook, vol. 3, *Alloy Phase Diagrams,* 10th ed., ASM Int., Materials Park, OH, 1992.

Villars, P., and L. D. Calvert: *Pearson's Handbook of Crystallographic Data for Intermediate Phases,* 3d ed., ASM Int., Materials Park, OH, 1991.

Hayes, F. H. (Ed.): *User Aspects of Phase Diagrams,* The Institute of Metals, London, England, 1991.

Massalski, T. B. (Ed.): *Binary Phase Diagrams,* 2d ed., ASM Int., Materials Park, OH, 1990.

chapter

6

Precipitation Hardening

The alloy of aluminum containing about 4.5% Cu and 0.5% Mg was discovered in Germany around 1912 and was found to be hardenable by quenching the solid alloy from a temperature close to the eutectic, i.e., from within the single-phase, solid-solution field, followed by aging at ambient or slightly elevated temperatures. This very useful phenomenon known as age hardening, or precipitation hardening, is now known to exist in a large number of alloy systems.

6.1 GENERAL MECHANISM OF PRECIPITATION HARDENING

Hardening of a metal or alloy results when obstacles to the free motion of dislocations through the material are established. An obstacle, as far as a dislocation is concerned, is any immobile part of the structure that exerts a force on the dislocation. Forces on dislocations arise from externally applied stresses and also from the interaction of the stress field that surrounds every dislocation with any inhomogeneity that produces a distortion in the crystal structure. For example, a solute atom of different size from the atoms of the solvent causes a local distortion of the crystal structure that exerts a force on an approaching dislocation. Dislocations also interact with each other because a region of distorted crystal structure surrounds each.

During work hardening, obstacles to dislocation motion are generated during the deformation process itself. These obstacles are arrays of other dislocations whose motion has been blocked by grain boundaries or which are cutting across the plane on which slip is occurring. Other hardening mechanisms involve the establishment of obstacles in the form of individual or clusters of solute atoms or point defects. The former are introduced by alloying, perhaps followed by heat treatment, the latter by irradiating the metal at low temperatures with particles, such as neutrons, sufficiently energetic to cause damage in the arrangement of atoms on the crystal lattice.

The rest of this discussion is confined to hardening effects achieved by certain types of alloying.

The effectiveness of the barriers introduced by alloying in hardening a metal depends on two factors:

1. The strength of the obstacle formed, i.e., its ability to withstand the passage of a dislocation through itself
2. The spacing between obstacles

The importance of the first factor is clear, for if dislocations can cause slip to occur within an obstacle, they can force their way through the barriers to their motion. The importance of the spacing between obstacles can be seen in this example. Suppose that the obstacles are small and randomly dispersed, solute atoms in a solid solution, for example. Then, for every obstacle, a given distance ahead of a dislocation line, there will be another nearby about the same distance behind. The forces that these two obstacles exert on the line very nearly balance, and the stress required to make the dislocation move through such an array of obstacles is not much greater than if the obstacles were not there. This situation is illustrated in Fig. 6.1. Figure 6.2 shows the situation when the same amount of solute is present but is collected into

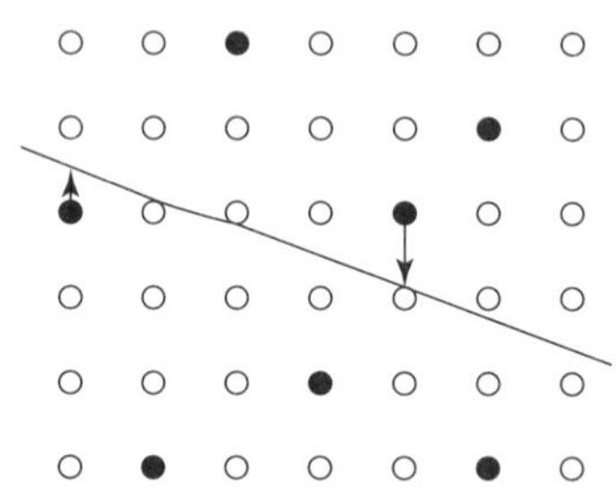

Figure 6.1 A dislocation line moving through a structure containing very small, randomly dispersed obstacles, e.g., atoms of different size in solid solution. The forces exerted on the dislocation by the obstacles balance out for the most part.

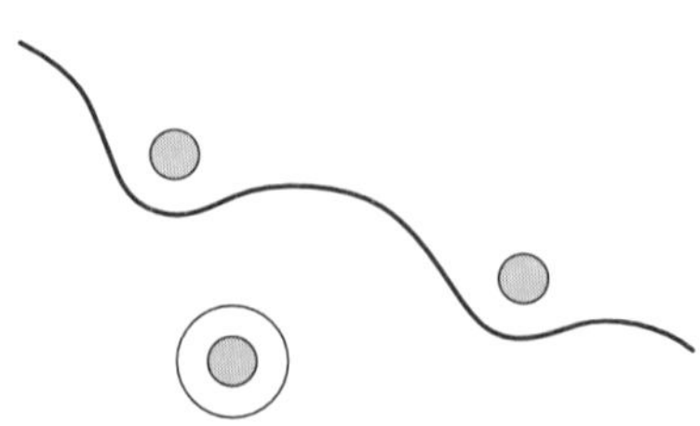

Figure 6.2 A dislocation passing large obstacles spaced far apart. After bending around each obstacle, it leaves a dislocation loop behind. The scale in this diagram is very different from Fig. 6.1, being microscopic rather than atomic.

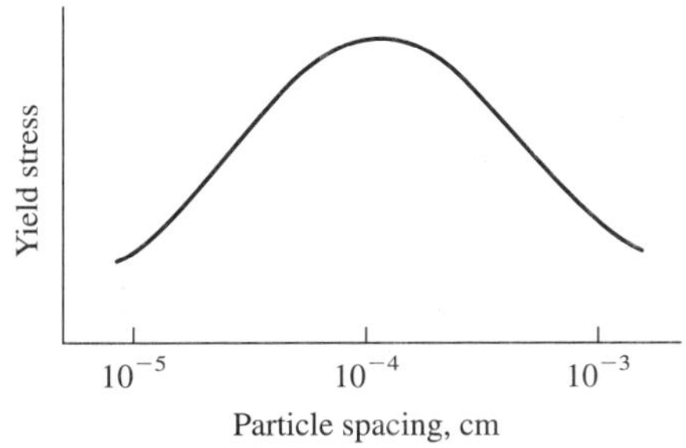

Figure 6.3 The dependence of the hardening effect of particles on the interparticle spacing for a structure such as that of Fig. 6.2, assuming a constant alloy content.

obstacles (second-phase particles) of larger size. In this case it turns out that the dislocation can bend around the obstacle and then pass on its way, after leaving a small dislocation loop behind. Again, the force restraining the dislocation motion is small. For some intermediate spacing of the obstacles, the hardening effect must be a maximum. This is illustrated in Fig. 6.3, which shows the effect of particle size on the yield stress of a metal.

For a given concentration of solute dissolved in a metal, the particle spacing depends on the particle size. The solute may be concentrated in a few large particles or in many small ones. In a solid solution, the solute atoms are dispersed as individuals in the solvent crystal and their hardening effect is relatively weak. To achieve optimum hardening from a given amount of an alloying element, some method of controlling its distribution in the alloy is required. In many alloy systems this can be realized by a heat treatment, which causes a controlled precipitation reaction.

6.2 PRECIPITATION FROM SOLID SOLUTION

Suppose that an alloy in equilibrium in a single-phase field (point 1, Fig. 6.4) is cooled to a temperature corresponding to a two-phase region of the phase diagram (point 2). The alloy will be supersaturated with solute that will begin to precipitate out of solution. The precipitating phase will, in general, have a different crystal structure and specific volume from that of the parent phase. Formation of precipitate particles within a homogeneous solid solution will result in severe local distortion of the matrix, creating concentrations of elastic-strain energy in the vicinity of the precipitating particles. This elastic energy plays an important role in determining the form of the precipitate particles.

The change in free energy resulting from precipitation of a second phase from a supersaturated solid solution is the sum of three separate changes:

1. A decrease in the free energy per unit volume of the precipitate
2. An increase in free energy due to the surface energy of the precipitate–parent-phase interface

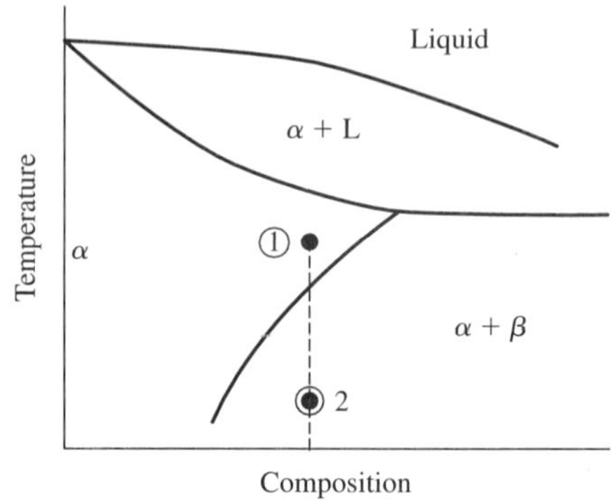

Figure 6.4 The partial phase diagram of an alloy system in which precipitation from solid solution may occur.

3. An increase in free energy due to local elastic distortion in the vicinity of the precipitate particles

The first two factors are similar to those determining the free-energy change during nucleation of a solid from a liquid; in fact, solid-state precipitation is, like solidification, a nucleation and growth process. There is, however, the additional contribution of the elastic-strain energy. It is this factor that primarily determines the form of the precipitating particles: they may be either coherent or incoherent with the surrounding matrix.

A coherent precipitate is a region of the solvent structure where solute atoms have concentrated to the degree required to give, in this region, the composition of the second phase. There is no true interface between the "particle" and the surrounding matrix. Instead, there is a more or less sharp difference in concentration and usually a change in crystal structure. Because the solute atoms will generally be of different size from the solvent atoms, there will be a large amount of elastic distortion of the structure around the precipitate "particles" (Fig. 6.5). In a coherent precipitate the elastic-strain-energy term in the total free energy will be large, but the surface-energy term will be zero, since there is no true interface between precipitate and matrix.

An incoherent precipitate particle is truly a distinct particle of a second phase. It has its own crystal structure and is separated by an interface from the surrounding matrix (Fig. 6.6). The elastic-strain energy around the particle is relatively low in comparison with the incoherent case.

In general, coherency will be favored in the early stages of the growth of a precipitate and incoherency in the later stages. This is because the surface-to-volume ratio for a small particle is large: the surface-energy term is relatively important, and the system will assume the form that minimizes the surface energy, i.e., a coherent particle. As the particle grows, the strain energy increases rapidly, and eventually a point is reached where coherency breaks down, the strain energy is

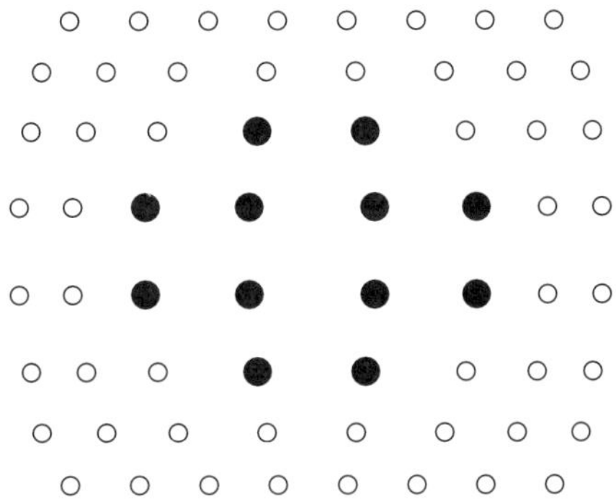

Figure 6.5 A two-dimensional view of a coherent precipitate particle formed, in this case, of atoms larger than the atoms of the matrix. There is no true interface separating the particle from the matrix, at least in the plane shown here.

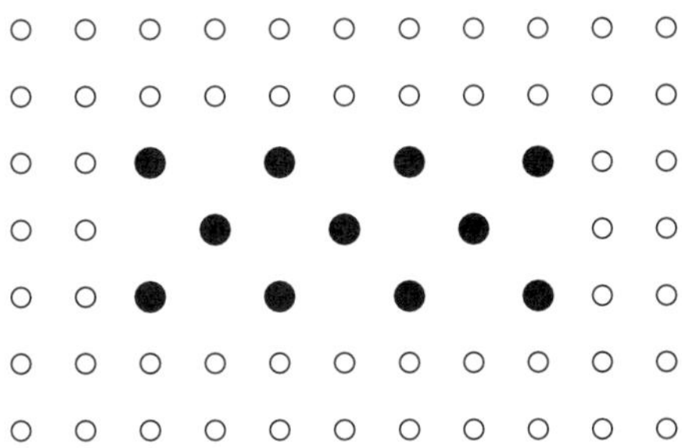

Figure 6.6 A two-dimensional view of an incoherent precipitate particle. It has a distinct crystal structure and is separated from the matrix by a true interface on the *X*, *Y* directions shown here.

decreased, and an interface is created, the increase in free energy due to formation of the interface being less than the decrease in the strain energy. Whether precipitate particles are coherent or incoherent determines to a considerable degree the effect of the presence of the precipitate on the mechanical properties of the material. Coherent precipitate particles are particularly powerful obstacles to the motion of dislocations, because the large elastic distortion of the matrix around the particles interacts strongly with the stress field of the dislocations.

6.3 STAGES OF PRECIPITATION HARDENING

The first requirement for age hardening is that the system has in its phase diagram the condition shown by Fig. 6.4, namely, a terminal solid solution where the solubility of the second component or solute is appreciably greater at an elevated temperature

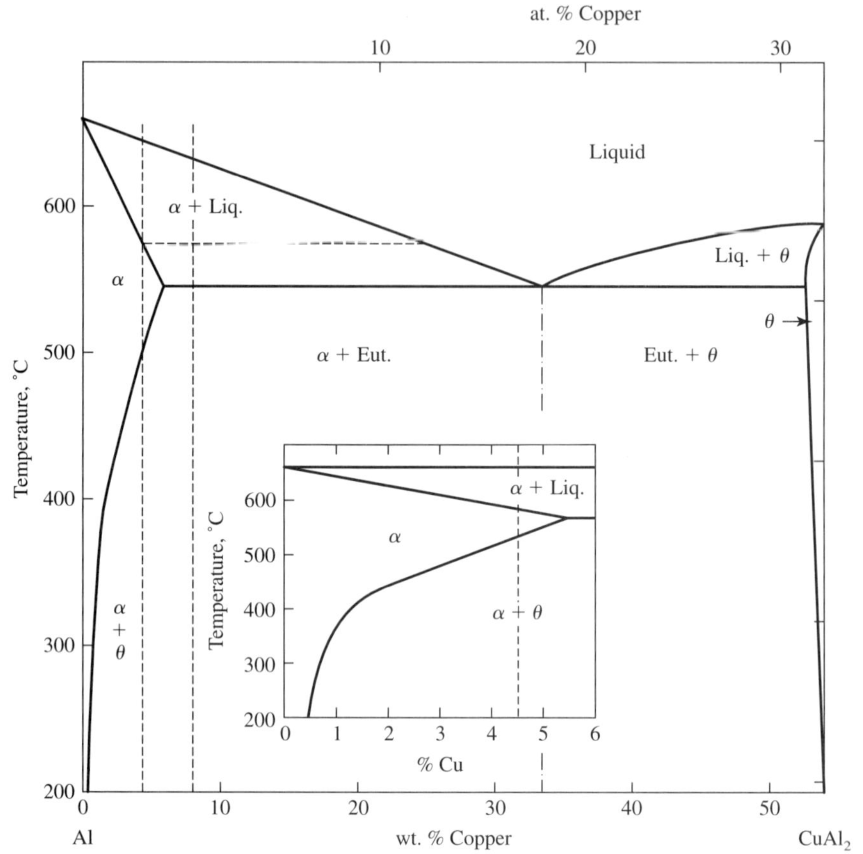

Figure 6.7 Phase diagram of the Al-Cu alloy system, from 0 to 54 wt % Cu showing a eutectic between α and θ and a decreasing solubility Cu in α with decrease in temperature below the eutectic.

than at room temperatures. Then, as in the case of a 4% Cu alloy (Fig. 6.7), the structure can be changed from a one-phase to a two-phase structure, or vice versa, simply by change of temperature. When a new or second phase forms within a previously single-phase structure, a host of fascinating variables such as nucleation and growth as functions of temperature (diffusion rates, etc.) make possible the technical control of properties over a wide range, from soft and plastic, to hard and plastic, to hard and brittle, together with all intermediate states.

Annealing It may be desired to put the alloy in a soft condition for fabrication purposes. In this case, the material is heated to under the solvus line, e.g., at 410°C (Fig. 6.7) and held long enough for a coarse precipitate to form. Then the alloy is cooled slowly to permit further growth of the precipitate, reducing the copper content of the matrix α. This greatly reduces solid-solution hardening and minimizes second-phase hardening by having a few coarse particles of θ rather than many very fine ones.

Solution Treatment The alloy is heated into the single-phase field and held a sufficient time for the second phase, θ, to dissolve and the atoms of copper to become randomly dispersed in the solid solution. A cast alloy with coarse, eutectiferous θ requires a longer solution time than a forged or rolled alloy, where θ has been broken up. The temperature must be below that of the eutectic to avoid any chance of localized eutectic melting.

Quenching The alloy must be cooled rapidly to prevent the precipitation of θ during cooling. Immersion of the hot metal in agitated cold water (or iced brine) provides the most rapid cooling but results in distortion of the part and high residual stresses. Immersion in hot water gives quite slow cooling, since the metal part in the quench is immediately covered with a steam blanket. As a result of this slow cooling in steam, some precipitate forms. This always occurs first at grain boundaries, where an interface already exists and little new surface energy is generated. Grain-boundary precipitates deplete the boundary areas of copper, impairing strength and corrosion resistance. Mixtures of 40°C water and glycols give intermediate cooling rates and satisfactory results.

Aging Hardening the quenched alloy involves the eventual formation of a precipitate of θ in the supersaturated solid solution. Although θ is the final form of the precipitate, several intermediate forms appear before the equilibrium phase appears.

Copper atoms are distributed randomly, not uniformly, in the 4.5% Cu solid-solution matrix. This means that in a sample of 1 million atoms, there would be 4.5 wt % Cu atoms but in a sample of only 1000 atoms, there could be from 1 to 20 wt % Cu. The sequence of precipitation involves first the formation of what are called GP[1] zones, i.e., areas of highly localized concentrations of copper atoms on cube planes of the matrix. With further aging, these regions of higher Cu concentration grow laterally to assume an ordered structure of Cu atoms on cube or (100) planes of the matrix, and these structures are called GP[2] zones.

The θ phase eventually to be formed has a tetragonal structure where the edge of its base is within 1% of the same dimension as the face diagonal of the cubic α. Thus the two different lattices can match when the basal plane of θ is oriented parallel to the cubic plane of α. When the GP[2] zones grow to a size where they have the tetragonal structure but distorted in base plane dimensions by coherency, the precipitate is called θ'. The θ' can grow to be fairly large plates, e.g., several hundred atoms across, before the matching plane-elastic-strain distortions cause loss of coherency. The growth is as plates simply because the coherency is two-dimensional and exists only on the stipulated matching planes. In the third dimension, normal to the matching plane, no coherency exists.

At the stage of maximum strength, it is difficult to resolve the particles of θ' by optical microscopy (Micro. 6.1). Overaging, as in Micro. 6.2, results in readily visible precipitates, here preferentially along slip planes resulting from severe quenching strains. Severe overaging is shown by the structure of Micro. 6.3. Accidental overaging is repairable but only by starting anew with a solution treatment.

Burning in aluminum alloys means overheating so that partial melting occurs. Micrograph 6.4 shows such burning and solidified eutectic formed upon cooling of

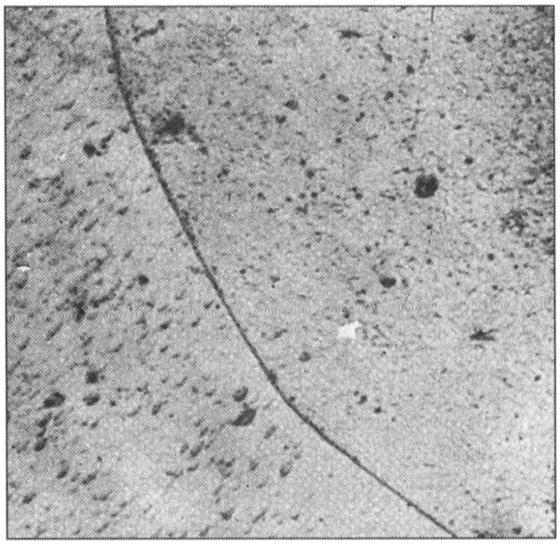

Micrograph 6.1 Al + 5% Cu; alloy of pure materials (less than 0.05% Fe + Si); quenched from 540°C, reheated 30 min at 200°C, ×1000; 0.5% HF etch. The heat treatment given this alloy resulted in maximum hardness (see Table 6.2), but there is no readily visible precipitate. The grain boundary has etched more deeply than it would in the as-quenched (from 540°C) structure, and there are a few markings within the grains, indicative of some change in the solid solution. Special etches have revealed more positive indications of precipitation in equivalent structures, but this photomicrograph is representative of the structure revealed by the usual micrographic technique.

Micrograph 6.2 Al + 5% Cu; ×1000; 0.5% HF etch. A specimen similar to Micro. 6.1 was quenched in cold water from a solution treatment at 540°C and reheated 1 h at 250°C. Distortion from the cold-water quench caused a slight plastic deformation of the alloy matrix, and the subsequent aging treatment resulted in precipitation of the θ phase, which occurred preferentially on the slip planes that were active in the plastic movement. Nucleation of the precipitation (or, more generally, of any solid phase change) occurs first at the least stable part of the matrix lattice, normally at grain boundaries but in the case of prior or simultaneous plastic deformation at slip planes. Note that precipitation is not uniform in all grains (in slight deformations, plastic movement is not uniform in all crystals). Particles of the precipitated θ are very fine and not clearly resolved here.

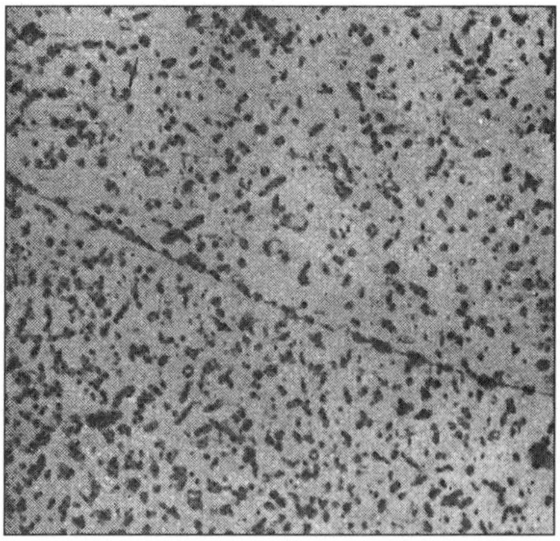

Micrograph 6.3 Al + 5% Cu; ×1000; 0.5% HF etch. A specimen similar to Micro. 6.1 was quenched from 540°C and reheated 1 h at 400°C. At this high aging temperature, the precipitated particles are very coarse, and partly as a result of their size and partly as a result of re-solution of θ (1.5% Cu is soluble at 400°C and only 0.6% at 250°C), there are fewer particles. Upon close examination, some needles are visible, and these, in the third dimension, would of course be plates. The plates show the genetic relationship between matrix and precipitate lattice orientations, as mentioned in the previous section on theory of age hardening. These visible plates, however, are tremendously larger than the platelets present at an early stage of aging. Notice the lineup of θ particles along the α_{Al} grain boundary.

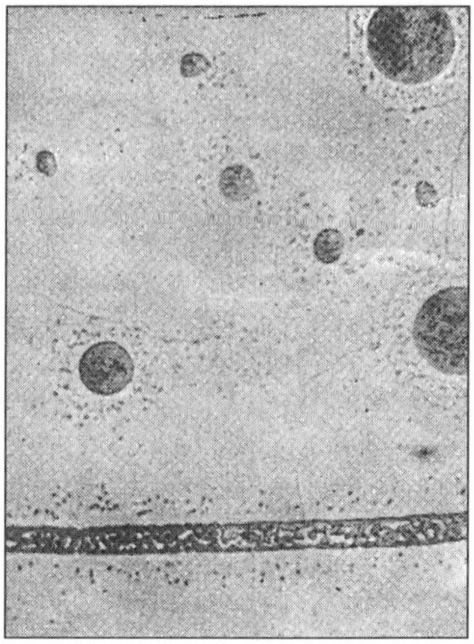

Micrograph 6.4 Same Al + 5% Cu alloy quenched from 620°C and reheated at 400°C; ×1000; 0.5% HF etch. The high-temperature treatment was well above the solidus temperature for this alloy (see Fig. 6.7), and "burning" occurred. Not only did a eutectiferous liquid form at the grain boundary (the horizontal eutectic structure), but also a similar liquid formed in spherical globules within the grains, resulting in the circular eutectic rosettes shown above the boundary. The precipitate reveals coring in the solid solution adjacent to the boundary eutectic and the rosettes; the $CuAl_2$ was precipitated in these regions upon the subsequent reheating to 400°C. The thin lines visible in parts of the structure are new grain boundaries, formed by recrystallization, at 400°C, of metal plastically deformed during the previous quench.

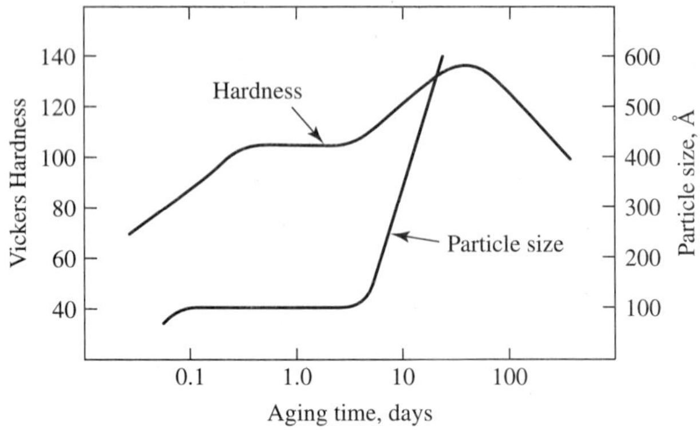

Figure 6.8 The hardness of an Al + 4.5% Cu alloy quenched from 500°C and aged at 130°C. Also shown is the diameter of the precipitating particles as determined by x-ray diffraction experiments.

the overheated alloy. Since such a structure, inherently brittle, is not repairable by heat treatment, the alloy must be remelted.

During aging, hardness and tensile strength increase, reach a maximum when the coherent precipitate particles attain optimum size, and then decrease. The aging process is ordinarily stopped when the optimum properties are reached, the alloy then being fully hardened. Figure 6.8 shows the variation of hardness with aging time for the Al + 4.5% Cu alloy held at an aging temperature of 130°C. Also shown is the average diameter of the precipitating "particles." The hardness curve is not entirely smooth, showing a preliminary maximum at about 1 day's time. This behavior reflects the complexity of the precipitation process in the age-hardenable Al-Cu alloys, i.e., the existence of intermediate forms of the precipitate.

In Fig. 6.8 the first hardness maximum is associated with the formation of the GP[1] zones. With further aging the regions of copper concentration assume an ordered structure and are designated as GP[2]. At the second hardness maximum of Fig. 6.8, the coherent form of the precipitate of $CuAl_2$ forms, designated θ'. Finally, as the alloy becomes overaged, the incoherent precipitate θ forms as microscopically visible particles of $CuAl_2$ in the aluminum solid-solution matrix. The continuously decreasing hardness of the alloy upon further aging corresponds to a continuing growth of the θ precipitate and a decrease in the number of θ particles.

6.4 VARIABLES AFFECTING PRECIPITATION HARDENING

Time and Temperature

Some increase in strength of Al-Cu alloys occurs at room temperature (natural aging); therefore, if the relatively soft, plastic, as-quenched structure must be preserved, the alloy is kept at dry-ice temperatures, for example, −40°C or less. However, attainment of maximum strength requires aging at elevated temperatures (artificial aging), for example, 75 to 150°C. Precipitation of the second phase involves diffusion and is

Table 6.1 Aging of Al + 5% Cu alloy as quenched from 540°C

Temp., °C	Time	Rockwell Hardness, F Scale	Temp., °C	Time	Rockwell Hardness, F Scale
25	0	61	200	5 h	87
	2 h	72		24 h	83
	1 day	85	250	1 h	84
	1 week	87	300	¼ h	83
	1 month	88		1 h	73
	1 year	88		5 h	65
150	1 h	80		24 h	43
200	¼ h	78	350	1 h	56
	1 h	90	400	1 h	44

time- and temperature-dependent. An increase in aging temperature of 10°C results in halving the time for a specific result, e.g., attainment of maximum strength.

Both isochronal and isothermal age-hardening data in Table 6.1 indicate the generality of aging time-temperature relationships on hardness. For example, the Al + 5% Cu alloy is approximately equally overaged in 14 h at 200°C, 1 h at 250°C, and 15 min at 300°C. Figures 6.9 and 6.11 for aging of an Al-Cu and Cu-Be alloy show more specific aging data of this type, respectively.

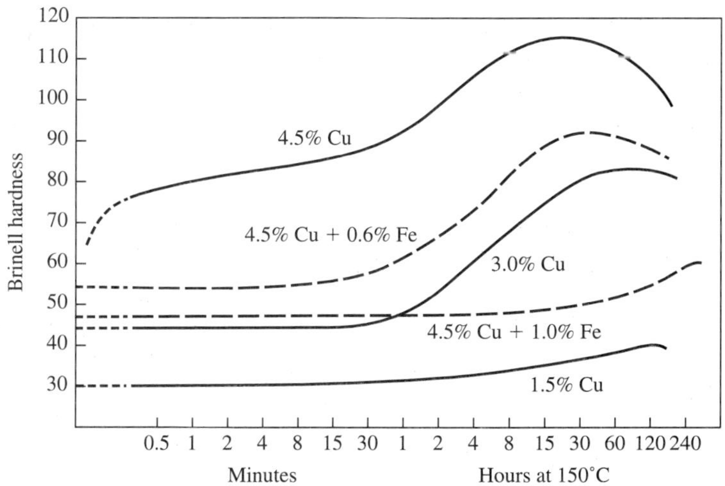

Figure 6.9 Effect of iron impurity in alloy of Al + 4.5% Cu on the response to age hardening at 150°C, following a solution heat treatment and quench. The aging response indicates that 0.6% Fe makes the ternary alloy age harden like a binary 3% Cu alloy and that 1% Fe reduces the aging response to that of a 1.5% Cu alloy.

Concentration of Solute and Presence of Other Elements

A higher concentration of Cu in Al-Cu alloys up to the solubility limit of 5.6% increases the volume of precipitant θ' and therefore naturally increases the maximum hardness, as is evident by the comparison in Fig. 6.9 of the 1.5, 3.0, and 4.5% Cu alloys. The 4.5% Cu level is a practical maximum, since higher contents of Cu have a very restricted solution heat-treating temperature range.

Iron as an impurity in Al-Cu alloys ties up the Cu as an insoluble Al_x-Cu_y-Fe_z compound. Thus the addition of 1.0% Fe to the 4.5% Cu alloy reduces its aging incremental response to that of a 1.5% Cu alloy (see Fig. 6.9). Even though there is some multiphase hardening from the Al-Cu-Fe compound, iron impurity has a very detrimental effect (Table 6.2).

Magnesium is an additive to Al-Cu age-hardening alloys employed to enhance the response to age hardening, particularly upon aging at room temperature, as shown by the data of Table 6.3. Micrographs 6.5 and 6.6 show the structure of alloy 2024 in the 190°C aged condition. A few particles of undissolved θ [or $Al(Mg)Cu_2$] are

Table 6.2 Effect of iron impurity in Al-Cu alloys quenched from the solid-solution field

% Cu	% Fe	Al Lattice Parameter	% Cu in Solution	% Cu in Compound	Max. BHN
4.48	0.01	4.0310	4.48	—	114
4.44	0.18	4.0321	4.08	0.36	109
4.35	0.32	4.0330	3.67	0.68	104
4.47	0.47	4.0335	3.45	1.02	101
4.46	0.61	4.0343	3.08	1.38	92
4.40	0.74	4.0351	2.73	1.67	75
4.43	0.90	4.0358	2.41	2.02	65
4.43	1.05	4.0364	2.16	2.27	60

Table 6.3 Effect of Mg in Al-Cu alloys upon age-hardening response

			Tensile Strength, MPa		Elongation in 50 mm, %	
Alloy	Cu, %	Mg, %	Aged Room Temp.	Aged 190°C	Aged Room Temp.	Aged 190°C
2014	4.4	0.4	425	480	20	13
2024	4.4	1.5	470	500	20	13

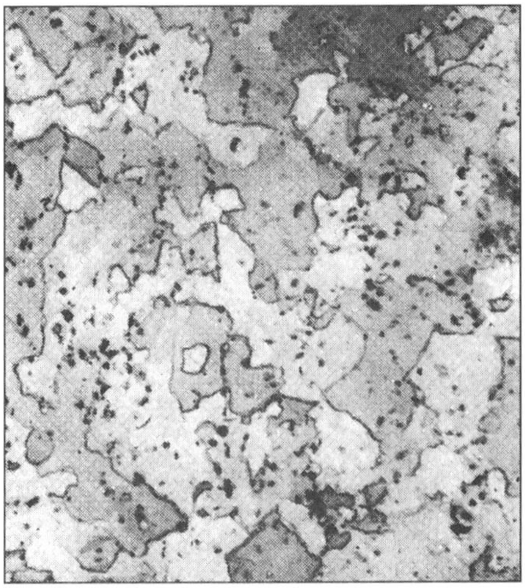

Micrograph 6.5 Al + 4.4% Cu, 1.5% Mg, 0.6% Mn, and about 0.5% Fe and Si as impurities (alloy 2024-T6); the alloy in sheet form was quenched from 500°C and aged at room temperature; ×75; HCl, HNO_3, and HF in water (Keller's etch). The etch employed in this case differentially attacks the aged α_{Al} solid-solution matrix to show the typical irregularly shaped grains of the structure on a section parallel to the rolling plane. Insoluble Al-Cu-Fe-Mn intermetallic compounds are again extended in the direction of flow during hot working. Only traces of copper or magnesium compounds are visible, since they were almost completely dissolved during the heat treatment and now cannot be seen in the extremely fine precipitated form characteristic of the aged alloy.

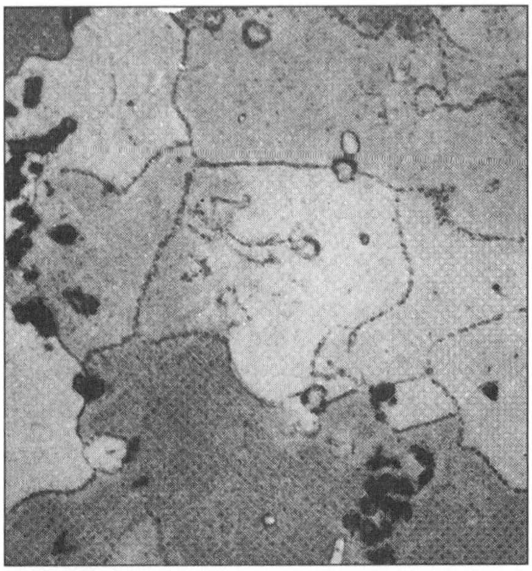

Micrograph 6.6 Same as Micro. 6.5 ×500. At the higher magnification, the clear white particles of residual $CuAl_2$ are visible, as well as the black and dark gray Al-Mn and Fe-Si compounds. In addition, a series of small dark particles along the grain boundaries are now resolved. They probably originated during cooling the alloy somewhat too slowly from the solution treatment, i.e., by quenching in hot water. The hot-water quench diminishes distortion and quenching stresses, but the grain-boundary precipitate (of θ or $CuAl_2$) results in a susceptibility to intergranular corrosion. The alloy may be protected against corrosion attack by coating the alloy with pure aluminum (Alclad 2024-T6).

visible in Micro. 6.6 along with insoluble impurity compounds. These micrographs are of rolled sheet examined on the sheet surface and show basically equiaxed α grains. A cross section would show that these grains are really of pancake shape in three dimensions. The impurities impede α-grain growth in a direction normal to the rolled surface during the solution heat treatment.

The effects of aging time on mechanical properties can be seen in Fig. 6.10 for a 2024 aluminum alloy. A sharp drop in ductility between 180 and 220°C with little change in tensile strength is caused by a heavy and almost continuous grain-boundary precipitation of θ with adjacent soft precipitate-free zones. Reheating to 500°C shows essential restoration of initial properties. The data of Fig. 6.10 also show that in this precipitation-hardening alloy, tensile properties do not move as ordinarily expected. For example, aging at 150°C reduces the yield strength while not affecting tensile strength or ductility. Therefore, this aging treatment could give better fabrication properties than aging at lower or higher temperatures.

Cold Work Cold work superimposed on age hardening is employed to obtain maximum strength of Al-Cu wrought alloys. Cold rolling can be done after solution quenching and room-temperature aging or, for maximum effect, after elevated-temperature aging. For example, some cold rolling after 150°C aging will increase the tensile strength of 2024 alloy from 480 to 525 MPa but with a related decrease in tensile ductility from 13 to 6% in 50 mm.

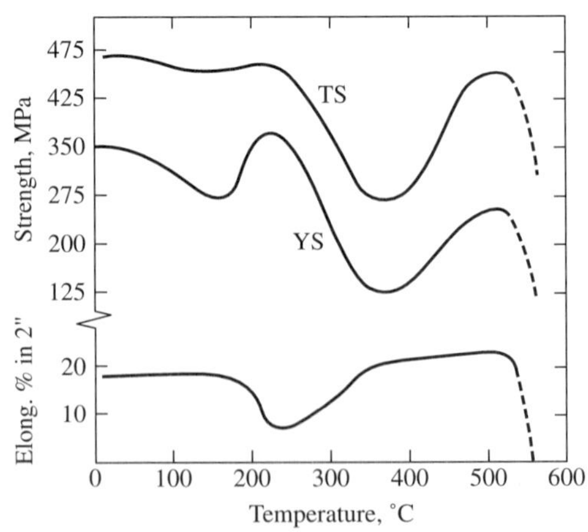

Figure 6.10 Tensile strength, yield strength, and tensile-test elongation values for aluminum alloy 2024 sheet specimens. They were reheated to varying temperatures for 1 h, quenched, and aged at room temperature. The difference between initial values at the left and the equivalent condition of 500°C reheat is presumably due to a slower original mill quench, e.g., in hot water, and the slight working resulting from a pass through straightening rolls. Both factors would increase yield-strength values particularly.

6.5 PRECIPITATION HARDENING OF Cu-Be ALLOYS

There are many popular references to "the lost art of hardening copper." Unknown is the means of hardening copper by this art, but also unknown is the degree of hardening attained. In any event, modern technology has developed an age-hardenable Cu-Be alloy that by heat treatment can be made as hard and strong as most heat-treated steels, namely, a Rockwell hardness of about C40. Although popularly known as beryllium copper, it is copper with about 2 wt % Be added plus a minor amount of nickel or cobalt.

The pertinent phase diagram (Fig. 6.11) shows the classical basis for age hardening, i.e., a solvus line with sharply decreasing solubility of Be from 866 to 500°C and below. The horizontal line at 605°C represents a three-phase reaction, $\beta \rightleftharpoons \alpha + \beta'$, but this is not significant to the aging process. It occurs only in slowly cooled alloys. The phase β' in this figure is approximately 50 at. % each of Cu and Be, that is, the intermetallic compound CuBe.

The vertical dashed line at 1.9 wt % Cu in Fig. 6.11 represents the commercial alloy. Because of the great difference in atomic weights of these two elements, 1.9 wt % Be corresponds to almost 12 at. % Be so that one atom out of every eight is Be.

The alloy is solution-annealed by heating at 790°C in an inert atmosphere for a time sufficient to dissolve all α phase present from earlier processing. It is then quenched in water to preserve at room temperature a supersaturated solid solution. In contrast to the Al-Cu age-hardenable alloys, this Cu-Be alloy will

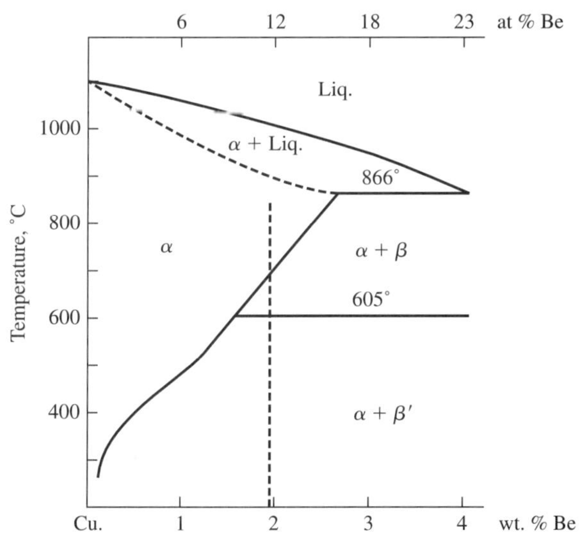

Figure 6.11 Copper end of the Cu-Be phase diagram showing the potential for age hardening. The dotted vertical line at 1.9 wt % Be represents the composition of a commercial Be-Cu alloy of very high strength.

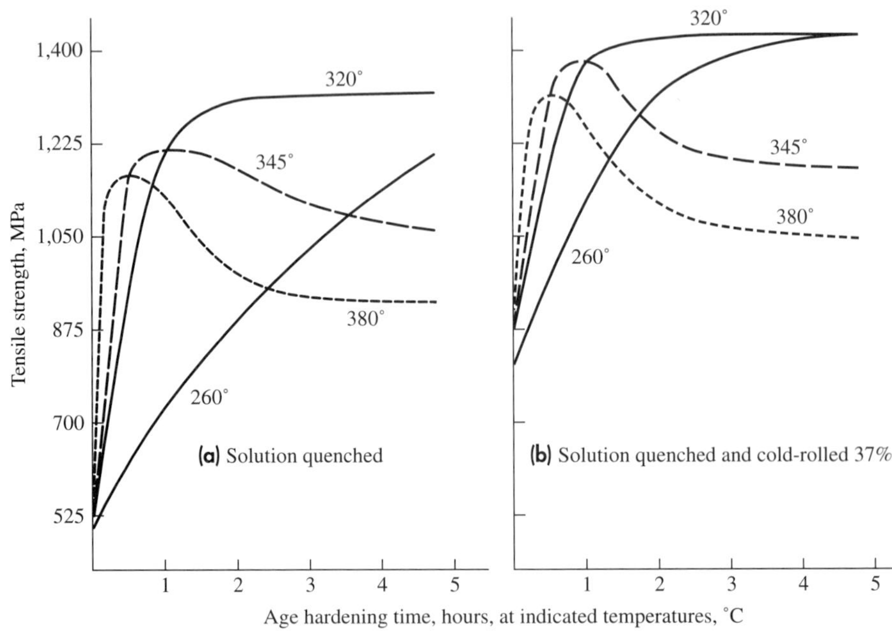

Figure 6.12 Age-hardening curves for the Cu + 1.9% Be alloy; tensile strength vs. aging time at four temperatures: (*a*) alloy quenched from 790°C, (*b*) alloy quenched from 790°C and then cold-rolled 37%.

remain soft indefinitely at room temperatures. Furthermore, this is the softest state for the alloy. As solution-annealed, it can be cold-worked by drawing or rolling, which of course superimposes work hardening on top of solid-solution hardening.

Reheating the alloy in the $\alpha + \alpha'$ field causes precipitation of the α' phase within the α and a marked increase in strength. Representative age-hardening curves are shown by Fig. 6.12 for both the solution-annealed and annealed + cold-worked states. An aging temperature of 320°C permits development of maximum strength in 3 h for the solution-annealed alloy or in 2 h for the annealed + cold-worked alloy. Ductility, in terms of percent elongation in 2 in, drops from 50% for the solution-annealed metal to about 15% as fully hardened. The corresponding change for the cold-worked metal is a drop from 5 to 2%. The electrical conductivity, corresponding to a decrease in solute content of the α, rises from 15% of that of pure Cu to 24% for the annealed and age-hardened alloy or from 10 to 22% for the cold-worked and age-hardened alloy.

The diversity of properties attainable in age-hardenable alloys is illustrated by a resistance-welding electrode of Cu + 1% Be and 1% Co. While it cannot attain the hardness of the alloy of Fig. 6.12, it is heat-treatable to a very respectable hardness of Rockwell C30 and at the same time shows an electrical conductivity 50% of that of pure Cu.

PROBLEMS

1. Describe the processing steps required to precipitation-harden a metal. Sketch and briefly describe the changes in microstructure that occur during each step.
2. Illustrate the portion of a typical equilibrium diagram that would be amenable to a precipitation process.
3. Sketch the typical microstructures for a Cu-Be alloy with 1.9 wt % Be as it is slow-cooled from the α phase at 800°C to room temperature.
4. The Ni-Be system shows a eutectic at 1157°C between α_{Ni} containing 2.7 wt % Be and α, the compound Ni-Be. The solubility of Be in α drops continuously from 2.7% at 1157°C to 0.20% at 400°C. Specify the alloy composition and heat treatment to be used for age-hardened Ni-Be alloy.
5. Compare and contrast the mechanisms of annealing and age hardening in an Al-Cu alloy using both mechanical properties and microstructure to aid your explanation.
6. Describe precipitate coherency and its effect on tensile properties in nonferrous alloys.
7. In the aging process, explain from an atomistic viewpoint the formation of GP zones and their development into precipitates.

REFERENCES

Porter, D. A., and K. E. Easterling: *Phase Transformations in Metals and Alloys,* Van Nostrand Reinhold, Berkshire, England, 1983.

Reed-Hill, R. E., and R. Abbaschian: *Physical Metallurgy Principles,* 3d ed., PWS Publishing, Boston, MA, 1994.

Brooks, C. R.: *Heat Treatment, Structure and Properties of Nonferrous Alloys,* ASM Int., Materials Park, OH, 1990.

Martin, J. W.: *Precipitation Hardening,* Pergamon, New York, 1968.

Embury, J. D., and R. B. Nicholson: "The Nucleation of Precipitates: The System Al-Zn-Mg," *Acta Metallurgica,* vol. 13, pp. 403–417, 1965.

Hatch, J. E. (Ed.): *Aluminum Properties and Physical Metallurgy,* ASM Int., Materials Park, OH, 1984.

chapter

7

Martensitic Transformation

All the metallurgical phenomena discussed to this point were a result of nucleation and growth mechanisms. This includes solidification of pure metals, solid solutions, and eutectics. It also includes recrystallization of a cold-worked metal as well as precipitation of a second phase during age hardening. In all these cases, some diffusion of atoms is required for the reaction to occur, so that both time and temperature are involved. Furthermore, the reactions will occur isothermally whenever the free-energy change is negative.

The final, major strengthening mechanism will now be discussed: the martensite transformation. This involves a change in crystal structure that occurs not by diffusion but by shear movements of atoms. The reaction usually occurs athermally, while the temperature is decreasing. The martensite reaction is found in many alloy systems, but the principal and most important system is the Fe-C alloys, or steels. However, before analyzing the martensite reaction, it is necessary to examine the eutectoid reaction, which is an alternative multiphase strengthening reaction in steels.

7.1 THE Fe-Fe_3C PHASE DIAGRAM

The phase diagram (Fig. 7.1) has been reproduced with the carbon-concentration scale plotted to expand the region including commercial steels, 0 to 1.4% C, and compress the area of cast irons, 2 to 4% C, in which the exact carbon content is of less importance. The basis of the differentiation between steels and cast irons is found in the phase diagram and may be expressed in two ways. From a practical standpoint, steels have such a high melting point (above 1440°C) that special and expensive equipment is required to melt them, whereas cast irons melt at 1350°C or less, a temperature much more readily attained with inexpensive equipment.

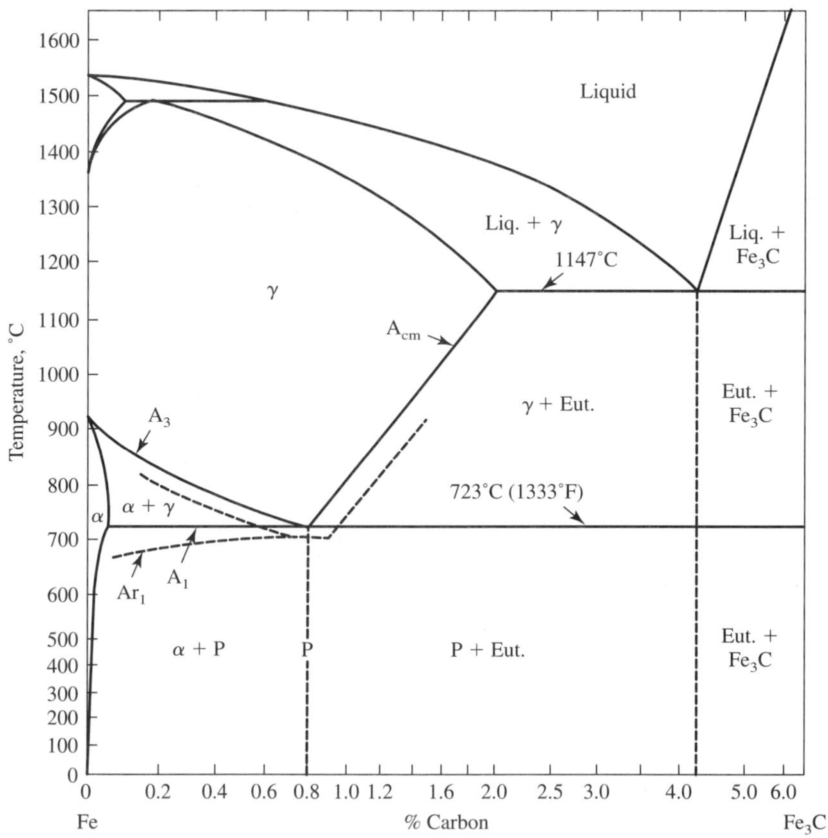

Figure 7.1 Phase diagram of the Fe-Fe_3C (metastable) alloy system; γ is fcc, α is bcc Fe, and P is pearlite, a two phase α + Fe_3C structure.

However, since the decrease in melting point (or liquidus temperature) with increase of carbon content is continuous from 1525 to 1147°C, this reasoning, although of practical importance, leads to no specific demarcation limit. A second means of distinguishing between the two classes of materials is on the basis of structure and properties; iron containing up to 2.06% C can be heated to a temperature at which it will show only one phase, fcc γ, while alloys containing more than 2.06% C always contain γ + Fe_3C eutectic structure. The eutectic has some 50% by weight (more by volume) of the brittle Fe_3C phase which tends to be continuous, and thus alloys above 2.0% C, containing eutectic at all temperatures, are somewhat brittle. The eutectic in alloys of less than 2.06% C, resulting from undercooling, can be dissolved in γ, which being fcc, is plastic. This distinction, that steels can be hot-rolled and cast irons cannot, is not absolute, however nicely it ties up with the phase diagram. For example, cast irons (white) of about 2.25% C, which contain only 20% of the eutectic and 80% of the γ phase at about 1100°C, can be hot-rolled if the initial soaking at the high temperature before rolling is effective in breaking the continuity of the carbide in the eutectic (by agglomeration) and if the initial breakdown passes are moderate. Alloys in the range 1.5 to 2.5% C are intermediate between steels and cast irons.

The phase diagram shows three horizontal lines, each representing a reaction involving three phases and occurring at a constant temperature. The reactions may be represented as:
At 1492°C:

$$\delta\ (0.08\%\ \text{C}) + \text{liquid}\ (0.55\%\ \text{C}) \rightleftharpoons \gamma\ (0.18\%\ \text{C}) \tag{1}$$

At 1147°C:

$$\text{Liquid}\ (4.2\%\ \text{C}) \rightleftharpoons (2.06\%\ \text{C}) + \text{Fe}_3\text{C}\ (6.7\%\ \text{C}) \tag{2}$$

At 723°C:

$$\gamma\ (0.08\%\ \text{C}) \rightleftharpoons \alpha\ (0.025\%\ \text{C}) + \text{Fe}_3\text{C}\ (6.7\%\ \text{C}) \tag{3}$$

The solidification process, through the peritectic range (0.1 to 0.5% C), solid-solution range (0.5 to 2.06% C), or hypoeutectic range (2.06 to 4.2% C) of carbon content, proceeds as it does in any alloy system showing comparable features. For example, an alloy of iron and 1.2% C upon cooling from the liquid state to 723°C behaves like an Al + 4% Cu alloy; it solidifies in the form of a solid-solution γ, and as this cools past the line marked A_{cm}, the decrease of solubility of Fe_3C causes the carbide phase to precipitate from solid solution in γ, which becomes correspondingly depleted in carbon. The carbide precipitation occurs chiefly at the γ grain boundaries, although if a large amount of Fe_3C must separate, it forms Widmanstatten plates within the grains. When the γ phase reaches a carbon content of 0.80% at 723°C (under equilibrium conditions), reaction (3) occurs. This reaction is missing from the Al-Cu diagram although, again, the construction of the phase diagram is identical if the γ field is taken to represent liquid alloys. Because of the essential similarity of reaction (3) to the eutectic (2), the former is known as a eutectoid. Whereas a eutectic represents the formation of a mechanical dispersion of two new solid phases, here $\gamma + Fe_3C$, from a liquid, a eutectoid represents the formation of a dispersion of two new solid phases, $\alpha + Fe_3C$, from a solid γ. In both cases, the two-phase dispersed structure has a distinctive appearance compared with either phase that may have formed in a different manner.

The term *hypoeutectoid* has the same significance as *hypoeutectic*. On cooling a 0.4% carbon steel past the line analogous to the liquidus in eutectics, here called A_3, crystals of α begin to form at the γ grain boundaries and continue forming until, at 723°C, they represent

$$\frac{0.80 - 0.40}{0.775} = 52\%$$

of the structure (under equilibrium conditions), after which the residual γ, now of 0.80% carbon content, goes through the eutectoid reaction. Hypereutectoid alloys behave in a similar manner, except that the phase separation from γ at its grain boundaries is Fe_3C.

Microstructures of the various phases can be seen in specimens that have been mounted and polished for metallographic examination. Polishing of steels is quite simple, with one important qualification. Surface flow usually considerably distorts the lamellar pearlite structure, and if this is relatively fine, as in normalized steels, flow may completely mask the lamellar characteristics. The surface flow is reduced but not eliminated by the use of light pressure during polishing. However, too light a pressure so increases the required polishing time that greatly increased surface pitting is encountered. The problem can be solved by etching after the first polish, repolishing on the last lap, and re-etching. This removes the distorted surface layer. Two repolishings may be required to reveal the structure clearly.

The common etching reagents are picral, usually a 4% solution of picric acid in alcohol, and nital, 2 to 10% concentrated HNO_3 in alcohol. Picral is somewhat superior in revealing carbides and is best employed to bring out details of carbide structures. The nital will also reveal pearlitic structures, although not quite so nicely, and since picral stains the fingers or hands, nital is more generally used. The more concentrated nital is used for lower-carbon-content steels or for quenched structures.

7.2 ALLOYS OF IRON AND CARBON

Both the α and γ phases of the iron-carbon system are interstitial solid solutions, α being bcc and γ fcc. In the fcc γ iron the interstitial "hole" positions are the midpoints of the cube edges (including the equivalent point at the center of the cube); in the bcc α iron they are at the midpoints between the cube edges and the face centers. In both cases the size of the interstitial openings between the iron atoms is substantially smaller than the size of the dissolved carbon atoms, and there is appreciable local distortion of the structure in the vicinity of each carbon atom. In γ iron the interstitial holes are, however, larger than they are in α iron, which may explain why the solubility of carbon in the fcc iron structure is so much greater than it is in the bcc structure (Fig. 7.2).

An important characteristic of interstitial solutes is the relatively high speed with which they can move about through the host structure. This is illustrated by comparing D_0 and the activation energy Q in the expression

$$D = D_0 e^{-Q/RT}$$

for diffusion of iron and carbon in α iron (Table 7.1). Also shown in this table are values of D calculated for a temperature of 600°C from D_0 and Q and the mean penetration distance x' resulting from diffusion occurring at 600°C for 1 day. The very high mobility of the carbon atoms relative to the self-diffusion of the iron is evident. Other interstitial solutes in α iron, boron, nitrogen, and oxygen also display similar high mobility.

The distortion of the structure surrounding a carbon atom dissolved in α iron makes possible a very sensitive experimental method for observing the mobility of the interstitial solute atoms. Figure 7.2 shows two of the interstitial sites in

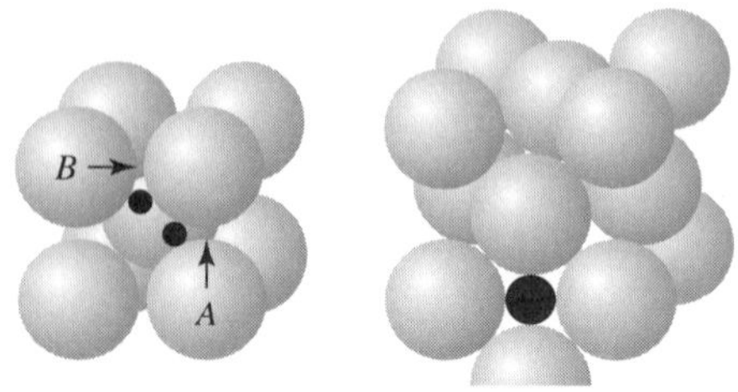

(a) (b)

Figure 7.2 Interstices of (*a*) bcc and (*b*) fcc lattices. The maximum-diameter foreign sphere (black) that can enter the bcc lattice is indicated by the black atom, with two of the four possible positions on one face shown here as filled. The fcc lattice has far fewer interstices, one per face, but as shown by the black sphere, the hole is much larger. (*a*) bcc lattice: holes at $\frac{1}{2}$, 0, $\frac{1}{4}$; $\frac{1}{4}$, 0, $\frac{1}{2}$, etc., for atom radius (α Fe) = 1.23 Å: edge of cube = 2.86 Å; radius, interstitial hole: 0.36 Å. (*b*) fcc lattice: holes at $\frac{1}{2}$, 0, 0; 0, 0, $\frac{1}{2}$, etc., for atom radius (γFe) 1.26 Å: edge of cube = 3.56 Å; radius, interstitial hole = 0.52 Å. Application of a tensile stress in the vertical direction causes the *A* site at 0, 0, $\frac{1}{2}$ of the bcc lattice to be a favored location for a carbon solute atom.

Table 7.1 Data for diffusion in α iron

Solute	D_O	Q, kcal/mol	D at 600°C	x' (1 day at 600°C), cm
Fe	5.8	59.7	6×10^{-16}	7×10^{-6}
C	2×10^{-2}	19.8	2×10^{-7}	0.1

the bcc structure, one marked *A*, which has its closest neighbors in the vertical position, and one marked *B*, with its closest neighbors horizontal. Since the carbon atom is too big to fit comfortably into its interstitial hole, its neighboring atoms will be displaced outward, vertically at the *A* site or horizontally at the *B* site. Suppose that a tensile stress in the vertical direction is now applied to the metal. This stress will help the neighboring atoms at the *A* site move outward when this site is occupied by a carbon atom but will hinder the outward motion of the atoms at the *B* site. The carbon atoms will therefore prefer to occupy the *A* sites. When the tensile stress is applied, there is a partial migration of carbon atoms from *B* to *A* sites; this causes an additional strain, over and above the elastic strain, to occur in the vertical direction. This is known as an anelastic strain. The speed with which the anelastic strain sets in is a measure of the mobility of the carbon atoms, while the magnitude of the anelastic strain is a measure of the concentration of carbon atoms in solution. Experimental study of this anelastic effect can therefore provide quantitative information about the behavior of carbon atoms dissolved in α iron.

Terminology

Fe-Fe_3C was one of the first metallic systems to be studied. This fact, plus the predominant commercial production and use of steels among all alloys, has led to the application of special names for phases and structures, in addition to the basic Greek-letter designations. The common ones are as follows:

Austenite = γ, the fcc structure that can dissolve up to 2.06% C at 1147°C

Ferrite = α bcc iron dissolving a maximum of 0.025% C at 723°C

Cementite = Fe_3C = iron carbide, or simply carbide

Eutectic = the reaction at 1147°C, liquid $\gamma + Fe_3C$; or the liquid composition participating in the reaction, that is, 4.3% C; the eutectic alloy, or the structure resulting from the reaction on cooling, known as ledeburite

Eutectoid = the reaction $\gamma = \alpha + Fe_3C$, or the alloy whose composition is that of the austenite in the reaction, that is, 0.8% C; the eutectoid steel, or the lamellar structure resulting from the reaction on cooling, pearlite

Pearlite = the distinctive two-phase lamellar structure resulting from the eutectoid reaction upon relatively slow cooling

A_{cm} = line showing thermal arrest resulting from precipitation of carbide from austenite (on cooling)

A_3 = line showing thermal arrest resulting from formation of ferrite in austenite upon cooling

A_2 = line at about 768°C, showing a magnetic change in ferrite; not a phase change and not shown on the phase diagram of Fig. 7.1

A_1 = horizontal line representing eutectoid reaction

In the field of heat treatment, we have the additional terms:

Normalizing = heating into the austenitic field followed by air cooling

Annealing (full) = heating above the Ac_3 for hypoeutectoid steels or above the Ac_1 for hypereutectoid alloys, followed by furnace cooling

Annealing (process) = heating to a temperature below but close to the A_1 line (does not result in any phase change but will soften work-hardened ferrite and spheroidize lamellar Fe_3C)

Spheroidizing = prolonged holding at a temperature just below A_1 to produce a spheroidal or globular form of carbide

Hardening = heating above the Ac_3 for hypoeutectoid steels or above the Ac_1 for hypereutectoid steels followed by quenching, i.e., cooling fast enough to prevent the normal eutectoid reaction from occurring so that instead the austenite transforms to martensite at a relatively low temperature

Equilibrium and Nonequilibrium

The phase diagram shown for Fe-Fe_3C alloys cannot be called an equilibrium diagram, which would imply no change of any phase with time, because even relatively pure Fe-Fe_3C alloys will change; e.g., a low-carbon steel, 0.08% C with low silicon content, held several years at 650 to 700°C (in an oil-distillation system), was found to consist of ferrite and graphite, not ferrite and carbide as the diagram would lead one to expect and as it was when originally installed. Since a long time at an elevated temperature is known to be conducive to the establishment of equilibrium, it is concluded that iron carbide is always a transitional or metastable phase. However, extraordinary temperature and time conditions (or a relatively high silicon content) are required to decompose carbide into graphite and ferrite,[1]

$$Fe_3C \rightleftharpoons 3Fe(\alpha) + C \text{ (graphite)}$$

Although the Fe-Fe_3C diagram is known to represent metastable conditions, the temperature-composition limits of phase changes shown by various lines can be considered as having equilibrium positions under slow heating and cooling conditions. Liquidus and solidus lines will be depressed by undercooling in the same manner as those on true equilibrium diagrams. The lines on the diagram outlining the eutectoid transformation are particularly subject to displacement from overheating or undercooling, since transformations requiring diffusion in the solid state are necessarily slower than those involving a liquid. (For example, the "metastable" solidus in solid solutions is not usually accompanied by a detectable metastable liquidus, since atomic diffusion is so much more rapid in the liquid than in the solid state.)

The displacement of the A_3 and A_1 lines of the phase diagram has led to the use of the designation Ac (C = chauffage = heating) for their positions on ordinary heating cycles and Ar (r = refroidissement = cooling) on ordinary cooling cycles, as indicated under terminology. A more complete discussion of these effects is given later in this chapter, but it is desirable to point out here that the A_3 and A_{cm} lines are subject to greater displacement, as a result of reluctance by the ferrite or cementite to form in austenite, than the A_1 line at which pearlite forms from austenite. The positions of all three of these important lines upon air-cooling of moderate-sized sections are shown in Fig. 7.1. The result of undercooling is immediately evident in the composition of the eutectoid; it is no longer fixed at 0.80% C but

[1] Above the A_1 temperature, carbide will break down to graphite and austenite when the alloy is to the right of the A_{cm} line. The eutectoid composition of the iron graphite system is 0.69% C, and the eutectoid temperature is 738°C.

includes a range of concentrations of 0.7 to 0.9% C. Earlier books on this subject often show carbon contents for the eutectoid different from the 0.80% given here, and the basis for past confusion is related to undercooling. The 0.80% value was obtained by heating and cooling tests of alloys between 0.7 and 0.9% C in which the change of temperature was only ⅛°C/min. The effect of undercooling on the carbon concentration and temperature of this eutectoid should be compared with the effect on the silicon content and temperature of the Al-Si eutectic, although in the latter case, since only the formation of silicon is suppressed by undercooling, the eutectic concentration is displaced in only one direction.

The construction of that part of the Fe-Fe_3C diagram (Fig. 7.1) shown by the dashed Ar_1 line is in violation of the Gibbs phase rule. With three phases present, γ, α, and Fe_3C, the temperature of the system and compositions of the phases should all be fixed; yet the austenite shows a range of compositions. The answer to this apparent discrepancy is that the phase rule applies only to equilibrium conditions while the Ar_1, Ar_3, and Ar_{cm} lines represent nonequilibrium conditions.

The carbon content of the eutectoid is important in applying the lever rule to hypo- or hypereutectoid structures, i.e., in predicting whether a specific alloy should show excess ferrite or carbide and how much or, analogously, in estimating carbon content on the basis of the presence and amount of an excess of one of these phases. The lever rule can be useful in both these ways but is quantitatively applicable only when the structures are obtained under conditions approaching equilibrium, i.e., very slow furnace cooling, or when the nonequilibrium conditions are known and specified quite precisely.

7.3 MICROSTRUCTURE OF NONHARDENED STEEL

While the entire two-phase field between 0.007 and 6.7% C at room temperatures consists of the phases ferrite and carbide, the structural appearance of different alloys varies with the state of aggregation of the ferrite and carbide as a function of their origin. The following ranges in slowly cooled alloys can be differentiated, chiefly on the basis of five types of carbide distribution:

1. 0.007 to 0.025% C; ferrite with carbide precipitated in a very fine form, usually invisible at ordinary optical magnifications
2. 0.025 to 0.8% C; ferrite + pearlite
3. 0.8 to 2.06% C; pearlite + carbide precipitated from austenite
4. 2.06 to 4.2% C; pearlite in a dendritic structure (from hypoeutectic primary austenite) + carbide precipitated from austenite (usually attached to and indistinguishable from eutectic carbide) + ledeburite (eutectic with austenite and cementite as constituents that decomposes into ferrite and cementite on cooling below Ar_1)
5. 4.2 to 6.7% C; primary carbide crystals + ledeburite

Of these five structures, 1 is not reproduced here; 2 and 3 are shown in Micros. 7.1 to 7.8; 4 is given in Chap. 12; and 5 is not reproduced since no commercial alloys fall within this range.

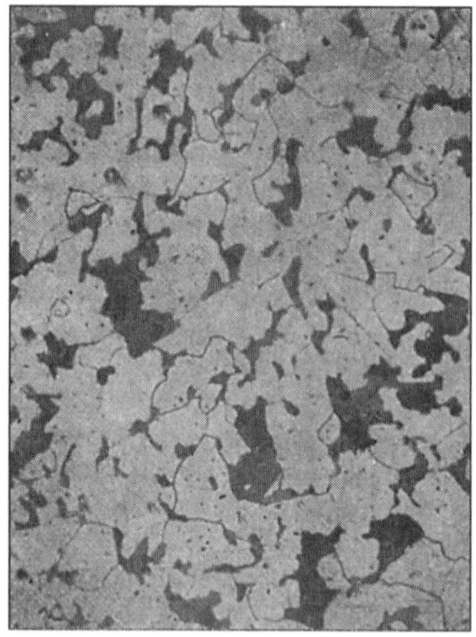

Micrograph 7.1 0.20% C + 0.42% Mn steel, annealed by furnace cooling from 1000°C; ×100; nital etch. At a low magnification the relative quantities and distribution of ferrite and pearlite are evident, as well as the ferrite grain size.

Micrograph 7.2 Same as Micro. 7.1 ×1000. At the higher magnification, the coarse lamellar carbide of the pearlite is resolved. Black spots are oxide inclusions.

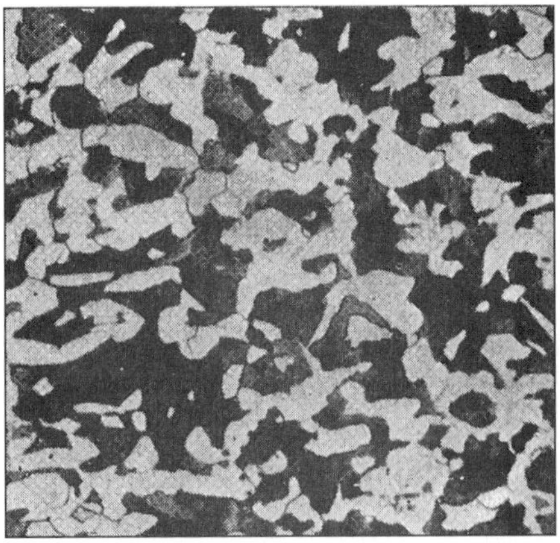

Micrograph 7.3 0.40% C normalized steel; ×100; nital etch. This structure is also composed of white ferrite crystals and dark areas of pearlite. (It is inaccurate to speak of grains of pearlite: it actually consists of two different types of crystals, α and Fe_3C, in a fine lamellar dispersion.) Ferrite here amounts to noticeably less than the expected 50% of the structure, because, upon air cooling of the specimen in the normalizing treatment, separation of hypoeutectoid ferrite is not complete and the carbon content of the pearlite consequently is diminished.

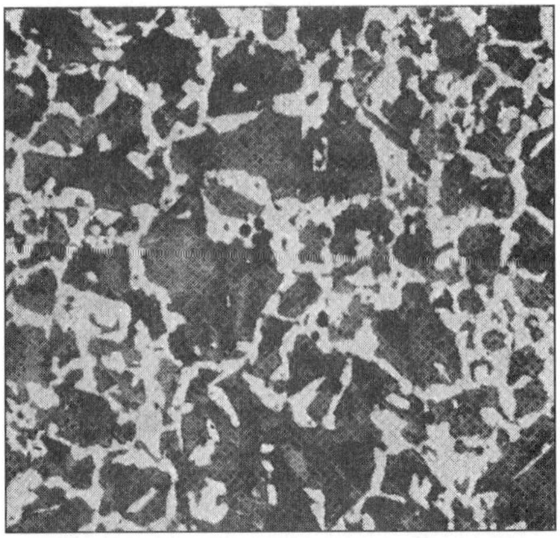

Micrograph 7.4 0.60% C normalized steel; ×100; nital etch. Moderately rapid air cooling of this specimen has resulted in somewhat less than the expected 25% ferrite. As the ferrite forms preferentially at the austenitic boundaries, with the center of the austenite grains later transforming to pearlite, the ferrite crystals outline the former austenite grains. This specimen showed considerable variation in the apparent size of the austenite grains; some are quite small, others are large. Ferrite formation is not confined completely to austenitic grain boundaries, since some white crystallites are visible inside the present pearlitic areas. Slow cooling from a very high temperature, as in castings of hypoeutectoid steels, sometimes gives a pronounced Widmanstätten pattern of ferrite plates in a pearlitic background. Color variations in the pearlite are associated with variations of the angular position of the ferrite-carbide lamellae with respect to the surface of polish.

Micrograph 7.5 0.8% C + 0.65% Mn steel, annealed by furnace cooling from 810°C; ×100; picral etch. The absence of any structural constituent other than pearlite results in a dark appearance at low magnification, with variations in shading that are associated with variations in the angle between carbide lamellae and the polished and etched surface.

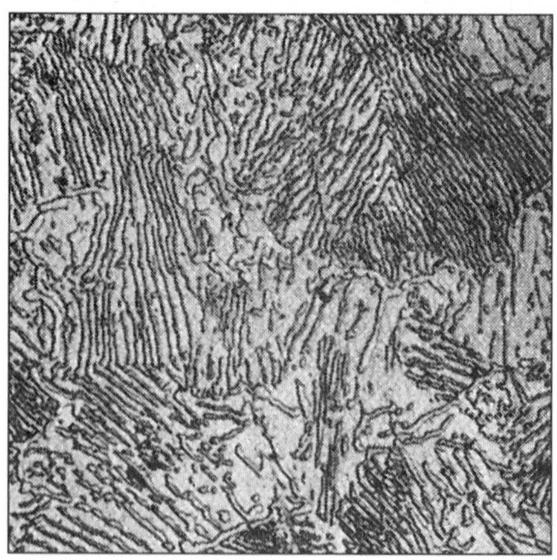

Micrograph 7.6 Same structure as Micro. 7.5 ×1000. Details of the lamellar pearlite show that ferrite, the background, is the continuous phase, representing (6.7 − 0.8)/6.7, or 88% by weight of the structure. When the carbide lamellae are normal to the surface (near upper right corner), the apparent spacing is less than when they approach parallelism to the surface. The actual spacing of lamellae must be the minimum measured on a plane section.

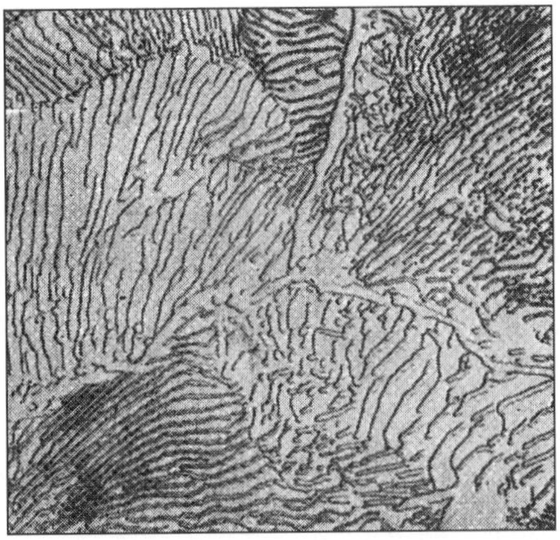

Micrograph 7.7 1.0% C annealed steel; ×1500; picral etch. For the reasons previously discussed, this slightly hypereutectoid steel would show little or no excess cementite if a relatively small section were air-cooled. When furnace-cooled from the austenitic field past the A_{cm} line, iron carbide forms at the grain boundaries of the austenite. This micrograph shows the intersection of three former austenite grains, outlined by a thin, continuous cementite envelope at their boundaries. The remainder of the structure is similar to the straight eutectoid steel. Note particularly that although ferrite and carbide lamellae are approximately parallel in some sections, in other areas within the same former austenite grain they are arranged in a different pattern. This demonstrates that in normalized or annealed steels the orientations of pearlitic areas do not reveal the austenitic grain size; some excess carbide (or ferrite) is required.

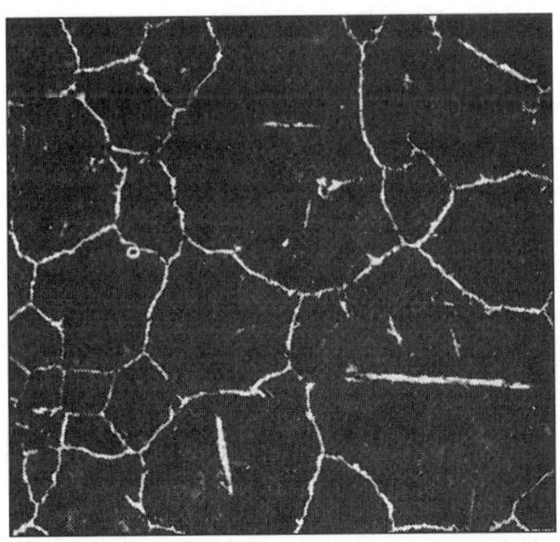

Micrograph 7.8 1.3% C annealed steel; ×100; nital etch. With a greater amount of excess carbide present, a high magnification is not required to show the structure of the steel. The white carbides outline former austenitic grain boundaries and reveal the size of the grains existing at the high temperature, namely, about ASTM 2. With the increased amount of excess carbides, some large plates of Fe_3C have formed within the former γ grains and outline the specific crystal plane on which they necessarily formed. The dark background structure is pearlite.

7.4 HEAT TREATMENT OF EUTECTOID STEEL

Formation of Austenite

Steels to be heat-treated are prepared for machining or fabrication by normalizing or annealing. Their structures consist, then, of pearlite mixed with either ferrite or cementite. The first step in heat treatment is to austenitize the steel, i.e., heat the steel to such a temperature that it is converted wholly or partially into austenite. The necessary temperature can be determined from the phase diagram; hypoeutectoid steels are usually heated to a temperature above the A_3 line, hypereutectoid steels to a temperature between A_1 and A_{cm}. Austenite forms by a process of nucleation and growth, and consequently a homogeneous austenite structure is not formed as soon as the steel reaches the austenitizing temperature. Even after the structure has completely transformed to austenite, it may not be homogeneous; areas that were formerly ferrite may be somewhat low in carbon, and those formerly carbide may be high in carbon concentration. The concentration gradients associated with these inhomogeneities can be eliminated only through diffusion, and if homogeneous austenite is desired, sufficient time for this must be allowed. The time required depends on the maximum temperature reached and the structural characteristics of the original ferrite-carbide matrix, as shown by the data in Fig. 7.3. A finely spheroidized

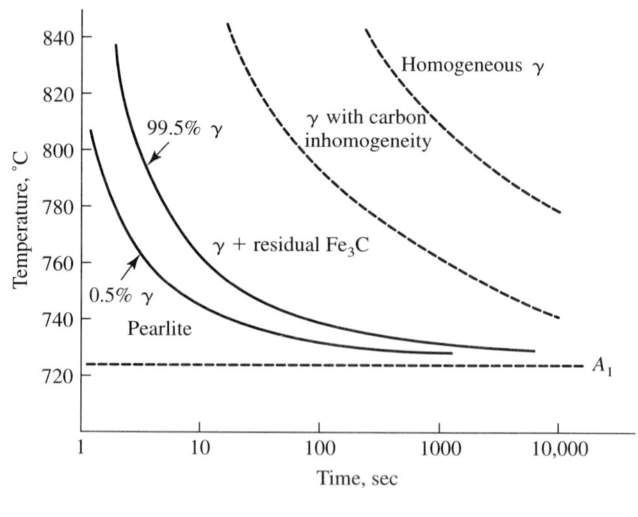

Figure 7.3 Graph showing time for austenitization of a plain carbon eutectoid steel as a function of austenitizing temperature, starting with a pearlitic structure obtained by normalizing from 875°C. The first curve (0.5% γ) represents the first visible evidence of austenite; the second curve (99.5% γ) represents the disappearance of pearlite, although some residual carbides remain undissolved. The third line, dashed, represents the approximate time-temperature limits for solution of all residual carbide; and the last dashed curve represents the probable attainment of homogeneity in the austenite.

carbide structure ("pinpoint" carbides) austenitizes most rapidly, next a fine pearlitic structure, then, most slowly, a coarse pearlitic or spheroidized structure.

Another factor which influences the time required for austenitizing is the composition of the steel. In a steel that contains any appreciable amount of alloying elements having an affinity for carbon, the carbide phase is not simply Fe_3C but is a complex carbide that may contain chromium, molybdenum, vanadium, or tungsten. Compared with Fe_3C, the complex carbides are relatively slow to dissolve, so much so that in the more highly alloyed steels it is not practical to attempt to get them all into solution.

In choosing the austenitizing temperature, several factors must be considered. If the temperature is too low, there may be incomplete solution of the carbides. If this is the case, the steel, after quenching, will not come up to its full hardness. If the temperature is too high, a number of complications may arise. At high temperatures the austenite grain size will become large, and this condition in heat-treated steel is often considered to be a cause of brittleness. A high austenitizing temperature also means that the quenching strains will be relatively great, and this may be a cause of cracking during heat treatment. Finally, there is danger of partial melting in highly alloyed steels when too high an austenitizing temperature is used. Since austenitizing must be accomplished in a reasonable length of time, the choice of the best temperature is always a compromise that, to a large extent, must be based on experience. Steel manufacturers generally publish recommendations for the best heat-treating conditions for their various products.

Transformation of Austenite to Pearlite

As soon as austenitized steel is cooled below the A_2 line, its structure should consist of pearlite and proeutectoid ferrite or cementite. In the heat treatment of steel it is important to know how much time is required for the formation of the equilibrium structure; to determine this time, isothermal-transformation experiments are used. The principle of these experiments can be explained most easily for the case of a steel of eutectoid composition. The transformation of steels of other compositions will be discussed in a later section. To perform an isothermal transformation experiment, samples of eutectoid steel are first austenitized and then quenched in a constant-temperature bath—usually molten lead or salt—held at a temperature below 723°C. Samples are withdrawn at successive time intervals, quenched, and examined under the microscope to see what progress the transformation has made. It is found that, for any given transformation temperature, a certain time t_s must elapse before any pearlite forms and that it is only after a time t_f that the transformation is complete, all the austenite in the sample having been converted to pearlite.

If isothermal-transformation experiments are carried out at a series of temperatures ranging downward from 723°C, t_s and t_f are determined as a function of temperature and the data necessary to construct a time-temperature-transformation (TTT) diagram, or C curve, for eutectoid steel are at hand. Such a transformation

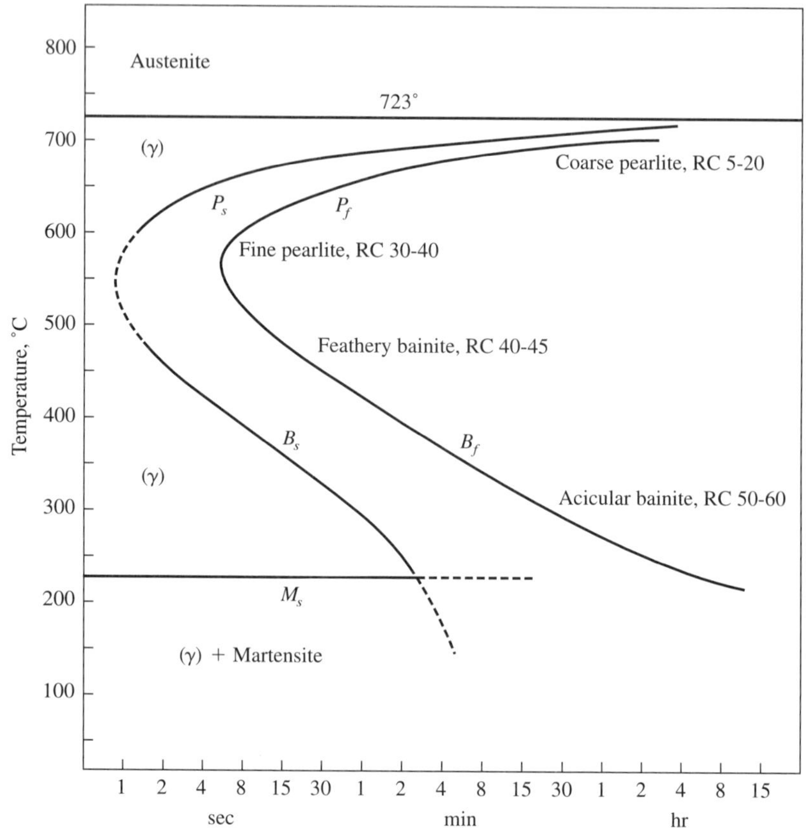

Figure 7.4 Isothermal transformation diagram for an 0.80% C steel containing 0.76% Mn, austenitized at 900°C with a No. 6 austenitic grain size.

curve is shown in Fig. 7.4. The various t_s points generate the pearlite-start curve marked P_s in the diagram, while the t_f points form the pearlite-finish curve. From the diagram, the time required for the pearlite reaction to begin and to be completed can be read off for any temperature. A sample represented by a point to the left of the P_s curve is all austenite, between the two curves it is a mixture of austenite and pearlite, while to the right of P_f its structure is all pearlite.

The reasons for the transformation diagram having the shape that it does can be understood in terms of the processes of nucleation and growth by which pearlite forms from austenite (see Chap. 3 on solidification). The amount by which the transformation temperature is below 723°C is the amount of undercooling, while t_s is the time required to nucleate pearlite in the undercooled austenite. The interval t_f–t_s is the time required for the growth of pearlite nuclei to complete transformation of the structure. Two factors control the speed of the transformation: the greater the undercooling, the greater the driving force (the free-energy difference between unstable austenite and its decomposition products) for pearlite formation. But as the undercooling is increased, the transformation temperature is lowered and the rate at which carbon diffuses to the cementite plates of the pearlite decreases, thereby tending to slow up the transformation. A degree or two below 723°C, the undercooling is so small, the driving force so weak, that t_s and t_f are very great.

As the transformation temperature is lowered, the increase in driving force is the dominant effect and t_s and t_f both become smaller. When a transformation temperature of about 550°C is reached, the decreasing diffusion rate of carbon begins to be more important than the increasing driving force and t_s and t_f begin to increase. In this way the characteristic C shape of the transformation curve is generated.

The mechanism by which the austenite-pearlite transformation takes place is not very well understood. Most of the experimental evidence on this subject comes from the examination of the microstructure of specimens that have been quenched to room temperature while the transformation was still in progress. One can never be sure with this technique that the structure existing at the transformation temperature is faithfully preserved. It appears, first of all, that pearlite always starts to form at austenite grain boundaries or, in an incompletely austenitized steel, at carbide particles as well as at the grain boundaries. Proeutectoid ferrite, on the other hand, does not seem to influence the initiation of the pearlite reaction. These observations suggest that it may be the formation of a carbide platelet that is the first step in the decomposition of austenite. Supporting evidence for this view is found in the observation that the crystallographic orientation relationships between proeutectoid ferrite and austenite are quite different from those for the pearlitic ferrite and austenite. Once a carbide platelet forms at a grain boundary in the unstable austenite, it grows edgewise into the austenite grain and at the same time carbon diffuses inward toward its flat faces, depleting the carbon content of the surrounding material. Ferrite plates then presumably nucleate in these carbon-deficient regions, rejecting carbon outward. The rejected carbon may then help to nucleate new, parallel carbide plates. Repetition of this process results in the creation of a colony of parallel carbide and ferrite plates. Growth of the colony is stopped only when it infringes on other colonies; it is not hindered by the austenite grain boundaries (Fig. 7.5). It is probable that during growth of a pearlite colony, the original carbide platelets branch to maintain a spherical or nodular shape of the colony.

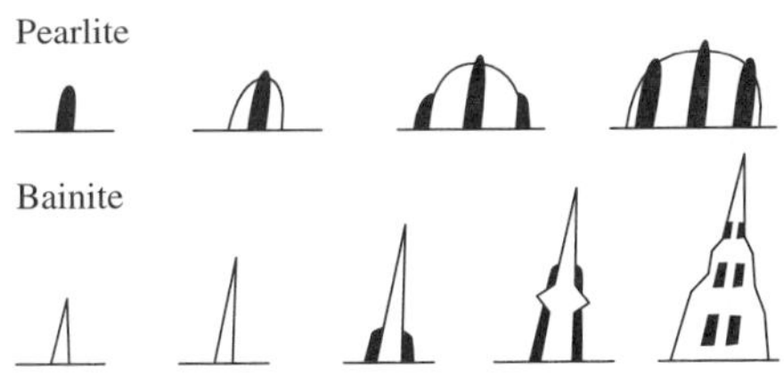

Figure 7.5 Growth of pearlite, nucleated by a carbide crystal, and of bainite, nucleated by a ferrite crystal and with adjacent rejected carbide, after reaching a critical concentration, as discontinuous small crystallites.

It is observed that the lower the temperature at which pearlite forms the finer the plate spacing. Presumably, at lower temperatures nucleation occurs rather more frequently while it is more difficult for diffusion to occur over great distances, resulting, then, in a tendency to form a fine structure. In examining photomicrographs of pearlite, it is important to remember that the true spacing in a colony will not ordinarily be revealed because the plates will generally not be normal to the specimen surface. The true spacing can be obtained from statistical analysis of the observed spacing in many colonies.

Formation of Bainite

If the isothermal transformation of a eutectoid steel is conducted at a sufficiently low temperature, the pearlite transformation is supplanted by a transformation which apparently proceeds by an entirely different mechanism and which produces as a product a different type of ferrite-cementite structure known as bainite. In bainite, as contrasted with pearlite, there is a definite orientation relationship between the ferrite and the parent austenite. Apparently in the bainite reaction ferrite is nucleated first, followed by precipitation of the carbide (Fig. 7.5). Because the reaction occurs at such low temperatures, the driving force for it must be very great; the nucleation frequency is high, but growth of the nucleated particles is slow because of the relatively low mobility of the carbon atoms. The result is a structure with a very fine particle size that ordinarily cannot be resolved in the optical microscope. The structure may have either a feathery or acicular, needlelike appearance, depending on whether it was formed at relatively high or relatively low temperatures.

7.5 THE MARTENSITE TRANSFORMATION

If a sample of eutectoid steel is cooled with such rapidity that its structure is all or partially austenitic when it reaches 230°C, a new and entirely different type of transformation occurs; the austenite transforms to martensite. Martensite is a metastable structure that has the same composition as the austenite from which it forms. It is a solution of carbon in iron having a body-centered tetragonal (bct) crystal structure. Because martensite forms with no change of composition, diffusion is not required for the transformation to occur. It is for this reason that martensite can form at such low temperatures. The most remarkable property of martensite is its potential of very great hardness. The hardness depends on the carbon content (Fig. 7.6), and at the eutectoid composition, martensite has a Rockwell C of about 65, which might be described as "glass-hard." In part, this hardness is an intrinsic property of martensite, but in part it is due to the very severe distortions that accompany the formation of the martensite. These distortions arise because martensite has a larger specific volume than the austenite from which it forms. This volume change also makes dilatometry (precise measurement of linear expansion and contraction) a common method to study martensite formation.

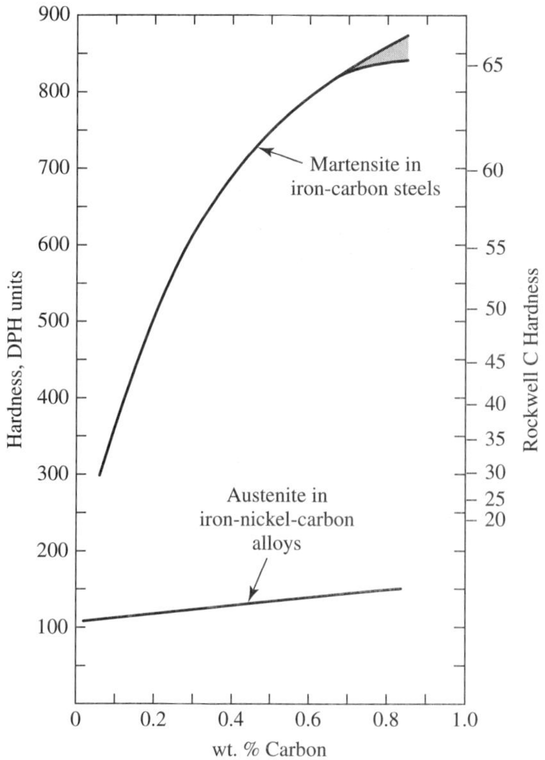

Figure 7.6 Effect of carbon on the hardness of austenite and on the hardness of martensite. The shaded area of the upper curve represents the effect of retained austenite.

The formation of martensite stands in marked contrast to the pearlite and bainite reactions. First, the martensite transformation is diffusionless and involves no change in composition. It does not occur by nucleation and growth. It cannot be suppressed by quenching. It is athermal. This means that austenite of a given composition begins to transform to martensite at a given temperature M_s and that as the temperature is lowered below M_s, the relative amount of martensite in the structure increases until at the M_s temperature the transformation is complete. At any temperature between M_s and M_f the amount of martensite characteristic of that temperature forms instantly, and holding at that temperature results in no further transformation.[2] The M_s and M_f temperatures are shown on the isothermal-transformation diagram as horizontal lines. Under the microscope, martensite gives the appearance of being formed of lens-shaped needles or laths. Each needle is a martensite crystal that forms by a complex two-stage shear mechanism. The shear apparently proceeds at a speed close to the speed of sound, for a microphone placed near a sample of transforming steel picks up a series of clicks.

Martensite transformations, i.e., diffusionless shear transformations, occur in several nonferrous alloys, brass containing 37.5% Zn, for example; and in several

[2] In some alloy steels a certain amount of isothermal martensite formation may occur.

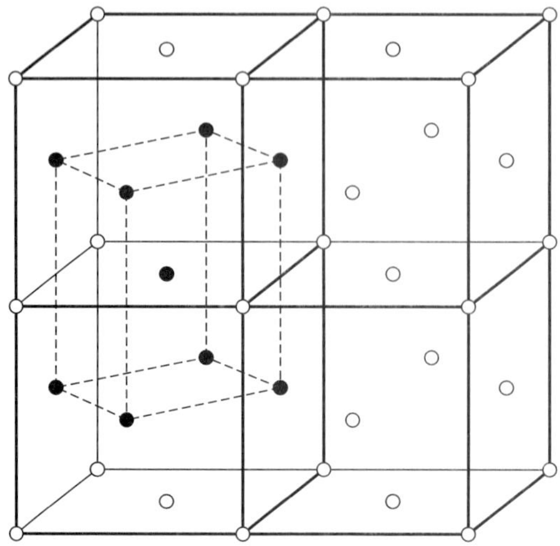

Figure 7.7 Illustration of the Bain model of the martensite transformation; the fcc structure (open circles) can be considered a bct structure (solid circles) with a *c/a* ratio of $\sqrt{2}/1$.

of these cases the mechanism of transformation has been completely worked out. The presence of carbon atoms in some of the interstitial holes of austenite, however, makes the transformation in steels not only more complex but also unique and of outstanding importance. One of the first mechanisms proposed is that due to Bain. It is not correct in its details but is relatively simple and is suggestive of how diffusionless transformations can occur. It is illustrated in Fig. 7.7. In this figure, four fcc unit cells of austenite are shown. Only the sites of the iron atoms are indicated; the carbon atoms occupy the interstitial sites. The dashed lines show how a tetragonal unit cell can be constructed within the austenite structure. As drawn, the *c/a* ratio of the tetragonal cells is $\sqrt{2}/1$. During the transformation to martensite it must change to a value between 1.00 and 1.08, depending on the carbon content.

In Fig. 7.8, the left-hand drawing represents the fcc austenite structure, and if one considers the iron atoms as contacting hard spheres, "holes" occur at the $00\frac{1}{2}$, $0\frac{1}{2}0$, $\frac{1}{2}00$, and $\frac{1}{2}\,\frac{1}{2}\,\frac{1}{2}$ positions in some of which carbon atoms are located. Iron atoms labeled 1 to 6 show the octahedral arrangement in the fcc lattice. When this transforms to martensite, which, with carbon present, is bct, the carbon atoms are trapped in these holes and cause the distortion from the normal bcc of ferrite to the tetragonal structure of martensite. Relative atomic diameters in these structures are shown in Table 7.2.

Table 7.2 Relative atom sizes

Atom	Form	Diameter, Å
Fe	γ	2.508
	α	2.478
C	Solution	1.410

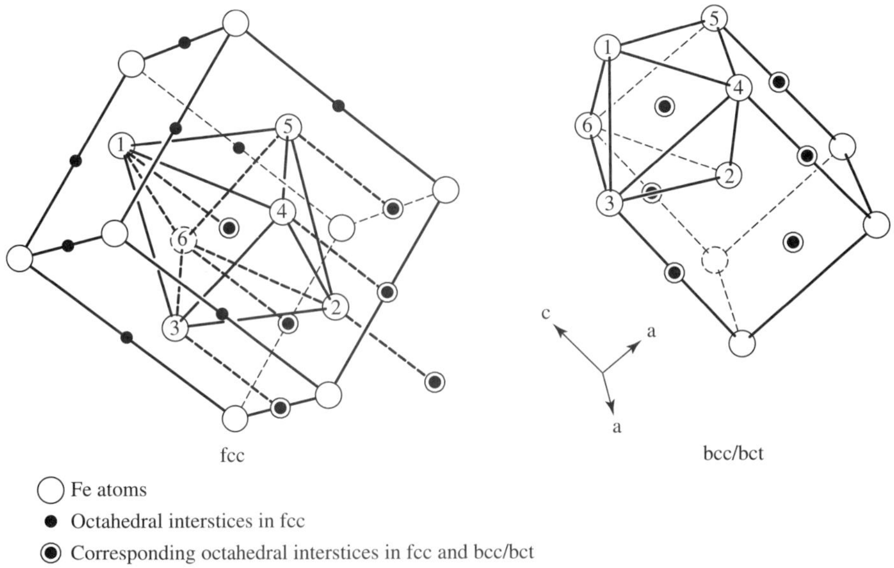

Figure 7.8 Correspondence of fcc and bcc/bct lattices, indicating relative orientations as they exist in steel. Note that all the octahedral interstices in the austenite are carried over into only one of the three possible sets in the martensite, thus defining the c direction of the unit cell as well as the alignment of the fourfold axes of the resulting nonregular octahedra.

In the absence of carbon atoms, the change from γ to α would result in the iron atoms being pushed apart symmetrically by 10% (the fcc lattice is more closely packed than the bcc). However, with carbon atoms present, the iron atoms 3 and 4 of Fig. 7.8 come closer together by 4%, and because the carbon atom is too large to fit between them, those in the 1 and 2 positions are pushed apart by 36%. These unsymmetric displacements result in the distorted tetragonal structure shown more clearly by Fig. 7.9. Cohen terms these dipole distortions.

The c axis of the tetragonal structure cannot be randomly oriented from unit cell to unit cell, or else x-ray diffraction would not clearly show the tetragonal structure. Therefore, within a given martensite crystal, constituted by millions of atoms even though microscopically small, carbon atoms are all trapped in the same crystallographic or xyz direction.

Upon cooling to the M_s temperature, martensite is nucleated by the shear formation of a unique interface that sweeps along, converting austenite into martensite. The shearing mechanism has been proved to result not only in translations that could be called slip but also in fine-scale (112) twinning.

The M_s temperature decreases with increasing austenitizing temperature. This is because M_s is quite sensitive to the carbon content of the austenite and a high austenitizing temperature generally means more complete solution of carbides. The formation of martensite is also sensitive to strain. If austenite is held somewhat above M_s, plastic deformation will often initiate the formation of martensite. This

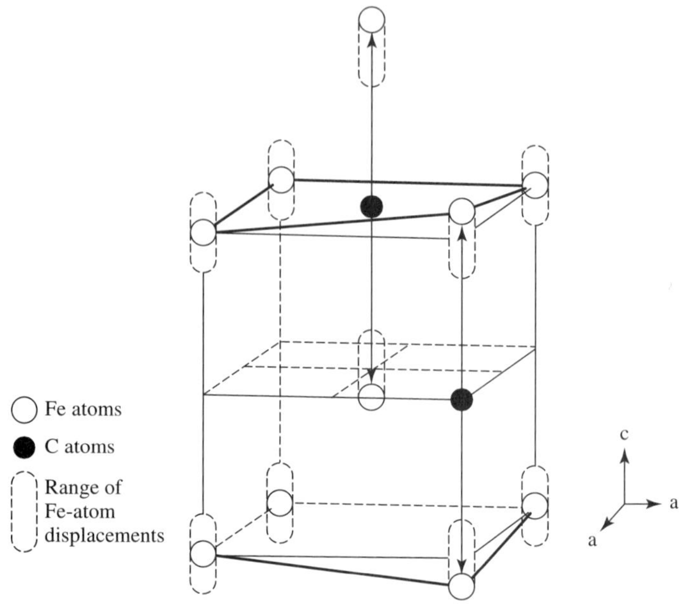

Figure 7.9 Schematic representation of iron atom displacements due to the presence of carbon in the bct lattice.

observation shows that M_s is not the temperature at which martensite first becomes stable relative to austenite; that temperature, called M_d, is always higher than M_s. The difference between M_d and M_s indicates that there is some sort of barrier to the formation of martensite. Apparently the nuclei from which the martensite needles form are already present in the austenite, and when the temperature becomes low enough—the driving force sufficiently high—individual needles of martensite begin to form. The transformation does not go to completion when M_s is reached, because each needle grows to only a limited size and, at a given temperature, only a few nuclei can overcome the energy barrier to transformation.

The temperature at which the martensite reaction is completed is known as M_f. As a practical matter, it is difficult to determine M_f. Most steels will, in fact, contain some retained austenite even after cooling to very low temperatures, for example, 80 K in the case of eutectoid steel. As will be discussed later, retained austenite may have an important influence on the properties of heat-treated steel.

The hardening of steel by the formation of martensite is a phenomenon essential to the well-being of mankind. Early in the eighteenth century the French mathematician and physicist Reaumur termed it "among the most wonderful phenomena of nature." It is only recently, however, that research has shown with any certainty the basic reason for the hardness of martensite. Figure 7.6 shows that carbon is an ineffective hardener of austenite but that it is essential to the hardness of martensite. Actually, the strength or hardness of martensite is attributable to the solid-solution hardening by carbon in interstitial sites of what would normally be a bcc lattice.

The related local, nonsymmetrical displacements of the neighboring iron atoms produce the dipole distortions of Fig. 7.9 that react strongly with and lock dislocations.

7.6 HEAT TREATMENT OF NONEUTECTOID STEELS

Shown in Fig. 7.10 are isothermal-transformation diagrams for a hypoeutectoid steel and a hypereutectoid steel. These diagrams differ qualitatively from that for eutectoid steel only by the presence of several additional lines. Thus the upper critical temperature A_3 is shown by a horizontal line above A_1. The lines marked F_s and C_s represent the start of the formation of proeutectoid ferrite and cementite, respectively. The corresponding finish lines have been omitted for the sake of clarity.

Comparison of the two diagrams shows that carbon content has only a minor effect on the time required for the pearlite reaction. However, dissolved carbon does tend to stabilize austenite at lower temperatures. It greatly retards the initiation and completion of the bainite reaction, displacing this part of the C by 200°C, whereas high-carbon steel retains considerable austenite even at room temperature.

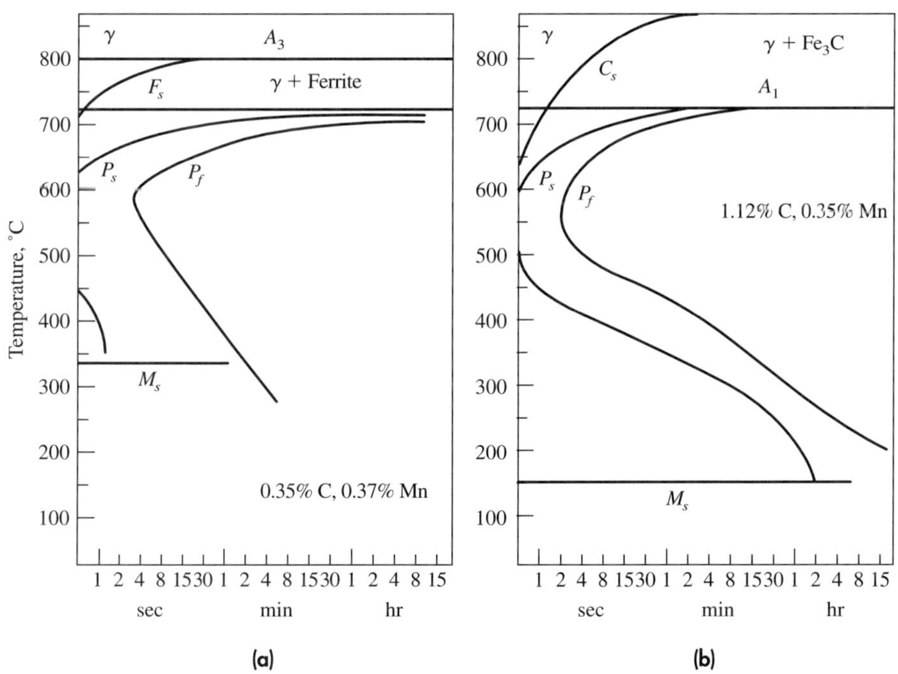

Figure 7.10 Isothermal transformation diagrams for (*a*) hypoeutectoid steel, (*b*) hypereutectoid steel.

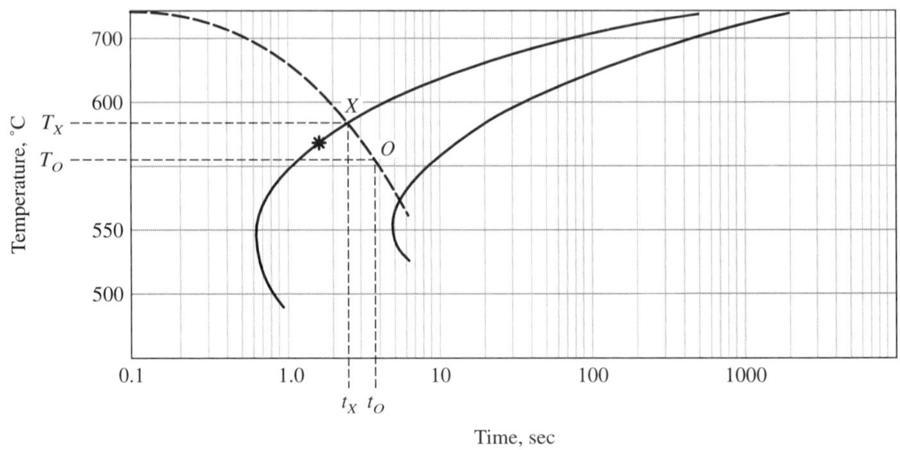

Figure 7.11 Estimation of time of start of pearlite reaction upon continuous cooling from the isothermal diagram. Upon cooling at rate shown by dashed cooling curve, assume nothing happens until point X is reached. Cooling to O requires time $t_O - t_X = 1.4$ s, and it is assumed that the average temperature for this period is halfway between T_X and T_O, that is, 620°C. Since 1.4 at 620°C is at the asterisk on the isothermal P_s line, then O is the P_s for continuous cooling.

Since the isothermal-transformation diagrams (Figs. 7.4 and 7.10) are fundamentally time-temperature charts, it should be possible to superimpose cooling curves in order to determine whether a given cooling rate would result in pearlite, martensite, or a mixture of the two. It is apparent that bainite could not form on continuous cooling of the plain-carbon steels, since upon starting at the A_1 line, no continuous-cooling curve could cross the B_s line from the metastable austenite field. However, it is not possible in any event to superimpose a cooling curve directly on an isothermal diagram. If the isothermal P_s time at 650°C is 5 s for a eutectoid steel, cooling from the A_1 to 650°C will not cause pearlite to start to form immediately. The isothermal diagram postulates an incubation time in this case of 5 s at 650°C. Upon making certain assumptions, it is possible to construct a continuous-cooling transformation diagram from an isothermal one.

Assume that, with reference to Fig. 7.11,

1. The extent of transformation at a time-temperature point X is not greater than if quenched to X; in other words, more time will be required for measurable transformation.
2. On cooling through a limited temperature range T_X to T_O the amount of transformation equals that on the isothermal diagram at the mean temperature $\frac{1}{2}(T_X + T_O)$ for the cooling interval $t_O - t_X$.

Using these assumptions, it is shown in Fig. 7.11 that, on continuous cooling to point O, conditions are equivalent to having isothermally reached the point * that lies on the P_s line; therefore, pearlite would just start to form. This point O was selected

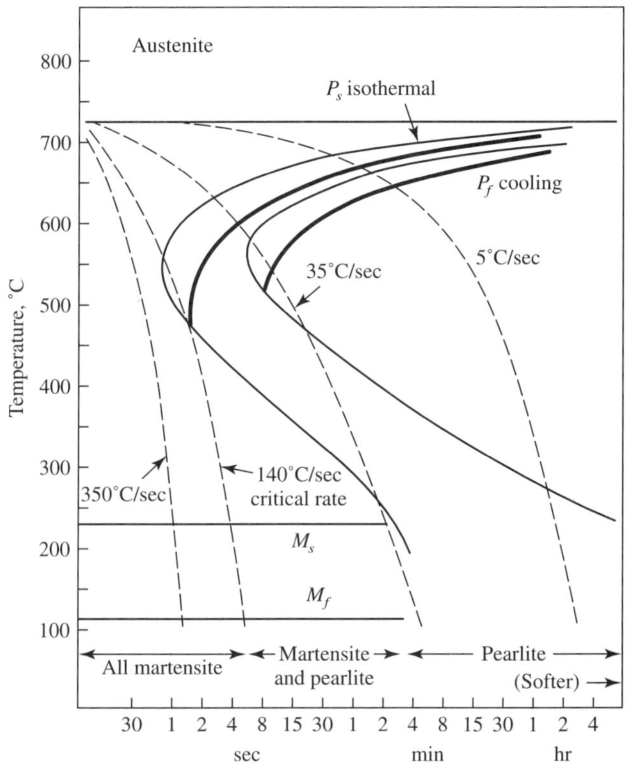

Figure 7.12 Transformation diagram for continuous cooling (heavy P_s and P_f lines) of a eutectoid steel in comparison with the isothermal transformation lines. Four different cooling-rate curves are superimposed, and the structures so obtained are labeled at the bottom.

by trial and error to obtain the asterisk point that lies on the P_s line. Later arbitrary times may be selected to find the P_f point for this cooling rate. Other cooling rates can be selected to locate continuous-cooling P_s and P_f curves. Then a derived diagram can be drawn, like Fig. 7.12, here applicable to a plain-carbon eutectoid steel.

It is also possible to determine continuous-cooling diagrams experimentally. One surprising result of such work is that bainite can form in a plain-carbon steel at certain critical cooling rates. An example of the difference between isothermal and continuous-cooling transformation diagrams is given by Fig. 7.12, representing a eutectoid steel, and Fig. 7.13, representing a typical medium-carbon steel.

Continuous-cooling transformation diagrams are generally more useful than isothermal diagrams because they approximate typical commercial practice. For example, suppose a thin strip of eutectoid steel is to be normalized; if it cools more rapidly than 35°C/s, Fig. 7.12 indicates that some martensite will be encountered and slower cooling would be necessary. On the other hand, normalizing a roll casting 1 m thick and weighing several tons might require an oil quench to obtain a cooling rate as fast as from 5 to 20°C/s.

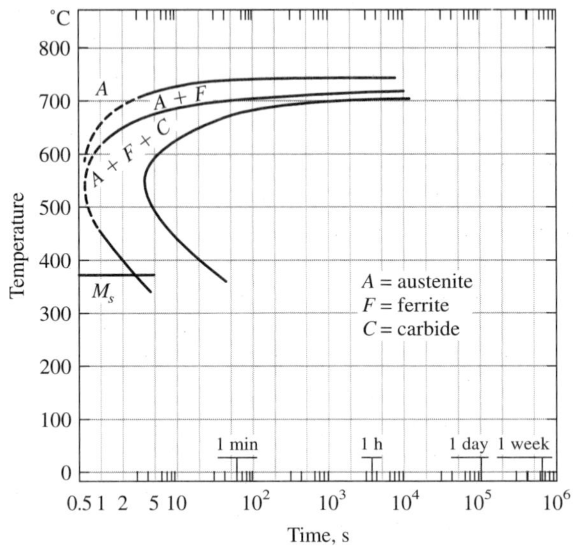

(a) I-T Diagram

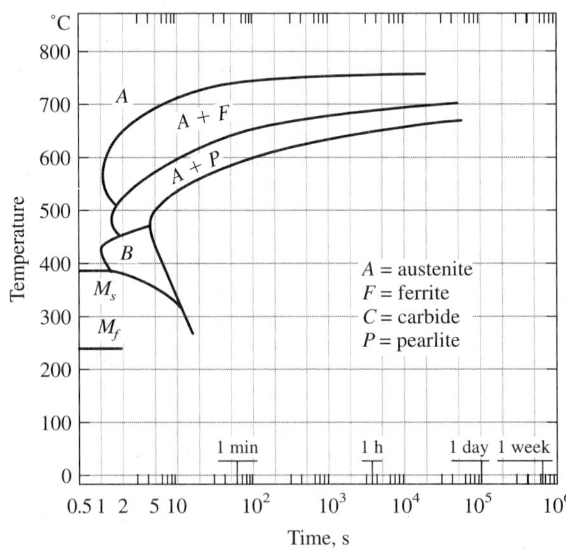

(b) C-T Diagram

Figure 7.13 Transformation diagrams for a medium-carbon steel, AISI 1040 (0.40% C, 0.5% Mn, 0.2% Si): (*a*) I-T, or isothermal, diagram; (*b*) C-T, or continuous-cooling, diagram.

The critical cooling velocity shown by this diagram is 140°C/s. If a fully martensitic structure were desired, that cooling rate must be met or exceeded to avoid some pearlite in the structure. An intermediate cooling rate, for example, 100°C/s, would give a pearlitic-martensitic mixture. From the earlier discussion, it should be remembered that the pearlite will be in the form of nodules centered along the former austenitic grain boundaries.

If only a small amount of pearlite is present, for example, 15 to 25%, the black etching nodules of pearlite in the white martensite nicely reveal the former austenitic grain size. This is one method of determining austenitic grain size and is particularly appropriate for eutectoid steels where no free ferrite or carbide can form to show the high-temperature grain size. Experimentally, the effective amount of pearlite is achieved by a gradient quench. The bar of the steel is austenitized and quenched from one end. Somewhere between the quenched and air-cooled ends a cooling rate will have existed that resulted in just the right proportion of pearlite nodules and martensite (see Micro. 7.16 on page 177).

7.7 PHYSICAL PROPERTY CHANGES DURING MARTENSITE FORMATION

The transformation of austenite into pearlite or martensite is accompanied by an expansion of the steel and the release of thermal energy. The critical temperatures can be determined by measuring dimensional changes with temperature (by use of a dilatometer) or by cooling or heating curves (thermal analysis). A simple demonstration of both dimensional and thermal changes can be made by stringing a piano wire, which is made of approximately eutectoid steel, between two supports with electrical contacts. (A 3-m length of 0.86-mm wire can be used directly across an ordinary 120-V power supply.) A current is passed through the wire to raise its temperature above the A_1, and then the current is shut off to permit the wire to cool in air. It sags during heating, but since the rate of heating is seldom uniform along its length, the transformation to austenite at the A_1 temperature is not readily detected. As it cools, however, the sagging wire rises steadily as the austenite contracts with falling temperature. Because of rapid cooling of the thin section in air, the steel does not transform until it reaches a dark-red color, corresponding to a temperature of about 550 to 600°C. As it then transforms to fine pearlite, the wire becomes visibly hotter, an effect commonly called recalescence. The expansion, resulting both from the transformation and from the rise in temperature, causes the wire to sag suddenly. Then, as the transformed structure continues to cool, the suspended wire resumes its slow contraction and steady rise to its original position.

Upon slow cooling, as in a furnace, there is no great temperature difference or gradient between the outside and center of even fairly heavy sections; all the transformation occurs at a high temperature (the Ar_1), and the expansion accompanying the transformation is accommodated by deformation of the plastic metal. Under these circumstances, there can be no appreciable residual stresses. When a heavy section, such as a steel roll casting 25 cm thick, is cooled in air or a liquid medium, the outside may reach the transformation temperature and expand while the center is much hotter and still completely austenitic. To accommodate the surface expansion, the metal in the central region is pulled toward the surface, and the ends may be forced in and become slightly concave. Well after the surface section has transformed and resumed its normal contraction, the central section reaches its transformation temperature and starts expanding. In order to conform to an expansion of the interior, the surface layers must expand. Since they are contracting at this time, the enforced expansion takes place by plastic flow, but at a much lower temperature and in metal having less plasticity than was the case during furnace cooling. If the roll has been quenched, this stage of the cooling cycle may find the surface in a brittle martensitic condition, not able to deform plastically, and surface sections may spall off, sometimes with explosive violence. Assuming that the surface plastically conforms to the expanding core, the final stages of cooling find the center, which transformed at a higher temperature and consequently is hotter, contracting more than the surface and now tending to pull it inward. This tensile pull of the center is balanced by the compressive elastic strength of the surface so that, when the metal is all at room temperature, there is a residual tensile stress in the center and a residual compressive stress at the surface. However, this state is reversed if the surface does not plastically deform as in the case of quenched, high-hardenability alloy steels.

Residual stresses are not confined to steel sections 25 cm in thickness; they may be found in many metals and in sizes less than 25 cm thick, if cooling from a high temperature was sufficiently rapid to produce a marked temperature differential between the surface and center sections. A high quenching temperature usually results in greater temperature gradients and thus introduces greater stresses, with a related increase in the danger of cracking or distortion. The stresses described above act in a circumferential direction on cylindrical specimens, but these are also accompanied by longitudinal (lengthwise on the surface) and radial (from surface to center) stresses. The stresses result in a displacement of the normal atomic spacing. If the crystal lattice is not too distorted, stresses can be detected by x-ray measurements. They can also be detected mechanically by machining off surface layers (or splitting tubes, etc.) since when part of the metal in a balanced elastic-stress system is removed, the remaining unbalanced stresses cause some deformation or distortion. This is frequently a problem of considerable magnitude in machine shop work.

Heating to a temperature sufficiently high to enable plastic deformation can relieve residual stresses. The elastic limit of most metals decreases rapidly at elevated temperatures, and since stresses are elastically balanced, they are diminished by internal flow. They are also reduced by an external or gross deformation; stretching a rod causes more flow in sections previously under tensile stresses and may eventually reverse the stress in sections under originally compressive forces, resulting in an equalization effect upon release of the external force.

Martensite, forming in relatively cold austenite, may be subject to stress of a much greater magnitude (since the elastic limit is higher) but on a much smaller scale. The first plates of martensite forming in plastic austenite may expand without trouble. When the adjacent austenite later transforms to martensite, the accompanying expansion is opposed by contact with the previously formed martensite. Thus the first martensitic needles are under a high localized elastic tensile stress, sometimes exhibiting transverse microcracks, with their newer neighbors balanced in compression.

7.8 TEMPERING OF MARTENSITE

A steel whose structure is freshly formed martensite is very hard but also very brittle. The brittleness is due in part to the intrinsic properties of martensite and in part to the internal stresses that accompany martensite formation. By tempering the quenched steel, i.e., reheating it to some temperature below 723°C, its ductility can be increased with usually some loss in hardness. Martensite, while stable relative to austenite at temperatures below M_D, is unstable relative to ferrite and cementite. There is some tendency for martensite to decompose into these constituents even at room temperature. Heating, as is done in tempering, accelerates the decomposition.

The tempering process can be studied most effectively by metallography, x-ray analysis, and dilatometry. These techniques show that the tempering process for ordinary times of reheating occurs in a series of well-defined stages in different temperature intervals:

1. 100 to 200°C Fresh martensite etches to a white color when prepared for metallographic examination, but after heating into this temperature range it etches dark. No individual particles can be resolved in the structure, which is known as tempered martensite. X-ray diffraction patterns of martensite tempered in this range show that marked changes in the structure, on too fine a scale to be resolved under the microscope, however, have occurred. This is illustrated by the data shown in Fig. 7.14. The c/a ratio of martensite is a measure of the amount of carbon in solution, and it is clear from the data in the figure that when 200°C (400°F) has been reached, the martensite is fully decomposed. No lines characteristic of cementite appear in the diffraction pattern, but the x-ray data indicate that in this temperature range, 75% of the carbon precipitates out of solid solution in some form of transition carbide structure. The formation of the transition carbide, called ϵ *carbide,* is accompanied by a slight increase in hardness.

2. 200 to 260°C In this region the steel begins to soften, but no marked change in structure occurs. If the steel contains retained austenite, this phase may begin to decompose, usually upon cooling.

3. 260 to 360°C X-ray diffraction patterns taken of steel tempered in this region begin to show diffraction lines characteristic of the cementite structure. Thus the transition carbide picks up additional carbon and transforms into very fine cementite particles. These cannot be resolved under the microscope, and the structure is

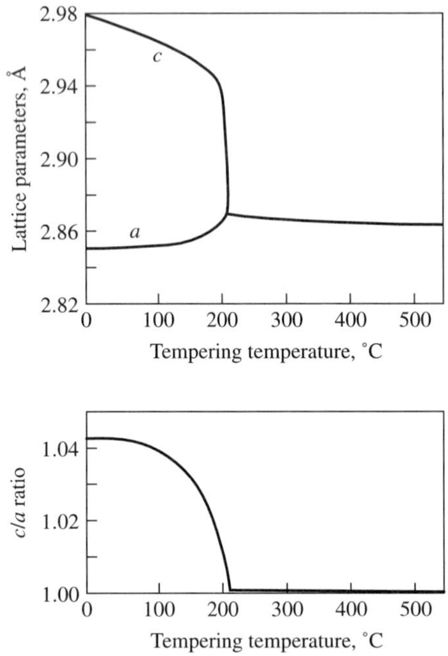

Figure 7.14 The lattice parameters *a* and *c* and the axial ratio *c/a* of martensite as a function of tempering temperature. When $c/a = 1.00$, the martensite has decomposed into ferrite and a carbide phase.

still called tempered martensite. There is, however, a marked decrease in hardness during tempering in this range.

4. 360 to 723°C The higher the temperature in this tempering range, the coarser the cementite particles. After tempering at 650°C or above, the particles are easily resolved under the optical microscope, and in this condition the structure is known as *spheroidite,* after the spherical shape assumed by the carbide particles. Near the high-temperature end of this range the steel becomes full softened.

The tempering of martensite involves the diffusion of carbon as particles of cementite grow from a homogeneous solid solution. Thus the amount of tempering depends on both the temperature and the time spent at temperature. The activation energy associated with the tempering mechanisms is high enough for temperature to be a more important factor than time; i.e., a small increase in temperature is equivalent to a large increase in time. For this reason tempering instructions are usually given in terms of temperature, it being understood that the time at temperature will be of the order of an hour or a few hours, the exact values usually not being very important. When it is desired to show how a property, such as hardness, changes with both temperature and time, the data may be plotted against the quantity:

$$T(c + \log t)$$

where T is absolute temperature, t time, and c a constant that depends on the nature of the steel. This parameter makes it possible to plot tempering data taken at different

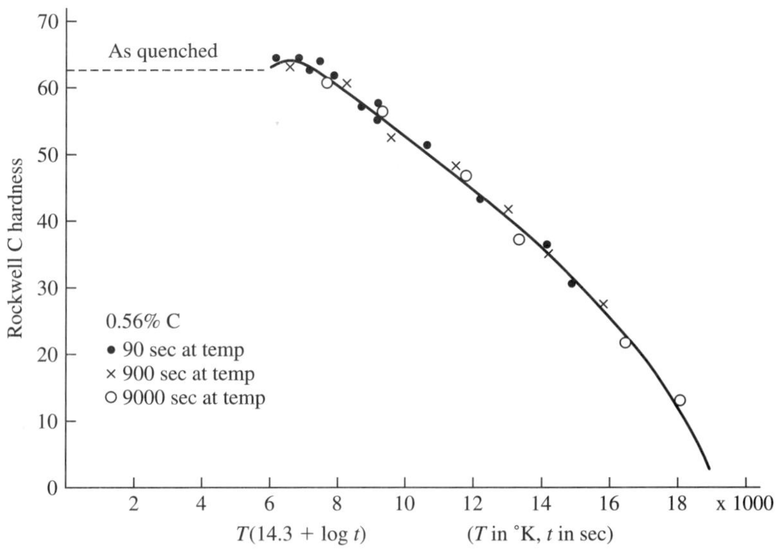

Figure 7.15 Tempering data for a 0.56% carbon steel, quenched to martensite and tempered over a range of temperatures for 90, 900, and 9000 s. The parameter $T(14.3 + \log t)$ permits all time-temperature data to be plotted as a single softening curve.

temperatures and times on a single curve, as illustrated in Fig. 7.15. Note that although tempering occurs in a series of distinct stages, the tempering curve shows a smooth decrease in hardness.

7.9 MICROSTRUCTURE OF ISOTHERMALLY TRANSFORMED STEEL

With proper equipment isothermal-transformation diagrams can readily be determined experimentally by using only a high-temperature austenitizing liquid salt bath, a moderate- to low-temperature transformation salt bath, and a cold-water tank. Hardness tests and a microscope can be used to estimate the time of start and of completion of the austenite transformation.

Austenite never appears as such to an appreciable extent in any microstructure of a plain-carbon steel. Any austenite present will largely transform to martensite upon quenching from the transformation bath to room temperature. The etching characteristics of martensite vary somewhat with the carbon content of the austenite from which it forms (Micros. 7.9 to 7.12).

Pearlite appears micrographically as described previously, the fine pearlite that forms near the nose of the transformation diagram being difficult to resolve except when the lamellae are nearly parallel to the surface of polish. Proeutectoid ferrite or carbide offers few or no difficulties as to polishing or identification (Micros. 7.9 to 7.20).

Micrograph 7.9 0.40% C + 0.71% Mn steel, austenitized at 1000°C, quenched to 684°C, held 10 s and quenched; Rockwell C50; ×1000; nital etch. After 10 s at the subcritical temperature, ferrite has started to form at the austenitic grain boundaries. The quench from 684°C did not affect the ferrite but changed the austenite to coarse, acicular martensite.

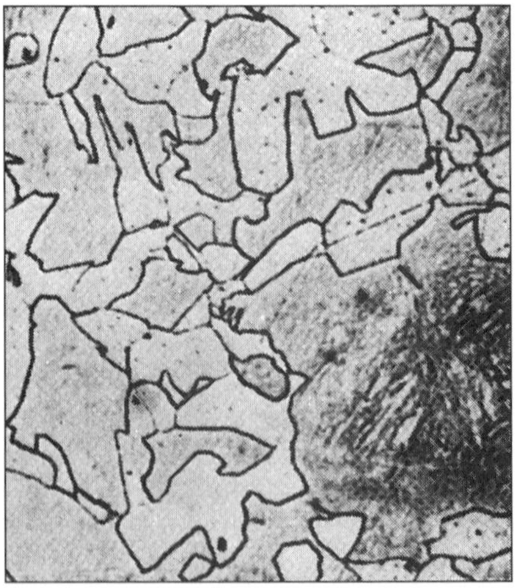

Micrograph 7.10 As in Micro. 7.9, but held 36 s at 684°C; Rockwell C48. During the interval from 10 to 36 s, more ferrite grains have appeared and grown in size at the austenitic grain boundaries. In areas adjacent to the ferrite, the remaining austenite has been enriched in carbon and the higher-carbon martensite formed there upon quenching etches much more slowly than the lower-carbon martensite on the right-hand side of the micrograph.

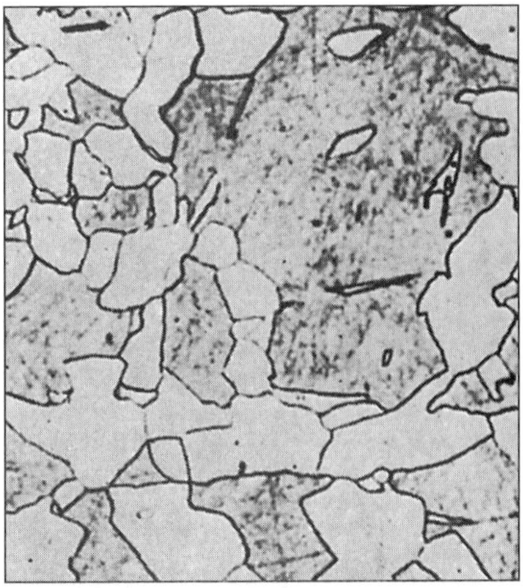

Micrograph 7.11 As in Micro. 7.9, but held 100 s at 684°C; Rockwell C34. The separation of ferrite from austenite has accelerated, as shown by the hardness change as well as by the microstructure. As ferrite separates, the remaining austenite becomes enriched in carbon. Again it transforms to martensite upon quenching, but this higher-carbon martensite (compared with that in Micro. 7.9) shows a less marked acicular, or needle, structure. It also etches to a lighter color.

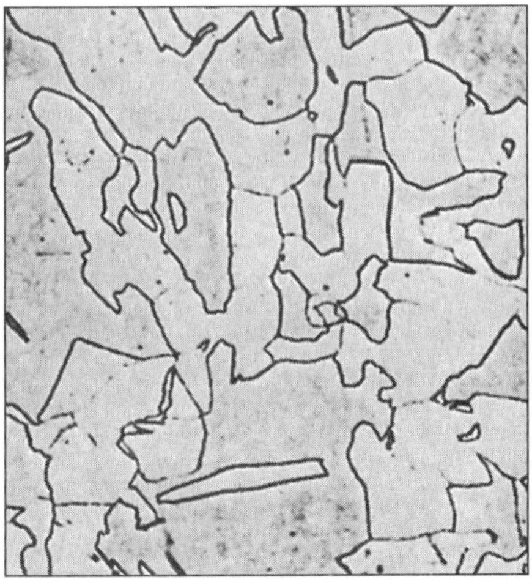

Micrograph 7.12 As in Micro. 7.9, but held 360 s at 684°C; Rockwell C31. Ferrite separation is nearing completion here, but no sign of transformation of the remaining austenite to pearlite is yet visible. In fact, after 3600 s this particular steel had a hardness of Rockwell C29 and a structure practically identical to this. Here, the effect of carbon content of the austenite on the etching characteristics of the martensite derived from it is more evident than in Micro. 7.11. In fact, it is difficult to distinguish between the soft ferrite and hard martensite.

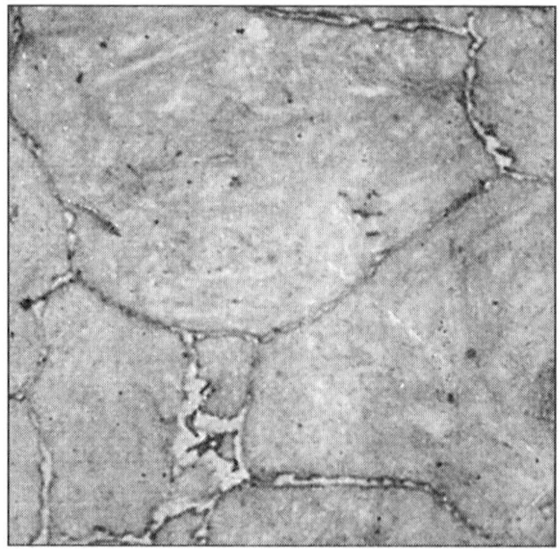

Micrograph 7.13 Razor-blade steel with 1.2% C, austenitized at 1000°C (1830°F), held 50 s at 710°C and quenched in water; ×1000; picral etch; hardness Rockwell C65. This specimen has been held long enough to cross the C_s line of the isothermal transformation graph (Fig. 7.10) but not to reach the line indicating the start of pearlite formation. Carbides outline the former austenitic grains, which transformed to white martensite in the water quench. Note the comparatively large size of the austenitic grains.

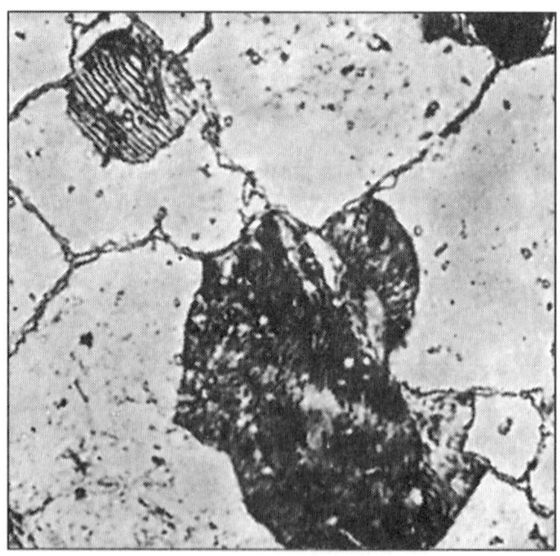

Micrograph 7.14 Same 1.2% C steel, austenitized at 1000°C, held 100 s at 710°C, and then quenched; ×1000; picral etch; hardness Rockwell C40. The transformation of austenite to moderately coarse pearlite is about one-third completed. Note in the upper left corner of this structure that pearlite is growing into an austenitic grain in two directions, at the sides and at the ends of the lamellae. Proeutectoid (excess) carbide is visible at the austenitic boundaries.

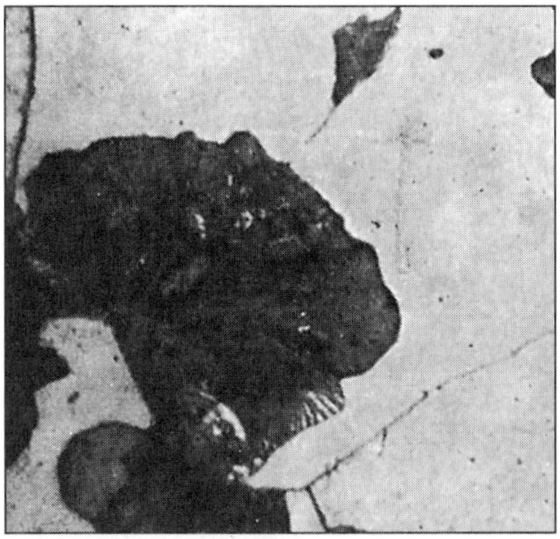

Micrograph 7.15 Same 1.2% C steel, austenitized at 1000°C, held 20 s at 680°C and then quenched; ×1000; picral etch; hardness C46. The reaction at this lower temperature, yielding a finer pearlite, is about one-third complete in a much shorter time. The front of the pearlite nodules shows that growth is chiefly at the ends of the pearlite lamellae at this temperature. Growth is mainly from the former austenite boundaries into one adjacent grain. At this temperature, only a trace of proeutectoid carbide has been able to form at the grain boundaries of the γ phase.

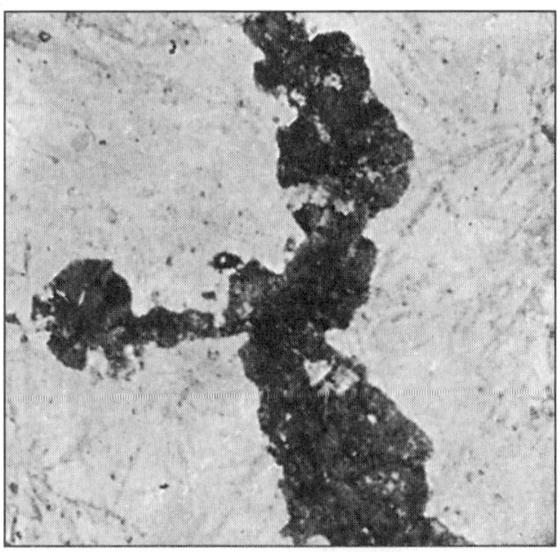

Micrograph 7.16 Same 1.2% C steel, austenitized at 1000°C, held 5 s at 610°C and then quenched; ×1000; picral etch; hardness C55. The transformation, here about 15% complete, has started much sooner at this temperature near the nose of the S curve, and the pearlitic structure is much finer, so fine that it is not resolved in this slightly overetched structure. The martensitic needles, in areas that were austenite before quenching, are more evident here because of the overetching. Note the nodular appearance of the fine pearlite areas and their characteristic radiation into two austenitic grains from the boundary. This structure is typical of a steel quenched at somewhat less than the critical cooling rate. The short time at 600°C has not permitted excess carbide to form at the austenitic grain boundaries, but the fine pearlite nodules at the boundaries reveal the austenite grain size.

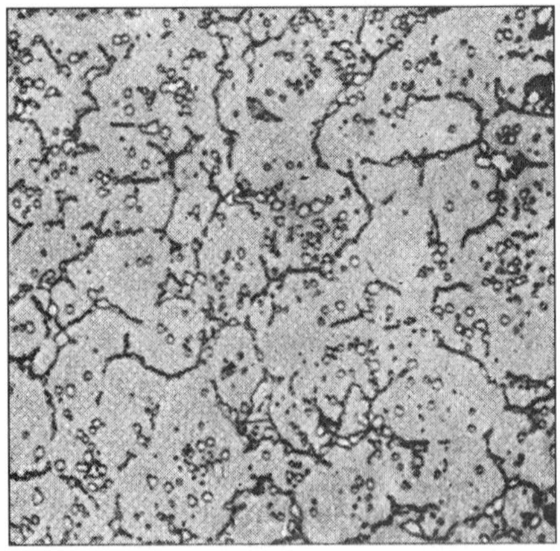

Micrograph 7.17 Same 1.2% C steel, austenitized at 850°C, held 5 s at 710°C and quenched; ×1000; picral etch; hardness C65. In this and the following structures, the same steel was austenitized below the A_{cm} line, in a commercial hardening temperature range at which the spheroidized carbides are not completely dissolved. Cementite has started to form at the austenitic grain boundaries much sooner than in the comparable specimen (Micro. 7.13), which was austenitized at a much higher temperature. Note the fine austenitic grain size in the present specimen, austenitized at a lower temperature.

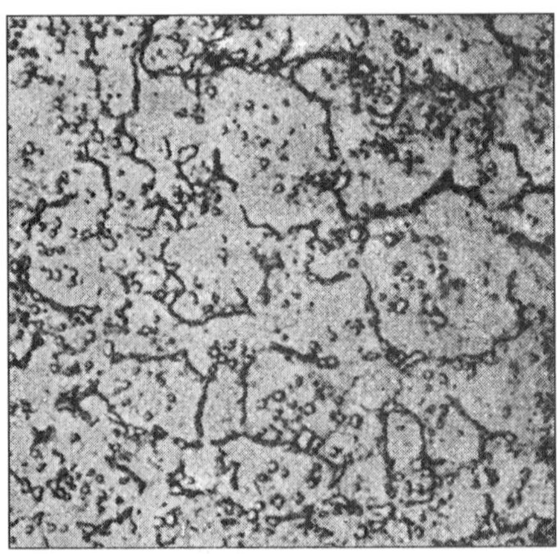

Micrograph 7.18 Same 1.2% C steel, austenitized at 850°C, held 10 s at 710°C and quenched; ×1000; picral etch; hardness C65. After a somewhat longer holding time, cementite at the grain boundaries and residual particles has thickened, but the pearlitic reaction has not yet started.

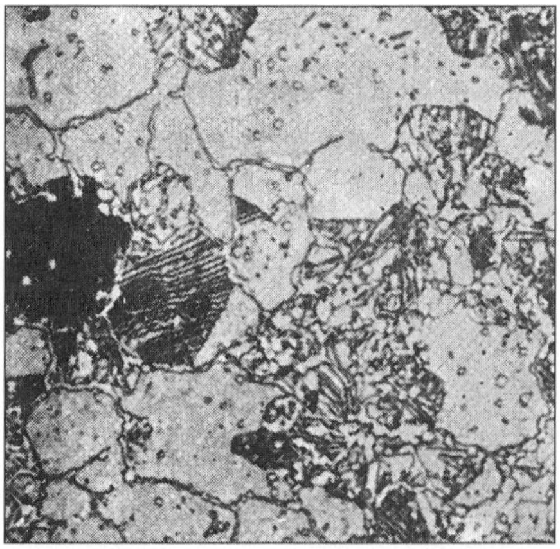

Micrograph 7.19 Same 1.2% C steel, austenitized at 850°C, held 30 s at 710°C and quenched; ×1000; picral etch; hardness C41. By now, the pearlite reaction has reached a stage of about 40% completion. Note that coarse pearlite has grown from the grain boundary into just one austenitic grain. Comparison of this structure with that of Micro. 7.14 shows that although the fine-grained austenite transforms more quickly the resulting pearlite spacing is similar.

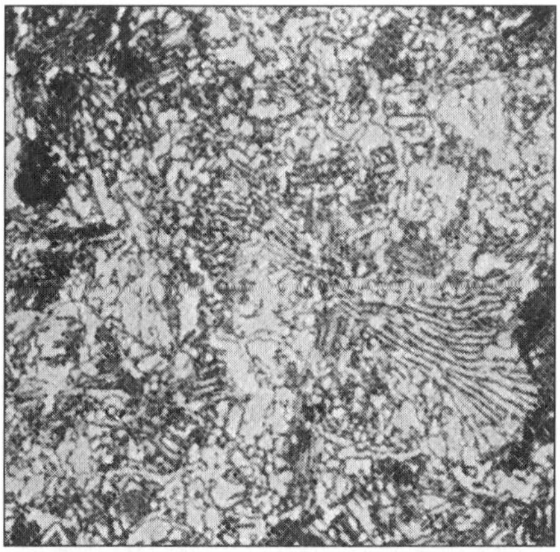

Micrograph 7.20 Same 1.2% C steel, austenitized at 850°C, held 100 s at 710°C and quenched; ×1000; picral etch; hardness C21. The transformation is complete, but not simply to pearlite. While some grains show this lamellar structure, others show merely coarse globular carbides. This is one form of a so-called abnormal, though very frequently encountered, type of structure. In the abnormal structure, carbides continue to form on carbides already present, which here were globular, so that the transformed structure contains large globular carbides and ferrite. Sometimes the excess grain-boundary carbides become very thick, with a correspondingly thick ferrite envelope next to them and with perhaps a little pearlite in the center of the former austenite grains.

Bainite formed at around 400°C has a feathery appearance. The parallel "fins" branching from a stem resemble pearlite but are straighter and finer than ordinary pearlite. At 250 to 300°C, bainite has an acicular structure much like tempered martensite. In specimens partly transformed to bainite, however, the quenched structure shows martensite needles to be white, while the bainite needles are black (Micros. 7.21 and 7.22).

The reason for martensite etching white while low-temperature bainite etches dark is related to the distribution of carbon. Martensite, forming by a diffusionless transformation, has the same composition as the parent austenite and is a solid solution, whereas bainite contains carbide particles in a highly dispersed form. These

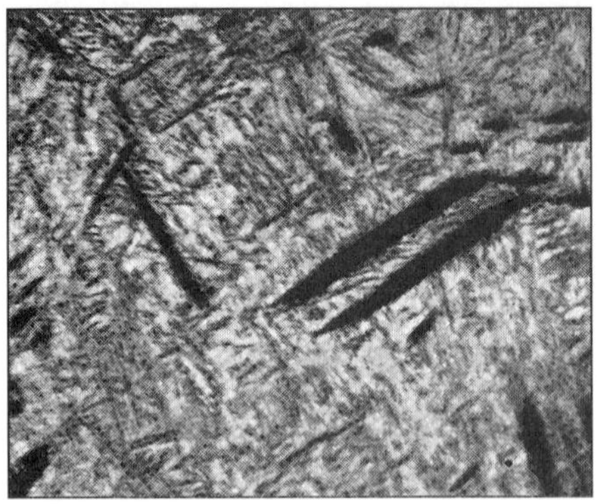

Micrograph 7.21 Thin section of 0.6% C steel, austenitized at 1000°C, quenched to 350°C, held 30 s at 350°C and quenched in water; ×1000, nital etch. The transformation of austenite to bainite has started with this structure showing about 2% bainite as scattered black needles in a martensitic background.

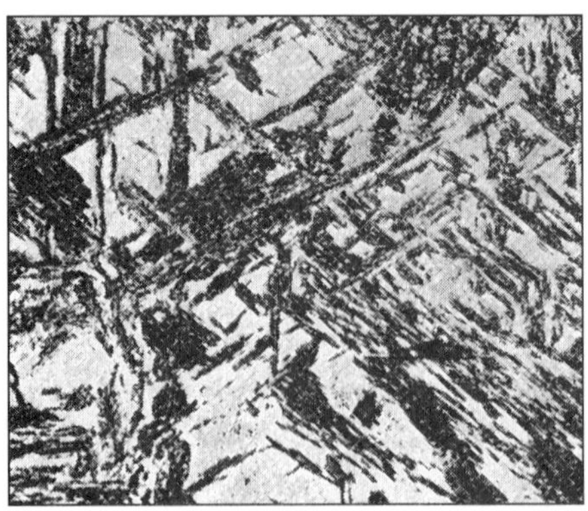

Micrograph 7.22 Same steel and treatment as Micro. 7.21 but held 100 s at 350°C and quenched (×1000, nital etch). The bainite reaction has progressed with time and is here about 60% complete.

carbide particles in bainite and also in tempered martensite are responsible for the dark etching of these constituents.

Bainitic and martensitic structures are quite easy to polish in comparison with pearlites. Their greater hardness practically eliminates difficulties from surface flow. However, whereas picral is preferred as an etchant for pearlite, nital is preferred for etching bainite and martensite. This is reasonable since nital was specified earlier as more desirable for delineating ferrite.

The micrographs in this section do not attempt to show the full development of all parts of the transformation graphs. Steps in the separation of ferrite from austenite of a hypoeutectoid steel are shown in Micros. 7.9 to 7.12. This series, in addition to showing ferrite separation in hypoeutectoid austenite as grain-boundary masses just below the A_1, beautifully shows the effect of variable carbon content on the etching characteristics of martensite. As ferrite separates, the remaining austenite becomes richer in carbon and the martensite formed upon quenching etches more slowly.

Similar partial-transformation structures for a completely austenitized hypereutectoid steel are shown in Micros. 7.13 to 7.16 and for the same steel, austenitized between the A_{cm} and A_1 in Micros. 7.17 to 7.20.

Tempered martensitic structures (Micros. 7.23 to 7.28) vary in etching characteristics according to the size of the carbide particles or the mean free path of ferrite between them. In a tempered series, it is desirable to etch all specimens for the same time, to obtain a specific comparative evaluation of "density of blackness" of the tempered martensite.

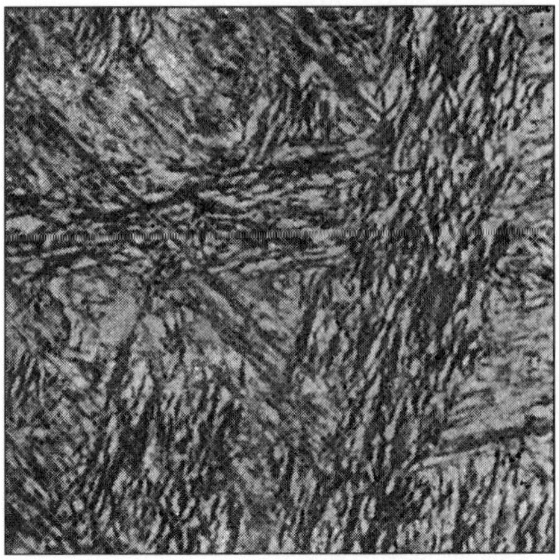

Micrograph 7.23 Forged steel of 0.70% C (SAE 1070), quenched from 925°C in cold water and tempered 1 h at 100°C; ×1000; etched 2 min in 4% picral; hardness C64. The structure after this very low tempering treatment is still essentially a white martensite. Only three directions of the acicular, or needlelike, martensitic plates are apparent in this micrograph, indicating that this entire field was only part of one austenite grain. The temperature from which this specimen was quenched was about 140°C above that ordinarily employed in commercial practice. The high temperature results in a coarse structure, well adapted to show the nature of martensite but too brittle for most service applications.

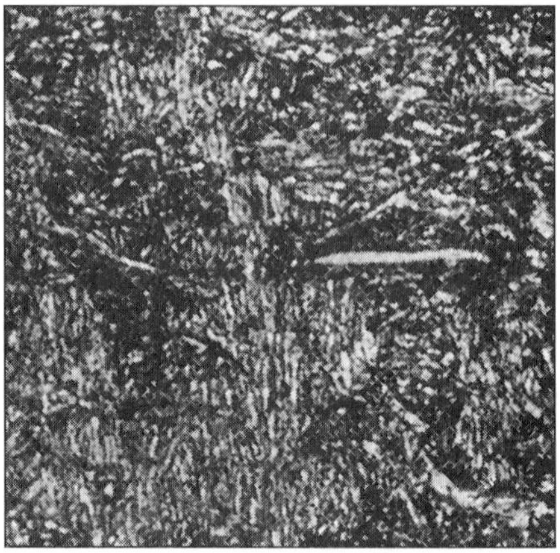

Micrograph 7.24 Same as Micro. 7.23, reheated 1 h at 200°C; ×1000; etched 40 s in 4% picral; hardness C60. The somewhat higher tempering temperature has caused the martensitic structure to etch more rapidly and darker. It is now called tempered martensite. The horizontal streak at the right center section represents a small oxide inclusion.

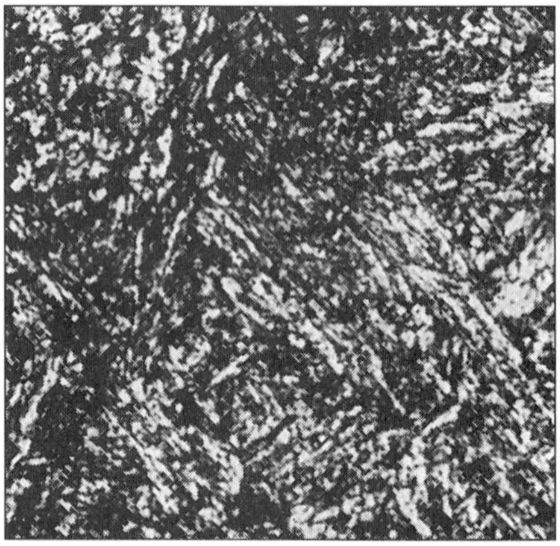

Micrograph 7.25 Same as Micro. 7.23, reheated 1 h at 350°C; ×1000; etched 25 s with 4% picral; hardness C50. General precipitation of fine carbides, below a resolvable size, causes the specimen to appear to be a black aggregate in which the martensitic plate directions are still evident. This structure is sometimes called troostite.

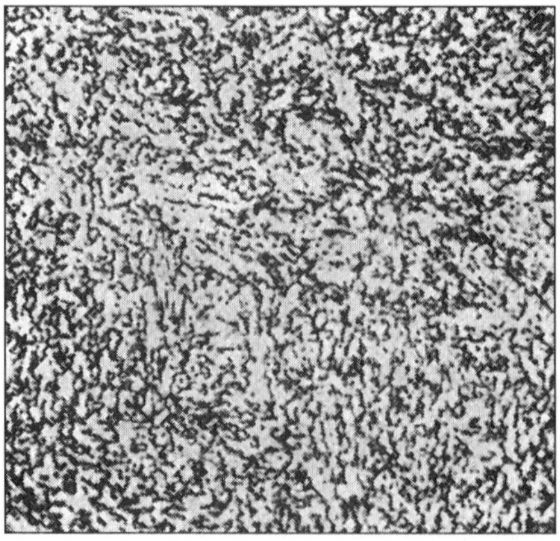

Micrograph 7.26 Same as Micro. 7.23, reheated 1 h at 600°C; ×1000; etched 25 s with 4% picral; hardness C30. Carbides have grown to a size just about resolved at this magnification, and the ferrite matrix is now evident. The structure is sometimes termed sorbite.

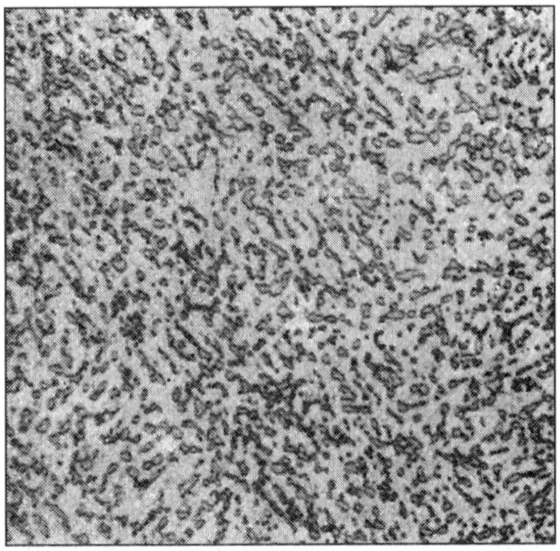

Micrograph 7.27 Same as Micro. 7.23, reheated 4 h at 720°C; ×1000; etched 25 s with 4% picral; hardness C8. The continued growth of carbides has made them clearly resolved so that this structure is much whiter and clearer than the preceding one. It would be called fine spheroidite. Note that alignment of the carbide particles still reveals the sites of the former martensitic needles.

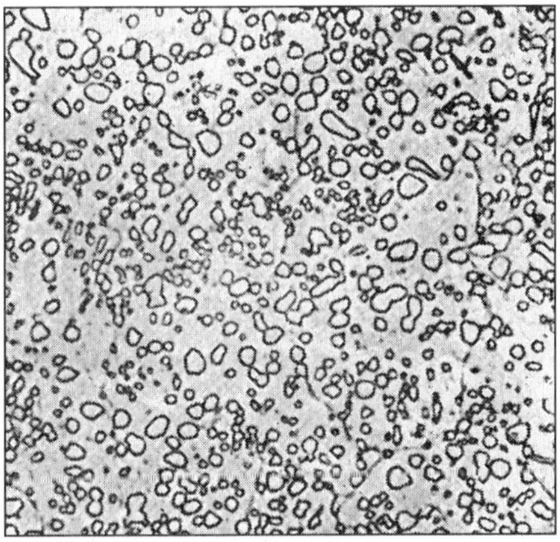

Micrograph 7.28 A commercial steel (SAE 52100) which gave machining difficulties; ×1000; nital etch. This structure, with its rather coarse carbide particles widely spaced in a ferrite matrix, could have originated only by an extremely long subcritical spheroidizing anneal.

7.10 GENERALIZED PROPERTIES OF HEAT-TREATED STEELS

Normalized and Annealed Steels

The mechanical properties of normalized steels, like those of any alloy, are determined by the phases present and their distribution. Ferrite, relatively pure bcc iron, has moderately good plasticity and strength, while the carbide is very hard and brittle.

In the aggregate or structural constituent called pearlite, the ferrite is nearly continuous. Consequently the eutectoid structure has some plasticity combined with moderately high hardness and strength. Hypoeutectoid alloys show a continuous ferrite grain structure containing islands of pearlite; therefore, these steels show good plasticity and strength, the plasticity decreasing and the strength increasing as the amount of ferrite decreases and the amount of pearlite increases with carbon content.

Hypereutectoid alloys, when slowly cooled past the A_{cm} line, exhibit carbide envelopes at the former austenitic grain boundaries, and, with this continuous brittle phase, structures such as those of Micros. 7.7 and 7.8 are inherently brittle. It is for this reason that, in commercial practice, hypereutectoid steels must be annealed below the A_{cm} line. The two structures shown would be very undesirable for an industrial alloy. Since the more rapid air cooling of a normalizing treatment suppresses the formation of hypereutectoid carbide envelopes, normalizing can be and is carried out above the A_{cm} line. Quantitative data showing these effects are presented in Table 7.3. For hypoeutectoid steels, the changes in properties are so nearly linear that they can be expressed with reasonable accuracy by a simple equation that relates the specific property to carbon content by means of the properties of ferrite and pearlite and the proportion of each present in the structure. Thus the tensile

Table 7.3 Mechanical properties of normalized and annealed steels

% Carbon	Yield Point, MPa	Tensile Strength, MPa	Elongation in 50 mm, %	Reduction in Area, %	BHN
Hot-rolled steel					
0.01	180	310	45	71	90
0.20	310	440	35	60	120
0.40	350	585	27	43	165
0.60	415	750	19	28	220
0.80	480	925	13	18	260
1.00	690	1050	7	11	295
1.20	690	1055	3	6	315
1.40	660	1020	1	3	300
Annealed					
0.01	125	285	47	71	90
0.20	250	410	37	64	115
0.40	303	517	30	48	145
0.60	340	660	23	33	190
0.80	360	800	15	22	220
1.00	360	745	22	26	195
1.20	350	700	24	39	200
1.40	345	685	19	25	215

Table 7.4 Tensile properties of the austenitic transformation products*

Reaction Temperature, °C	Hardness, HRC	Yield Strength (0.2% Offset), MPa	Tensile Strength, MPa	Elongation in 50 mm, %	Reduction Area, %	Pearlite Spacing, Å
700	19	345	830	13	20	6300
650	30	655	1070	16	35	2500
600	40	930	1310	14	40	1000
550†	38	910	1275	12	30	
500	36	900	1240	16	46	
450	40	1035	1310	18	54	
400	44	1170	1450	16	52	
350	48	1310	1585	13	44	

*Data obtained on a eutectoid plain-carbon steel (0.80% C, 0.74% Mn, 0.24% Si); corrected for recalescence during transformation.
†Data variable in this temperature range.

strength (MPa) of annealed hypoeutectoid steels (using the data of Table 7.3 and the lever rule) is related to carbon content by

$$\text{Tensile strength} = \frac{285\ (\%\ \text{ferrite}) + 115{,}000\ (\%\ \text{pearlite})}{100}$$

$$= 285\ (1\%\ \text{C}/0.8) + 115{,}000(\%\ \text{C}/0.8)$$

In pearlitic structures, although the ferrite is continuous and occupies 90% by volume of the structure, the carbide in the form of thin lamellae interrupts the continuity of the ferrite much more than the small carbides of bainite. Thus, although a fine pearlitic structure may approach a higher-temperature bainitic in strength, it will have significantly poorer ductility. The data of Table 7.4 for an isothermally transformed eutectoid steel show the variation of strength and ductility properties through the pearlite region, 700 to 550°C, and through the bainite region, 500 to 350°C.

Earlier, in Fig. 7.6, the strong effect of carbon content in martensitic hardness was shown. Theoretically, tensile- and yield-strength curves would essentially parallel the hardness curve. However, in actuality, the microstresses in quenched martensites of over 0.20% C are so high that premature fracture tends to occur at any minor defect. Thus low-carbon, for example, 0.10%, steels show good tensile strengths of about 1300 MPa and acceptable ductility, but these results are not commercially feasible. Such low-carbon steels have a high A_3 temperature and require very fast critical cooling rates so that only thin sections can be quenched to martensite. Parts so quenched show severe distortions. Therefore, in general, useful high strengths are best obtained by quenching to obtain a completely martensitic structure and

Table 7.5 Properties of a medium-carbon steel (0.39 C, 0.71 Mn), 25 mm rod, quenched and tempered 1 h as indicated

Temp., °C	Tensile Strength, MPa	Yield Strength, MPa	Elongation in 50 mm, %	BHN
Oil-quenched 850°C				
200	780	590	19.5	262
300	780	595	19.8	255
400	765	570	20.0	244
500	740	520	23.5	229
600	685	460	28.0	196
700	610	425	33.5	183
Water-quenched 850°C				
200	900	670	16.5	514
300	890	660	17.5	464
400	860	640	20.0	376
500	810	610	22.0	285
800	730	545	25.5	232
700	595	430	33.0	187

then tempering back to obtain the desired combination of strength and ductility (Table 7.5). The same strength can thus be obtained over a range of carbon contents, e.g., from 0.30 to 0.60% C. The selection of carbon content and heat treatment will be discussed in later chapters.

PROBLEMS

1. How might a spheroidized structure in an 0.8% carbon steel be converted to (*a*) a fine pearlite, (*b*) a coarse pearlite, or (*c*) a spheroidized structure appreciably finer than the original?
2. Why is a tempered martensite structure preferable to pearlite of equal hardness, e.g., Rockwell C30, for an automotive connecting rod?
3. What is the quantitative difference in the time required for the transformation at 710°C of coarse austenite vs. fine austenite plus residual carbides (see micrographs in this chapter)?
4. Of the two austenitic structures of Prob. 3, which, upon quenching, would show (*a*) greater quenched hardness and (*b*) faster critical cooling rate?

5. Specify two methods other than chemical analysis for checking the carbon content of a hypoeutectoid steel.
6. How might an annealed 0.40% carbon steel be heat-treated to show (*a*) large areas of free ferrite plus areas of fine spheroidized carbides in ferrite, (*b*) uniformly distributed fine spheroidized carbides?
7. Suppose a quenched but untempered (martensitic) 0.8% carbon-steel part was placed in a furnace at 800°C. What would be the effect of this structure on the austenization time at 800°C? If cracks in the part were found after the subsequent quench, explain their probable origin.
8. If a 1.2% carbon steel were hardened from 950°C instead of the usual 790°C, why is it most important that it be tempered twice rather than the usual once? (Assume a 190°C temper for a tool.)
9. Describe in detail an experimental technique by means of which the M_s temperature of a steel could be determined by using metallographic techniques.

REFERENCES

Olson, G. B., and W. S. Owen (Eds.): *Martensite: A Tribute to Morris Cohen,* ASM Int., Materials Park, OH, 1992.

Kaufman, L., and M. Cohen: "Thermodynamics and Kinetics of Martensite Transformations," *Prog. in Metal Physics,* vol. 7, p. 165 (1958).

Bain, E. C., and A. W. Paxtow: *Alloy Elements in Steel,* 2d ed., ASM, Materials Park, OH, 1961.

Krauss, G.: *Steels: Heat Treatment and Processing Principles,* ASM Int., Materials Park, OH, 1989.

Krauss, G.: "Tempering and Structural Change in Ferrous Martensites," in *Phase Transformations in Ferrous Alloys,* A. R. Marder and J. I. Goldstein (Eds.), The Metallurgical Society, Warrendale, PA, 1984.

Christianson, J. W.: *The Theory of Transformations in Metals and Alloys,* Pergamon, New York, 1965.

Pickering, F. B.: *Physical Metallurgy and the Design of Steels,* Applied Science Publishers, London, UK, 1978.

Honeycombe, R. W. K., and H. K. D. H. Bhadeshia: *Steels—Microstructure and Properties,* 2d ed., Wiley, New York, 1996.

Shewmon, P. G., *Transformations in Metals,* McGraw-Hill, New York, 1969.

VanderVoort, G. F., (Ed.): *Atlas of Time-Temperature Diagrams for Irons and Steels,* ASM Int., Materials Park, OH, 1991.

ASM Handbook, vol. 1, *Properties and Selection: Irons, Steels, and High-Performance Alloys,* 10th ed., ASM Int., Materials Park, OH, 1990.

SECTION 3

METALLIC MATERIALS ENGINEERING

chapter

8

Low-Carbon Steels

Steels are the most widely employed and generally useful of all engineering materials. The next four chapters of this book deal with both carbon and alloy steels: however, the low-carbon steels, the subject of this chapter, constitute the bulk of steels used in terms of tonnage. The products involved include the steels used in buildings, bridges, pipelines, ship plate, and the numerous sheet products employed in the transportation, food processing, and construction industries.

Commercial steels are never binary alloys of iron and carbon because some manganese, silicon, sulfur, and phosphorus are always present in the steels produced by any of the industrial refining processes. Plain-carbon steels, produced from basic oxygen converters using up to 25% scrap metal or in the open-hearth furnace using from 25 to 50% scrap, may also contain residual elements such as copper, nickel, chromium, and tin, to name a few, in amounts up to 0.1%. All these elements affect some properties of the steel, but in general the effects are small. The $Fe\text{-}Fe_3C$ phase diagram is not noticeably changed by the presence of any of these residual impurities or by deliberate additions, e.g., manganese or silicon, in the amounts usually found in carbon steels. Modifications in the diagram, in the properties of steels, or in their response to heat treatment as affected by larger additions of other elements are discussed in later chapters.

8.1 TERMS RELATED TO STEELMAKING PROCESSES

Most of the steel produced today is made by the basic oxygen steelmaking process, where liquid pig iron and scrap metal up to 25% are charged into a kettle-shaped converter and large quantities of oxygen are blown onto the surface of the liquid metal. The actual steelmaking process involves oxidation. The silicon, carbon, and phosphorus are oxidized from the molten bath to reduce their amount from the level typical of pig iron to that specified for the steel being made. Slag-forming materials, usually rich in lime, are added to combine with the oxidized impurities and remove them from the molten bath. This BOP process requires less than 1 h to produce 100 to 300 tons of steel.

An alternate but common steelmaking process is electric-furnace steelmaking, used most commonly for making alloy steels, particularly if the alloying elements tend to oxidize readily. The charge in this furnace, which is a shallow bath heated by electric arcs between the steel and large carbon electrodes, may be liquid pig iron or scrap in any proportion. In producing high-alloy steels, the charge is often selected high-quality scrap of controlled composition. One or sometimes two slags are used, and steels are usually produced with fewer oxide or sulfide inclusions than steels made by other methods.

The need for the utmost in combined strength and toughness for high-quality aircraft, missile, and nuclear components has led to a number of special techniques for further refining steels after primary processing. These methods include vacuum melting and heating, vacuum arc remelting, and electroslag refining. Since these processes are usually applied to medium-carbon or higher-carbon steels, they will be discussed in more detail later. Within the context of this discussion, however, it should be noted that such processing techniques are available to reduce the nonmetallic impurity content of the steel and are being increasingly applied to a large number of steels.

The liquid steel from the oxygen converter or electric furnace is poured into ladles and then teemed into ingots or most recently into continuous slabs. The deoxidation practice used in the ladle and ingot molds will put the steel into one of the following classifications.

1. **Rimmed steel.** With only minute additions of deoxidizer to the liquid steel, nearly pure iron begins to freeze at the ingot mold surfaces. After a thin skin, or "rim," freezes, the concentration of carbon and of oxygen in the liquid at the skin-liquid interface increases until the reaction C (dissolved in steel) + O (dissolved in steel) $\rightarrow$ Fe + CO (gas) begins. The CO gas so formed causes an apparent boiling action in the liquid. Gas blowholes are entrapped just under the skin and later near the center of the ingot.
2. **Capped steel.** After the mold is filled with liquid steel which is chemically similar to that for rimmed steel, a cap is put over the mold. This permits pressure to build up, which partially suppresses the C-O reaction and results in a somewhat more homogeneous ingot. Both rimmed and capped steels generally have under 0.15% C.
3. **Semikilled steel.** If deoxidation reduces the C-O reaction to the point where no boiling occurs, only late in the solidification process is sufficient gas liberated to develop blowholes. These have a sufficiently large volume to compensate for the volume change or shrinkage upon solidification so that the ingot has essentially a flat top. Semikilled steels typically are in the 0.15 to 0.25% C range.
4. **Killed steels.** In these grades, sufficient metallic deoxidizers are added for all oxygen in the steel to be converted into stable oxides, generally SiO_2, and Al_2O_3.

Aluminum is typically used alone for sheet products, while Si or Si + Al is used for other products. There is no reaction between carbon in the liquid steel and oxygen, and therefore no gas evolution. Solidification is quiescent, and normal shrinkage occurs, resulting in a loss in yield of the final product upon cropping or rejecting the top part of an ingot showing an open shrinkage cavity.

Electric-furnace steels, high-carbon steels, and high-alloy steels are usually killed steels. Open-hearth or converter steels of medium or low carbon content and/or low alloy content may be produced as killed steels where the higher qual ity and higher costs of the killed-steel method are warranted by service requirements.

8.2 GRAIN SIZE OF STEEL

Grains have been defined in this book as adjacent crystals of the same phase. In the Fe-Fe_3C system, it is possible to have ferrite grains or austenite grains. It is not possible to have carbide grains, since the carbide crystallites are not contiguous, except to a very minute degree, with other carbide crystallites. Likewise, it is not possible to have pearlite "grains," since ferrite crystals are in contact with carbide crystals, and vice versa (see page 152).

From a glance at Micros. 7.3 to 7.4, it is clear that in normalized or annealed steels, ferrite grain size is significant only in steels with less than about 0.4% C; above this value, the amount of free ferrite becomes so small that at most there are excess or proeutectoid ferrite crystallites in contact with pearlite areas.

Austenite grain size is never observed directly in the plain-carbon steels here considered, since austenite as a phase disappears at the Ar_1 temperature. Micrographs 7.4 and 7.8 show the grain size of the austenite that existed in the γ field by the evidence of the small amount of free ferrite or carbide that separated at the austenitic grain boundaries during the relatively slow cooling. There are other methods of determining at room temperature the grain size of austenite previously existing at a high temperature.[1]

Ferrite grain size is controlled in the same way as α-brass grain size, as discussed earlier. The grain size obtained on heating to 600°C may be increased by heating to a higher temperature. However, here there is a limit well below the melting point; on heating above 723°C, any pearlitic areas present will transform to austenite grains that, upon continued heating, will grow at the expense of ferrite grains even as the latter are growing at each other's expense. The process is complete, of course, at the Ac_3 temperature when all ferrite disappears.

In connection with ferrite grain size, there is another variable to be considered in addition to the ones previously discussed in connection with brass, namely, degree of cold work, annealing temperature and time, purity of ferrite, etc. The new variable is that a given set of ferrite crystals can come into existence in the solid state by a

[1]For methods to determine grain sizes in steels, refer to American Society for Testing and Materials Standard E112-96, Estimating the Average Grain Size of Metals.

process other than recrystallization. Ferrite crystals form by the γ-α transformation in pure iron or low-carbon steels upon cooling from the austenitic state. It is not possible to state the relationship mathematically, but since the ferrite grains are nucleated and grow, their size will depend on time and temperature or, specifically, on the cooling rate. Slow cooling of low-carbon austenite will give a much coarser ferrite grain size than fast cooling. The austenitic grain size itself is a variable, since, for a given cooling rate, a coarse-grain austenite will give a coarser ferrite grain size than a fine-grained austenite. Factors such as a finely dispersed insoluble phase are important; such a dispersoid would be conducive to a fine-grained austenite and, necessarily then, a fine-grained ferrite.

The usual rules of grain size apply to austenite as well as to ferrite. Although it is not possible to cold-work a plain-carbon-steel austenite, hot working involves recrystallization and grain growth to a degree dependent on the hot-working temperature. If the austenitic grains do not form in this way, they form in pearlite upon heating steel containing this structure. Since they form eutectoidally, both ferrite and carbide must be present at the site of nucleation, but obviously this is no limitation in a pearlitic area, where α and Fe_3C are everywhere in contact. This being the case, heating rate through the Ac_1 temperature only slightly affects the resulting initial austenitic grain size. The coarseness of the pearlitic lamellae will also have only a slight effect.

The newly formed austenitic grains, just above the Ac_1, are of minimum size, which increases slightly with time and markedly with temperature. Austenitic grain-growth curves are greatly influenced by the presence of a dispersoid. Relatively insoluble or slowly soluble excess carbides and finely dispersed oxides such as Al_2O_3 will retard grain growth in the usual heat-treating temperature range above the Ac_1 or Ac_3. Typical growth curves are shown in Fig. 8.1. It is evident that grain-size differences are not great at temperatures just above the Ac_1 range or at very high temperatures.

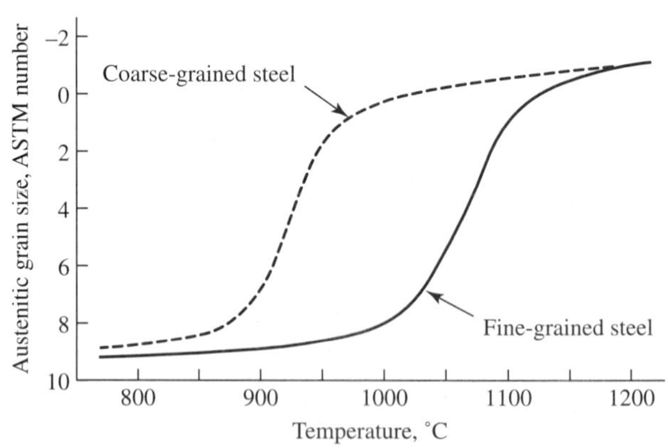

Figure 8.1 Typical austenitic grain-growth curves of a coarse-grained steel (not deoxidized with aluminum) and a fine-grained steel (deoxidized with aluminum). The steels differ materially in austenitic grain size only in the usual pack carburizing range, for example, 950°C, not at hardening temperatures (800 to 850°C) or at forging temperatures (1150 to 1200°C).

8.3 NONHARDENABLE LOW-CARBON STEELS

The great bulk of tonnage of steels produced and consumed by our industrial civilization is of relatively low carbon content and is never heat-treated to attain a martensitic structure. This, of course, does not mean that close metallurgical control is not employed but only that the costs of heat treatment are too great for the increase in properties attainable or that sizes of parts are too great. The grain size of these low-carbon steels can be controlled by composition or use of fine-grain practice. For hot-worked steels, grain size is also controlled by control of hot-working temperatures and particularly the temperature of finishing. Finally the cooling rate is controlled; in the case of hot-rolled flat strip, by controlling the temperature of the steel when it is coiled. Hot-rolled strip 2.5 mm thick cools quite rapidly when open but quite slowly when coiled. The qualitative effects of these variables can be deduced on the basis of an understanding of Chap. 7.

A great tonnage of structural steels and of sheets and plates is produced from plain-carbon semikilled OH or BOP steels; these are used in the hot-worked state. Apart from occasional gross and readily visible defects, properties of these steels are determined by the amount of hot working, the temperature at which it is completed, and the subsequent cooling rate. The amount of hot working controls the degree of elimination of the coarse structure and segregation characteristic of the cast ingot. Lower finishing temperatures, down to a minimum of about 732°C, result in a related lower temperature of recrystallization, and therefore a finer grain size. A faster cooling rate results in more pearlite colonies of a finer lamellar structure, and therefore harder and stronger steel. Since cooling is in air, the cooling rate is affected greatly by the thickness of the section. This can be modified, however; e.g., by coiling 3-mm strip while it is at about 732°C, the cooling rate can be greatly retarded.

The advent of continuous high-speed five-stand cold-rolling mills, which reduced the costs of cold rolling greatly and permitted improved control of gage together with improved surface smoothness, has resulted in a tremendous growth in the use of cold-rolled steels in gages of 0.75 mm down to 0.15 mm. Ordinarily, the steels are rimmed or capped steels of less than 0.25% C. Such steels seldom are used in the cold-rolled state but are process-annealed, i.e., below the A_1 critical temperature, in a reducing atmosphere, either in coil form or continuously as strip.

Process annealing results in the normal recrystallization of cold-worked ferrite and spheroidization of the Fe_3C, which, if in lamellar form prior to cold working, is broken up physically by the extensive deformations used in cold rolling. A critical element in process annealing of continuous strip is the cooling rate from 675 or 700°C. Too rapid cooling retains too much carbon (and nitrogen) in solution in the recrystallized ferrite and makes the annealed steel susceptible to quench aging, i.e., the precipitation of carbide and nitrides upon heating slightly above room temperatures, e.g., 150 to 230°C.

Process-annealed steel as annealed is generally too soft and has too pronounced a yield point; i.e., carbon and nitrogen atoms at dislocations greatly raise the stress

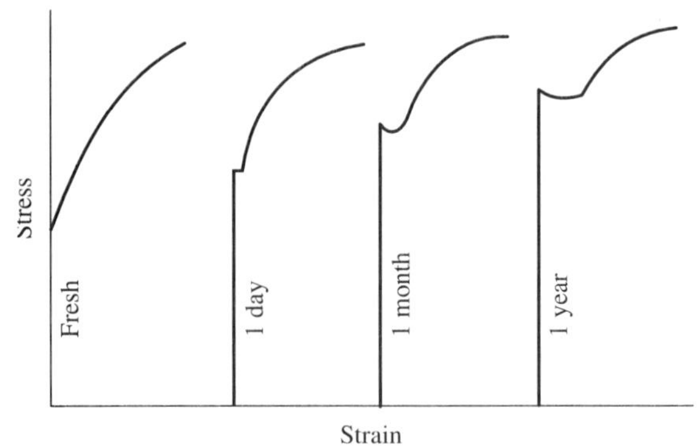

Figure 8.2 Stress-strain curves for low-carbon steel at various times after cold working. As strain aging occurs, a sharp yield point with yield flow develops in the steel.

required to initiate yielding, but once plastic flow starts in one locale, the stress required to cause it to continue there drops abruptly, thus causing continued flow in one location. This effect is not only interesting scientifically but very troublesome commercially, since it causes localized surface roughness called stretcher strains or worms. Because of these two considerations of softness and yield-point elongation, most process-annealed, cold-rolled steel is given a slight reduction of $^1/_2$ to 1-$^1/_2$ % by cold rolling, called a temper pass or skin pass. This not only slightly strengthens the steel but also eliminates the yield point, i.e., results in a smooth transition from elastic to plastic deformation rather than the abrupt drop in stress once plastic flow starts.

The yield point of temper-rolled, rimmed, or capped steel will return upon aging at room temperature (Fig. 8.2) at a rate somewhat dependent on the cooling rate from the prior anneal. The effect can be eliminated again if the rolled strip is roller-leveled by small-diameter rollers just before use. If the low-carbon steel is an aluminum-killed grade, however, the nitrogen is tied up structurally as aluminum nitride and therefore is not available to lock dislocations in place, with a resultant elimination of the yield-point effect.

Porcelain Enameled Ware

To produce the porcelain-coated steel shapes or panels familiar to us in homes and buildings, steel sheets are formed to the desired shape, coated with a siliceous gel, and heated to the temperature needed to melt the frit to a glossy state, usually about 815°C. Two coats, a ground coat and a finish coat, may be used with an inorganic pigment added to the latter to obtain the desired color. Low carbon and low hydrogen contents of the steel are desired to avoid the formation of carbon monoxide gas or hydrogen at the interface between the fused ceramic coating and surface of the metal at 815°C temperatures. A new technique for removing carbon from enameling steel is the so-called open-coil annealing process, based on a technique formerly used only in laboratories. A coil of cold-rolled carbon steel is wound to have a gap, or space, between successive turns. With the axis of such an "opened" coil vertical, the material is heated to the desired temperature of 675 to 700°C, with a strong fan blowing dissociated ammonia ($N_2 + H_2$) and water vapor through it. If P_{H_2}/P_{H_2O} is between 2 and 10 at 700°C, the resulting reaction of $C + H_2O \rightarrow H_2 + CO$ will remove carbon and nitrogen without oxidizing the iron. In relatively thin strip, hydrogen will diffuse out subsequently during slow cooling, particularly in the range 100 to 25°C.

Automobile-Body Stock

In this case, the critical considerations are the ability to be cold-formed into the desired shape in a minimum number of operations and the condition of the surface after forming. Cold forming of steel is still largely an art, but certain qualitative statements can be made about the steel properties needed (the design of forming dies, lubricants, etc., being ignored):

1. A combination of low yield point and high tensile strength is desirable.
2. Along with item 1 is the need for a considerable degree of uniform elongation (plastic stretching between yielding and maximum load in a tensile test). A high work-hardening rate results in delay of the localized "necking" which precedes fracture.
3. A high Olsen-cup-test value or ability to make a deep hemispherical indention (20 mm diameter) before fracture indicates biaxial ductility. This requires minimal inclusions in size and number, a higher strength normal to the surface[2] than in the plane of the sheet, and little directionality of properties in that plane.
4. Moderately fine grain size; if finer than no. 9, forming is impaired; if coarser than no. 7, orange-peel, or surface, roughening will occur on forming.
5. A low yield-point elongation is desirable and can be obtained by temper rolling to about 1% reduction or by roller flexing or "breaking" the grain, e.g., stretching each surface locally beyond the yield point. A high yield-point elongation does not reduce formability but does lead to surface roughening from worms or Luders bands of localized elongation.

[2]The relative strengths of a steel sheet in a direction normal to the surface and in directions within the plane of the sheet are indicated by the *R*-values, determined from standard tensile tests of sheet specimens. The *R*-value is the ratio of decrease in width of the specimen at fracture to the decrease in thickness at the neck. A high *R*-value means the metal resisted thinning in a direction normal to the surface more than it resisted stretching and relatedly narrowing in the test direction. Aluminum-killed, low-carbon steel sheet typically has much higher *R*-values than rimmed or capped steel.

Tin Plate

This material is the second largest standard tonnage item of steel production, representing the consumption of 10 million tons of steel a year in the United States. It is a low-carbon steel, ordinarily hot- and then cold-rolled to finished gage, process annealed, temper-rolled, and electrolytically plated with tin, generally of only 20 micrometer thickness, which is subsequently fused to brighten the surface coating. With most specific metallurgical details ignored, as in the two preceding cases, attention is called to a technological revolution during 1962–1963 in this field. Before that time, it was considered necessary for the steel to have at least 10% tensile ductility in order to flange the ends of can bodies so as to seam or crimp on end closures. When aluminum and plastic containers began to take away certain parts of the container market previously held by tin plate, it was necessary to go to thinner gage and therefore lower cost (per unit area) tin plate. This was accomplished by annealing at a gage perhaps 50 to 75% thicker than that finally to be used and then cold rolling to this final gage. Thus material could be rolled as thin as 0.13 mm and in the cold-rolled state be strong enough to be pulled through tin electroplating lines at normal high speeds. The economics are somewhat obscure but basically hinge on the fact that 50% more area of tin plate is produced per ton of steel, and containers are made in terms of units of area and not weight of tin plate.

The relative properties of commercial tin plates are shown in Table 8.1. Most notable technologically is the fact that the double cold-reduced tin plates can be flanged without cracking, which means a tensile stretch of 8%, even though the ordinary tensile-test elongation to fracture is recorded as 1%! Of course, the recorded 1% elongation is over a 50-mm gage length, and if one measures the strain just

Table 8.1 Properties of low-carbon* steels for automobile bodies and tin plate

Grade	Yield Strength, MPa	Tensile Strength, MPa	Elongation in 50 mm, %
Rimmed steel, annealed and temper-rolled	240	305	35
Killed steel, annealed and temper-rolled	170	285	41
MR† tin plate, box-annealed and temper-rolled	325	365	28
Continuous, annealed and temper-rolled	400	435	20
Box-annealed and rolled 40%	550	560	1.0
Transverse	610	615	0.8
Continuous annealed and rolled 35%	690	700	
Transverse	770	780	0.8

*Typically from 0.05 to 0.08% C.
†Composition: 0.12% C max, 0.2 to 0.6% Mn, 0.01% Si max, 0.05% S max, 0.02% P max, 0.2% Cu max.

before fracture over a gage length of 0.25 mm in the vicinity of localized deformation, the elongation will be found to be from 40 to 50%. (It must always be remembered that tensile-test elongations have two components, uniform elongation over the entire gage length and local elongation during the localized thinning preceding fracture; it is the latter which frequently makes gage length a critical element of percent elongation data.) A second gimmick in the discrepancy between ability to stretch by flanging vs. tensile-test elongation is that the stretch must be in the rolling direction; i.e., the tin-plate cans must have the rolling direction of the steel circumferential. Even though the transverse tensile-test elongation of cold-rolled tin plate is almost the same as the longitudinal, cracking always occurs if stretch flanging is attempted in the transverse direction. Cracks are nucleated by inclusions or by cut-edge damage.

Heavy Steel Plates for Ships and Tanks

About 20% of all steel produced in the United States in the recent past has been in the form of plates and shapes for structural applications. The non-heat-treated structural steel plates of this type are most commonly made to one of a number of American Society for Testing and Materials (ASTM) specifications such as A36, A285, A441, A515, A516, and A662. Table 8.2 lists the chemical compositions and mechanical property requirements for these grades. Most of these steels are given no heat treatment after hot rolling and thus develop their mechanical properties as a result of control of composition and grain size (through deoxidation practice). In some cases, for example, A515 and A516, plates greater than 50 mm thickness are normalized. Although only moderate in yield and tensile strength, these kinds of steels have the proper combination of strength, ductility, toughness, and weldability to perform satisfactorily in structural applications. The carbon content is rarely over

Table 8.2 Typical mechanical properties and applications of structural steels

ASTM Grade*	Use	Yield Strength, MPa	Tensile Strength, MPa	Elongation in 50 mm, %
A36	Bridges and buildings	310	450	28
A285 grade C	Pressure vessels	275	450	31
A441	Bridges and buildings	380	520	26
A515 grade 70	Pressure vessels, boilers	310	520	26
A516 grade 70	Pressure vessels, low temperature	310	520	26
A662 grade A	Pressure vessels, low temperature	345	450	27

*The steels as a group have 0.18 to 0.31% C, 0.60 to 1.25% Mn, 0.035% P, 0.040% S, 0.15 to 0.35% Si. Individual grades are more restricted within this range. A662 may have lower carbon and higher manganese than the above limits indicate.

0.25% (to increase it above this level may reduce toughness and weldability) and will rarely be below 0.15% for reasons of strength. The most common significant alloying element, besides carbon, is manganese, which is used to increase yield and tensile strength without reducing ductility. Copper is often added for improved corrosion resistance.

Steels for Low-Temperature Service

Occasional failures of ships and tanks during cold weather over the past 50 years have called attention to the brittle-fracture behavior of low-carbon steels and, more generally, of almost all materials other than those having the fcc structure. This area of brittle fracture was discussed in more detail in Chap. 2, but here it will suffice to say that metallurgical factors can be controlled to lower the brittle-to-ductile transition temperature (Chap. 1) or increase the fracture toughness so that brittle fracture will not be encountered in normal winter service temperatures. The basic control measure is to ensure fine austenite (and therefore ferrite) grain size by such procedures as using aluminum-killed steel, performing extensive hot work, controlling finishing temperatures (just above the A_1), and normalizing. Manganese additions above the normal level may also improve toughness. An example of such a material is ASTM A516 (which may be contrasted to A515). This material is specified to have a fine austenite grain size (finer than ASTM 5) and to be normalized for improved notch toughness. The intended application of such a steel is for "service in welded pressure vessels where improved notch toughness is important."[3] The microstructure of this steel, which is typical of the group, is seen in Micro. 8.1. A further step in improving notch toughness is illustrated by ASTM A662, which has lower carbon content and higher manganese content and for which impact toughness requirements between −25 and +25°C may be specified.

There has been an increasing trend in the last few years, particularly in Europe, to develop and utilize structural steels which have quite low carbon content, usually 0.15% C or less, sometimes as low as 0.06%, but high manganese content and V, Ti, or Nb as alloying elements. These steels rely on precipitation-hardening effects produced by carbide formation during rolling or subsequent heat treatments or by compounds of nitrogen with these alloys or with aluminum. These steels have attractively high yield and tensile strengths and retain good weldability and adequate toughness compared with the usual as-rolled or normalized carbon steels. They require more care in production and heat treatment than carbon steels, however, and are therefore somewhat more expensive.

Ship-plate steels are quite similar to the low-carbon steels listed in Table 8.2 but are normally made to the American Bureau of Shipping specifications. These plates are also usually as rolled but may be normalized when high toughness in service is desired.

[3]ASTM Specification A516-96, Pressure Vessel Plates, Carbon Steel, for Moderate and Lower Temperature Service, 1999.

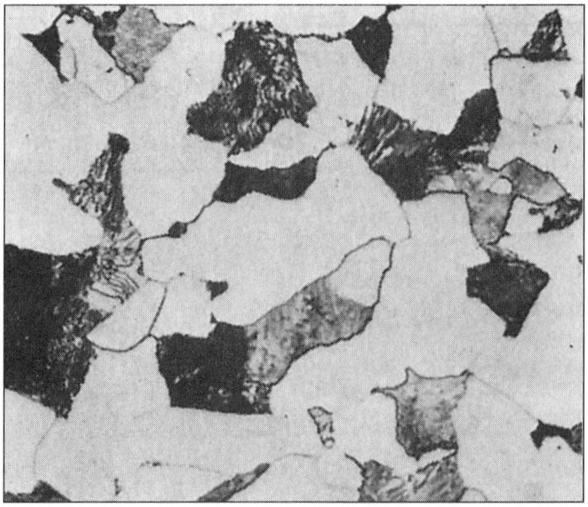

Micrograph 8.1 A516 Grade 70 normalized from 900°C; ×500, nital etch. The microstructure consists of ferrite and fine pearlite.

Structural Shapes and Pipe

A wide variety of structural shapes are also produced from low-carbon steel for bridge and building construction. They include such items as I beams, channels, angles, and wide-flange beams. The section thicknesses are normally about 12 mm but can include large H beams many inches thick and many feet in overall cross section. Larger structural shapes, such as box girders and plate girders used in bridge construction, are usually built up by welding of plate or may be made partly from rolled shapes with welded or bolted attachments. Many of these shapes fall under the same ASTM specifications as plate products, such as A36.

Pipeline steels are also quite similar to the plates and structural grades discussed above and are generally used in the as-rolled condition. Since large-diameter (100 cm and larger) pipelines are usually welded both during their manufacture into lengths and in the field, good weldability is a necessary characteristic. The manufacture of pipe, where section thicknesses seldom exceed 20 mm, can be accomplished by high-speed rolling (and welding) processes, and thus fine, closely controlled microstructures can be achieved, producing higher yield and tensile strengths than would be characteristic of plate or shapes of the same composition. Most of this pipe is manufactured to American Petroleum Institute (API) specifications. As with the plates and shapes, there is an increasing interest in pipeline steels with low carbon content but containing additions of Nb, V, and Ti. For a variety of reasons, these steels are economically more feasible in the form of pipe than plate, where thicknesses are typically greater, and their good weldability and high yield strength have made them attractive alternatives to more conventional steel pipe. Thus these pipe steels are finding an expanding market in the United States to a greater extent than the plate steels. These steels are sometimes called microalloyed steels because the amounts of alloy element used are small in comparison with those used in more conventional steels, normally less than 0.1%.

8.4 HIGH-STRENGTH, LOW-ALLOY (HSLA) STEELS

While the structural, ship-plate, and pipeline steels are often used in the as-rolled or normalized condition and contain little alloy addition beyond manganese, the low-alloy, high-strength steels can contain appreciable alloy content, up to 10%, and are often quenched and tempered to give high levels of both strength and impact toughness. Tables 8.3 and 8.4 list typical chemical analyses and mechanical properties for a number of commonly used low-alloy, high-strength steels. The steels are usually designated by an ASTM specification number. The ASTM system of numbering specifications is not related to the chemical composition, mechanical properties, or service application of the steel but to the chronological order in which they were developed and approved for inclusion within this system. Thus two steels next to each other in numerical sequence may be entirely different in character. The higher-numbered alloys have been developed most recently.

The steels in Table 8.3 exemplify the metallurgical approach used in developing these types of steels. Some grades, such as A517 type F and A543 class 1, have high yield and tensile strengths in combination with good toughness. The alloy element selection in these cases is based on promoting the formation of martensite

Table 8.3 Typical chemical composition of high-strength, low-alloy steel

Element	A533 Grade B	A517 Grade F	A543 Class 1	A542 Class 1	A203 Grade D	A553 Type 1
C	0.22	0.15	0.15	0.12	0.12	0.10
Mn	1.25	0.80	0.35	0.45	0.45	0.65
P	0.015	0.015	0.010	0.020	0.015	0.010
S	0.015	0.015	0.010	0.020	0.015	0.010
Si	0.20	0.20	0.25	0.25	0.25	0.25
Ni	0.50	0.85	3.25		3.50	9.00
Cr		0.50	1.75	2.25		
Mo	0.50	0.50	0.50	1.00		
V			0.02			
Zr		0.10				
B		0.002				
Cu		0.30				

Table 8.4 Typical mechanical properties and applications of high-strength, low-alloy steel

ASTM Grade	Use	Yield Strength, MPa	Tensile Strength, MPa	Elongation in 50 mm, %
A533 grade B	Nuclear vessels, steam-generation class 1 equipment	415	620	25
A517 grade F	Bridge, building construction, other heavy construction	760	860	21
A543 class 1	Pressure vessels, plates, forgings (similar composition)	655	760	23
A542 class 1	Chemical and refinery pressure vessels	655	760	22
A203 grade D	Low-temperature equipment	285	520	
A553 type 1	Cryogenic tanks and equipment	655	760	25

or bainite on quenching over a range of section thicknesses with good toughness developed by tempering at relatively high temperatures (600 to 650°C). These types of materials are used in plates, shapes, and forgings, and for welded construction from bridges to commercial nuclear pressure vessels. Other steels, such as A542, are quenched and tempered for high strength but are alloyed with Cr and Mo for high-temperature creep and corrosion resistance. The same composition is made in the normalized condition under a different ASTM specification number. Another possible option is the selection of steels such as A203 grade D and A553 type 1, which are alloyed with nickel for excellent cryogenic temperature toughness. The impact transition temperature of A203 grade D is below −60°C, and it is used in a variety of low-temperature service applications. The impact transition temperature for A553 type 1 is below −200°C, and thus it is suitable for use in pressure vessels for the transport of liquefied natural gas (−170°C). In addition to the examples listed above, there are many other low-alloy, high-strength steels that provide special properties appropriate to various kinds of service.

Because of their relatively high alloy content, these steels are not as easily welded as the normal carbon plate steels. With proper care, however, they can be and are used in wide-ranging service involving welded construction. The fact that many have a low carbon content is helpful in this respect. Although they are somewhat more expensive than the unalloyed low-carbon steels, their superior strength and toughness make them attractive in many instances where alternate materials would have to be much more massive to provide the same load-carrying capacity.

Typical microstructures of low-alloy, high-strength steels are shown in Micros. 8.2 to 8.4. The microstructure of the A515 steel (Micro. 8.1) reflects the fact that this material is too low in carbon or alloy content to produce transformation products other than ferrite and pearlite when quenched in section thickness over about 1.3 cm. In contrast, the A533 grade B steel has sufficient hardenability to produce mixed

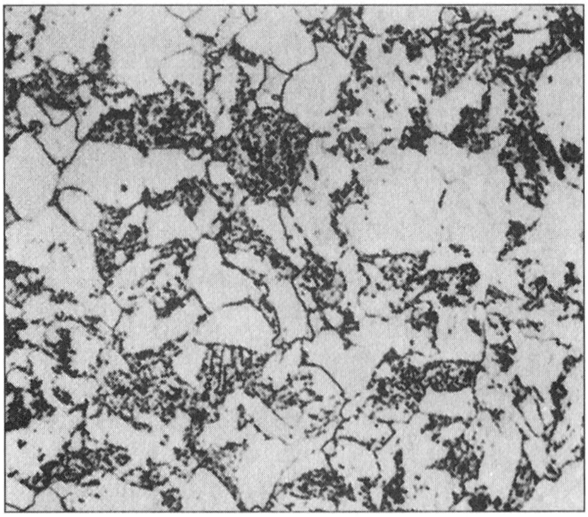

Micrograph 8.2 A533 grade B, quenched from 900°C and tempered at 620°C; ×500, nital etch. This is a heavy section plate; the microstructure consists of ferrite and tempered bainite.

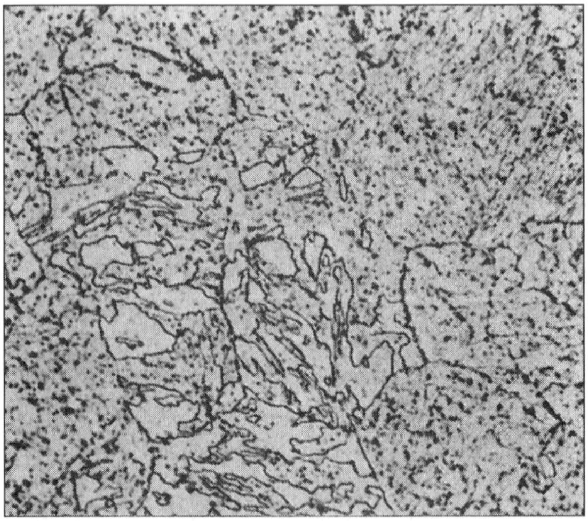

Micrograph 8.3 A543 class 1, quenched from 850°C and tempered at 650°C; ×500, nital etch. The microstructure in this sample is tempered bainite and martensite.

microstructures containing ferrite and carbide and, in thin sections, martensite. In the heavy-section microstructure shown in Micro. 8.2, this steel contains ferrite and tempered bainite. The higher-alloy materials, such as A517F, can be quenched to martensitic microstructures and are subsequently tempered for toughness. As seen in Micro. 8.4, its microstructure is predominantly tempered martensite.

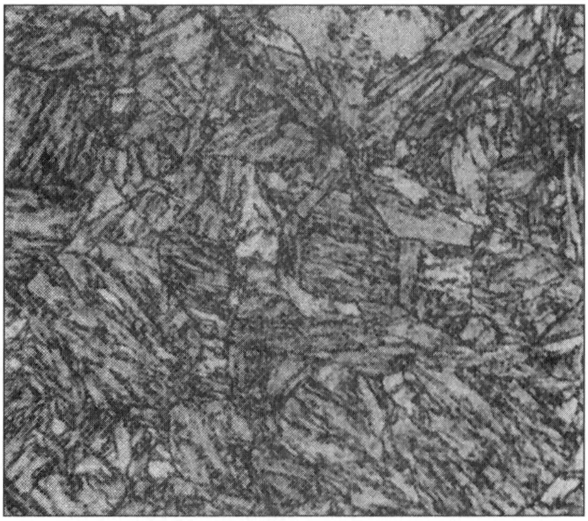

Micrograph 8.4 A517 grade F, quenched from 925°C, tempered at 650°C; ×500, nital etch. The microstructure is tempered martensite.

8.5 WELDING OF LOW-CARBON STEEL

Metallurgically, fusion welding is simply a melting and casting operation where the "cast" weld metal becomes integral with the "mold," the solid metal being welded. Necessarily, then, there will be a gradient of temperature from metal generally at ambient (room) temperature to liquid metal. For welding mild steel, a few of the metallurgical considerations implicit in this statement are as follows:

1. The weld metal should be low in gas content and low in oxides or carbon to avoid the liberation of excessive amounts of gases upon solidification of the liquid metal.
2. The weld metal will solidify very rapidly and will therefore be very fine-grained.
3. Metal adjacent to the liquid will be heated into the austenitic state and usually will be cooled very rapidly by adjacent cold metal.
4. The resulting quenching effect on the austenitized metal will result in brittle martensite unless the carbon content is very low (then the martensite is not brittle) or the hardenability is low (little of such alloying elements as Cr and Mo).
5. There will be a zone adjacent to the austenitized metal that has been heated to just below the A1 temperature. For an initially cold-rolled steel, this will be an annealed and locally softened zone subject to strain aging and (if the steel is hardened) tempering.

The wide variety of low-carbon steels currently available for use today almost all possess sufficient weldability to permit easy and economic fabrication by some current welding process. Variations in weldability between similar materials can also be accommodated by the proper choice of welding procedure. The welding process used and the procedure employed will therefore depend on the material welded, the joint geometry, and the conditions under which the weld must be made. Increasingly, the weld quality (actually weld-inspection technique) required will also influence this choice. The increasing use of sensitive nondestructive testing techniques in recent years has placed emphasis on welds of high quality.

In general terms, a high-quality weld is one in which the process, procedure, and/or filler metal will provide (1) a joint that is free of defects above a certain minimum size at the time of fabrication and (2) a joint that has acceptable mechanical properties in service. This first requirement, sometimes called fabrication weldability, is usually interpreted to mean that the joint is sufficiently free of slag inclusions, gas porosity, cracks, undercut, and other defects in weld size or contour to be able to meet code or specification requirements. The second requirement, sometimes called service weldability, is usually interpreted to mean that the properties of the weld metal and heat-affected zone are compatible with the base material. The means by which the two objectives are achieved differ for different materials, and sometimes a compromise is made to meet both requirements simultaneously.

Of the two, service weldability is usually not so difficult to achieve. Up to strength levels of about 860 MPa tensile strength in steels, the use of the appropriately alloyed weld metals and properly controlled welding processes will generally deposit weld metal and produce heat-affected zones that are compatible in properties to the base metals joined. This does not necessarily mean that the chemical composition of the weld metal will match the base metal or that the properties will be matching, but it does mean that they will be compatible. Unfortunately, the conditions that tend to produce increased strength in steels usually cause the fabrication and service weldability of the steel to be decreased. Except for the high-strength alloys, however, the most severe weldability requirement is usually the fabrication-weldability one.

Fabrication weldability is usually limited by the occurrence of weld-metal or baseplate hot shortness cracking and by baseplate heat-affected zone (cold) cracking. Hot cracking is cracking which occurs in weld metal (and to a lesser extent in the heat-affected zone) at temperatures close to the fusion temperature. It is usually caused by too high sulfur and/or carbon content or an inappropriate alloy-content level. It may be aggravated by weld sizes and geometries that are large and non-equiaxed. Cold cracking in steels occurs at temperatures below about 300°C only when (1) hydrogen gas, (2) restraint, and (3) a hard martensite microstructure are present. The higher the strength and the alloy (or carbon) content of a steel, the more likely it is to cold crack on welding. Micrograph 8.5 shows a cold crack in low-alloy, high-strength steel. As is typical, it lies in the weld-heat-affected zone next to the fusion line. For this reason such cracks are often called underbead, i.e., under the weld bead, cracks. In addition to these phenomena, the tendency to produce defects such as porosity, lack of fusion or penetration, and entrapped slag may also tend to increase with steel strength because the welding-energy input used for higher-strength steels must be held to lower levels to maintain service weldability.

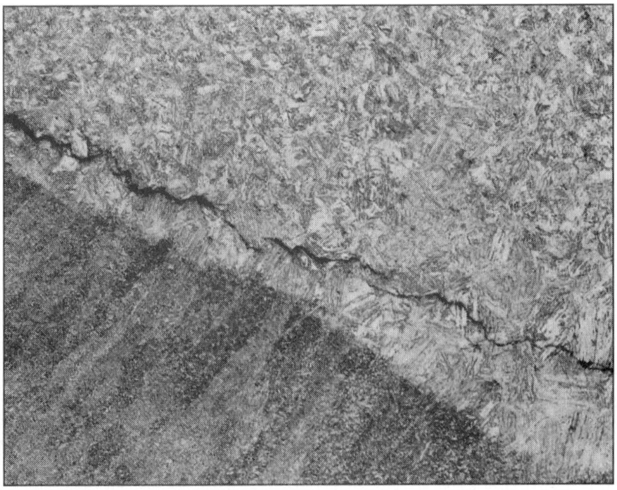

Micrograph 8.5 Cold (underbead) crack in low-alloy, high-strength steel; ×50, nital etch. This hydrogen-induced welding crack occurred at temperatures below 300°C in the heat-affected zone of low-alloy steel. Typically very fine-grained, columnar solidified weld metal is observed in the lower part of the micrograph.

A wide variety of welding processes are available for structural pipe and pressure-vessel construction; however, those used commercially in substantial amounts are limited to about half a dozen.

The currently applied welding processes have some common parameters that can be used to describe their characteristics. Probably the most widely used parameter, although not necessarily one on which there is universal agreement, is the welding heat input,

$$\text{Heat input } H = \frac{\text{welding current } I \times \text{arc voltage } V \times 60}{\text{travel speed } S}$$

where I is in amperes, V in volts, and S in inches per minute. The heat input has the units joules per inch of weld. This parameter is helpful because it can be used to define (1) the minimum and maximum operating conditions possible for a process, (2) the optimum range of welding conditions suitable for meeting fabrication-weldability requirements by a given process, and (3) the optimum range of welding conditions suitable for meeting service-weldability requirements for a given material. Unfortunately, the heat input is not the only parameter that influences the behavior of the steel or process: such things as preheat, shielding gas, and/or flux composition, joint geometry and thickness, restraint, and electrode size are also important. As a first approximation, however, welding heat input is probably the most useful parameter to characterize welding.

In general, steel weld metals are lower in carbon and higher in alloys, particularly manganese, than the base metal, which gives them good strength and toughness. The consumables such as the shielding gas may also influence the chemical composition of the weld metal, particularly with respect to oxygen, nitrogen, and hydrogen. Thus they become an "alloying" addition and must be controlled. Some welding processes, such as resistance welding, do not require the use of an external filler metal. These welds depend on the composition of the base metal to produce satisfactory weld metal.

8.6 SURFACE HARDENING OF LOW-CARBON STEEL

If a hard steel surface is desired, high-carbon steel can be used or much cheaper low-carbon steel can be heated in a carbonaceous atmosphere to increase the carbon content of the surface layers. The latter process, called carburizing,[4] can be carried out by heating the steel while it is packed in a mixture of charcoal and a catalyst such as $BaCO_3$. At elevated temperatures, for example, 900°C, the following reactions occur:

$$C + O_2 \text{ (initial air in charcoal)} \rightarrow CO_2 \qquad CO_2 + C \rightarrow 2CO \tag{1}$$

$$Fe + 2CO \rightarrow Fe \text{ (C in solution)} + CO_2 \qquad CO_2 + C \rightarrow 2CO \tag{2}$$

$$BaCO_3 \rightarrow BaO + CO_2 \qquad CO_2 + C \rightarrow 2CO \tag{3}$$

It is apparent that some oxygen atoms are required to transport the carbon atoms from solid pieces of charcoal to iron via gaseous CO. The catalyst, here $BaCO_3$, functions to increase the rate of supply of carbon atoms or the amount of the active agent, CO.

Carburizing of a large number of parts can be done more cheaply by heating in a closed furnace containing an atmosphere of hydrocarbon gas such as natural gas or methane. The reactions are

$$Fe + CH_4 \rightarrow Fe(C) + 2H_2 \qquad CH_4 \rightarrow C + 2H_2 \tag{4}$$

The gas must flow over the work to remove the hydrogen, or the reaction would go to equilibrium and stop. Too rapid a flow might result in the decomposition of the gas at the iron surface at a rate faster than the carbon is dissolved, in which case free carbon deposits as a soot. This effect can be eliminated by controlling excess hydrogen.

The equilibria of reactions (2) and (4) at varying temperatures are shown in Figs. 8.3 to 8.5.

A most significant fact revealed by a comparison of these graphs pertains to the relative effect of temperature on reactions (2) and (4). As the temperature is increased, the equilibrium constant for reaction (2) decreases; i.e., the carbon concentration in iron would decrease for a specific CO_2/CO ratio. In other words, to

[4]Not carbonizing, which is a process driving the hydrocarbons from bituminous coal and thereby converting it to relatively pure carbon.

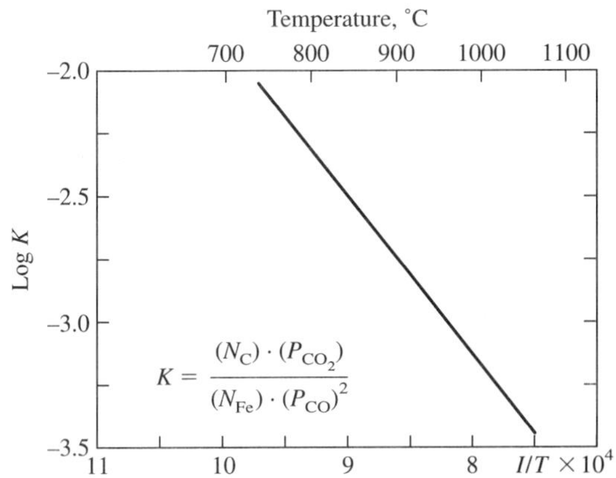

Figure 8.3 Relation between temperature T and the equilibrium constant K for the reaction $2CO + Fe \rightarrow C$ (in Fe) + CO_2. The decrease in K with increase of temperature means that the concentration of C in iron at equilibrium decreases for a fixed CO_2/CO ratio.

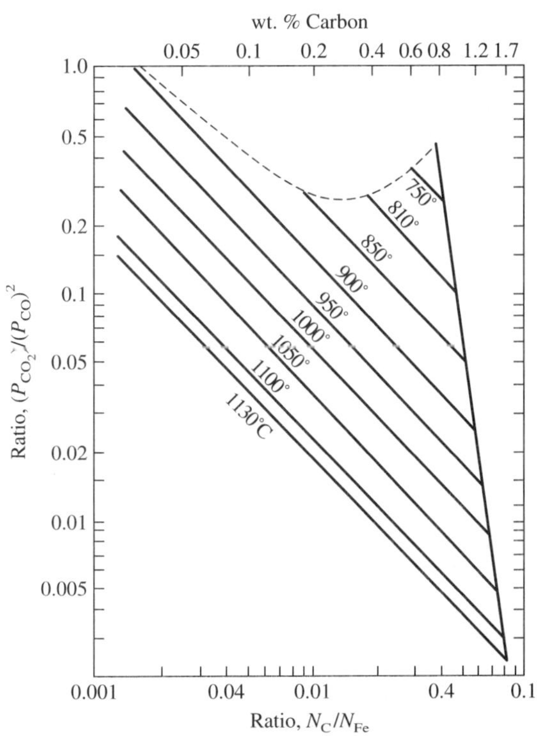

Figure 8.4 Equilibrium between carbon content of iron at austenitic temperatures and ratio of partial pressure of CO_2 to the square of the partial pressure of CO. Carbon content of iron is expressed as molal concentration (bottom) or weight % (top).

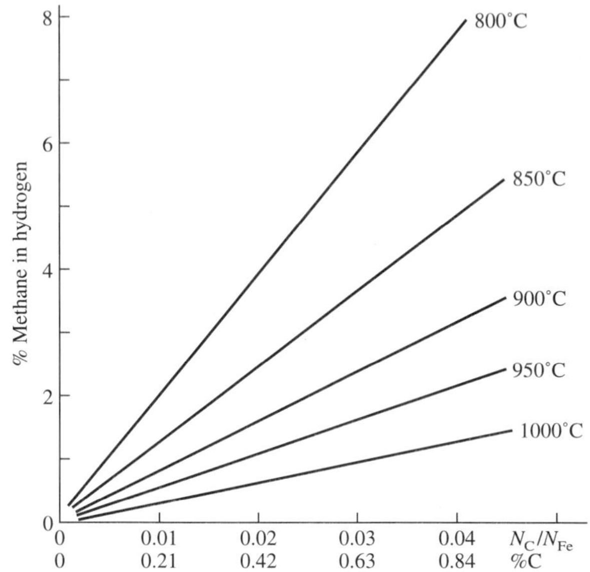

Figure 8.5 Equilibrium relationships between concentration of CH_4 in H_2 and C in Fe at different austenitic temperatures. At a specific methane content, the carbon content of iron increases with the temperature (opposite to temperature effect when CO is the carburizing agent; see Fig. 8.3).

maintain the same carbon concentration in iron, the proportion of CO in the gas must increase. The effect of temperature on carburization with methane is the reverse. Figure 8.5 shows that, for a specific CH_4/H_2 ratio, the carburizing power or the carbon concentration in iron increases with temperature.

If the steel is almost entirely ferrite, i.e., pure iron or a low-carbon steel below the Ac_1 temperature, the ferrite can absorb a maximum of 0.025% C and the external carbon supplied to the surface can form only a very thin surface layer of Fe_3C. If the steel is above the Ac temperature but below the Ac_2, the surface carbide can eutectoidally react with any adjacent ferrite that is present, to form austenite, but this intermediate reaction slows up carbon absorption, and carburizing in this temperature range results in thin, shallow cases with a very high surface-carbon content (up to 3.0% in the outer 0.1-mm layer). However, if the steel is initially in the austenitic condition, carbon at the surface is soluble to an amount exceeding 0.80% and is free to diffuse into the steel. The higher the temperature attained the deeper the carbon penetration and, correspondingly, the lower the surface-carbon concentration, since the carbon diffusion rate into the steel generally increases more rapidly than the increase in rate of carbon supply to the surface. In most commercial carburizing, the surface may contain about 0.80 to 1.0% C, with the concentration tapering off until, at a depth (usually 0.25 to 1.0 mm) determined by the time, temperature, and other carburizing-practice variables, the carbon content is that of the original low-carbon steel.

ASM Handbook, Volume 4, *Heat Treating* (see References), contains articles devoted to carburizing as well as to other surface treatments. However, it should be emphasized here that the limitations of these surface treatments are to a considerable extent explicable on the basis of the pertinent phase diagrams, particularly the solid-solution fields, since diffusion is impossible in the absence of solid solubility. Thus the concentration limits away from the immediate surface of the steel being treated are set by the phase solubility limits. The structures and properties attained after heat treatment of carburized steels follow the generalizations covered in Chap. 7 for steels of the appropriate range of carbon contents.

An important aspect of carburizing is the growth of the surface layer that results from the addition of carbon atoms. Even though the carbon dissolves interstitially, it is still too large for the interstices and somewhat expands the iron structure. Necessary dimensional adjustments with the core are readily made by austenite, which is plastic at the carburizing temperatures. On relatively slow cooling, however, the surface layer of higher carbon content expands more on transforming to pearlite than the core does, which transforms to ferrite, largely, and to pearlite. As a result, the surface layer is left with residual compressive stresses that are a desirable feature of carburized parts. Actually, carburized parts are practically always quenched, and the same types of residual compressive stresses are then obtained.

The surface of steel may lose carbon to its surrounding atmosphere as well as gaining carbon. Reaction (2) is reversible and may proceed to the left, removing carbon from the surface layer if the steel is heated in CO_2. Whether this decarburizing reaction or carburizing occurs depends on the ratio of CO_2/ CO or their partial pressures, as follows:

For the chemical reaction, $2CO + Fe \rightleftharpoons Fe(C) + CO_2$; the mass-action law states that, at a temperature T, an equilibrium state is reached as expressed by the equilibrium constant

$$K_T = \frac{[C(\text{in Fe})] \times [CO_2]}{[Fe] \times [CO]^2}$$

On the assumption that solid-state concentrations are constant and that the active concentrations of the gases CO and CO_2 are proportional to their partial pressures, the critical ratio becomes

$$\frac{P_{CO_2}}{P_{CO}^2}$$

Some other possible *decarburizing* reactions are

$$Fe(C) + H_2O \rightarrow Fe + CO + H_2 \tag{5}$$

$$Fe(C) + O_2 \rightarrow Fe + CO_2 \tag{6}$$

Decarburization is chiefly a problem with higher-carbon-content steels that are heat-treated after being fabricated to final shape. At the moment, it is sufficient to say that if the surface has low carbon content, it will have the properties of low-carbon steel.

PROBLEMS

1. Calculate the percentages of microstructural constituents in slow-cooled steels containing (*a*) 0.08% C, (*b*) 0.15% C, (*c*) 0.30% C, (*d*) 0.70% C, and (*e*) 1.3% C.
2. Why is a low-carbon-steel sheet preferred for automobile fenders, and under what conditions is it desirable to have this steel a killed type?
3. Why will a 0.20% carbon steel carburize more quickly at 875°C than ingot iron?
4. What would be the difference in structure of 0.20% carbon steel if it were (*a*) furnace-cooled from above the A_{cm} and (*b*) normalized from above the A_{cm}, reheated just above the A_{cm}, and then furnace-cooled?
5. If the hypoeutectoid steel of Micro. 8.1 had been deoxidized with aluminum, carburized, and then cooled very slowly from above the A_{cm}, the resulting microstructure would be abnormal, showing heavy carbide envelopes at the former austenitic grain boundaries and then a zone of free ferrite between the carbide envelopes and pearlitic areas. Give a possible explanation of the mechanism of the development of this abnormal structure.
6. At austenitic temperatures, a decarburized 0.8% carbon steel has a smooth gradient of % C vs. distance, increasing from 0.0% at the immediate surface to 0.8% C at some depth, for example, 250 micrometers. Describe, in terms of the A_3 line of the Fe-C diagram, the redistribution of carbon on very slow cooling that results in a free columnar ferrite band to perhaps 100 micrometers and, abruptly, a virtually complete pearlitic structure of nearly 0.8% C.

REFERENCES

ASM Handbook: vol. 1, *Properties and Selection: Irons, Steels, and High Performance Alloys*, 10th ed., ASM Int., Materials Park, OH, 1993.

ASM Handbook: vol. 4, *Heat Treating,* 10th ed., ASM Int., Materials Park, OH, 1995.

1999 Annual Book of ASTM Standards: vol. 01.04, *Steel—Structural, Reinforcing, Pressure, Railway*, ASTM, West Conshohocken, PA, 1999.

Dieter, G. E.: *Mechanical Metallurgy,* 3d ed., McGraw-Hill, New York, 1986.

Bain, E. C., and H. W. Paxton: *Alloying Elements in Steel*, 2d ed., ASM, Metals Park, OH, 1966.

chapter

9

Medium-Carbon Steels

Typically, iron may have a tensile strength of about 275 MPa, but the introduction of fractional percentage points of alloying elements such as carbon followed by heat treatment can produce as much as a tenfold increase in strength. Different heat treatments can produce different combinations of strength and ductility within these limits. While such alloying and heat treatment may be applied to steels of any carbon content, they are most widely applied to steels of medium and high carbon content to produce alloy steels and tool steels of high strength and hardness. The tool steels will be considered in the next chapter. This chapter is devoted to the medium-carbon alloys that are so widely used for machine parts and high-strength structural-component applications. These steels are often called engineering alloy steels.

Steel may be heat-treated for purposes other than the modification of its strength and ductility. Because the principles involved in such treatments have been discussed in previous chapters, this chapter is primarily concerned with heat treatments whose aim is the marked improvement of the strength of steel. Such heat treatment generally involves three distinct operations: (1) heating the steel to a relatively high temperature so as to convert it to austenite, (2) quenching (rapid cooling) of the hot steel to form martensite, and (3) tempering the martensitic steel by heating to a relatively low temperature so as to obtain the desired reduction in hardness and increase of ductility. As with most other hardening mechanisms, one usually cannot "have one's cake and eat it too"—in this case, high strength and high ductility simultaneously in the heat-treated steel. The proper combination of strength and ductility is critical to the usefulness of the engineering alloy steels. In the heat-treated medium-carbon steels, the balance of strength and ductility can be closely controlled and is one of the most satisfactory to be found in engineering alloys.

9.1 CLASSIFICATION OF MEDIUM-CARBON STEELS

While the low-carbon structural and alloy steels are usually, although not exclusively, referred to by the ASTM designation system, medium-carbon steels are almost universally referred to by their American Iron and Steel Institute (AISI) or equivalent Society for Automotive Engineers (SAE) designations. The system covers a broad spectrum of steels from very low carbon content (0.06% maximum) with only modest amounts of manganese as an alloy element up to high-carbon (1.1% maximum)

Table 9.1 AISI-SAE standard carbon steels

	Composition, %		
No.	**C**	**Mn**	**S**
Free-cutting (resulfurized)			
1110	0.08–0.13	0.30–0.60	0.08–0.13
1118	0.14–0.20	1.30–1.60	0.08–0.13
1119	0.14–0.20	1.00–1.30	0.24–0.33
1144	0.40–0.48	1.35–1.65	0.24–0.33
Free-cutting (resulfurized and rephosphorized)			
1211*	0.13 max	0.60–0.90	0.10–0.15
12L14†	0.15 max	0.85–1.15	0.26–0.35
Nonresulfurized grades‡			
1010	0.08–0.13	0.30–0.60	
1015	0.13–0.18	0.30–0.60	
1020	0.18–0.23	0.30–0.60	
1025	0.22–0.28	0.30–0.60	
1030	0.28–0.34	0.60–0.90	
1035	0.32–0.38	0.60–0.90	
1040	0.37–0.44	0.60–0.90	
1045	0.43–0.50	0.60–0.90	
1050	0.48–0.55	0.60–0.90	
1055	0.50–0.60	0.60–0.90	
1060	0.55–0.65	0.60–0.90	
1070	0.65–0.75	0.60–0.90	
1080	0.75–0.88	0.60–0.90	
1090	0.85–0.98	0.60–0.90	

* 0.07 to 0.12 P; = all others, P = 0.040 max.
† 0.15 to 0.35% Pb.
‡ P = 0.040 max. and S = 0.050 max.

Table 9.2 AISI-SAE standard alloy steels

No.	Composition, %*				
	C	Mn	Ni	Cr	Other
1330	0.28–0.33	1.60–1.90			
1340	0.38–0.43	1.60–1.90			
4023	0.20–0.25	0.70–0.90			0.20–0.30 Mo
4032	0.30–0.35	0.70–0.90			0.20–0.30 Mo
4130	0.28–0.33	0.40–0.60		0.80–1.10	0.15–0.25 Mo
4140	0.38–0.43	0.75–1.00		0.80–1.10	0.15–0.25 Mo
4320	0.17–0.22	0.45–0.65	1.65–2.00	0.40–0.60	0.20–0.30 Mo
4340	0.38–0.43	0.60–0.80	0.65–2.00	0.70–0.90	0.20–0.30 Mo
4422	0.20–0.25	0.70–0.90			0.35–0.45 Mo
4620	0.17–0.22	0.45–0.65	1.65–2.00		0.20–0.30 Mo
4720	0.17–0.22	0.50–0.70	0.90–1.20	0.35–0.55	0.15–0.25 Mo
4820	0.18–0.23	0.50–0.70	3.25–3.75		0.20–0.30 Mo
5120	0.17–0.22	0.70–0.90		0.70–0.90	
5140	0.38–0.43	0.70–0.90		0.70–0.90	
5150	0.48–0.53	0.70–0.90		0.70–0.90	
6150	0.48–0.53	0.70–0.90		0.80–1.10	0.15 V
8620	0.18–0.23	0.70–0.90	0.40–0.70	0.40–0.60	0.15–0.25 Mo
8630	0.28–0.33	0.70–0.90	0.40–0.60		0.15–0.25 Mo
8640	0.38–0.43	0.75–1.00	0.40–0.70	0.40–0.60	0.15–0.25 Mo
8650	0.48–0.53	0.75–1.00	0.40–0.70	0.40–0.60	0.15–0.25 Mo
9255	0.51–0.59	0.70–0.95			

*Small quantities of certain elements are present which are not specified or required. Considered as incidental, they are acceptable to the following maximum amounts: 0.35 Cu, 0.25 Ni, 0.20 Cr, and 0.06 Mo. All steels contain 0.035% P and 0.040% S max., and 0.20 to 0.35% Si.

steels and those containing over 4% alloy element. The low- and medium-carbon steels designated by this system are listed in Tables 9.1 and 9.2. In both cases, the system utilizes a four-digit number to designate the carbon and alloy content of the steel. The first two digits indicate the alloy content, and the last two indicate the carbon content. The steels with the initial digits 10, as in steel 1040, are often called plain-carbon steels and have only carbon and manganese as deliberate major alloy elements. They also contain minor amounts of other elements, such as silicon, and impurities such as phosphorus and sulfur, but these are set either at deliberate low levels or are as low as the particular steel-refining process used can produce.

The alloy steels are those that have initial digits other than 10, for example, 5140. In this case the digits are intended to designate that the material is a chromium alloy steel. The last two digits again indicate the carbon content, in the case of 5140, 0.40%.

Both single- and multiple-alloy steels can be designated using this system. The actual combinations of alloys used may seem arbitrary, but they have been established according to experience and to some rules of thumb concerning alloying. For example, it was felt in times past that an ideal balance of nickel to chromium was 2.5:1 for best-quenched and tempered properties. Thus the nickel-chromium alloy ratios tend to follow this formula in some compositions, for example, 43xx, 47xx.

There are other elements usually found in these steels that are not normally determined in a typical chemical analysis for the grade. Examples of such alloys are copper, tin, arsenic, and antimony. Although they may actually be present in amounts equal to or greater than deliberate alloy additions, e.g., copper, which is introduced in steel scrap and not removed in steelmaking, they are not considered alloys but impurities. As will be described later, some applications of these steels require that the impurities be held to quite low levels.

9.2 HARDENABLE CARBON STEELS

Low-carbon steels, e.g., with about 0.15% C but without alloys, are not usually considered to be hardenable by quenching from the austenitic state. However, plain 0.15% C steel sheet, quenched in water from 900°C, will have a tensile strength of 1030 MPa and a tensile elongation of 5%, whereas in the annealed and 50% cold-rolled state, it will have a tensile strength of only 700 MPa with less than 1% elongation. Thus by control of structure through heat treatment, the metal can be made stronger and more ductile simultaneously.

The hardenability of steel is determined by three factors, the grain size, homogeneity, and composition of the austenite.

1. **Austenite grain size.** Since pearlite is nucleated at austenite grain boundaries, fine-grained austenite tends to transform to pearlite more rapidly than coarse-grained. Steels with fine-grained austenite then tend to have less hardenability than coarse-grained steels.

2. **Austenite homogeneity.** Since the pearlite reaction is nucleated by carbide formation, residual carbides or localized areas of austenite rich in carbon cause the pearlite reaction to start sooner and thus contribute to shallow hardening. A high austenitizing temperature, by tending to produce a coarse-grained, homogeneous austenite, will thus give steel a higher hardenability than a low austenitizing temperature will.

3. **Austenite composition.** Alloying elements, which, in addition to carbon, must diffuse during the pearlite reaction, in most cases slow up the formation of pearlite and thereby increase hardenability. In the plain-carbon steel illustrated on p. 167, the nose of the C curve is relatively far to the left, and only a water quench of relatively small pieces will avoid pearlite formation. For this reason the steel is said to be water-hardening. The addition of a percent of manganese, chromium, or nickel slows up the pearlite reaction and displaces the P_s curve well to the right. This

displacement may be sufficient for a quench in oil to give a cooling rate fast enough to form a fully martensitic structure. The result is an oil-hardening steel. Addition of more of these alloying elements with perhaps vanadium or molybdenum can slow down the pearlite reaction to the point where the steel becomes air-hardening; i.e., even air cooling results in the formation of a large proportion of martensite in the structure. One of the principal reasons for using alloy steels is to obtain the desired hardenability.

Carbon steels containing more than 0.25% C are nearly fully hardenable by quenching; i.e., they can be put in a hard and strong martensitic state. Steels with from 0.25 to 0.55% C are generally used in the quenched and tempered state. A wide range of properties can be attained, between 48 HRC or tensile strength of about 1585 MPa and 20 HRC or tensile strength of about 760 MPa. Within this range, a heat-treated plain-carbon steel is chosen for heat treatment if:

1. Service temperatures are close to ambient, i.e., not above 260°C.
2. The section is thin enough to quench to martensite throughout, i.e., less than 12 mm thick.
3. Water quenching is permissible; i.e., quench distortion or sharp section-thickness changes are not factors.

Item 2 implicitly refers to hardenability, which is very sensitive to grain size and to chemical composition, namely, manganese, silicon, and residual elements such as chromium or molybdenum when these elements are in solution in the austenite. When these elements are all on the high side of the usual range, plain-carbon steels in thin sections may harden satisfactorily upon oil quench or will harden in thicker sections upon water quench.

At times a martensitic surface may be desired, for wear resistance, with a fine ferritic-pearlitic core structure for toughness. This is obtained of course by a water quench of a thicker section, e.g., a 25-mm-diameter bar, after heating the entire structure to the proper temperature just above the Ac_3 line. Alternatively, it can be obtained by heating only the surface layers of a thicker section to this temperature by using high-frequency induction or intense flames.

Not only can the surface alone be heated for hardening; it also is possible to heat just a part of an assembly, i.e., the bearing sections of a small crankshaft. Alternatively, it is possible to water-quench only a part of a section, e.g., wrenches, pliers, etc., for a time and then oil-quench the entire assembly, again obtaining a martensitic part and a fine pearlite part.

In all cases, the maximum martensitic hardness is determined by the carbon content of the steel, on the assumption that the austenitizing treatment brings substantially all the Fe_3C into solution in the austenite. In order to assure this, many structures, before the hardening heat treatment, are normalized to obtain a fine initial carbide structure.

After quenching of carbon steel, the structure must be tempered, the carbon content being assumed above 0.20%. (Low-carbon martensites are relatively ductile and therefore no tempering is required for most kinds of engineering service.) The reduction of hardness and tensile strength is a nearly linear function of temperature up to 650°C. There is a corresponding nearly linear increase of ductility as indicated by tensile elongation and reduction of area at fracture. The scatter of properties observed when thousands of tests are conducted will generally be an indication of variability of structure associated with marginal hardenability. Small amounts of pearlite will always be detrimental.

Even though the quenched and tempered steel may have the same tensile strength and hardness as a fine-grained normalized steel, the quenched and tempered structure is much superior for certain applications requiring toughness, i.e., the ability to absorb large amounts of energy without brittle fracture. The fine spheroidal carbide particles developed upon tempering martensite do not initiate internal cracks, whereas lamellar Fe_3C crystals in pearlite of the same carbon content will have a greater surface area and, being brittle, will initiate cracks. Hence, for certain services, the expenses of hardening and tempering are warranted.

9.3 HARDENABLE ALLOY STEELS

The mechanical properties of medium-carbon (0.25 to 0.55% C) alloy steels at room temperature are not distinguishable from those of plain-carbon steel having the same ferrite structure and same carbide particle size, amount, and distribution. The major reason for the use of most alloy steels is that the most desirable and toughest microstructure, tempered martensite, is attainable with carbon steels only in thin sections. To obtain such structures in heavier sections, the hardenability of the steel must be increased, and the only practical means of doing this to a marked degree is by introducing alloying elements.

While carbon steels have been considered as binary Fe-C alloys, they actually are much more complex, since up to 1.00% Mn and 0.30% Si may be present. These change the eutectoid reaction from being invariant to being bivariant; i.e., either the composition of the reacting phases or the temperature can change from those shown by the Fe-C phase diagram. While this is true, the changes for carbon steels are small. However, for alloy steels, the changes frequently are significantly large, as is shown by the graphs of Fig. 9.1 for Mn, Si, Ni, Cr, Mo, and W. It will be observed that all these elements reduce the carbon content of eutectoidal austenite. Nickel and manganese lower the eutectoid temperature, while the others raise it. All of them of course change the eutectoid temperature from a single value (at equilibrium) to a range of temperatures.

All these elements will be present in at least two of the three phases shown in the Fe + C diagram, namely, the austenite, ferrite, and carbide. Below the A_1 temperature, nickel and silicon will be present solely in the ferrite and will cause, among other effects, some solid-solution strengthening. Manganese, chromium, and molybdenum form carbides and under equilibrium conditions will be partitioned between the ferrite and carbide phases. Within the limits of the amounts of these elements employed in the standard alloy steels of Table 9.2, however, the alloying elements

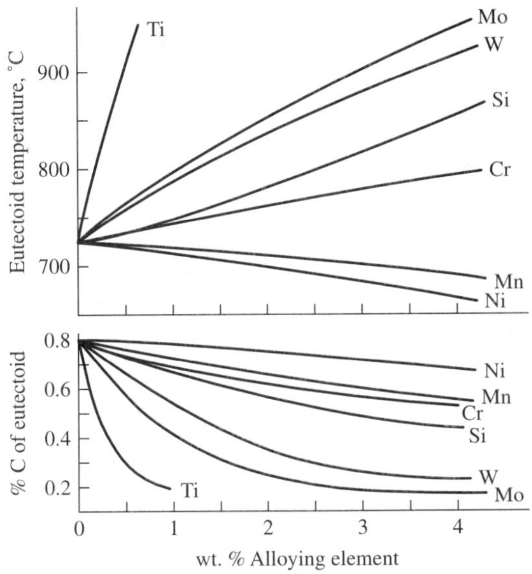

Figure 9.1 Effect of some common alloying elements on the eutectoid temperature and eutectoid carbon content of the Fe-C system alloys.

do not change the crystal structure or basic formula of the carbide, Fe_3C: atoms of the alloy element merely replace a small fraction of the iron atoms.

Because atoms of the alloying elements are randomly distributed in solid solution in homogeneous austenite and must be allowed some diffusion time to be redistributed and permit ferrite and pearlite to form, all these elements cause substantial changes in the pertinent alloy-steel transformation diagrams and, relatedly, substantial increases in hardenability. Typical effects are reproduced for a 3½% Ni steel (formerly AISI 2340) (Fig. 9.2); 5140 or Cr steel (Fig. 9.3); and 4340 or Cr-Ni-Mo steel (Fig. 9.4).

It will be observed that nickel does not change the appearance of the transformation diagram but merely lowers the A_3 and A_1 temperatures, slightly retards the initiation of the pearlite reaction, and also slows its rate; i.e., more time is required to go from start to finish at any temperature.

For the Cr or 5140 steel, the transformation diagram is changed in general appearance. Instead of a single C curve, there appear to be two C curves, with one at higher temperatures for the pearlite reaction and a second C curve at 600 to 300°C for the bainite reaction. For steels with this type of transformation diagram, it is possible to form some bainite upon direct cooling of austenite without isothermal treatment. However, because of the slope of the bainite start B_s and finish B_f lines in Fig. 9.5 an isothermal treatment would still be necessary if a completely bainitic structure were desired.

The Cr-Ni-Mo, or 4340, steel transformation diagram (Fig. 9.4) shows features similar to Fig. 9.3, but the pearlite reaction is much more delayed and, therefore, the probability of getting some bainite for a continuous, moderately slow cool is increased.

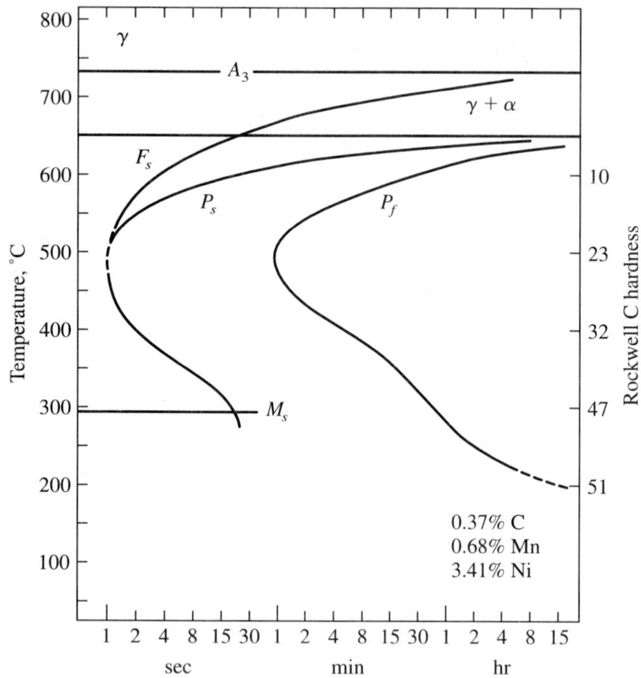

Figure 9.2 Isothermal transformation diagram for a 3½% Ni steel (formerly AISI 2340) austenitized at 790°C; grain size 7 to 8. Rockwell hardnesses of the completely transformed structures are shown.

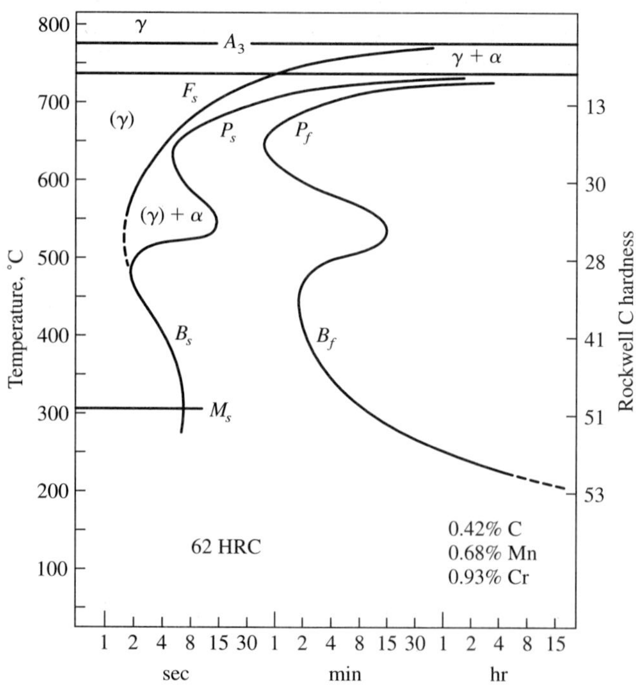

Figure 9.3 Isothermal transformation diagram for steel AISI 5140; austenitized at 845°C; grain size 6 to 7. Rockwell hardnesses of the completely transformed structures are shown.

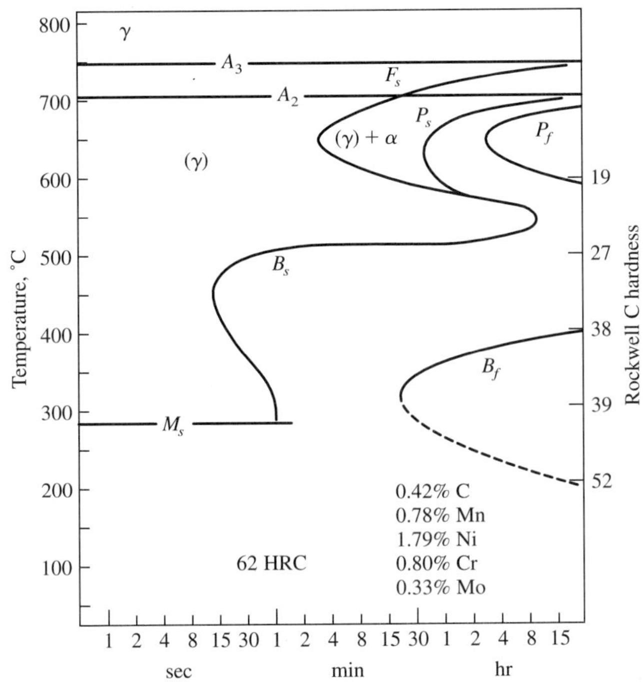

Figure 9.4 Isothermal transformation diagram for steel AISI 4340; austenitized at 845°C; grain size 7 to 8. Rockwell C hardnesses of completely transformed structures are shown.

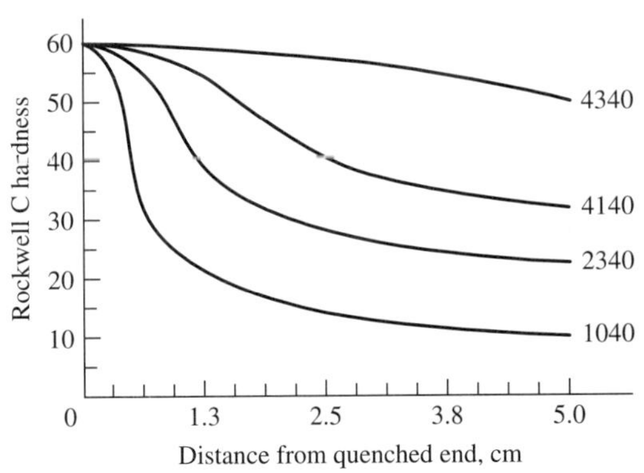

Figure 9.5 Typical hardenability curves for four medium-carbon steels austenitized at 845°C from an initial normalized condition.

Typical end-quench curves for these steels, in comparison with plain-carbon 1040 steel, are given in Fig. 9.5. Note that the martensitic end hardness is the same for all steels. However, it differs appreciably away from the quenched end, and whereas the structure 5 cm from this end would be ferrite and pearlite in the 1040 and nickel steels, it would be ferrite, pearlite, and bainite in 4140 steel and martensite plus bainite in the 4340 steel.

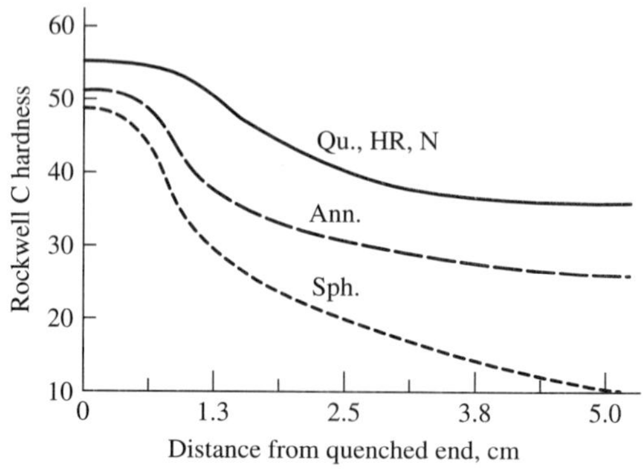

Figure 9.6 Effect of prior structure on Jominy hardenability of AISI No. 4140 steel (0.40% C, 0.90% Cr, 0.20% Mo). The prior structures were: (Qu) quenched from 945°C; (HR) hot-rolled in austenitic state to 25-mm rod and air-cooled; (N) normalized by air cooling from 845°C; (Ann) annealed by furnace cooling from 845°C; (Sph) spheroidized by normalizing and reheating at 700°C for 24 h. All bars were then heated 10 min at 845°C and end-quenched. (When bars were heated 4 h at 845°C and end-quenched, their hardenabilities were the same and corresponded to the top curve.)

The Cr-Mo and Cr-Ni-Mo steels are more susceptible to incomplete austenitization than the other steels, as shown in Fig. 9.6. The carbides containing chromium and molybdenum, if initially in coarse form in spheroidized steel or, to a lesser degree, in pearlite annealed steel, may not dissolve completely if held too briefly or at too low an austenitizing temperature. If the carbides are not completely dissolved, full martensite hardness is not obtained nor is the normal hardenability found because of the lower carbon and alloy content of the austenite.

As in the case of plain-carbon martensite, the alloy-steel martensites must be tempered to reduce microstresses and increase ductility to a serviceable level. Typical tempering curves for several alloy steels at the 0.45% C level are shown in Fig. 9.7. Elements dissolved in ferrite have little effect on the martensite-softening curve, and any small effect observed is associated with solid-solution strengthening of ferrite by dissolved nickel, silicon, and manganese.

Chromium, molybdenum, and vanadium atoms, on the other hand, tend to diffuse to the carbide phase when the temperature rises to the point where substitutional atoms can diffuse at an appreciable rate, in the vicinity of 300 or 400°C. At lower temperatures, softening rates are not too different from those of carbon steels, as carbon diffusion permits some Fe_3C particles to grow and others necessarily to dissolve and disappear. However, when the carbide is $(Fe, Cr, Mo)_3C$, chromium and molybdenum atoms must also diffuse for the carbide to grow. The slower diffusion of the substitutional solutes causes a retardation in the rate of softening. Therefore, for specific tempered hardnesses of under 40 HRC, the alloy steels require somewhat higher tempering temperatures or much longer tempering times. This means that the alloy steels will show more complete relaxation of

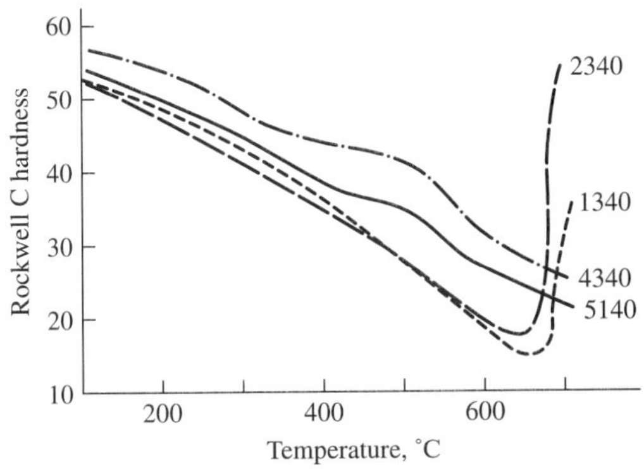

Figure 9.7 Tempering curves for four medium-carbon alloy steels (1340, 1.7% Mn; 2340, 3.0% Ni; 5140, 1.0% Cr; 4340, 1.7% Ni, 0.8% Cr, 0.3% Mo). Tempering time for all curves was 20 h. The sharp rise in hardness of 1340 and 2340 steels above 660°C is because Mn and Ni lower the A_C temperature and since the alloys were quenched from the tempering temperature, new martensite formed from whatever austenite was present above 660°C.

microstresses and a greater toughness than a quenched carbon steel tempered to the same hardness.

Beyond considerations of hardenability, specific steels are selected for various types of service on the basis of the contributions of their individual alloy elements for the special conditions under which the steels must perform. These considerations are secondary to the strength of the steel in most cases but may nonetheless be important. For example, chromium in solution in ferrite or austenite contributes improved corrosion resistance. While the chromium levels used in the engineering alloy steels are well below those used in stainless steels, there is a substantial decrease in corrosion rate because of the presence of chromium, and this can be a reason for selection of chromium-containing engineering alloy steels over others. Similarly nickel is often used because it improves toughness in the quenched and tempered steel.

Molybdenum is used because it improves machinability at high hardness levels, improves creep resistance, and slows development of temper embrittlement. This last phenomenon is of considerable importance because the tempering process is intended to improve ductility and toughness in martensitic steels. During the tempering process, however, some steels begin to show an increasing impact transition temperature with time, i.e., loss of toughness, especially when held in the 425 to 525°C range. Steels already tempered (at higher temperatures) will show this effect when slowly cooled through this range. It is not detectable by a change in tensile properties or in hardness. The mechanism for embrittlement has been studied for over 75 years, but no final single mechanism has been determined. It is known that it occurs because of interactions between major alloy elements like Cr, Mn, and Ni and impurity elements such as P, As, Sn, and Sb, which enter the steel with the scrap during steelmaking. The embrittlement is reversible and can be eliminated by heating the embrittled steel above 535°C. Retempering at 500°C will restore the

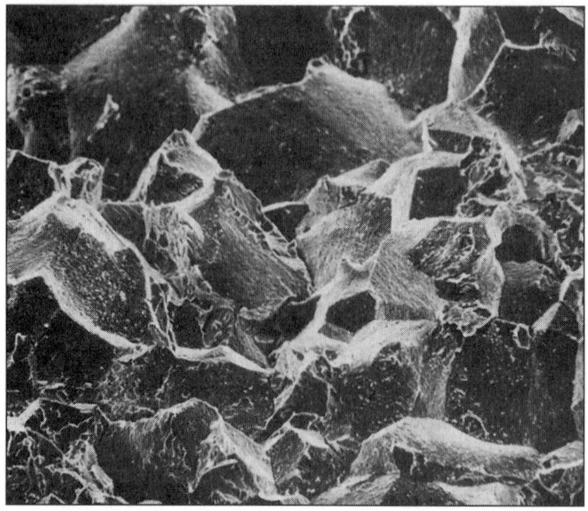

Micrograph 9.1 Fracture surface of impact-broken specimen of temper-embrittled low alloy steel; scanning electron micrograph at ×500, no etch. Fracture occurred along prior austenitic grain boundaries of this quenched and tempered steel.

embrittlement. Ambient-temperature toughness can be reduced to low values, and thus the effect of Mo in slowing the embrittling process makes it a valuable addition to steels tempered in the critical range. The fracture surfaces of a temper-embrittled steel tested below its transition temperature show almost complete intergranular fracture rather than cleavage. The fracture surface of an embrittled steel is shown in Micro. 9.1.

To summarize, medium-carbon alloy steels are industrially important, not because they are necessarily stronger in the quenched and tempered state than plain-carbon steels, but because:

1. They can be cooled more slowly, namely, in oil rather than in water when quenched, which minimizes the distortion and cracking tendencies associated with large temperature gradients and the related volume contractions of cooling vs. volume expansions during the austenite-to-martensite transformation.
2. They can be made fully martensitic (or have a high martensite content, for example, 90%) at the center of relatively thick sections; thus in such sections the toughness advantages of a tempered martensitic structure are gained.
3. Those grades containing Cr, Mo, and/or V are more resistant to tempering, which means that they will not further soften at service temperatures that would soften carbon steels. It also means that at the same tempered hardness in the range 30 HRC to 45 HRC the alloy steel will be more completely stress-relieved and therefore somewhat tougher.
4. Implicit in the above is that an alloy steel of somewhat lower carbon content than a comparable carbon steel may be used in order to obtain the same quenched and tempered hardness and still permit adequate tempering because of the retardation of softening upon tempering.

9.4 AUSTEMPERING AND MARQUENCHING

Austempering is the name of a special heat-treating process that involves isothermally transforming austenite at a temperature between the range of fine pearlite and martensite formation. The structures developed in this range are called *bainites*. The transformation graphs show the range of hardness of bainitic structures. In the lower part of the temperature range, bainite has a hardness directly comparable with that of tempered martensite. Since this structure forms directly from austenite at a higher temperature than martensite, the microstresses that develop in bainite are much lower.

Bainitic structures with hardnesses exceeding 50 HRC have high plasticity for hardened steel. For example, a 0.5-cm section may have a 35% reduction of area (RA) and an absorbed impact energy of 48 J as bainite. On the other hand, a tempered martensite with equivalent hardness would only have a 1% RA and 4 J. However, it is only in this hardness range that bainites show superior properties and only in the case of carbon steels. In the range of 40 to 45 HRC, the tempered martensite is superior for the same reasons it is better than fine pearlite. Bainitic structures cannot be obtained in the hardness range required for cutting tools, 60 to 65 HRC. Other limitations of the process involve size requirements. Sections must be thin enough to cool past the nose of the C curve and rapidly enough to avoid the formation of fine pearlite. More expensive equipment is required to quench into a 250 to 400°C bath and hold at a given temperature. Also, variations in different heats of steel result in variable transformation conditions.

Marquenching

The ductility and toughness of tempered martensite is superior to direct transformation structures in the important hardness range of 25 to 45 HRC. Hardenability also becomes an important issue since it is desirable to be able to quench a thick section directly to martensite. This effect increases the frequency of problems associated with distortion, residual stresses and cracking, which result from the thermal gradients necessary for rapid cooling.

Examination of Fig. 9.2 reveals a method of reducing thermal gradients and the associated problems. Immediately above the M_s line, austenite is metastable for a period of time. Therefore, instead of quenching to room temperature, it is possible to quench in a liquid at a temperature above M_s, until the piece is thermally stable. Then the steel may be removed from the hot liquid and slowly cooled below the M_s temperature. No thermal macrostresses are superimposed on the microstresses induced by the austenite to martensite transformation. Distortion, residual stresses, and cracking are almost eliminated. This process is termed *marquenching,* although it was called martempering at one time.

The word martempering was chosen by the originator of the process, presumably because it results in low stresses analogous to a tempered martensite. However, the term is misleading since it is structurally different than tempered martensite. There are no precipitated carbides and, while the macrostresses are eliminated, there are still significant microstresses that must be reduced with the usual subsequent tempering operation.

9.5 ULTRA-HIGH-STRENGTH STEELS

In this section, attention is directed to the metallurgical means of obtaining maximum useful strength from steels for use in a moderate temperature range, for example, −110 to +185°C, that is, excluding the high-temperature requirements of 3220 to 4830 km/h aircraft or of space vehicles subject to atmospheric reentry. There are structural requirements within the above more limited temperature range that are now being met by alloy steels treated to have 1725 to 2415 MPa strengths with sufficient ductility to be used safely.

AISI 4340 has been known and used for years because of the high-strength properties resulting from standard oil quenching and tempering treatments. AISI 4330V is essentially the same steel modified with vanadium, which raises the coarsening temperature so that the heat-treated steel tends to have a finer grain size.

AISI H11 is a hot-worked die (tool) steel with a carbon content sufficiently low to show acceptable ductility; it is now being used for rocket-booster cases and other structural uses. It contains enough Cr and Mo to be air hardening and to show secondary hardening upon tempering (Chap. 10). Welding of cases of this steel requires preheat and postheat operations to minimize cracking associated with martensite formation and thermal stresses.

The 18Ni (300) is one of the newly developed maraging steels that basically are low-carbon high-nickel-iron martensites, further hardenable by cold work and precipitation during aging. With 18% Ni, the steel does not need to be quenched from the austenitic state at 820°C but may be slowly cooled, with a resultant mixed martensite-austenite structure at room temperature. Being a very low-carbon martensite, this structure can be cold-rolled to 80 or 90% without cracking. Subsequent aging at 475°C results in precipitation of a phase based on the Co-Ti components of the structure. Since the final structure is obtained by a relatively low temperature treatment, is immune to decarburization, appears to be readily weldable, is insensitive to notches, and has high fracture toughness, it is particularly suitable for high-strength pressure vessels, components, and other structures.

The HP9-4-30 is a quenched and tempered steel like the H11 but with high hardenability and toughness. In this case the cobalt serves to adjust the M_s and M_f temperatures to values above the ambient as well as having a solid-solution strengthening effect in the tempered structure.

The 17-7PH is essentially a stainless steel that has had increasing use for its combination of strength and corrosion resistance. It is precipitation-hardened by the formation of nickel-aluminum compounds and requires a three-step heat treatment. The 17-4PH is also a high-strength precipitation-hardening stainless steel that is popular for its strength and corrosion characteristics. These two alloys are discussed in more detail in Chap. 11.

Thermomechanical Processing A type of extremely high-strength steel has appeared as a result of the concept of deforming unstable austenite of moderately alloyed steels at a temperature below the A_1 but above the lower bainite reaction temperature. The technique is applicable to many steels but most readily to the Cr-Mo steels such as the

H11 hot die steel (listed in Table 9.3). Ausforming is the process of plastically deforming austenite at a temperature within the "bay" that exists between the pearlite and bainite reactions (for example, Fig. 9.4), followed by cooling to room temperature. This will give much higher strengths than are otherwise obtainable with such steels.

Table 9.3 Compositions and properties of certain ultra-high-strength steels

	Composition, %						
Steel	**C**	**Mn**	**Si**	**Ni**	**Cr**	**Mo**	**Other**
4330V	0.30	0.90	0.30	1.8	0.80	0.40	0.07 V
4340	0.40	0.85	0.20	1.80	0.75		0.25
300M	0.40	0.75	1.60	1.85	0.85	0.40	0.08 V
H11	0.40	0.35	1.0		5.00	1.40	0.45 V
18 Ni	0.03	0.10	0.11	18.5		4.50	7.0 Co, 0.22 Ti, 0.003 B
9-4-30	0.31	0.25	0.10	7.50	1.00	1.00	0.11 V, 4.5 Co
17-7PH	0.09	1.00	1.00	7.10	17.0		1.0 A1
17-4PH	0.07	1.00	1.00	4.00	16.5		4.0 Cu, 0.30 Ti + Nb

	Properties				
AISI	**Yield Strength, MPa**	**Tensile Strength, MPa**	**Elongation, %**	**Charpy at 25°C, J**	**Endurance Limit 10^6 cycles, MPa**
4330V	1415	1620	12	34	
4340	1860	1980	11	145	
300M	1670	1995	10	30	
H11	1660	2035	6.6	20	900
D6AC	1725	1960	7.5	14	760
18 Ni	1850	1900	11	31	
17-7PH	1415	1655	9	8	795
17-4PH	1275	1380	13	26	620

	Heat Treatments		
Steel	**Austenitized, °C**	**Tempered, °C**	**Aged, °C**
4330V	870, oil quench	315	
4340	845, oil quench	205	
300M	870, oil quench	315	
H11	925, oil quench	315	
D6A	900, oil quench	315	
18 Ni	815, air cool		480
17-7PH	1065, air cool	760 air cool	565 air cool
17-4PH	1035, oil quench		480

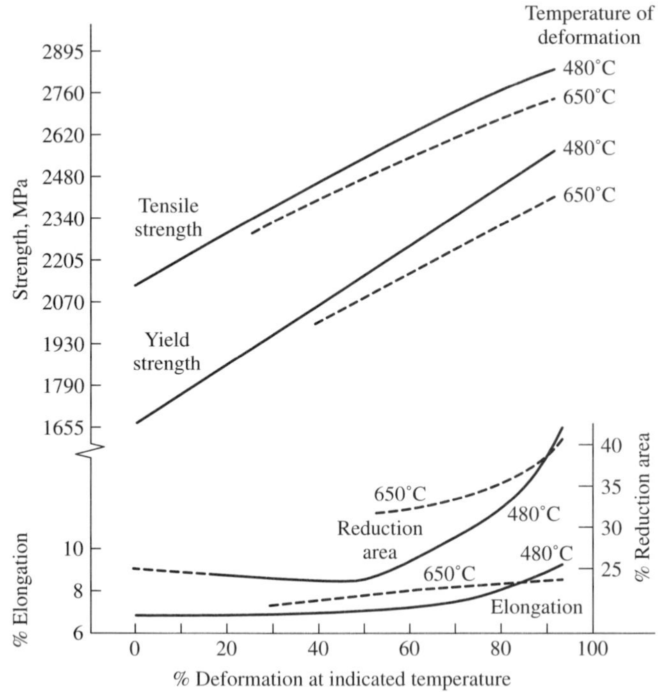

Figure 9.8 Effect on yield strength and tensile strength of deforming austenite of H11 Cr-Mo die steel, austenitized at 1040°C, cooled to the indicated temperature, deformed 50, 75, or 90%, cooled to room temperature, and tempered twice (double-tempered) at 510°C.

Figure 9.8 shows that the tensile properties attainable for H11 steel in sheet form may be increased from the 2070-MPa level to the vicinity of 2760 MPa, maintaining a useful level of ductility, by a 94% deformation in the vicinity of 475°C. Other data support the indication given by this figure that the exact temperature of deformation is not critical; the steel analysis is such that the TTT graph has a metastable austenite area with an appreciable time lag before the pearlite or bainite reactions start. Likewise, it has been found that austenitizing temperature and austenite grain size are not significant variables in this ausforming process. On the other hand, carbon content in the range 0.20 to 0.60% plays the same vital role here as in the case of conventionally processed steels, with increase in carbon linearly increasing the strength and decreasing the ductility determined after tempering. Tempering of ausformed steels is not noticeably different from that for the same steels conventionally hardened, except for some evidence of the effects of a predictable decrease in the amount of retained austenite prior to tempering.

Ausforming of high-hardenability steels is actually a particular case of a more general processing and heat-treatment sequence known as thermomechanical treatment. The types of thermomechanical treatments available for use on ferrous materials are listed in Table 9.4. Conventional treatments are of the Class Ia type. The

Table 9.4 Classification of thermomechanical treatments

Class I Deformation Before Austenite Transformation
- (*a*) Normal hot-working processes
- (*b*) Deformation before transformation to martensite
- (*c*) Deformation before transformation to ferrite-carbide aggregates

Class II Deformation During Austenite Transformation
- (*a*) Deformation during transformation to martensite
- (*b*) Deformation during transformation to ferrite-carbide aggregates

Class III Deformation After Austenite Transformation
- (*a*) Deformation of martensite followed by tempering
- (*b*) Deformation of tempered martensite followed by aging
- (*c*) Deformation of isothermal transformation products

ausforming treatment described above is a procedure falling under the category of Class Ib, and is one of the more dramatic of the thermomechanical treatments in terms of property enhancement. The improvement apparently results from interaction of carbon atoms with lattice defects induced in the austenite and inherited by the martensite. Treatments of the Class Ic type result in fine austenite grain size and thus a favorable balance of strength and ductility only if hot-working conditions are carefully controlled. Some plate and pipe materials are processed by rolling using Class Ic procedures for enhancement of notch toughness, especially in Europe.

Class IIa is most often applied to austenitic stainless steels. The strengthening results from both transformation to martensite and strain hardening of both austenite and martensite. Class IIb results in improved strength through microstructure refinement and perhaps some dispersion strengthening. Class IIa and IIb treatments produce strengthening by work hardening of martensite followed by precipitation phenomena, either as solute atmospheres around structure defects or as a more dispersed carbide structure. Class IIIc strengthening occurs primarily by dispersion strengthening.

Some of the thermomechanical treatments have been used for a number of years, such as parenting of wire, which is a treatment of the Class IIIc type, while others, such as ausforming, are more recent. The requirement for high-strength steels in advanced technology applications will probably promote increasing use of thermomechanical processing in the future.

9.6 SPECIAL PROCESSING OF STEEL

Wrought steels at ordinary temperatures contain, in addition to the normal metallic alloying or impurities, oxygen in the form of microscopic oxides, nitrogen present in interstitial solution and as very fine nitrides, and hydrogen also in interstitial solution. The coarse oxides, elongated by hot working, are inherently weak and brittle, forming internal discontinuities that weaken the steel. This is especially true in a direction normal to their long axes. During cyclic stressing or vibration, these discontinuities can initiate cracks that propagate until the material fails by fatigue. Therefore, while oxides may not noticeably impair normal longitudinal tensile strength and ductility, they affect fatigue strength, transverse ductility, and especially fracture toughness.

Relatively coarse oxides cannot be removed by any practical process from solid steel. Hydrogen will diffuse out if the metal is supersaturated at a low temperature. To avoid the embrittling effect of hydrogen at room temperature, it should be removed at around 150 to 300°C, where the solubility is quite low. Unfortunately, the diffusion rate is also low, so the hydrogen can only be removed in a reasonable time (a day or two) from thin sections.

The best way to eliminate gaseous elements is by vacuum processing of the steel while in a liquid state. In a vacuum, the oxygen in liquid steel will react with carbon to form carbon monoxide gas. As long as the gaseous product of the reaction is removed by continued evacuation, the reaction will go to completion. Almost all of the oxygen will be removed and the solid steel will be free of oxide inclusions. Obviously, this degree of removal will not be achieved except with impractically high vacuums and long times, especially when metallic elements such as aluminum are present, which have a high affinity for oxygen. However, there are several practical processes for vacuum-processed steel.

Air-melted steels can be poured into an evacuated ladle, effectively "spraying" small globules having large surface-to-volume ratios. When the vacuum is below 1 torr pressure, the material will be effectively, if not completely, degassed. The process can be used for large heats (100 to 200 tons) of liquid steel.

Another approach is to move the liquid steel hydraulically upward from a ladle into an evacuated refractory chamber using the pressure of an inert gas and then reverse. This efficient Dortmund-Horder (D-H) process is repeated several times, exposing new surfaces and degassing continuously during the cycling. The similar R-H vacuum degasser causes the liquid steel in a ladle to flow continuously through a refractory pipe into the vacuum chamber and then back through a refractory pipe into the ladle. By this process, the oxygen in a 300-ton heat can be reduced from 400 to 50 ppm in 20 minutes.

Other processes may result in more complete degassing but are more costly and are limited to smaller volumes. They include melting the steel under vacuum in an induction-heated crucible, by a process known as vacuum induction melting (VIM), or by a secondary technique, vacuum arc remelting (VAR). In the latter, an arc is struck between the VIM electrode and the bottom of a water-cooled copper crucible, allowing the electrode material to drip into the crucible. Before the molten material solidifies, inclusions and oxides may be removed by flotation, chemical, or physical processing. The resulting VIM-VAR steel has superior cleanliness and microstructural homogeneity. One drawback is that metallic elements with high vapor pressures such as chromium and manganese, may be significantly reduced through evaporation in these high vacuums.

With some structural steels, the transverse ductility and toughness are decreased by the presence of sulfides in the steel. This becomes more prominent when the sulfides can be flattened during rolling processes, as in plates. To reduce this effect, steelmaking processes have been improved to the point where phosphorus and sulfur levels below 0.010% are commonly produced. As a further step, it is possible to reprocess the steel by remelting it into a water-cooled copper crucible, similar to the VAR process. The difference is that this process is conducted in air under a molten slag. The energy is electrical and the steel is melted through the slag by resistive heating with the steel ingot and copper crucible acting as electrodes. The process is called electroslag remelting (ESR) and is applied to sizable heats, resulting in high-quality slabs.

An alternate approach to remelting is to restrict the sulfides to less harmful forms using small additions of rare earth elements. These additions strengthen the sulfide phase and, as a result, do not readily flatten during rolling. Thus, the sulfides present are not as detrimental as they would be in the form of elongated stringers.

PROBLEMS

1. Justify the use of heat-treated 4340 steel for jet aircraft landing gear components in preference to the alternative of strong aluminum alloy forgings. Use strength-weight ratios and elastic deflection under impact loading as criteria.
2. What is the difference in the shape of inclusions in hot-rolled alloy-steel round bars vs. the same hot-rolled steel in flat plates? For a plate, which is the direction of least ductility: rolling direction, transverse in the rolling plane, or normal to the rolling plane?
3. What are the relative problems of electric arc welding of liquid-oxygen rocket cases of standard heat-treated 4340, vacuum-melted heat-treated 4340, and the 18% Ni maraging steel of Table 9.3?
4. Why would a normalized 2.5-cm bar of 4340 steel be difficult to machine and what is probably the lowest-cost process to impart relatively good machinability to such a rod?

REFERENCES

SAE Handbook, vol. 1, *Materials,* Society of Automotive Engineers, 1989.

Steel Products Manual, "Alloy, Carbon and High Strength Low Alloy Steels: Semifinished for Forging, Hot Rolled Bars and Cold Finished Bars, Hot Rolled Deformed and Plain Concrete Reinforcing Bars," AISI, March 1986.

ASM Handbook, vol. 1, *Properties and Selection: Irons, Steels, and High-Performance Alloys,* ASM Int., Materials Park, OH, 1990.

ASM Handbook, vol. 3, *Heat Treating,* ASM Int., Materials Park, OH, 1991.

The Metals Black Book, 3d ed., CASTI Publishing, distributed by Society of Automotive Engineers, Washington, DC, 1998.

chapter

10

High-Carbon Steels

The production of practically everything in our industrial economy requires tools of an almost infinite variety, including cutting tools, dies for forming or shaping many kinds of materials, and, necessarily, gages to ensure satisfaction of dimensional tolerances. The specific requirements for tools vary with the service demands on them, but generally they include high hardness to resist deformation, resistance to wear in order to achieve an economic tool life, dimensional stability, etc.

The basic tool materials are high-carbon or tool steels, of which there are at least twenty, when classified by types of composition. While the total tonnage of tool steels produced per year is relatively small, their value per pound may be fifty or more times that of the high-tonnage steels and their economic value is incalculable.

In addition to steels used for tools, other materials, particularly cemented carbides, have been developed for special service. Their structures and properties are also important to those concerned with tool materials.

10.1 CLASSIFICATION OF HIGH-CARBON STEELS

Steels used for tools generally have at least 0.60% C in order to assure attainment of a martensitic hardness of at least 60 HRC. Carbon in excess of this minimum is employed only to have undissolved carbides in the martensitic structure to increase the resistance to wear. Alloying elements may be added for specific structural or property effects.

Table 10.1 lists the composition of the principal types of tool steels. The water-hardening plain-carbon grades have such low hardenability that sections thicker than 1.3 mm can be quenched to martensite only in the surface layers, so that the interior has a softer but relatively tough fine pearlitic structure. The shock-resistant grades have small amounts of chromium or molybdenum which somewhat increase

Table 10.1 Classification of high-carbon steels

AISI-SAE Designation	Composition, %							Typical Uses
	C	Mn	Cr	V	W	Mo	Co	
Water-Hardening Grades								
W1	0.6–1.4							Cold-heading dies, woodworking tools, etc.
W2	0.6–1.4	...	...	0.25				
Shock-Resisting Tool Steels								
S1	0.5	...	1.5	...	2.5			Chisels, hammers, rivet sets, etc.
S5	0.5	0.8	...	...	...	0.4(2.0 Si)		
Oil-Hardening Cold-Work Tool Steels								
01	0.9	1.0	0.5	...	0.5			Short-run cold-forming dies, cutting tools
02	0.9	1.6						
Air-Hardening Medium-Alloy Cold-Work Tool Steels								
A2	1.0	...	5.0	...	...	1.0		Thread rolling and slitting dies, intricate die shapes
A5	1.0	3.0	1.0	...	...	1.0		
High-Carbon High-Chromium Cold-Work Steels								
D2	1.5	...	12.0	...	...	1.0		Uses under 480°C, gages, long-run forming and blanking dies
D3	2.25	...	12.0	...	...	1.0		
D4	2.25	...	12.0	...	...	1.0		
Chromium Hot-Work Steels								
H12	0.35	...	5.0	0.4	1.5	1.5		Al or Mg extrusion dies, die-casting dies, mandrels, hot shears, forging dies
H13	0.35	...	5.0	1.0	...	1.5		
H16	0.55	...	7.0	7.0				
Tungsten Hot-Work Steels								
H21	0.35	...	3.5	...	9.5			Hot extrusion dies for brass, nickel, and steel, hot-forging dies
H23	0.30	...	12.0	...	12.0			
Tungsten High-Speed Steels								
T	0.70	...	4.0	1.0	18.0			Original high-speed cutting steel. Most wear-resistant grade
T15	1.50	...	4.0	5.0	12.0	...	5.0	
Molybdenum High-Speed Steels								
M1	0.80	...	4.0	1.0	1.5	8.5		85% of all cutting tools in United States made from this group
M2	0.85	...	4.0	2.0	6.25	5.0		
M3	1.0	...	4.0	2.4	6.0	5.0		
M10	0.85	...	4.0	2.0	...	8.0		
M15	1.50	...	4.0	5.0	6.5	3.5	5.0	Most wear-resistant grade

hardenability so they can be quenched in oil. They have lower carbon contents in order to increase impact strength. All other steels in the table have sufficient alloying elements to make their hardenability great enough to permit quenching either in oil or in air (or in a nonoxidizing atmosphere) and still develop a martensitic structure. Metallurgically, the functions of the various alloying agents are as follows:

Chromium is a relatively low-cost element that increases hardenability and, in sufficient excess together with carbon, forms a chromium-rich carbide, $Cr_{23}C_6$, for wear resistance.

Molybdenum and *tungsten* have generally similar effects and when they are present in large percentages, form with carbon a hard carbide, $(Mo\text{-}W)_6C$, which, upon tempering a quenched alloyed austenite, precipitates as fine particles in martensite and resists growth at low temperatures. This is the basis of red hardness, or secondary hardening, of high-speed types of steel. Molybdenum is less expensive than tungsten and has a greater volume per unit weight; hence M-type tool steels are less expensive than the equivalent T types.

Vanadium forms a very hard carbide of the type V_4C_3, which resists solution in austenite and therefore generally remains in the microstructure unchanged through heat-treatment cycles. It is the hardest of all carbides, and therefore vanadium, when sufficient carbon is present as in the M15 steel, gives the greatest wear resistance. It is an expensive element, though, and therefore high-vanadium steels are more costly.

10.2 HEAT TREATMENT OF HIGH-CARBON STEELS

Transformation Diagrams

The transformation diagrams of two of the more frequently used tool steels are presented in Figs. 10.1 and 10.3. Since these are being reproduced for their value in understanding the normal heat treatments, the diagrams are drawn for austenitization conditions employed in industrial heat treatments but with some discussion of the effects of overheating or underheating, i.e., austenitizing at too high or too low a temperature.

The upper portions of Fig. 10.1 indicate that plain-carbon steel must be cooled rapidly past 500°C to avoid the formation of fine pearlite. This is an essentially water-hardening steel, although a somewhat higher manganese content, for example, 0.50 instead of about 0.25%, may permit thin sections such as saw blades to be hardened by oil quenching. The effect of higher austenitizing temperatures on the martensite reaction is given in Table 10.2. These data show that higher quenching temperatures give a coarser grain structure, fewer residual carbides to resist abrasion, and more retained austenite upon cooling to room temperature.

The value of these transformation diagrams can be illustrated by considering the hardening of a 15-cm block of carbon tool steel. Suppose that it is quenched by a high-pressure spray that will cool the surface layers rapidly to about 65°C,

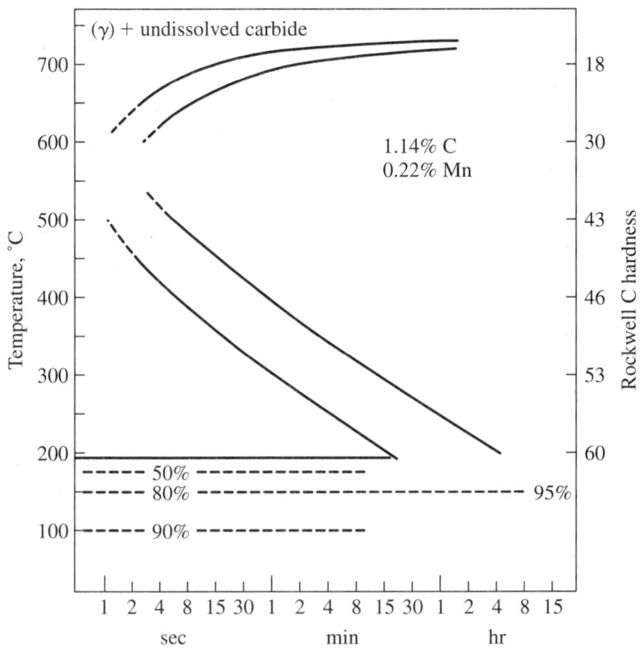

Figure 10.1 Isothermal transformation diagram for a water-hardening plain-carbon tool steel austenitized according to the usual hardening treatment at 790°C, therefore from the $\gamma + Fe_3C$ field with undissolved carbides in the structure. Rockwell hardnesses of transformation structures are shown. The degree of transformation of austenite to martensite, shown by dashed lines, may vary with cooling rate (or cooling stresses) in large sections. The 95% transformed at the right end of the 80% horizontal (150°C) means 80% martensite + 15% bainite, since long holding here would cross the bainite transformation line if that were extended below 200°C.

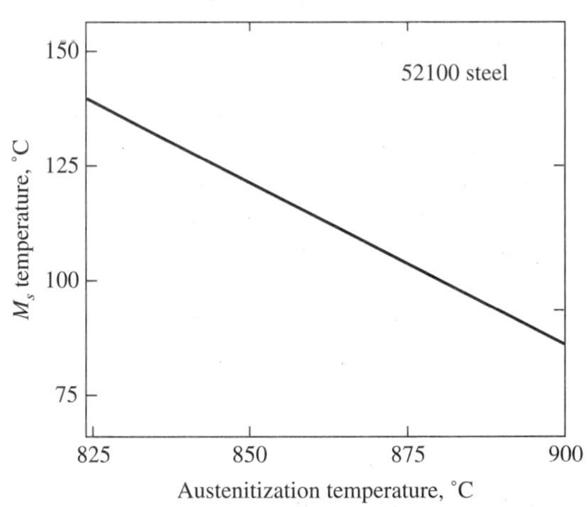

Figure 10.2 The M_s temperature for 52100 steel (1% C, 1¼% Cr) as a function of the austenitization temperature; increased carbon solution lowers the M_s temperature.

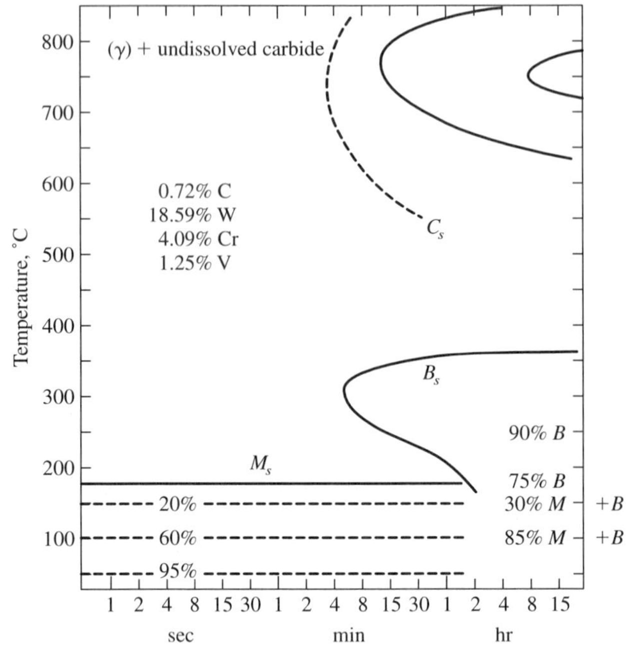

Figure 10.3 Isothermal transformation diagram for 18-4-1 high-speed steel as austenitized at the usual hardening temperature of 1290°C from an annealed structure. In the martensitic transformation range, long-time holding permits bainite to form as indicated by the increased percentage transformation at long times (15 h). The dashed C_s line indicates approximate start for carbide precipitation from supersaturated austenite prior to the eutectoid reaction.

Table 10.2 Effect of austenitization treatment on martensite reaction of a 1.2% C plain-carbon tool steel

	Austenitization Treatment		
	790°C, 1 h	**830°C, ½ h**	**870°C, ½ h**
γ grain size, ASTM	No. 9	No. 8	No. 6
Undissolved carbides	Very many	Many	Some
% martensite, at 150°C	80	40	5
At 100°C	90	80	50
At 30°C	100	95	85

permitting their complete transformation to martensite. The center of the block will cool much more slowly and transform at about 550°C to fine pearlite. Somewhere between the center and the surface will be a zone which at a specific time will have missed the pearlite reaction but which would not have cooled below the M_s line. In this zone, the austenite will remain untransformed for some time. If the block is transferred to the tempering furnace while the surface is still warm (the recommended procedure) and tempered 1 h at 200°C, some austenite will remain in this subsurface zone and finally transform upon cooling of the entire block to room temperature.

The volume expansion accompanying the transformation if it occurs at this time, upon cooling from tempering, will result in high surface tensile macrostresses. These may well cause the corners of the block to chip off, sometimes many hours after final cooling from the tempering furnace.

This difficulty could be avoided if thermal gradients were eliminated during original martensite formation or if the block were allowed to transform completely before tempering. However, marquenching (martempering) of such a large block of carbon tool steel could not be accomplished because of the rapid cooling required past 550°C. If the quenched steel were allowed to stand around too long at room temperature to permit complete transformation before tempering, the surface would become too cold and might chip or crack from quenching stresses combined with transformation stresses. A successful compromise is to quench as before to about 65°C and then, if a heavy mass is involved, hold at this temperature in a warm oil bath until the temperature is uniform.

For oil-hardening steels, slower cooling means smaller thermal gradients and more uniform transformations throughout the cross section. Oil-quenched 52100 steel is not strictly a tool steel, having been developed for ball bearings. It is about 95% transformed to martensite by the time it reaches 95°C upon quenching if austenitized at the normal temperature. The effect of higher austenitizing temperatures on the martensite reaction of the 52100 ball-bearing steel is shown by Fig. 10.2. As in the case of the plain-carbon steel, higher austenitization temperatures result in more complete solution of hypereutectoid carbides, austenitic grain growth, and a related depression of the M_s and M_f temperatures.

The transformation diagram of an air-hardening tool steel is reproduced in Fig. 10.3. Notable is the degree to which the knee of the reaction, corresponding to fine pearlite formation in carbon steels, is displaced upward and to the right, to about 780°C and 600 s, compared with 550°C and less than 1 s. In addition, there is a temperature interval, of about 600°C to about 350°C, in which metastable austenite shows no indication of transforming even when held a period of weeks! However, the carbide precipitation that may occur upon holding austenite at these temperatures (see C_s line in Fig. 10.3) will change the M_s and M_f temperatures and affect tool properties.

The transformation diagram (Fig. 10.3) does not specifically identify the carbides present in annealed high-speed tool steels. There are three types: a chromium-rich carbide, $M_{23}C_6$, a harder tungsten-molybdenum-rich carbide, M_6C, and a very hard vanadium-rich carbide, M_4C_3. These do not all dissolve at the same temperature or rate, as is shown by the data of Fig. 10.4. It is evident that although at higher austenitizing temperatures more carbides dissolve, those remaining undissolved will be the harder and more wear-resistant M_6C and M_4C_3 types.

Transformation diagrams like Fig. 10.3 permit a quenching process equivalent to marquenching, although much older in usage, known as hot quenching. After austenitizing, the steel is quenched in a liquid salt bath at about 550 to 500°C, held

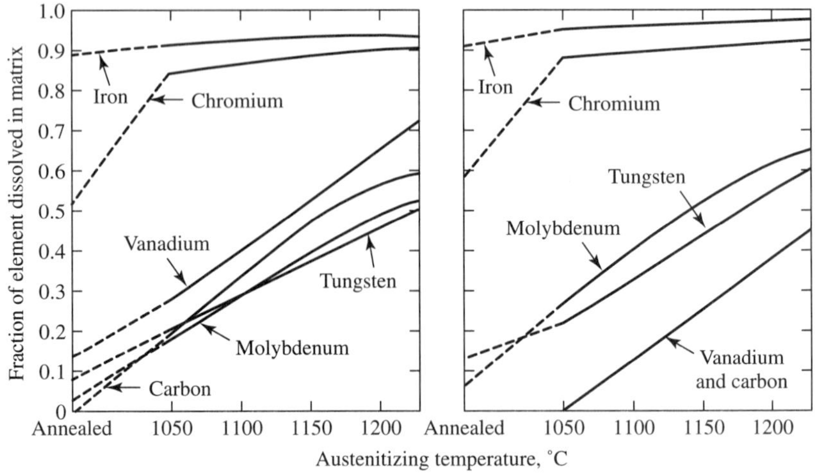

Figure 10.4 Partition of elements between matrix and carbides in high-speed steels, M1 (left) and M4. Most of the chromium carbides are dissolved at 1050°C; vanadium, molybdenum, and tungsten carbides dissolve much more slowly.

until the temperature is uniform, and then quenched in oil. The chief advantage of doing this as contrasted with air cooling is economic. Moderate-sized sections of the steel will harden quite successfully upon air cooling, as is obvious from the transformation diagram. However, if a heat-treating shop is processing a large number of parts in this way, a room of normal size is soon filled up with hot steel and no further processing, i.e., tempering, is possible until the parts are nearly at room temperature. On the other hand, an oil quench directly from the high austenitization temperature is certain to give more distortion and also result in severe thermal gradients, with accompanying macrostresses and potential cracking. The hot quench is a successful compromise between the economic and the metallurgical difficulties.

The effect of increased austenitization temperature on the formation of martensite is shown in the case of the high-carbon high-chromium steel by the data of Table 10.3.

Table 10.3 Effect of austenitization treatment on martensite reaction in air-hardening high-carbon high-chromium steel

	Austenitization Treatment			
	1010°C, 1 h	**1070°C, ½ h**	**1110°C, ¼ h**	**1150°C, 5 min**
γ grain size, ASTM	No. 11	No. 10	No. 9½	No. 9
% martensite, at 95°C	95	20	Trace	0
At 30°C	100	85	30	Trace
Rockwell hardness, as air-cooled	C65	C59	C47	C36

The pronounced drop in M_s and M_f temperatures upon overhardening is typical of the higher-chromium-content steels. The related large increase in the amount of retained austenite upon cooling to room temperature is the obvious cause for the decrease in as-quenched hardness. If, immediately after quenching, such steels are cooled to well below room temperature, for example, −80°C, the austenite retained upon cooling to room temperature almost completely transforms to martensite. When returned to room temperature, the steel will be harder than before the subzero treatment, and if careful dimensional measurements are made, it will be found that the steel has "grown" or increased in volume. These effects, to be anticipated upon the transformation of austenite to martensite, would not be observed if the steel were underhardened with only a relatively small amount of carbide dissolved in austenite.

If these higher-alloy-steel tools warp or distort upon heat treatment and must be straightened, this plastic deformation should be performed while the steel is relatively soft, i.e., partly austenitic. The transformation diagrams show that while the steel is still warm, immediately after quenching, a normally hardened high-speed steel may contain 30 or 40% austenite. This is the stage when slight deformation for straightening can best be performed.

Tempering

The tempering curves for a plain-carbon tool steel and 52100 ball-bearing steel, initially in the martensitic state, are shown in Fig. 10.5. The sequential structural changes for the plain-carbon steel have already been described, namely, precipitation of carbide in martensite and change from bct to bcc matrix lattice, followed by precipitation of carbide in the retained austenite and conversion of this austenite to

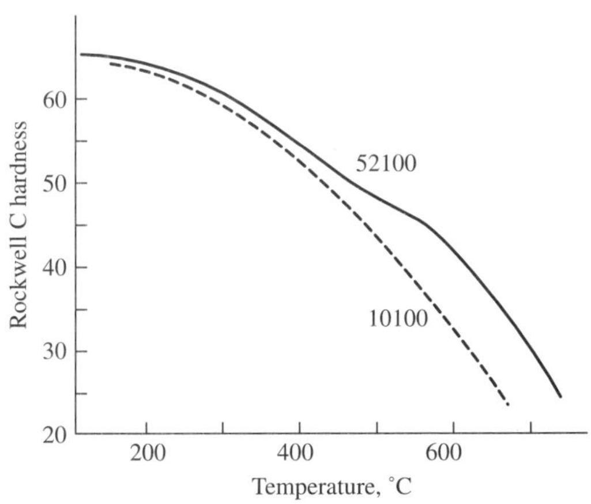

Figure 10.5 Tempering data for a plain high-carbon steel, water-quenched from 790°C, and for 52100 (1% C and 1.2% Cr) after being quenched in oil from 845°C. Tempering time for both curves was 1 h.

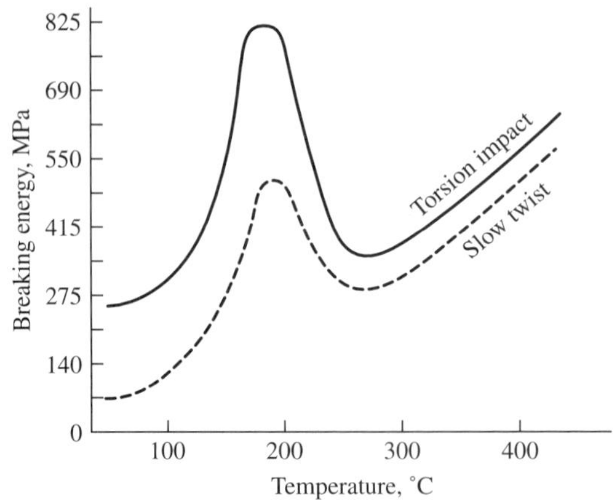

Figure 10.6 Energy required to break tempered martensitic structures as a function of the tempering temperature. The 1.1% C tool steel was quenched in brine from 790°C and tempered for 1 h at the indicated temperatures. The breaking energy for the specimens given a slow twist was obtained by integrating the area under the torsional-stress-vs.-torsional-strain curve for each test specimen. The peak at about 190°C presumably corresponds to stress relief of martensite, with no change of retained austenite; the low values at about 260°C result from the stresses generated by retained austenite transforming to martensite upon cooling from the tempering operation.

martensite (upon cooling), formation of regular Fe_3C from ε carbide, and, finally, growth or coarsening of Fe_3C carbides in ferrite, with a corresponding decrease in their number, a concomitant increase in spacing, and a pronounced softening.

Microstresses associated with the transformation of retained austenite and the related localized expansions cause a decrease in torsional impact strength of plain-carbon tool steel at about 230°C, as shown by the data of Fig. 10.6. In properly hardened carbon steel or 52100 steel, there is insufficient retained austenite to cause any strong inflections in the hardness-tempering curve. The slight inflection at 500°C and the subsequent slower softening of the 52100 steel (Fig. 10.5) are associated with formation of the stable carbide, $(FeCr)_3C$, since for its formation and growth chromium must substantially diffuse as well as the interstitial carbon.

Three tempering curves of a high-speed steel are shown in Fig. 10.7. Of the three curves, the dashed line represents an underhardened steel, i.e., austenitized at 1150°C. Its hardness is highest initially because, on the assumption that the steels were all cooled just to room temperature after hardening, this steel would have the least dissolved carbon and alloy in the austenite and therefore the least amount of soft retained austenite in the structure before tempering. Conversely, the overhardened specimen, represented by the dotted line for austenitization at 1250°C, would have the most retained austenite and be softest after quenching. This steel would be even softer, initially, if it were tempered before being cooled to room temperature. If, for example, it were cooled only to 100°C, almost half the structure would be austenite and the hardness might be in the Rockwell C50 to C60 range.

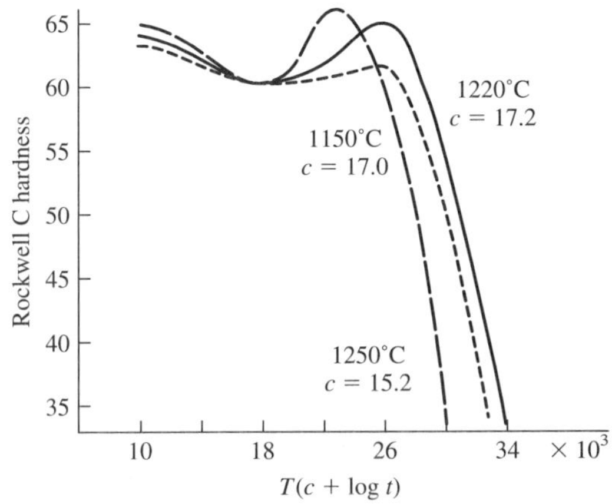

Figure 10.7 Tempering curves for 6-5-4-2 high-speed steel austenitized at three different temperatures. Tempered hardness is plotted against the parameter $T(c + \log t)$, where T is the tempering temperature, t the tempering time, and c a constant for a specific steel and austenitization treatment.

Before tempering, these steels consist of highly alloyed martensite, highly alloyed retained austenite, and undissolved carbides of the type $(MoW)_6C$ and V_4C_3. The changes upon tempering can best be shown by the schematic diagram below:

Tempering Temperature		Structural Components	
	Highly alloyed γ	Highly alloyed martensite	$(MoW)_6C$ and V_4C_3
		ppt ε carbide	
205°C		Forms Fe_3C	
430°C	ppt $(FeCr)_{23}C$ + lower alloy γ	Forms $(Fe_3Cr)_{23}C_6$	
540°C	ppt $(MoW)_6C$ + lower alloy γ	ppt $(MoW)_6C$ (Secondary hardening)	
	To martensite on cooling	Growth of carbides	No change

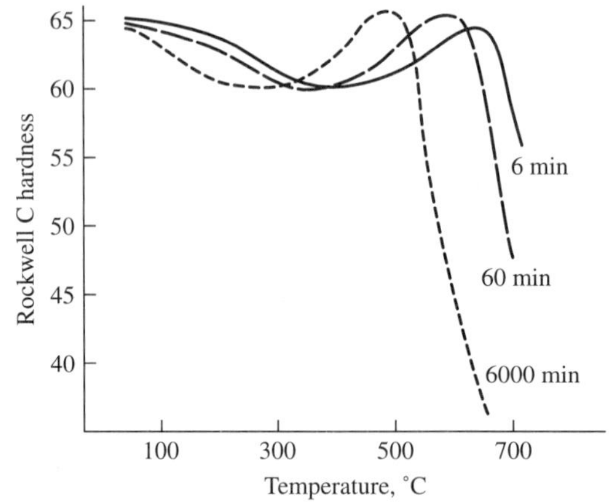

Figure 10.8 Tempering curves for 18-4-1 high-speed steel, quenched in oil from the usual hardening temperature of 1290°C curves for hardness after tempering 6, 60, and 6000 min are reproduced. Almost identical tempering curves are obtained for the 6-5-4-2 high-speed steel when oil-quenched from its usual hardening temperature of 1220°C.

The formation of new martensite in a hard and moderately brittle tempered martensite, upon cooling from the tempering operation, results in high microstresses. If the steel is quenched from the tempering operation or high macrostresses are present from other sources, e.g., sharp corners, sharp changes in section size, or an originally too drastic quench, the steel may crack upon cooling from the first tempering operation. It is more likely to crack at this time than during the hardening quench, because there is no soft cushion of retained austenite to deform and reduce the stress level. For these reasons, it is the practice of most shops to employ two tempering treatments. The second heating tempers the martensite formed from the first tempering and, more importantly, reduces the micro- and macrostresses in the tool and also the likelihood of brittle failure in service.

The effect of time at tempering temperature has not been discussed, but it is obviously a significant variable since diffusion is the primary requisite for carbide precipitation and growth. The graphs of Fig. 10.8 showed the effects of tempering time of an 18-4-1 high-speed steel. It is clear that the character of the tempering curve is unchanged: only the temperature of the secondary hardness peak is lowered or displaced to the left by longer tempering times. As in the case of medium-carbon steels, a single parameter has been found to express the two variables of tempering time and temperature. This parameter, of the form

$$T(c + \log t)$$

where T = temperature
t = time
c = constant whose value depends on composition of austenite

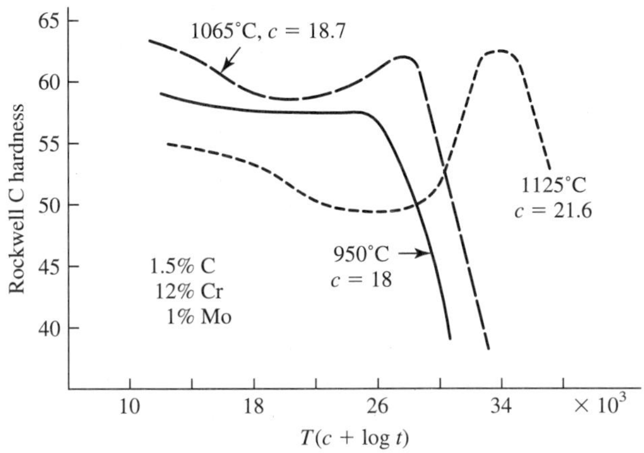

Figure 10.9 Tempering curves for an air-hardening die steel austenitized at three different temperatures, as indicated on each curve. Tempered hardness is plotted against the parameter $T(c + \log t)$, where T is the tempering temperature, t is the tempering time, and c is a constant for each curve whose value depends on the austenitization temperature. Hardening from the lowest temperature results in no secondary hardening; hardening from the highest temperature results in low initial hardness because of retained austenite and then a related pronounced secondary hardening.

was used as the abscissa in the tempering curves of Fig. 10.9. The graph of this figure shows in a particularly noteworthy way the effect of austenitization temperature. The 950°C treatment gives a quenched hardness of C64, since, with more carbides dissolved, the martensite is harder, although some retained austenite is now encountered, as is shown by the secondary hardening upon tempering. The 1125°C austenitization gives a low-quenched hardness of only C55 because too many carbides have dissolved, resulting in a very large amount of retained austenite. On tempering, a pronounced peak results from transformation of retained austenite upon cooling from tempering. The displacement of the peak to the right is somewhat misleading in this graph, since the temperature of maximum secondary hardening does not increase noticeably; the high austenitization temperature changes the constant c used in plotting abscissa values.

The use of the parameter for abscissa is this. Suppose that the hardness desired shows a $7(c + \log t)$ value of 30,000. A convenient time could be chosen and the corresponding required temperature immediately calculated if the constant c were known. Thus, data of this sort enable the heat treater to choose convenient times, overnight or otherwise, and calculate the exact temperature required to achieve the desired hardness.

Surface Effects upon Hardening

The working parts of a tool are the edges or the surfaces. It is quite possible unintentionally to carburize or decarburize this part of the tool during heat treatment.

Annealed stock usually will have a surface decarburized during hot-working and annealed treatments at the steel mill. This is not objectionable in most cases, since machining of the tool will expose a fresh, representative surface. Decarburization during the hardening heat treatment, however, will require grinding of the surface to remove the soft, low-carbon surface layer. Grinding of a hardened tool requires leaving enough stock on the tool before hardening to permit meeting specified dimensions. Furthermore, grinding is expensive and in some cases, e.g., hacksaw blades, impractical. Finally, grinding of a martensitic structure, if not performed with care, may cause the formation of cracks. Local overheating and local expansions followed by contractions, all within a hard and brittle structure, result in cracks. Whenever cracks are observed on a tool at right angles to the direction of grinding, it is fairly certain that they originated in grinding, not in the heat treatment.

It has been shown that heating 18-4-1 high-speed steel at the usual austenitizing temperatures for short times in most gaseous atmospheres results in surface carburization followed by decarburization over longer periods. This is true whether the atmosphere is oxidizing, i.e., contains oxygen from excess air in the combustion mixture, is neutral (gas completely burned to CO_2 and H_2O with no O_2 or CO), or is reducing (contains some CO and H_2). Actually, equilibrium at 1290°C in the reaction $C + CO_2 \rightarrow 2CO$ is far to the right; that is, CO_2 will burn carbon and also oxidize iron. How then can carburization occur in oxidizing or neutral atmospheres?

Presumably CO_2, H_2O, or O_2 present in the gas oxidizes iron very rapidly at 1290°C. As oxygen atoms can diffuse through the thin oxide scale more quickly than carbon atoms, carbides are less rapidly oxidized. Rapid oxidation of the metal surface raises the temperature locally and permits solution in austenite of carbides that were in the scale. Thus the austenite adjacent to the scale may briefly contain up to 1.0% C although the steel contains only 0.7% C.

Oxygen diffusion through the scale, required for continued scaling, decreases in rate very rapidly with thickening of the scale. However, carbon can diffuse interstitially, and its rate is less affected by scale thickness. Therefore, increased time and thicker scales permit carbon to diffuse to the surface and oxidize at a rate faster than scaling of the iron. The result is decarburization.

Decarburization produces a soft surface layer on the hardened steel, often best detected by a file test after tempering. Carburization at the high temperature may also cause the surface to be file-soft by reason of the large amount of retained austenite that would be found there.

Steels of the high-speed type are usually preheated at 700 to 850°C and held there to attain a uniform temperature and thereby reduce the time of exposure to the high temperature. Surface oxidation or scaling during the preheat may cause slight decarburization, but severe decarburization can occur only at the high temperature. Although no data have been published showing decarburization during the tempering operation at about 550°C, it apparently can also occur at this relatively low temperature.

Protection from surface chemical changes is most necessary at the high hardening temperature. It may readily be achieved by packing the steel part in cast-iron chips, but this slows up the heating and prevents close control of the time at the high temperature, since it is difficult to determine when the steel reaches the proper temperature. Protection is also achieved by using a carbon block to hold the steel. Above 1000°C, the atmosphere within a carbon block is nearly 66% N_2 + 34% CO, and scaling and decarburization will not occur if no oxide scale was formed during the preheat.

Suitable liquid baths adequately prevent compositional changes at the metal surface if the bath is properly protected from oxygen. At high temperatures, this can be achieved by a covering of silicon carbide or by comparable conditioners on the surface of the liquid bath.

Extremely hard cutting edges can be obtained by nitrating the hardened high-speed steel in ammonia or in a liquid cyanide bath. The basis for these treatments was discussed earlier.

Dimensional Changes

The specific volumes of ferrite + carbide, of austenite, and of martensite are all somewhat different (see Table 10.4). The annealed structure of alloyed ferrite and carbide contracts upon austenitization, since the fcc lattice has a greater packing density of atoms. There is an expansion when austenite changes to martensite on cooling, but even when this transformation is complete, the expansion does not exactly coincide with the prior contraction upon heating and formation of austenite. The atomic dispersion of carbon in martensite and perhaps the presence of microstresses result in a greater volume than corresponds to the original ferrite and carbide. Therefore, tempering of martensite is accompanied by a slight contraction. Dimensional changes for an 18-4-1 high-speed steel upon cooling through the M_s temperature are shown in Fig. 10.10. This graph also shows that holding at room

Table 10.4 Specific volume of phases present in carbon tool steels*

Phase or Phase Mixture	Range of Carbon, %	Calculated Specific Volume at 26°C, cm^3/g
Austenite	0–2	0.1212 + 0.0033 (% C)
Martensite	0–2	0.1271 + 0.0025 (% C)
Ferrite	0–0.02	0.1271
Cementite	6.7 ± 0.2	0.130 ± 0.001
ϵ carbide	8.5 ± 0.7	0.140 ± 0.002
Graphite	100	0.451
Ferrite plus cementite	0–2	0.1271 + 0.0005 (% C)
Low-carbon martensite plus ϵ carbide	0–2	0.1271 + 0.0015 (%C − 0.25)
Ferrite plus ϵ carbide	0–2	0.1271 + 0.0015 (% C)

*B. S. Lement, "Distortion in Tool Steels," American Society for Metals, 1959.

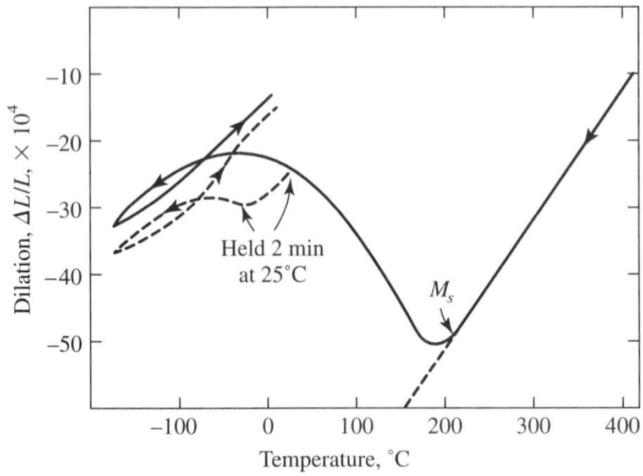

Figure 10.10 Volume changes as indicated by length changes upon cooling of an 18-4-1 high-speed steel from 1290°C. The solid line shows the length changes upon continuous cooling from 1290 to −190°C. Expansion begins at 207°C, the *M* point, and continues until at about 75°C, all austenite has transformed to martensite. The dotted line represents a case where cooling was interrupted by holding the specimen at room temperature for 2 min. This noticeably stabilized the austenite still present so that the austenite-martensite transformation did not again start until the specimen cooled to a temperature of about −75°C. In this latter case, the dimensions upon reheating to room temperature show that not all austenite was transformed to martensite even upon cooling to −190°C.

temperature before subzero cooling stabilizes the retained austenite and reduces the effectiveness of the subsequent cooling.

If the dimensional changes in a specific tool take place at greatly different times as a result of thermal gradients, not only macrostresses but also greater volume changes result. The maximum thermal gradients encountered upon water, oil, and air cooling of steel rods of several diameters have been calculated by Scott and are shown in Table 10.5. It is evident that the difference between oil quenching

Table 10.5 Calculated maximum temperature difference between center and surface of steel cylinders quenched from 930°C into medium at 30°C

	Temperature Difference, °C		
Cylinder Diam., cm	**Water**	**Oil**	**Air**
10	846	567	55
5	801	414	30
2.5	734	277	16
0.1	617	171	9

Table 10.6 Increase in volume of alloy tool steels on hardening

Type	Composition, %						Heat Treatment, °C	Increase in Volume, %
	C	Mn	Cr	Mo	W	V		
01	0.9	1.1	0.5	. . .	0.5	. . .	780 to water	0.69
	1.0	0.2	1.3	. . .	. . .	. . .	800 to water	0.52
A2	1.0	0.6	5.2	1.0	. . .	0.2	945 to gas*	0.30
D2	1.5	0.2	11.5	0.8	. . .	0.2	1010 to gas	0.11

*Quenched in dissociated ammonia to preserve a bright, scale-free surface.

and air cooling is far greater than the difference between oil and water quenching. Table 10.6 shows the increases in volume that were found upon hardening four common tool or die steels. It would be expected that distortion and stresses would be greatest in the steel with the greatest volume change.

Shrinkage occurs upon tempering of the air-hardening chromium die steel. Complete stability of dimensions is not obtained until all martensite has been tempered to the point of precipitation of the stable carbide, here $(CrFe)_7C_3$ and some $(FeCr)_3C$. Transitional carbides that form at temperatures below 600°C may cause some shrinkage but not to the dimensions of the original annealed part.

10.3 CEMENTED CARBIDES

The harder a material is the more easily it can indent other materials and the better it will resist abrasive wear. However, when extreme hardness is accompanied by brittleness, the material will not be useful as a tool. The technical solution to this problem is to surround particles of the hard, wear-resistant material with a continuous thin film of a soft and plastic substance, a "cement." The relative Knoop hardnesses of several relatively hard materials are shown in Table 10.7.

Cemented tungsten carbides are produced by taking WC particles, usually in the size range of 1 to 3 μm, blending them with soft cobalt powder so that each WC grain is coated, pressing the mixture to form a "green" low-strength compact, and then sintering in hydrogen at a temperature above the melting point of the cobalt. The liquid cobalt wets the WC grains and dissolves a small amount of WC. The sintered material shrinks appreciably, leaving a small amount of residual porosity. Upon cooling, the Co solidifies and the dissolved WC precipitates on the grains already present, leaving a final structure of WC cemented together with thin films of Co. The data of Table 10.8 show the effect of increasing cobalt content on hardness, elastic modulus, and compressive strength. Although these properties decrease, shock resistance increases with increase in Co content.

Table 10.7

Material	Knoop Hardness
Fe-C martensite (C-65)	700
Fe_3C	1100
Cemented carbides	1400
$Cr_{23}C_6$ (in high-speed steel)	1800
WC, TiC, TaC	2100–2400
Sapphire	2800
Diamond	8000

Table 10.8 Classification and properties of cemented carbides*

Carbide Group	Composition %Co	Composition Carbide Type	Rockwell A Hardness	Elastic Modulus, GPa	Compressive Strength, GPa
1	3	WC	92.7	645	4.7
2	6	WC	91.5	615	4.3
2	9	WC	90	590	3.6
3	13	WC	89	550	2.7
4	5	WC/TiC	93		
5	8	WC/TiC	91		
7	6	WC/TaC	92		
8	9	WC/TaC	91		

*Classes 1 to 3 are used for machining cast iron and most nonferrous metals. The higher-cobalt-content carbides are used where increased shock resistance is required. Classes 4, 5, 7, and 8 are used for machining ferritic steels since the TiC or TaC present with WC increases resistance to cratering.

PROBLEMS

1. It is desired to test the machinability of specimens of a coarsely spheroidized 1.0% carbon steel with the following structures: (*a*) coarse pearlite + some spheroidized carbides, (*b*) entirely fine pearlite, (*c*) entirely fine spheroidal carbides, (*d*) moderately coarse pearlite + ferrite, (*e*) partially spheroidized moderately fine pearlite. Describe the heat treatments required, in terms of temperatures and cooling rates, to produce these structures (if they can be obtained).

2. What quenching mediums are used for cooling steels from the austenitic range? (Arrange in order of decreasing cooling rates.)

3. What advantages would be gained by quenching a 1.3-cm section of carbon steel into brine, then, as soon as its exterior has cooled below 540°C, transferring the section into an oil bath? After removing from the brine, holding in air a few seconds, and then replacing in the brine, what might the final surface structure be?
4. How might a large screwdriver be treated, an external source of heat being used only once, to develop a fine pearlitic structure in the shank and a tempered martensitic structure at the tip?
5. What difficulties are associated with the hardening of dies used for cutting threads? How would you suggest heating a large high-speed steel wedge to bring the heavy section within the proper hardening temperature without burning the thin edge?
6. Since neither the hardness nor the toughness of a high-speed steel cutting tool is usually increased by tempering at 565°C, why should it be given this treatment?
7. Specify in detail a heat treatment of hardened high-speed steel, based on the transformation diagram of Fig. 10.3, designed to achieve maximum softness for remachining.
8. Why does 52100 steel, after a spheroidizing anneal, sometimes show columnar ferrite grains at the immediate surface, then a zone of pearlite between the pure ferrite and the spheroidized carbide structure? Explain why the ferrite is columnar in structure and why the intermediate zone is pearlite rather than spheroidite.

REFERENCES

Cahn, R. W., P. Haasen, and E. J. Kramer (Eds.): *Materials Science and Technology: A Comprehensive Treatment,* vol. 7, *Constitution and Properties of Steels*, Wiley, New York, 1992.

Smith, W. F.: *Structure and Properties of Engineering Alloys,* 2d ed., WCB/McGraw-Hill, New York, 1993.

ASM Handbook, vol 1, *Properties and Selection: Irons, Steels, and High-Performance Alloys,* 10th ed., ASM Int., Materials Park, OH, 1990.

Metals and Alloys in the Unified Numbering System, 5th ed., SAE/ASTM, Warrendale, PA, 1989.

chapter

11

Stainless Steels

The stainless steels are a branch of the family of ferrous alloys designed for extremely high levels of corrosion resistance. This effect is achieved by alloying primarily with chromium but may also be enhanced by the addition of elements such as molybdenum and nickel. Moreover, these alloy elements may significantly alter the phase relationships in the steel and produce a wide spectrum of possible microstructures. The range of microstructures serves to qualify some stainless steels for special types of service beyond their use in corrosion service.

Austenitic alloys are well suited for cryogenic service because of their toughness and for high-temperature service because of their creep resistance. Some martensitic stainless steels are sufficiently high in strength and hardness to be used as tool and die materials or as high-strength aerospace components. The stainless steels are therefore found in applications that utilize all their properties, not their corrosion resistance alone. However, it is their corrosion characteristics that form the basis for their present grouping in this chapter as stainless steels.

11.1 PHASE DIAGRAMS OF STAINLESS STEELS

Stainless steels contain chromium or chromium and nickel or manganese. Present also will be some carbon and other elements, deliberately added or as unavoidable impurities. The Fe-Cr diagram of Fig. 11.1 illustrates that bcc chromium tends to stabilize bcc α-iron and its high-temperature counterpart δ that merge to form what is called the closed γ loop. Thus beyond 16% Cr, no austenite will be found in binary alloys, and, for this reason, they resemble most nonferrous solid-solution alloys. They display no transformation hardening by quenching and no grain refinement by heat treatment. Thus, binary Fe-Cr alloys, free of carbon, are not properly called steels but stainless irons. The σ phase of the binary system, which may be formed by ordering of Fe and Cr atoms to attain the structure of FeCr in high-chromium-content alloys, is of specific importance in certain stainless steels.

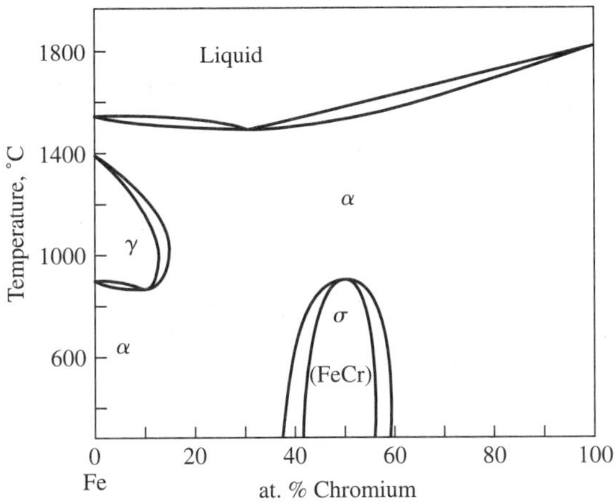

Figure 11.1 Binary Fe-Cr phase diagram illustrating the closed γ loop as body-centered chromium and body-centered iron form ferrite solid solutions. The sigma phase centered at the 50/50 composition is an ordered structure of the CsCl type.

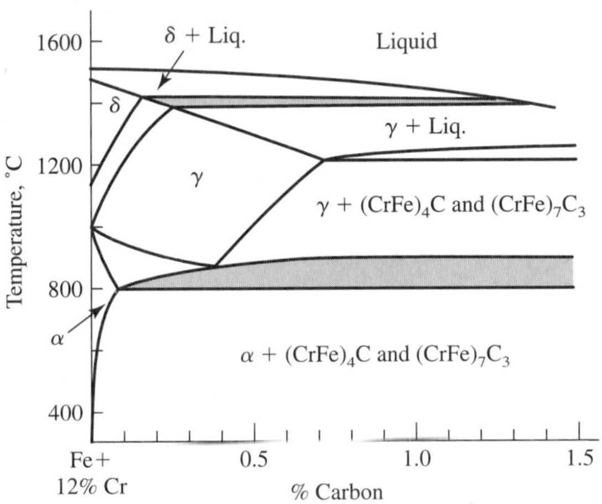

Figure 11.2 Quasi-binary phase diagram of iron plus 12% Cr, with varying percentages of carbon. Tie lines in two-phase fields do not show compositions of the two phases or relative proportions. The crosshatched areas are three- or four-phase fields.

The binary Fe-Cr diagram does not show the effect of carbon that is soluble in austenite and increases the chromium limit of the γ loop. For a discussion of the hardenable chromium steels, e.g., the so-called stainless cutlery grades, the pseudobinary diagram for (Fe + 12% Cr)-C (Fig. 11.2) is useful. Chromium constricts the γ field of the Fe-C diagram, decreasing the eutectoid composition to about 0.35% C and the maximum carbon solubility in austenite to 0.7%. The eutectoid temperature is considerably higher, and the eutectoid reaction is no longer represented by a line but by a field (dashed) since Gibbs's phase rule for this three-component system permits one degree of variability for the $\gamma \rightarrow \alpha +$ carbide reaction; that is, $F = 3 - 3 + 1 = 1$.

The chromium in fully annealed steels is present in both the ferrite and the carbide phases. The distribution in terms of the identity of the carbide phase is shown

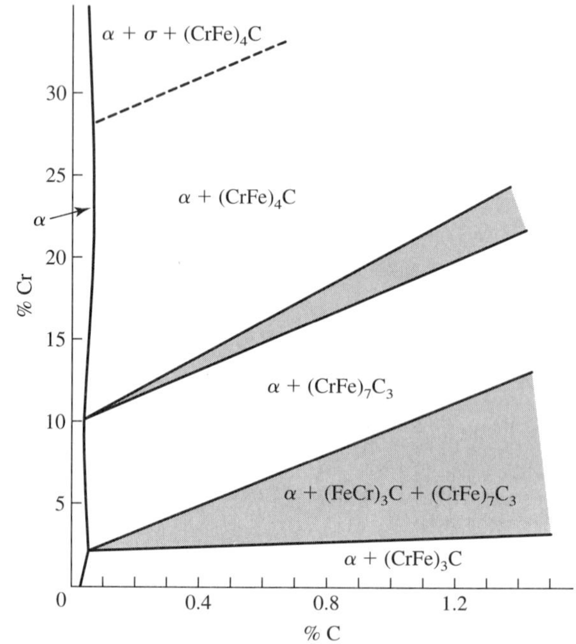

Figure 11.3 Diagram showing the phases present in slowly cooled Fe-Cr-C alloys as a function of chromium and carbon content. Chromium can enter Fe_3C in amounts up to 15% Cr without changing its structure, and this carbide is designated as $(FeCr)_3C$. The next carbide is $(CrFe)_7C_3$, which contains a minimum of 36% Cr. The $(CrFe)_4C$ is the carbide usually found in stainless steels and contains over 70% Cr.

in Fig. 11.3 as varying with chromium and carbon content. The medium-carbon alloy steels discussed in Chap. 9 contain normal orthorhombic Fe_3C with chromium replacing iron up to 15% by weight of the carbide, that is, $(FeCr)_3C$. In the high-C–high-Cr tool steels of Chap. 10, the carbide phase is $(CrFe)_7C_3$, which contains at least 36% Cr. The 12% Cr stainless steels may contain this same carbide, but most stainless grades have the next carbide in the series, $(CrFe)_4C$, which contains a minimum of 70% Cr by weight.

When fcc nickel is added to iron, it tends to depress the A_3 temperature, as shown in Chap. 8, and stabilize fcc austenite. With 30% Ni or more, binary Fe-Ni alloys are completely austenitic at all temperatures. The effect of increasing the nickel content in Fe-Cr alloys is shown by the diagrams of Figs. 11.4 and 11.5. Like Fig. 11.3, these are pseudobinary diagrams of (Fe + 18% Cr + 4% Ni)-C and (Fe + 18% Cr + 8% Ni)-C, respectively. They show that, with 4% Ni, the stainless steels containing less than about 0.4% C can become a mixture of ferrite and carbide, but only in the vicinity of room temperature. With 8% Ni present, the three-phase eutectoidal field is pushed to still lower temperatures and carbon contents. As the formation of ferrite is further restricted, the equivalent high-temperature phase δ is also similarly restricted. Diagrams for higher-nickel contents are not given, but the

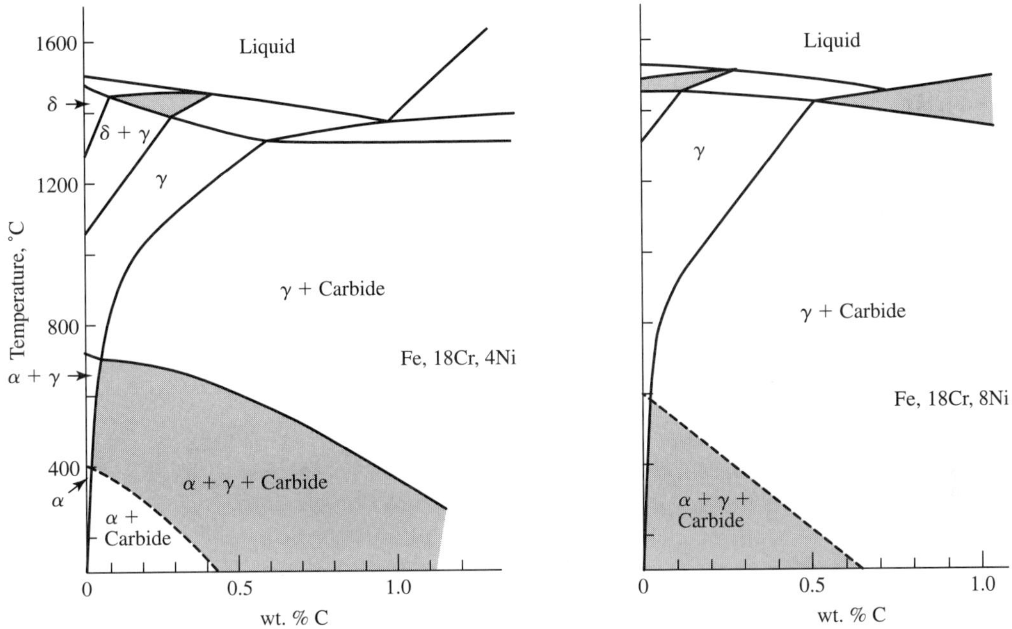

Figures 11.4 and 11.5 Quasi-binary phase diagrams Fe + 18% Cr + 4% Ni vs. varying carbon content (11.4) and Fe + 18% Cr + 8% Ni vs. varying carbon content (11.5). The carbide in both cases is $(CrFe)_4C$, at least up to 0.5% C. Note how increasing nickel content stabilizes austenite, restricting the high-temperature δ and room-temperature α. The three-phase fields are crosshatched. Phase compositions and relative proportions cannot be obtained from tie lines in two- or three-phase fields in these diagrams.

diagrams shown clearly indicate that low-nickel content in an Fe + 18% Cr base would be conducive to hardening by martensite formation while high-nickel contents would tend to give a more stable austenite.

The carbide solubility line of Fig. 11.5, equivalent to the A_{cm} line of the Fe-C diagram, is of great importance with regard to the heat treatment and properties of the 18-8 types of stainless steel. Unfortunately, these pseudobinary diagrams do not indicate compositions of the various phases, but Fig. 11.3 shows that the stable carbide in an 18% Cr, 0.10% C steel is $(CrFe)_4C$. Since nickel is not a carbide-forming element, this may be presumed to be the carbide in 18-8 stainless steel. With a minimum of 70% Cr in the carbide phase and with only 18% Cr in the austenite at temperatures above the A_{cm}, precipitation of the carbide phase inevitably results in depletion in chromium content of the remaining austenite. As in almost every case yet studied, the new phase precipitating from a solid solution does so preferentially at the solid-solution grain boundaries. This commonly occurs in the 18-8 steels heated in the range 500 to 800°C after the steel has been rapidly cooled from above the A_{cm}. An exception is when the structure has been cold-worked; then precipitation is just as rapid on planes of slip as at grain boundaries. The significance of grain-boundary precipitation of carbide and localized depletion in chromium content of the austenite is discussed on page 259.

11.2 STAINLESS-STEEL ALLOY DESIGNATIONS

The types of stainless steels most commonly used are listed in Table 11.1 by classification based on structure. The 300 series are all austenitic Cr-Ni alloys, where the austenite structure is obtained by nickel additions. Problems resulting from grain-boundary precipitation of chromium carbide are minimized in grade 304L by

Table 11.1 Wrought stainless steels

	Nominal Composition, %				
AISI Type	**C**	**Mn**	**Cr**	**Ni**	**Other**
Austenitic Grades					
201	0.15 max	7.5	16–18	3.5–5.5	0.25% N max
202	0.15 max	10.0	17–19	4.0–6.0	0.25% N max
301	0.15 max	2.0	16–18	6–8	
302	0.15 max	2.0	17–19	8–10	
304	0.08 max	2.0	18–20	8–12	
304L	0.03 max	2.0	18–20	8–12	
309	0.20 max	2.0	22–24	12–15	
310	0.25 max	2.0	24–26	19–22	
316	0.08 max	2.0	16–18	10–14	2–3% Mo
316L	0.03 max	2.0	16–18	10–14	2–3% Mo
321	0.08 max	2.0	17–19	9–12	(5 × % C) Ti min
347	0.08 max	2.0	17–19	9–13	(10 × % C) Nb-Ta min
Martensitic Grades					
403	0.15 max	1.0	11.5–13		
410	0.15 max	1.0	11.5–13		
416	0.15 max	1.2	12–14		0.15% S min
420	0.15 min	1.0	12–14		
431	0.20 max	1.0	15–17	1.2–2.5	
440A	0.60–0.75	1.0	16–18		0.75% Mo max
440B	0.75–0.95	1.0	16–18		0.75% Mo max
440C	0.95–1.20	1.0	16–18		0.75% Mo max
Ferritic Grades					
405	0.08 max	1.0	11.5–14.5	0.1–0.3 Al	
430	0.15 max	1.0	14–18		
446	0.20 max	1.5	23–27		
Precipitation-Hardenable Grades					
17-4PH	0.07	1.0	15.5–17.5	3–5	0.15–0.45% Nb; 3–5% Cu
633*	0.07–0.11	0.5–1.2	16–17	4–5	0.07–0.13% N

*Similar designation as AISI but not in that system.

maintaining carbon content at less than 0.03% or in grades 321 and 347 by the addition of the strong carbide formers titanium or niobium and tantalum, respectively.

The 200 series are also austenitic, but manganese and nitrogen are substituted for a part of the nickel, the cost of the alloy being thus lowered.

The martensitic types basically contain chromium, but their chromium content is low enough so that austenite will form at high temperatures and transform, upon rapid cooling, to martensite at or near room temperatures. These grades are used for stainless-steel cutlery.

The ferritic grades range from the low-cost and increasingly popular type 405 to the higher-chromium grades such as type 446. Type 405 has low carbon content and aluminum additions that prevent hardening to martensite in spite of the low chromium content.

One ferritic type, 430, has sufficient chromium, i.e., about 17%, to prevent austenite formation at any temperature. This alloy can be hardened only by cold work. Cast stainless steels are also widely used in alloy compositions similar to the wrought alloys of Table 11.1, although they are usually identified by the Alloy Casting Institute (ACI) system. (The High Alloy Product Group of the Steel Founders Society of America has replaced ACI administration of the system.)

11.3 HEAT TREATMENT OF STAINLESS STEELS

The austenitic steels are annealed differently than carbon steels: slow cooling from the annealing temperature is never practiced for the austenitics. The $(FeCr)_4C$ phase must be dissolved by heating to the vicinity of 1035°C and then, to prevent the carbide from precipitating at austenite grain boundaries during cooling, the stainless steel is quenched in air or water. The only other thermal treatments used are a stress-relief anneal at 345 to 455°C to improve the elastic properties of cold-worked stainless or a stabilization treatment at 870 to 900°C to form the stable titanium or niobium carbides in grades 321 or 347 and prevent subsequent precipitation of $(FeCr)_4C$. It is the intermediate temperature range of 485 to 815°C which not only "sensitizes" austenitic stainless to corrosion but also reduces its impact or notch strength, i.e., embrittles the steels.

The martensitic grades are annealed by slow cooling from around 760 to 815°C or hardened by oil quenching from about 1000°C. Stress-relief heating at 315°C causes little softening of the alloyed martensite. Since these alloys show temper embrittlement upon heating at 400 to 500°C, tempering must be carried out at 535 to 600°C, if softening is desired.

Some stainless steels are subject to the formation of the brittle σ (sigma) phase (see Fig. 11.1) at grain boundaries upon prolonged service at elevated temperatures. The tendency is found only in type 316 steel among the austenitic grades, which indicates that molybdenum promotes δ formation. The high-chromium ferritic steel 446 may also contain δ after service at elevated temperatures. In both cases, δ causes an embrittlement that can be removed only by a high-temperature solution anneal.

Precipitation-Hardenable Stainless Steels

The precipitation-hardenable stainless steels are a group of mostly proprietary alloys that combine corrosion resistance with high strength. The 17-4PH alloy is transformed to low-carbon martensite on cooling from austenite and subsequently hardened by precipitation. The 17-7PH, referred to as a semiaustenitic alloy, requires a more complex series of treatments to produce a precipitation-hardened martensitic structure. The critical characteristic of these steels is that nickel content is reduced, the stability of the austenite thus being lowered, and elements such as aluminum or titanium are added which lead to the potential formation of coherent alloy precipitates.

The nominal composition of 17-7PH is 0.07% C, 17% Cr, 7% Ni, and 1% Al. When it is solution-annealed at 1065°C and rapidly cooled, this steel will contain about 5 to 20% δ ferrite in its structure, largely because of the aluminum present, which is a ferrite former. In this annealed state, the alloy is soft and easily formed. Like the 301 grade, it will harden rapidly upon cold work because of the relatively low nickel content. The unique characteristic of the 17-7PH, however, is its capacity to harden greatly by thermal treatments alone.

11.4 MECHANICAL PROPERTIES OF STAINLESS STEELS

Stainless steels are never chosen for a given use solely because of their strength or ductility. However, if the corrosion resistance of a stainless type of steel is required, the mechanical properties may become important with regard to ease of fabrication or service requirements.

Stainlessness requires that 12% Cr be in solution in the structure, be this austenite, martensite, or ferrite. This amount of chromium induces a solid-solution strengthening effect. If the steel is austenitic by reason of the presence of at least 7% Ni, the alloy will be more ductile than a ferritic alloy steel; e.g., compare the ductility of annealed 302 with that of annealed 410 alloy (Table 11.2).

Recall from the pseudobinary phase diagrams (Figs. 11.4 and 11.5) that the austenite in the 300 grades is not completely stable at room temperatures. On this basis, one would expect the 300 type of austenitic alloy steels to transform to harder structures, perhaps martensitic upon deformation at room temperature and, therefore, to work-harden much more than ferritic alloys. This is shown by the data in Table 9.4 for 25 and 45% cold-rolled 301 alloy. Not only do these alloys significantly work-harden, they also become strongly magnetic at the same time.

Since nickel stabilizes austenite, it would be predicted that higher nickel content would reduce the work hardening of 300 series stainless. This is shown to be the case by the comparative work-hardening data for the 301 and 302 stainless steels in Table 11.2. The straight (nonhardenable) ferritic stainless steels 405, 430, and 446 are similar to ferritic low-carbon steels because they have low work-hardening capacity. During cold working, the yield strength approaches the ultimate tensile strength, and the elongation goes to zero. Hence shapes that must be produced by

Table 11.2 Typical mechanical properties of wrought stainless steels

Grade	Condition	Tensile Strength MPa	Yield Strength MPa	Elongation, % in 50 mm	Rockwell Hardness
Austenitic Grades					
301	Annealed	817	230	68	B85
	25% cold-rolled	1150	890 min	24 (min)	C38
	45% cold-rolled	1970	1400 min	7 (min)	C46
302	Annealed	660	250	61	B80
	20% cold-rolled	970	845	22	C29
	50% cold-rolled	1240	1055	6	C38
304L	Annealed	560	210	55	B76
316	Annealed	595	245	55	B80
321	Annealed	610	245	55	B80
347	Annealed	645	245	50	B84
Ferritic or Martensitic Grades					
405	Annealed	490	280	30	B80
410	Annealed	525	280	30	B82
	985°C, qu, 315°C	1260	980	15	C39
420	Annealed	665	350	25	B92
	1035°C, qu, 315°C	1610	1365	8	C54
440B	Annealed	750	430	18	B96
	1035°C, qu, 315°C	1955	1835	3	C55
Precipitation-Hardenable Grades					
17-4PH	Solution-annealed	1050	7700	C33	
	Hardened 485°C	1400	825	C44	

cold forming are more easily produced from austenitic grades than from ferritic stainless steels.

While the 400 series of alloys are all ferritic in the annealed state, those with chromium contents below 14% may be hardened by heating between 980 and 1035°C and quenching to form austenite. These alloys have transformed to a martensitic state for full resistance to corrosion because of chromium depletion of the matrix upon formation of $(FeCr)_4C$ by annealing. When the steel is hardened, the carbide dissolves and the chromium remains atomically dispersed in the martensitic structure. The 475°C temper embrittlement observed in these alloys causes a marked drop in notch impact strength. This detrimental effect increases rapidly with chromium content, reaching a maximum in type 446.

The precipitation-hardenable types like 17-4PH do not have better mechanical properties than the comparable cold-worked 300 series. They do have the unique advantage of being capable of easy forming to shape while in the soft, solution-annealed state and the capacity subsequently to be hardened by a simple thermal treatment without distortion.

Table 11.3 Typical properties of 17-7PH stainless steel after various stages of thermal treatment

Condition	Treatment	Yield Strength, MPa	Tensile Strength, MPa	Elongation in 50 mm, %
A	Mill anneal, 1065°C	280	910	35
TH(1)	Conditioned, 760°C, 90 min			
	Cooled to 15°C	690	1010	9
(2)	Aged 565°C, 90 min	1360	1445	9
RH(1)	Conditioned, 955°C, 10 min,			
	room temperature	295	930	19
(2)	Cooled in 1 h to −75°C, 8 h	805	1220	9
(3)	Aged 510°C, 1 h	1460	1570	6
LH(1)	Conditioned, 635°C, 2 h			
(2)	Cooled to −75°C, 8 h			
(3)	Aged 510°C, 1 h	1460	1570	6
CH	Cold-rolled 60% and aged			
	1 h at 510°C	1820	1850	2

The so-called TH (temper, hard) hardening sequence, which gives lower strength but better ductility than other sequences, requires an initial treatment of 90 min at 760°C, which precipitates chromium carbide at high-energy points such as grain boundaries and active slip planes (of cold-formed shapes). This precipitation, by reducing the carbon and chromium content of the austenite, makes it susceptible to transformation on subsequent cooling. Variations in temperature and time from that specified will somewhat affect the properties obtained but will not change the phenomenon. On cooling, martensite begins to form the M_s temperature of 95°C, and cooling must continue, to 15°C at least, in order to obtain the amount required for desired strength levels.

A bigger increment in strength is obtained by the next step in treatment, i.e., aging at 565°C for 90 min. This combination of temperature and time gives the optimum combination of strength and ductility. Lower temperatures, for example, 485°C, give higher strength but poorer ductility, whereas aging at 595°C has reverse effects.

These property effects are shown in Table 11.3, as well as the results of other treatment sequences. The RH[1] sequence with the higher conditioning temperature of 955°C results in more carbon in solution in the austenite, a lower M_s temperature, and the need to cool to −75°C in order to form martensite. The LH sequence uses a conditioning treatment of 635°C that minimizes scaling and distortion during treatment. The CH treatment eliminates any conditioning treatment but uses cold

[1] RH, LH, and CH treatments are given in Table 11.3.

working and aging to develop high yield strength. The cold-rolled strip, of course, does not have the formability that is required for many parts.

The mechanical-property data discussed so far are all for ambient temperatures. Actually, the austenitic stainless steels have exceptionally good strength and ductility properties at very low or cryogenic temperatures and at elevated temperatures, i.e., approaching low red heats or 650°C.

The low-temperature properties have become important in the space program. Liquid oxygen (LOX) at −185°C is widely used as the oxidizer in liquid-fueled rockets, and this means large, insulated Dewar vessels for storage and also large vessels as a part of the rocket casing to hold the liquid oxygen. Austenitic stainless steel, e.g., type 304, has proved to be satisfactory for this service. It does not embrittle at low temperatures, whereas ferritic types of steel lose all ductility at cryogenic temperatures. It is much stronger or has a better strength-to-weight ratio than weldable aluminum alloys, and it also has a lower thermal conductivity.

11.5 CORROSION RESISTANCE OF STAINLESS STEELS

Stainless steels owe their name and uses to the phenomenon of passivity, which is a condition of negligible corrosion in spite of a chemical tendency of the metal to react with its surroundings. The good corrosion resistance of aluminum, in spite of its great affinity for oxygen, is a result of the oxide film that quickly forms, is adherent, and is nearly impervious to the passage of more oxygen atoms. Likewise, the passivity of stainless steels is always found in association with oxidizing conditions and the oxide-film theory of corrosion resistance here again is useful. In connection with stainless steels, several broad generalizations are possible:

1. Corrosion resistance depends on passivity.
2. Chromium is the basic element for attaining passivity.
3. Corrosion resistance generally increases with chromium of the matrix phase.
4. Strongly reducing conditions, i.e., an absence of oxidizing conditions, cause a susceptibility to corrosive attack.
5. Strongly oxidizing conditions cause extraordinary resistance to attack.
6. The chlorine ion is destructive to chromium steels.
7. Nickel not only improves the engineering properties but increases resistance to corrosion in neutral chloride solutions and acids of low oxidizing capacity.
8. Molybdenum expands the passivity range and improves resistance to corrosion in hot sulfuric and sulfurous acids, as well as neutral chlorides, including seawater.
9. Intergranular attack of austenitic types is one of the foremost characteristics of these steels. It is avoided by low carbon contents, by proper heat treatment, or by alloying with niobium or titanium plus tantalum.

To expand a few of these generalizations, the requirement of chromium in the matrix is exemplified by the improved corrosion resistance of the 410 and 420 straight chromium alloys when in the martensitic condition. The grain growth that may occur in the completely ferritic 446 (28% Cr) alloy does not affect its corrosion resistance, which is comparable with that of the austenitic steels. However, grain growth, e.g., upon welding, makes the straight chromium alloy less desirable from the standpoint of mechanical properties.

In contact with salt solutions that are freely exposed to air, surface oxidation of stainless steels produces passivity and substantial immunity to corrosion. However, stainless steels cannot be recommended for storage tanks or other applications in contact with stagnant solutions. The original oxygen in the solution may be quickly used up, and if its oxidizing capacity is not maintained, for example, by aeration, pitting corrosion is possible. A small activated (anodic) region and a large passivated (cathodic) region results in destructive pitting at the anodic spots. A high nickel content and the presence of molybdenum, as in type 316, are conducive to minimizing general corrosion and pitting, respectively, in neutral chlorides. However, this steel cannot be relied upon for all service conditions. For example, in the least oxidizing acids, HF and HCl attack all stainless steels at almost all concentrations and temperatures.

A limitation of the effective use of some austenitic stainless steels has been their sensitivity to stress-corrosion cracking. This phenomenon occurs when austenitic and some ferritic alloys are under stress (residual or applied) in a specific corrosive environment. These may be different environments for different types of steels. The effect of this type of corrosion is cracking rather than general weight loss or pitting, and it may lead to failure of the stainless-steel part in a very short time. The cracks generally have a highly branched transgranular appearance, although they are intergranular under some conditions also. This type of attack is usually controlled by changes in composition. In the most common case, the austenitic alloys, the nickel content is increased.

The chief service difficulty encountered with 18-8 stainless steel has been intergranular corrosion, made possible by precipitation of $(CrFe)_4C$ at austenitic grain boundaries. The basis of this condition is shown by the schematic drawing of Fig. 11.6. Chromium diffuses from the immediate vicinity of grain boundaries to form the carbide phase at these sites. The resulting depletion of the matrix lowers chromium content below the 12% required for corrosion resistance. The low-chromium-content grain boundary metal is anodic to the rest of the grain, and attack in a corrosive medium is localized at the grain boundaries. If a strip is sensitized, i.e., treated to obtain grain-boundary carbide precipitation, and then boiled for 72 h in the Strauss test solution (47 ml H_2SO_4 and 13 g $CuSO_45H_2O$ in 1 liter of solution), the individual austenitic grains in the corroded zone are almost completely disconnected. If the strip is bent, the surface grains may fall out as individual crystallites.

The effect of temperature and time is illustrated in the graphs of Fig. 11.7. An 18-8 steel with 0.08% C was solution-treated at 1050°C and quenched. Different specimens were reheated for 3 min, 1 h, and 1000 h at constant temperatures from

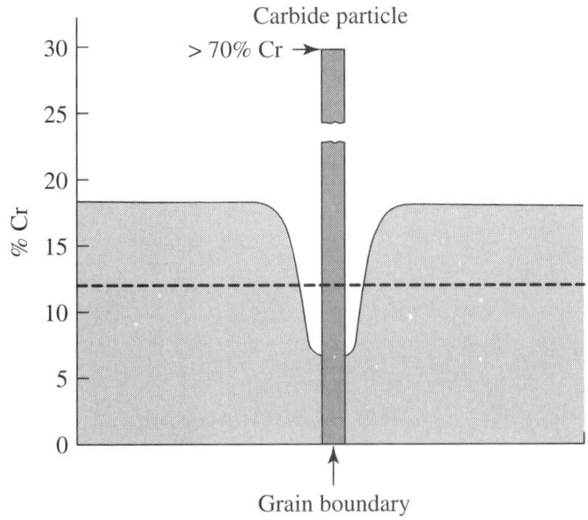

Figure 11.6 Schematic section through boundary of two austenitic grains of 18-8 stainless steel, with a carbide particle, $(CrFe)_4C$, precipitated at the boundary. This plot of chromium content through the section shows chromium depletion of austenite adjacent to the carbide particle. The amount of chromium left in solution at the boundary is below the 12% Cr (dashed horizontal) required for corrosion resistance.

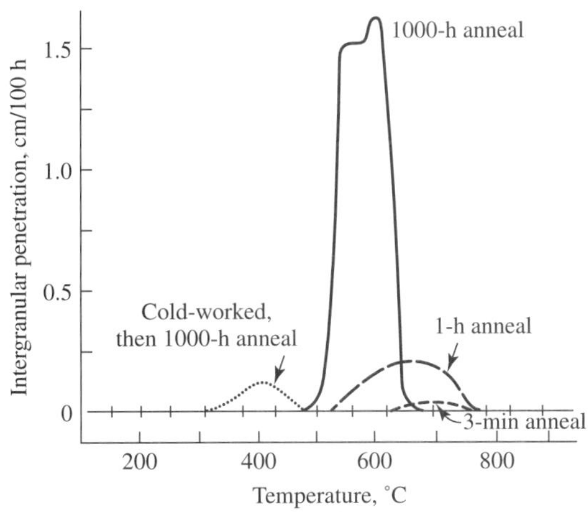

Figure 11.7 Effect of variation in annealing or reheating time of stainless steel at temperatures in the carbide precipitation range, as shown by intergranular attack in modified Strauss solution (47 ml H_2SO_4 and 13 g $CuSO_45H_2O$ diluted with water to 1000 g). The solid line represents annealed stock, whereas the dashed lines represent metal cold-rolled 40% before reheating. Intergranular attack is greatest and occurs at lower temperatures for long heating times. Cold working before heating induces precipitation at a lower temperature and more uniformly through the structure, reducing relative intergranular penetration.

200 to 800°C. Intergranular penetration after boiling in the Strauss solution for 100 h was measured by the decrease in electrical conductivity, a sensitive measurement since the zone of penetration tends to become a nonconductor.

Deterioration is greater and occurs at a lower temperature for longer times of heating. In the same graph, a comparison is made between sensitivity of annealed and cold-rolled metal upon precipitation treatments of 1000 h. The cold-rolled metal is only slightly affected, in comparison with the annealed, and the temperature of sensitization is lower. This should be anticipated in view of the discussion of precipitation in cold-worked Al-Cu alloys. Precipitation occurs sooner and is more general in cold-worked structures.

The curves of Fig. 11.7 also show that at 650°C, for example, little intergranular penetration occurs after 3 min, a large amount occurs after 1 h, and almost none is found again after 1000 h. This does not mean that the carbide precipitate redissolves after 1000 h at 650°C. Rather, it should be interpreted as (1) carbide-particle growth, which increases continuity of the boundary matrix, and (2) diffusion of chromium from the matrix into the impoverished boundary zone (Fig. 11.6), raising the chromium content there to at least the 12% minimum requirement and diminishing the anodic-cathodic difference between boundary and matrix.

Despite our knowledge of solutions to the corrosion problems associated with uses of stainless steels, some field failures are still encountered. Some examples are:

1. Despite the use of niobium to prevent carbide sensitization by tying up carbon as niobium carbide, it is possible during fabrication of 18-8 stainless type 347 to carburize the surface and, after using up all niobium, to form chromium carbide and thereby deplete the surface of chromium to the degree that corrosion can occur. This has been observed in hot spinning using a reducing flame and a greased spinning tool.

2. Entrapped degreasing agents containing chlorine will slowly liberate HCl, which will attack the passive film on which stainless steel depends for its corrosion resistance, with pitting or rusting resulting. There are many other materials such as chlorinated lubricants and plastics that in time can cause similar difficulties.

3. One of the common mechanisms that initiate corrosion of stainless steel is the formation of an oxygen-concentration cell, resulting in "crevice" or "contact" corrosion in material meeting all standard corrosion-resistance tests. If the surface of the stainless steel is covered with a moisture film and particles of sand or other solid material are present, the general surface remains passive because the moisture film is saturated with oxygen. However, the moisture film under the particle of adherent material (or in crevices) is not in contact with air and thus becomes starved of oxygen. In seacoast or equivalent locations, local attack can begin by breakdown of passivity in these oxygen-starved areas, resulting in corrosion products, generally ferrous and ferric chlorides. This small area becomes an active anode that, with the very large adjacent cathode (passive) area, is rapidly attacked. A pit forms that leads to a hole in sheets or tubing of appreciable thickness.

4. Airborne steel or iron "dust" which is deposited on stainless-steel surfaces will be even more effective in creating active anodic spots than sand particles. Such contamination may be common in industrial environments. Similar effects may result from plain-steel particles transferred to the surface of the stainless steel from a dull cutting tool or embedded during grinding, tumbling, polishing, drawing, forming, or buffing. The best cure for these problems is electropolishing that removes the contaminants and gives a smoother finish having less tendency to pick up surface contaminants.

PROBLEMS

1. Explain the formation of a sigma phase in stainless steel with excess carbon and how it might be detrimental in a welded structure.
2. Describe how a stainless steel may be made hardenable, and compare the hardening method with that of an Al-Cu alloy.
3. Why is high-carbon, high-chromium tool steel (Chap. 10) not stainless and, despite its chromium content, not capable of being made stainless by heat treatment?
4. Carbon forms carbides with Fe, Mn, Cr, Mo, Ti, and Nb. On what basis would the relative affinity of C for each element, which determines the distribution of C in the structure, be expressed, or how could this be estimated from literature sources?
5. Can a stainless steel, which has been sensitized, be returned to an unsensitized condition? If so, how?
6. Describe at least one method of stabilizing a stainless steel against sensitization.
7. Draw a curve that shows how the oxidation resistance of steel at elevated temperatures may be affected by different amounts of chromium.

REFERENCES

Davis, J. R.: *ASM Specialty Handbook: Stainless Steels,* ASM Int., Materials Park, OH, 1994.

Sedricks, A. J.: *Corrosion of Stainless Steels,* Wiley, New York, 1979.

Lula, R. A.: *Stainless Steel,* ASM Int., Materials Park, OH, 1986.

Potts, D. L., and J. D. Gensure: *International Metallic Materials Cross-Reference,* Genium Publishing, Amsterdam, NY, 1989.

Krauss, G.: *Stainless Steels, Steels: Heat Treatment and Processing Principles,* ASM Int., Materials Park, OH, 1990.

VanderVoort, G. F.: “The Metallography of Stainless Steels,” *Journal of Metals,* vol. 41 (3), pp. 6–11, March 1989.

Schaeffler, A. L.: “Constitution Diagram for Stainless Steel Weld Metal,” *Met. Prog.,* vol. 56 (11), pp. 680–680B, 1949.

Blair, M.: Cast Stainless Steels, *Properties and Selection: Irons, Steels, and High-Performance Alloys,* vol. 1, *ASM Handbook,* ASM Int., Materials Park, OH, 1990.

Davison, R. M., T. DeBold, and M. J. Johnson: Corrosion of Stainless Steels, *Corrosion,* vol. 13, *ASM Handbook,* ASM Int., Materials Park, OH, 1987.

chapter

12

Cast Irons

In a quiz show, a presumably well-informed inquisitor asked which of the following were alloys and which metals: copper, brass, bronze, cast iron, and aluminum. The contestant named brass, bronze, and cast iron as alloys, only to be informed that he was wrong—cast iron was a metal! Engineers know cast iron as a cheap structural material; materials scientists consider it to be a large family of eutectic ferrous alloys, consisting mainly of iron, carbon (typically more than 2%), and silicon. However, it has also become, with metallurgical control, an alloy that can be produced with high strength and several other desirable properties not readily obtained with other alloys.

12.1 CAST IRON (Fe-C-Si) PHASE DIAGRAM

The chemical composition of the liquid cast iron and its nucleation potential are two factors that determine the alloy's ability to form a stable or metastable eutectic. If this so-called graphitization potential is high, the carbon-rich second phase that forms will be graphitic. If it is low, the metastable austenite–iron carbide (Fe_3C) eutectic will form with significantly different mechanical properties. As one reviews the classifications of cast irons, it becomes apparent that controlling the structure of the eutectic is the essence of cast iron engineering.

It is evident that silicon as well as carbon is an important alloying agent. Rather than present the complex ternary phase diagram, a vertical section at a 2% Si content has been reproduced in Fig. 12.1. As shown in previous chapters, a third component changes the eutectic and eutectoid from an invariant reaction at a single temperature to an invariant reaction occurring over a range of temperatures, even under equilibrium conditions. The carbon contents of austenite at the eutectoid and the eutectic are reduced by the silicon from 0.8 and 1.7 to about 0.6 and 1.5% C, respectively. The carbon content of the eutectic is also reduced from 4.25 to 3.65%. Since

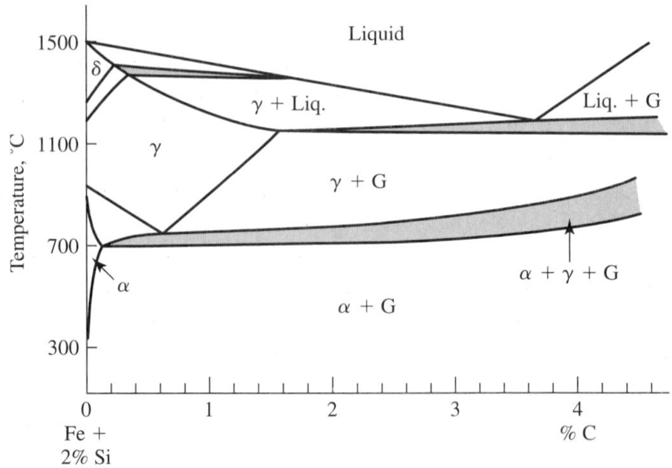

Figure 12.1 A vertical section of the Fe-C-Si system at a constant 2% Si concentration. Crosshatched fields are three-phase areas.

this reduction in eutectic carbon content is linear with silicon content over the range of interest here, it can be expressed as

$$\text{Eutectic carbon }\% = 4.25 - 0.30\ (\%\ \text{Si})$$

This relationship is in common use as a means for determining the closeness to eutectic composition of a gray cast iron.[1]

12.2 GRAY CAST IRON SOLIDIFICATION

When cast iron is melted, poured, and allowed to solidify in a mold and then broken, the fractured surface may be gray, white, or a mixture of the two. A sooty gray appearance of the fracture means that the structure consists of graphite flakes in a metallic aggregate. The graphite being weak, fracture progresses from flake to flake, with ultimately a broken surface consisting mostly of graphite. If the fracture is white, then the structure consists solely of carbide and ferrite, and fracture has occurred along or through the brittle white masses of eutectiferous Fe_3C. A mottled color means that, in some places, graphite flakes are present and, elsewhere, eutectiferous carbide. Thus, historically, cast iron names were based on the appearance of their fracture surfaces. Today, classifications based on microstructural features are common such as compacted graphite or bainitic (austempered) cast iron.

[1] J. Rehder (*Trans. AFA,* 1947) showed that, if it is assumed that one silicon atom ties up three iron atoms as Fe_3Si, then removing the proper quantity of Fe as Fe_3Si for a specific silicon concentration leaves the carbon content of the eutectic in the remaining free iron at 4.25%. The Fe_3Si is not a separate phase but is in solution. Other data cited by Rehder support the postulation of Fe_3Si as a molecular compound in solution.

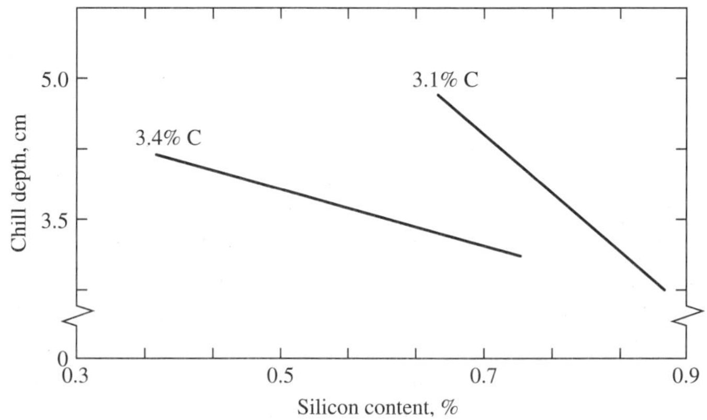

Figure 12.2 Chill depth (depth of white cast iron from a chilled surface) upon solidification of a cast iron as diminished by increasing silicon content for two constant carbon contents.

Since the stable Fe-C system involves iron and graphitic carbon, it is reasonable that slow solidification favors the stable iron-graphite system and rapid solidification favors the metastable system, iron and iron carbide. Iron foundrymen are familiar with the practice of using metal chill plates at certain parts of a sand mold. These plates induce rapid solidification at these points, which may result in a hard wear-resisting structure of Fe-Fe_3C eutectic. Elsewhere, slower freezing may result in a softer iron-graphite structure.

The most important elements present in cast irons of ordinary composition are carbon and silicon, and a high content of either element or both is conducive to solidification of the iron according to the stable system, i.e., with graphitic carbon. Figures 12.2 and 12.3 show that an increase in the concentration of either element, the other being kept constant, decreases the depth of white or carbidic iron when the alloy is solidified under conditions of constant surface chilling (somewhat analogous to the Jominy end quench for hardenability of steels). Other elements that promote graphitization if present in the cast iron are nickel, aluminum, titanium, zirconium, and copper.

Manganese is in itself a moderately strong carbide-forming element, and its presence in cast iron tends to stabilize carbide or prevent graphitization. For example, if just enough silicon is present to give a completely graphitic structure under specific controlled conditions, a slight increase in manganese may make the iron mottled (partly carbide in the eutectic). A large increase in manganese may also cause the iron to solidify completely in the metastable or carbide condition. This effect is contingent on the absence of sulfur.

Sulfur chemically acts to stabilize iron carbide, although it does not participate in the carbide formation. It has a very strong influence: it is ordinarily considered that each 0.01% S is sufficient to neutralize the graphitizing influence of 0.15% Si. However, sulfur has a strong affinity for manganese to form a manganese sulfide compound, which has little influence on carbide or graphite formation. Therefore, the first

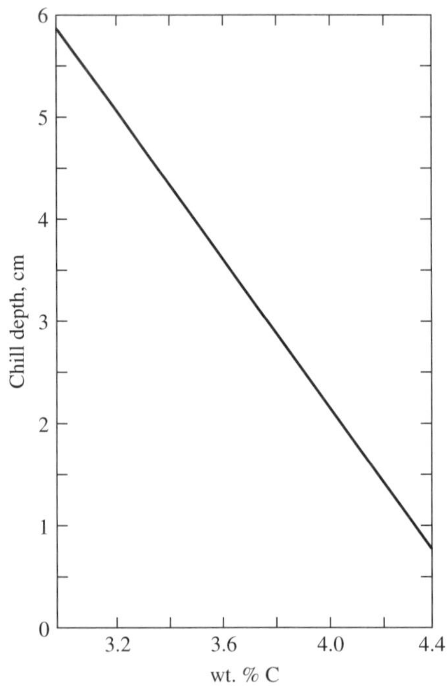

Figure 12.3 Chill depth upon solidification of a cast iron as diminished by increasing carbon content for an approximately constant silicon content of 0.81 to 0.89%.

additions of sulfur to an iron with moderately high manganese content have an indirect graphitizing tendency by removing the carbide-stabilizing manganese. On the other hand, the first additions of manganese to a moderately high-sulfur iron remove some of the sulfur from an active to an inactive role and thus promote graphitization.

Phosphorus chemically acts to promote carbide formation. Physically, it forms a phosphide eutectic with a melting point below that of iron and carbon. This causes the $\gamma + Fe_3C$ eutectic to solidify over a temperature range which increases the critical time available for silicon to promote graphitization. With moderately low phosphorus contents, the physical effect predominates, and graphitization is encouraged, but large amounts of phosphorus cause it to act chemically as a carbide stabilizer.

Gaseous elements, particularly hydrogen and oxygen, may enter cast iron during melting and affect the cast structure. Hydrogen seems to stabilize carbides and when combined with oxygen as steam or moisture is quite active in preventing graphitization during solidification, but it does not seem to have any effect on graphitization of the solid iron. Oxygen, as iron oxide, seems to promote graphitization during solidification and retard the process in the solid alloy (during malleabilizing, page 467).

Of the alloying elements, nickel, which like silicon dissolves completely in ferrite, also acts as a graphitizer, while the carbide-forming elements, specifically chromium and molybdenum, tend to stabilize the carbide phase. Thus, by adding these elements in proper relative quantities, the graphitizing characteristics of the original unalloyed cast iron will be relatively unaffected. The desirable properties obtained by the addition of alloying elements will then be related to their effect on graphite flake size and on the transformation characteristics of the austenite present, with graphite, after solidification.

Graphitization has so far been used to refer to eutectic solidification, a choice between one of the two following eutectic reactions:

$$\text{Liquid} \rightarrow \text{austenite} + Fe_3C \qquad [1]$$

or

$$\text{Liquid} \rightarrow \text{austenite} + \text{graphite} \qquad [2]$$

A third possibility is that the reaction is according to the metastable system, i.e., to $\gamma + Fe_3C$, with an almost immediate decomposition of the Fe_3C. That is,

$$\text{Liquid} \rightarrow \gamma + Fe_3C \quad \text{then} \quad Fe_3C \rightarrow \gamma + \text{graphite} \qquad [3]$$

Graphite nuclei may exist in liquid cast iron and if present at the time of solidification will promote solidification directly as graphite. Superheating destroys these graphitic nuclei and induces greater undercooling. Since the γ-Fe_3C eutectic temperature is a few degrees below the γ-graphite eutectic, the undercooled liquid solidifies according to the metastable system. However, a high silicon and carbon content induces immediate subsequent graphitization. Inoculation of the superheated iron may permit attainment of the normal and usually more desirable graphite structure type A of Fig. 12.4.

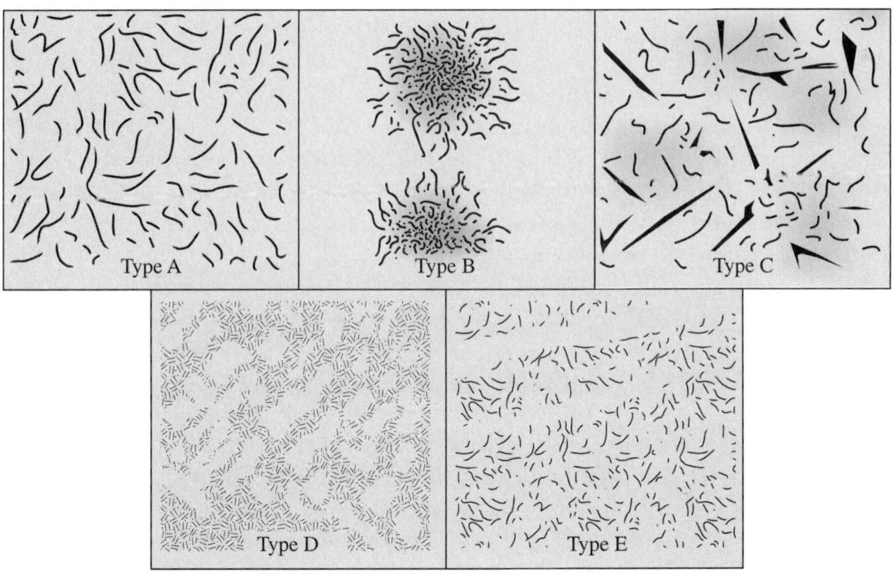

Figure 12.4 Graphite flake types in gray cast iron according to AFS-ASTM standards: type A, uniform distribution, random orientation; type B, rosette groupings, randomly oriented; type C, superimposed different flake sizes, random orientation; type D, interdendritic flakes, random orientation; type E, interdendritic flakes, preferred orientation.

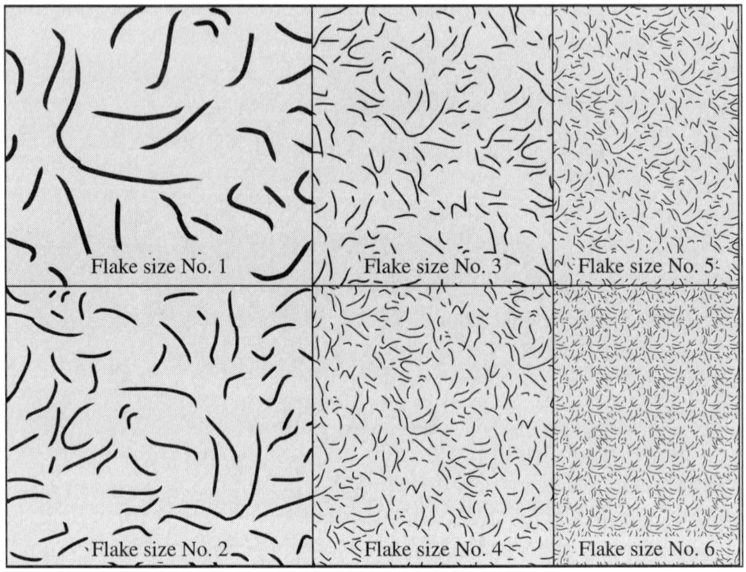

Figure 12.5 Graphite flake size in gray cast iron according to AFS-ASTM standards (reduced 50% in reproduction): no. 1, longest flakes greater than 100 mm at ×100; no. 2, longest flakes 50 to 100 mm at ×100; no. 3, longest flakes 25 to 50 mm at ×100; no. 4, longest flakes 12 to 25 mm at ×100; no. 5, longest flakes 6 to 12 mm at ×100; no. 6, longest flakes 3 to 6 mm at ×100; no. 7 (not shown), 1.5 to 3 mm at ×100; no. 8 (not shown), less than 1.5 mm at ×100.

Rapid solidification in general results in a finer grain size and a finer eutectic structure, and cast iron is no exception to this rule. Figure 12.5 shows the AFS-ASTM graphite flake-size chart. The finest flakes, size 8, are not shown here, but size 6 represents a structure where the longest flakes are ⅛ to ¼ in long at ×100. At the other extreme, size 1 represents a structure where the longest flakes are 2 to 4 in long at ×100.

12.3 DUCTILE CAST IRON SOLIDIFICATION

There is an important cast iron alloy that, without heat treatment, shows appreciable tensile ductility. During solidification of this iron, the graphite forms as tiny balls or spherulites rather than as flakes during solidification of gray iron, or as the compacted aggregates ultimately formed by decomposition of eutectic Fe_3C during heat treatment in the case of compacted graphite iron. Because of its morphology, the alloy was originally termed nodular iron or spheroidal graphite (SG) cast iron. Today, it is predominantly referred to as ductile cast iron.

This unique spheroidal graphite structure is obtained by taking a liquid iron of composition similar to that of unalloyed gray cast iron. However, undesirable residual elements such as sulfur and phosphorus must be held below specified limits. Magnesium, cerium, and other elements are controlled to produce the spheroidal graphite structure.

In any event, nodular spherulitic graphite (SG) or ductile cast iron when properly produced will have no more than 0.01% S and, retained from the addition agent, 0.03 to 0.08% Mg and 1.0 to 1.5% Ni. The presence of magnesium changes the nucleation and growth of graphite during solidification of the Fe-C eutectic from the formation of flakes to the formation of nearly perfect spherulites. Elimination of flakes, of the internal notches associated with flakes, and of the weakening effect of large surfaces of graphite, i.e., substitution of spherulites having a low surface-to-volume ratio, results in the cast iron being ductile.

12.4 CONCEPTS OF GRAPHITIZATION IN CAST IRON

An iron-carbon alloy may solidify as a white carbidic cast iron or as a gray graphitic iron. In either case, the major part of the structure is austenite, primary and eutectiferous. Upon slow cooling in a mold from the eutectic to the eutectoid temperature, the austenite will reject excess carbon. In a hypoeutectic cast iron, there are no or comparatively few austenitic grain boundaries since the structure consists of primary dendrites completely surrounded by interdendritic eutectic. However, the eutectic carbon exerts a powerful preferred nucleation effect, and the precipitation of excess carbon from austenite, as required by the slope of the A_{cm} line, results in the growth of eutectic Fe_3C in the white iron or of graphite in the gray iron.

Upon cooling in air or in the mold to the eutectoid or A_1 temperature, the austenite will transform, and, as in the case of the eutectic, two alternative eutectoid reactions are possible:

$$\gamma \rightarrow \alpha + Fe_3C \quad [1]$$

$$\gamma \rightarrow \alpha + \text{graphite} \quad [2]$$

The normal pearlite reaction [1] will generally occur in white cast irons (Micros. 12.1 and 12.2), although in the eutectiferous austenite no pearlite may appear. Again, the massive eutectic Fe_3C exerts a nucleation effect, and the eutectoidal Fe_3C may form integrally on the massive carbide, leaving a structure of ferrite and eutectic carbide.

In a gray cast iron, the type of eutectoid reaction will depend on the carbon and silicon content, on the possible presence of other alloying elements in the austenite, and, of course, on the cooling rate. A given composition may result in a completely graphitic eutectic structure and a regular pearlitic carbide eutectoid structure. Therefore, of a total 3.50% C, 2.70% may be in the form of graphite and 0.80% as Fe_3C in pearlite.

A fairly strongly graphitizing composition, e.g., one containing more silicon, and less rapid cooling may cause the eutectoid to form ferrite and graphite. This may occur only at points of preferred nucleation adjacent to graphite flakes, with

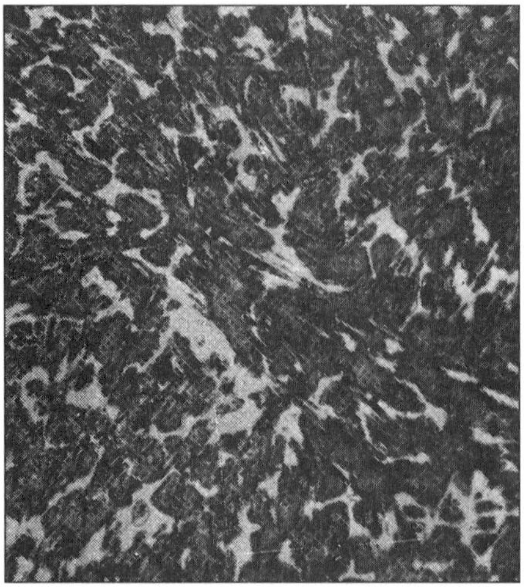

Micrograph 12.1 Commercial white cast iron (about 2.50% C); ×50; picral etch. This specimen shows a hypoeutectic structure in which the gray background was chiefly primary austenite that transformed on further cooling to pearlite. The white masses are iron carbide. The eutectic structure of γ and Fe_3C is not evident, since the austenite part of the eutectic was formed upon, and indistinguishable from, the primary austenite. Although white iron is ordinarily considered to be brittle, hard, and unmachinable, this structure is sufficiently low in carbide for the pearlite to be nearly continuous. This white cast iron showed a slight ductility in the tension test and could be machined.

Micrograph 12.2 Same white iron as Micro. 12.1; ×1500; picral etch. At a high magnification, details of the pearlitic background and massive carbides become readily visible. Although in this hypoeutectic structure three different forms of carbide should exist, specifically, eutectic Fe_3C, Fe_3C separating out from γ along the A_{cm} line, and eutectoid carbide, only the first and last are distinguishable. Presumably, Fe_3C separating out from austenite along the A_{cm} line formed on the massive, eutectiferous Fe_3C already present, rather than at the austenitic boundaries. This form of preferred nucleation, or, in reality, growth of large nuclei already present, is rather frequently encountered in all alloys where a comparable condition exists. Here, the eutectoid reaction was normal, and all the eutectoid carbon apparently formed pearlitic carbide.

the eutectoidal graphite a part of the flakes and ferrite bands alongside the flakes. Elsewhere a pearlitic structure may be present so that, of a total of 3.50% C, 3.10% may be present as graphitic flakes and only 0.40% as Fe_3C in pearlite. This structure would naturally be softer than a completely pearlite-graphite structure. Maximum softness is achieved by a composition and cooling rate that result in complete graphitization of the eutectoid carbide as well as that of the eutectic, resulting in a structure of ferrite (containing silicon in solid solution) and graphite flakes.

The discussion thus far is concerned with graphitization upon continuous cooling from the original solidification. It is possible, of course, to have a composition which would result in graphitization upon extremely slow cooling but which would form a completely carbidic or white cast iron upon ordinary cooling, as of a sand casting. In this case, reheating of the casting and holding at an elevated temperature will result in a decomposition of the iron carbide. For example, a 2.25% C–1.0% Si melt would probably solidify as a white iron. Upon heating the casting to 900°C, the structure would be austenite, with about 1.1% C in solution plus eutectiferous Fe_3C. This would, with time, graphitize as

$$Fe_3C \longrightarrow 3Fe\ (\text{austenite, } 1.1\%\ C) + \text{graphite}$$

There is a considerable difference in the structure of graphite formed in a solid alloy previously free of graphite and graphite formed in a solidifying iron or in an iron already containing flakes. Flakes are brittle and disrupt the continuity of plastic ferrite, and their edges constitute sharp internal notches. Graphite formed by the above reaction with no flakes present grows in all directions, thereby forming compact aggregates. These disrupt the ferrite much less and greatly diminish the internal notch effects. Therefore, the alloy with compact aggregates of graphite is malleable and shows some ductility in contrast to a structure with flake graphite that is fairly brittle.

Cooling from the graphitizing temperature of about 900°C again permits carbon to separate from austenite and form on the aggregates already present. At the A_1 temperature, very slow cooling or a strongly graphitizing composition will result in a graphitic eutectoid reaction, with the eutectoidal graphite forming on the aggregates already present. Faster cooling or a less graphitizing reaction can give a pearlitic eutectoid and a final structure of graphite aggregates in pearlite.

The factors affecting the rate of graphitization of a white cast iron are:

1. Composition, particularly carbon and silicon.
2. Rate of cooling of the original casting upon solidification. The finer grain size and eutectic structure of a more rapidly solidified iron give more interfacial surfaces for the initiation of graphitization (Fig. 12.6).
3. Rate of heating to the annealing or graphitizing temperature.
4. The temperature of annealing (Fig. 12.6).
5. The atmosphere in the annealing furnace.
6. Hereditary effects from the original pig iron or the melting process.

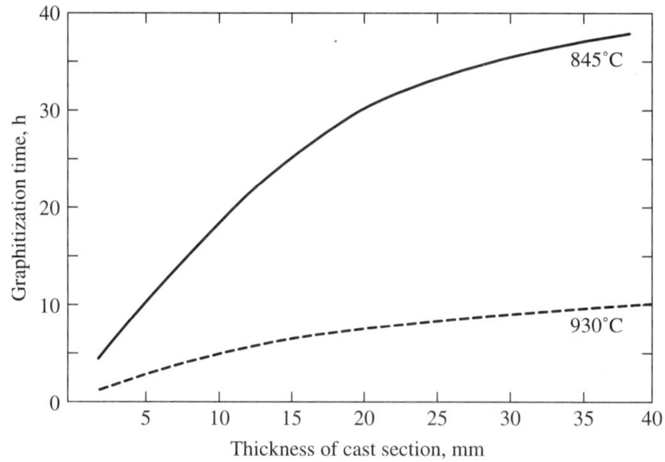

Figure 12.6 Effect of annealing temperature and prior solidification rate, i.e., thickness of cast section, on the time for complete graphitization of a white cast iron at two annealing temperatures.

This includes compositional effects not ordinarily measured, i.e., the oxygen and hydrogen contents, etc. Some production foundries add 0.25% graphite to the melt charge, and it appears that this results in some graphite nuclei in the melt that are retained in the white iron and speed up malleabilizing upon annealing.

12.5 PROPERTIES OF CAST IRONS

Structurally, the matrices of gray cast irons resemble steels in that they contain varying proportions of ferrite and pearlite. The ferrite may be a little stronger than that of most carbon steels because of the dissolved silicon, but the pearlitic part of the structure may be softer as a result of its somewhat greater coarseness. Overbalancing both these factors is the weakening and embrittling effect of a relatively large amount (3% by weight corresponds to 12% by volume) of the soft, brittle graphite flakes that disrupt the continuity of the plastic matrix. The edges of the flakes are likely to be comparatively sharp, and each acts as an internal notch that, upon deformation, tends to initiate a crack in the plastic matrix. For this reason, gray cast irons break with a brittle, sooty gray fracture and at stresses of only 140 to 420 MPa (Table 12.1). Some foundries have been pouring iron for several years with no test bars in the 420-MPa class ever breaking below this value. Such higher strengths have been achieved in two ways, by greatly refining the graphite flake size and by attaining a fine, completely pearlitic matrix. Success in achieving this structural condition depends on close control of the chemical composition of the iron and of pouring temperatures.

Table 12.1 Typical compositions of gray cast iron based on strength and section thickness

Type*	%C	%Si	%P	%S	%Mn	Carbon Equivalent†	Metal Section	Brinell Hardness	Tensile Modulus, GPa	Tensile Strength, MPa
Class 20 (L)	3.65	2.50	0.50	0.10	0.60	4.56	0.25	180	84	170
Class 20 (M)	3.50	2.40	0.40	0.10	0.60	4.34	1.00	170	84	150
Class 30 (L)	3.30	2.20	0.22	0.10	0.65	4.03	0.50	200	105	225
Class 30 (H)	3.05	1.90	0.20	0.10	0.60	3.68	1.00	217	105	225
Class 40 (L)	3.10	2.05	0.17	0.10	0.55	3.77	0.25	230	120	310
Class 40 (M)	3.05	1.85	0.15	0.08	0.60	3.65	1.00	225	120	300
Class 60 (L)	2.85	2.05	0.15	0.09	0.60	3.51	....	250	135	435
Class 60 (M)	2.65	2.00	0.10	0.07	0.85	3.37	....	270	135	435

* (L), light section; (M) medium section; (H) heavy section.
† % C + 0.3 (%Si + %P). These cast irons do not obey Hooke's law, so the modulus is arbitrarily expressed as the slope of the stress-strain curve at a stress equal to one-fourth the tensile strength.

Aside from strength properties, gray cast irons have several other features that fit them particularly well for certain applications. Their relatively low melting point and ready castability make them relatively cheap, although, naturally, costs will be increased if high-strength specifications require use of alloying elements and laboratory control. More important in some applications is the fact that the internal structural discontinuities offer sites for the local dissipation of vibrational energy. This is equivalent to saying that gray cast irons have a high internal friction or damping capacity (Fig. 12.7). Used as a base for machinery or any equipment subject to vibration, the structure of the iron permits the vibrations to be absorbed internally. Machine bases, or piano frames, could be made of welded steel assemblies, but these assemblies would not so readily absorb external vibrations. At frequencies approaching the natural vibration period of the structure, the amplitude of vibration

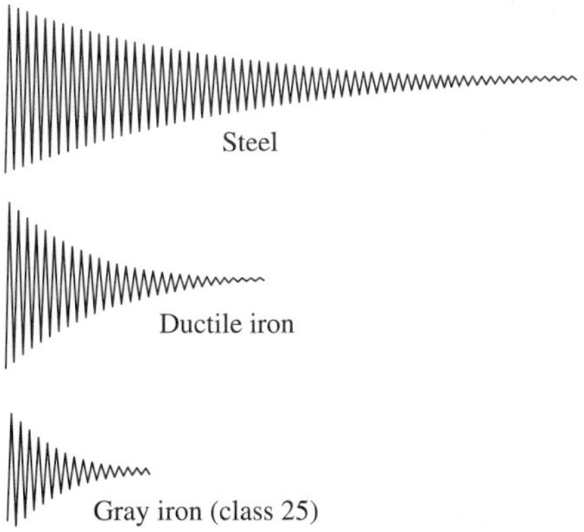

Figure 12.7 Relative damping capacities of steel, ductile, and class 25 gray iron. The ordinate is amplitude of vibration, which decreases with time along the abscissa to the right.

might well increase to the point where the structure would break by fatigue stressing. The great importance of this feature of cast irons is coming to be recognized and more fully utilized.

The data of Fig. 12.8 show the relationship between tensile and compressive strengths of cast iron. On the average, an iron with a tensile strength of 140 MPa has a compressive strength of 560 MPa. Therefore, it is evident that the notch effects of graphite flakes are limited to tensile loading. In compression, the properties of the iron are basically determined by the matrix structure.

Other properties of gray cast iron that make it specifically more suitable than steel for many uses include machinability, corrosion resistance, and wear resistance. Because of the discontinuities offered by graphite and the lubricity of graphite, gray cast iron can be machined much more easily than steel. This, for example, is the major reason it is universally used for glass molds for which machining costs are generally high. Because of the high silicon content and perhaps other factors, gray cast iron is more resistant to atmospheric and some other types of corrosion than mild steel. Finally, gray cast iron is notch-insensitive; i.e., the presence of sharp surface notches does not noticeably decrease strength. The reason, of course, is that there are so many internal notches from the edges of graphite flakes that an external notch is relatively harmless.

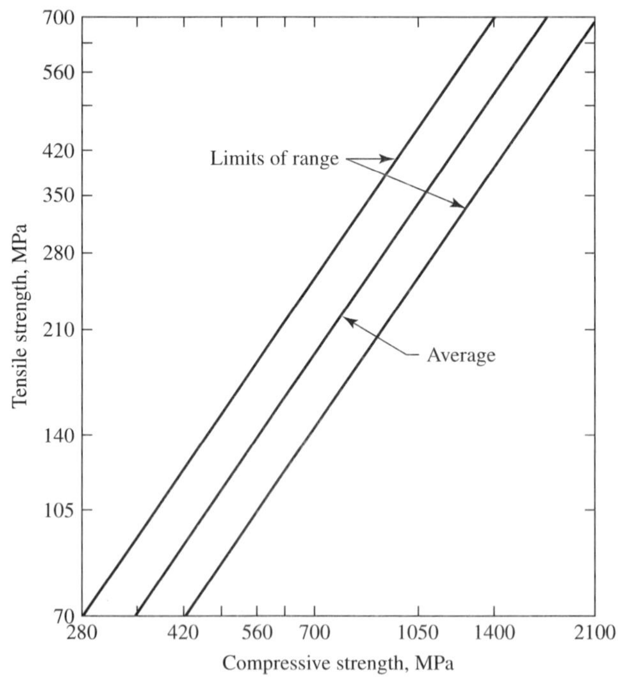

Figure 12.8 Relation between tensile and compressive strength of gray iron.

Compacted graphite iron castings are of intermediate cost and properties between gray iron and steel. The compacted aggregate form of graphite, or temper carbon, does not interrupt the continuity of the ferritic matrix (Micros. 12.3 and 12.4), and the combined structure may show strengths of around 385 MPa, combined with elongation values in the vicinity of 12 to 18%. Close control of composition and pouring temperatures (sometimes achieved by the use of special melting furnaces) gives a metal that is consistently white in the as-cast form and yet graphitizes quickly upon reheating.

The ductile cast iron, formerly called nodular or spherulitic graphite cast iron, in the as-cast state contains varying proportions of ferrite and pearlite in the matrix, depending on composition, inoculation practice, and cooling rate, the latter normally being determined by section size. The graphite spherulites distributed in the matrix have a range of diameters or sizes, those first formed being largest and therefore the most readily observed in microstructures (Micro. 12.5). These spherulites influence the properties of the iron less than flakes, and the properties of nodular iron therefore more closely follow those of the matrix. To obtain the softest iron (Table 12.1), i.e., one with a completely ferritic matrix (Micro. 12.5), the as-cast structure must be annealed by heating to 900°C and cooling, the rate being 20°C/h between 800 and 650°C. This permits carbon in solution in austenite to diffuse to the graphite spherulites rather than forming Fe_3C in pearlite.

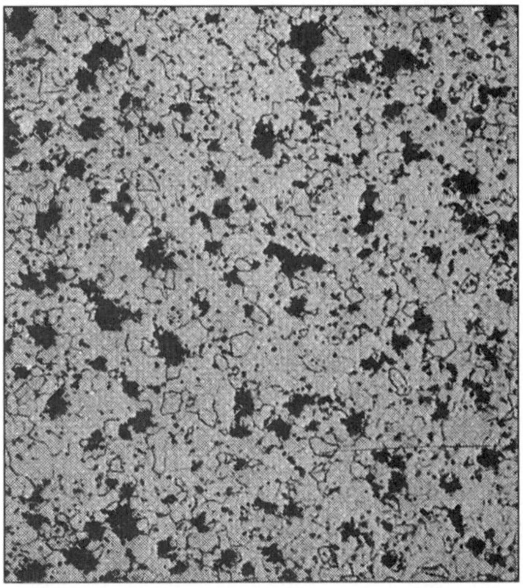

Micrograph 12.3 Compacted graphite cast iron; ×50; nital etch. If the white cast iron of Micro. 12.1 were heated long enough below the eutectic temperature, the carbide would decompose to graphite by the reaction $Fe_3C \rightarrow 3Fe\ (\gamma) + C$ (graphite). The graphite forming in a solid structure grows in all directions from nuclei in the carbide to form compacted aggregates of graphite or temper carbon particles in austenite. If sufficient amounts of dissolved silicon are present, very slow cooling through the eutectoid causes the eutectoid reaction to be the reaction $\gamma \rightarrow \alpha + C$ (graphite), and this additional graphite forms on the nodules already present. The final structure shown here consists of a continuous, moderately fine-grained ferrite containing irregular, randomly dispersed compact graphite aggregates.

Micrograph 12.4 Compacted graphite cast iron; ×300; nital etch. At a higher magnification, details of the ferritic matrix and compacted aggregates of graphite or temper carbon particles are more evident.

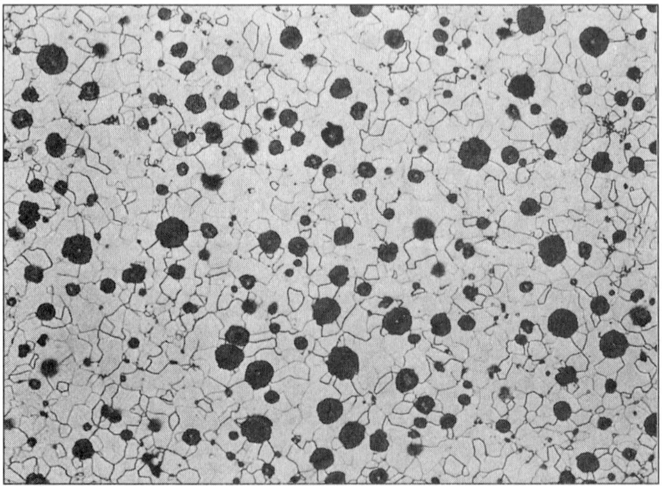

Micrograph 12.5 Ductile cast iron; ×100; nital etch. This is the structure of a commercial casting which has been given a ferritizing anneal by heating to 900°C, cooling at 20°C/h from 800 to 650°C, and then normal furnace cooling. The spheroidal graphite is dispersed in a completely ferritic matrix.

Any cast iron, i.e., gray, compacted graphite, or ductile type, can be heated above the A_1 temperature and form austenite with the carbon content approaching 1.0% and with no noticeable change in the prior graphite structure. Of course, the more pearlite rather than ferrite that is originally present in the cast iron structure, the more easily and quickly is eutectoidal austenite formed. Subsequent cooling—in the furnace (annealing), in air (normalizing), or in oil or water (quenching)—will have the usual effects discussed in prior chapters (with the limitation that, upon slow cooling, the graphite present will provide nuclei that will alter the usual eutectoid structure). Thus one can produce a range of matrix structures from soft to very hard. However, in quenched or martensitic cast iron, although the matrix may have a hardness of C65, the aggregate or composite of matrix and graphite will show a hardness of only C50 since graphite is so soft. Tempering of martensitic cast irons follows the same laws as those applicable to tempering of hardened carbon steels, with regard to the property changes of the matrix.

Both compacted graphite and ductile cast irons have some of the inherent virtues of gray cast iron such as machinability, corrosion resistance, and wear resistance superior to mild steel. Either type of iron can replace steels for many applications. However, the choice between these two is not easily made, since their mechanical and other properties are quite similar (Table 12.2). Both types have reasonable ductility and shock resistance. Lacking the internal notches of graphite flakes, both are more susceptible to external or surface notch embrittlement than gray cast iron. Neither one can readily be welded, but both can be brazed to make assemblies. The final criterion for selection of one instead of the other tends to be cost: because of the absence of a lengthy malleabilizing anneal, ductile cast iron tends to be slightly lower in cost.

The service requirements and economics dictate the choice between gray iron, ductile iron, and compacted graphite iron for a specific service, as in all materials problems. Gray iron is the lowest in cost and also the easiest to cast and obtain a sound, essentially nonporous structure. Ductile iron is basically similar but is subject to more shrinkage on solidification and requires bigger risers.

Table 12.2 Compositions and properties of compacted graphite and ductile cast iron

	Compacted Graphite		Ductile	
	Ferritic	Pearlitic	Ferritic 60-45-10*	Pearlitic 80-60-3†
Carbon, %	2.25	2.25	3.50	3.40
Silicon, %	1.15	1.30	2.40	2.15
Manganese, % 0.40	0.75	0.50	0.50	
Sulfur, %	0.10	0.10	0.01	0.01
Nickel, %	—	—	1.00	1.00
Magnesium, %	—	—	0.06	0.06
Tensile modulus, GPa	175	195	160	160
Tensile strength, MPa	390	560	455	665
Yield strength, MPa	260	350	335	455
Elongation in 50 mm, %	20	6	12	5

*Annealed at 900°C and furnace-cooled.
†As cast.

PROBLEMS

1. Compare the similarities of steel and cast iron and contrast their differences.
2. Describe the influence of graphite shape and distribution on the properties of cast iron.
3. Other than carbon, explain the effects of element additions, such as silicon and manganese, on cast iron.
4. Why is welding of ductile cast iron more practical than welding of compacted graphite cast iron? Explain the desirability of preheating the casting of ductile iron to 260°C before are welding. Why is full annealing after welding necessary to obtain ductility at the welded joint?
5. Give the reason in terms of microstructure for chromium additions of 0.5 to 1.0% having the effect of increasing the dimensional stability or minimizing the growth of gray cast iron subjected to service temperatures of 450 to 550°C.

REFERENCES

Bates, C. E.: "Alloy Element Effects on Gray Iron Properties," Part II, *Trans. AFS,* 1986, p. 889.

C. F. Walton and T. F. Opar (Eds.): *Iron Castings Handbook,* Iron Castings Society, Des Plaines, IL, 1981.

Gundlach, R. B., and J. F. Janowak: "A Review of Austempered Ductile Iron Metallurgy," in *Proceedings of the First International Conference on Austempered Ductile Iron: Your Means to Improved Performance, Productivity and Cost,* ASM, Materials Park, OH, 1984.

ASM Handbook, vol. 1, *Properties and Selection: Irons, Steels, and High Performance Alloys,* Materials Park, OH, 1990.

Foundryman's Handbook, Facts, Figures, and Formulae, 9th ed., Elsevier Science, New York, 1988.

Courtney, T. H.: *Mechanical Behavior of Materials,* 2d ed., McGraw-Hill, New York, 2000.

chapter

13

Aluminum Alloys

For over fifty years, aluminum has been a tonnage metal, second only to steel as a major factor in the metal industry. The growth has been based on characteristics such as light weight, nonrusting properties, reasonably good strength and ductility, easy fabrication, modern metallurgical control of structure and properties, and favorable economics.

The major alloying elements used in aluminum are silicon, magnesium, and copper. The Al-Si phase diagram (page 114) is a simple eutectiferous system between aluminum containing a maximum of 1.65% Si in solid solution and nearly pure silicon. Supercooling, achieved partially by chill casting or more completely by small additions of metallic sodium which suppresses the nucleation of silicon crystals, lowers the eutectic temperature from 577°C to 550 to 560°C and increases the silicon in the eutectic from 11.6% to 13 to 14% (see Micro. 13.13).

Inspection of the Al-Si phase diagram leads to certain useful qualitative conclusions. Since the aluminum solid solution makes up 85 to 90% of the eutectic, this phase should be and is continuous in the microstructure and therefore the eutectic structure shows some ductility. However, since silicon is hard and brittle, hypereutectic alloys containing large primary crystals of silicon would not be expected to be as useful as eutectic or hypoeutectic alloys.

The aluminum end of the Al-Mg system (page 115) shows a eutectic at 450°C between aluminum containing 15.35% Mg in solid solution and a β phase which in composition approaches the stoichiometric ratio Al_3Mg_2. This intermetallic phase is quite hard and brittle. By use of the lever rule, one can see that the eutectic contains so much β phase that the eutectic structure could be expected to be brittle—and is. Therefore, useful alloys must contain less magnesium than the maximum soluble in the solid solution, i.e., less than 15.35%.

The aluminum end of the Al-Cu phase diagram (page 132) is somewhat similar to the above in showing a eutectic between a solid solution containing 5.6% Cu and an intermetallic compound, here designated T, which approaches the stoichiometric ratio $CuAl_2$. Again, the lever rule indicates that the T phase, which is hard and brittle, is dominant in the eutectic structure, which is therefore brittle. Consequently, most useful wrought alloys contain less copper than the 5.6% maximum amount soluble in solid aluminum.

Wrought aluminum alloys, i.e., those used for rolled sheet or shapes, forgings, or extrusions, have standard alloy composition and temper designation systems. A four-digit numerical system developed by the Aluminum Association has traditionally been used to identify composition; it is still the common language among the industry. The international Unified Numbering System has adopted a five-digit designation for aluminum with an A prefix. Examples are given in tables throughout this chapter. Rationale for the four-digit system is given below.

1XXX Pure aluminum: 99.00% aluminum or greater. The principal alloy is 1100. AA1060 is 99.60% aluminum.

2XXX Age-hardenable alloys containing about 4.5% Cu. AA2014, containing 0.8% Si, 0.8% Mn, and 1.5% Mg besides the copper, used principally for forgings.

3XXX Cold-work hardenable manganese alloys, AA3003 contains 1.2% Mn; the stronger 3004 alloy contains an additional 1.0% Mg.

4XXX Silicon alloys. Only one is commonly used, AA4032, Al + 12% Si, and 1% each of Mg, Cu, and Ni. This is an age-hardenable, forgeable alloy used for elevated-temperature service, e.g., for automotive pistons. Age hardening in this case comes from decreased solubility with temperature of the phase Mg_2Si.

5XXX Work-hardenable Mg alloys. These include:

5052: 2.5% Mg + 0.5% Cr

5056: 5.2% Mg + 0.5% Cr + 0.1% Mn

5186: 4.5% Mg + 0.5% Cr + 0.8% Mn

In each case, chromium is present as a grain-refining agent, and Mn simply enhances the strength of the cold-rolled alloy. These alloys exhibit good weldability when the magnesium content is below 5%.

6XXX Heat-treatable Mg-Si alloys. AA6061 is used widely for hot extrusions, e.g., window or door frames.

7XXX Heat-treatable zinc alloys. The aircraft structural alloy 7075, containing 5.5% zinc, 2.5% magnesium, 1.5% copper, and 0.4% chromium.

The basic temper designations for wrought products are:

F As fabricated.

O Annealed.

H Work-hardened.

H1X Strain-hardened only. The X designates the degree of hardening; H12 is quarter hard, H18 full hard, and H19 extra hard (usually from an 80+% cold reduction).

H2X Strain-hardened and then recovery annealed. After the slight anneal softening, the strength will be as high as the H1X equivalent but ductility will be higher.

H3X Similar to H2X except this is called stabilized and applies only to Mg (5XXX) alloys that otherwise gradually soften at room temperatures.

T Used for age-hardenable alloys that usually are solution-treated and aged to a stable condition. The digits after the T represent aging as follows:

T1 Naturally aged at room temperature following an elevated-temperature operation, e.g., hot extrusion.

T4 Solution treated, quenched, and naturally aged.

T6 Solution treated, quenched, and aged at a slightly elevated temperature.

T8 Same as T6 except that the alloy is cold-worked before aging.

13.1 WORK-HARDENABLE WROUGHT ALUMINUM ALLOYS

The major alloys in this classification and their properties in three tempers are listed in Table 13.1. It is evident that in the cold-worked tempers, for example, H18, tensile strengths of from 275 to 415 MPa can be obtained with respectable ductilities,

Table 13.1 Properties of principal work-hardenable aluminum alloys as 1.6-mm sheet in three common tempers

Alloy	Composition, %	Temper	Tensile Strength, MPa	Yield Strength, MPa	Elongation in 50 mm, %
1100	99.00 Al	O	90	35	35
		H14	125	120	9
		H18	170	155	5
3003	1.2 Mn	O	110	40	30
		H14	150	55	16
		H18	200	190	4
3004	1.2 Mn, 1.0 Mg	O	180	70	20
		H14	245	200	9
		H18	290	250	5
5052	2.5 Mg, 0.2 Cr	O	195	90	30
		H34	265	220	14
		H38	295	260	8
5056	5.2 Mg, 0.1 Mn, 0.1 Cr	O	295	155	35
		H18	440	410	10
		H38	420	350	15

eliminating the need for heat treatments with their extra costs. At the same time, particularly for cold-rolled sheet and strip, costs have been kept low. Thus these products have become large-tonnage items for the specialized mills set up to produce them.

13.2 HEAT-TREATABLE ALUMINUM ALLOYS

These alloys are appreciably higher in cost than the work-hardenable types, yet they are the backbone of the aerospace industry. Here, savings in weight made possible by higher strength far outweigh additional costs of more scrap generated in-plant, lower volume, and costs of heat treatments.

The principal alloys in this category used for sheet are listed with compositions and strength properties in Table 13.2. Alloys used for forgings are listed in Table 13.3.

Among the forging alloys, 6061 is so soft at high temperatures that very intricate forgings can be produced; although moderate strength values are obtained by heat treatment, they are low in comparison with other alloys. Alloy 2014 represents the other extreme. It develops very good strength properties but is not suitable for complicated shapes because of its relative stiffness, even at forging temperatures. The other alloys are intermediate in these regards.

Table 13.2 Properties of principal age-hardenable aluminum alloys for 1/16-in sheet in several tempers

Alloy	Composition	Temper	Tensile Strength, MPa	Yield Strength, MPa	Elongation in 50 mm, %
2014	4.4 Cu, 0.8 Si, 0.8 Mn, 0.4 Mg	O	190	100	22
		T4	440	280	18
		T6	490	420	10
2024	4.5 Cu, 0.6 Mn, 1.5 Mg	O	190	80	20
		T4	475	330	20
		T6	480	400	10
		T86	525	495	6
6061	1.0 Mg, 0.6 Si, 0.2 Cr	O	125	55	25
		T4	245	150	25
		T6	315	280	12
		T91	410	400	6
7075	5.5 Zn, 2.5 Mg, 1.5 Cu, 0.3 Cr	O	230	105	17
		T6	580	510	11

Table 13.3 Typical mechanical properties of aluminum forging alloys

Alloy	Tensile Strength, MPa	Yield Strength, MPa	Elongation in 50 mm, %	BHN	Endurance Limit, MPa
4032-T6	385	320	9	120	110
2014-O	190	100	18	45	90
2014-T4	435	310	20	105	125
2014-T6	490	420	13	135	125

The data of Table 13.2 hold several points of interest. Alloy 6061 is relatively low in cost, is easily formable, and has a high corrosion resistance. The 2024 alloy, available also in extruded forms, etc., offers a tensile strength five or six times that of annealed, commercially pure aluminum, with fairly good ductility in the ordinary tensile test. It does have poor local plasticity and is sensitive to notches so that a sheet test specimen will frequently break at a scratched gage mark. The alloy, except when quenched in cold water or cold-worked and aged, also is sensitive to intercrystalline corrosion from grain-boundary precipitation of the second phase. The matrix phase at grain-boundary areas, after precipitation, is depleted in solute atoms and in a corrosive solution acts as an anode. The rest of the grain serves as a cathode, and as current passes, the grain-boundary areas are selectively dissolved. Under these circumstances, a small amount of corrosion (small in terms of total amount of metal corroded away) may be accompanied by a large decrease in strength and ductility. The ductility is especially reduced by the notch effect of grain-boundary attack.

For use in corrosive environments, a surface layer of pure aluminum, which is integrally attached by hot rolling and pressure welding, protects the 2024 alloy. The deformation breaks the aluminum oxide layer always present to some degree on aluminum alloys and permits atomic structural continuity to be developed across the interface. Although slightly reducing strength and endurance properties, particularly in bending fatigue, the pure aluminum is anodic to the alloy and prevents intercrystalline corrosion, even at cut edges. Furthermore, the highly plastic aluminum surface layer eliminates notch sensitivity so that minor scratches are no longer sources of brittle failure.

The atomic continuity across the alloy-aluminum interface permits diffusion to occur, and if copper diffuses to the surface, most of the advantages of the clad form are lost. The diffusion may be minimized by restricting the heat treatment to the shortest effective time (see Table 13.4). If the diffusion rate of copper in aluminum at the heat-treatment temperatures is known and certain assumptions about the initial state of the alloy are made, it is possible to calculate the maximum time the clad alloy could be held at the solution heat-treatment temperature. Practically, it is customary to take test specimens of clad annealed sheet and heat-treat in a liquid salt bath for increasing increments of time, for example, 2, 5, 10, and 20 min, and determine the minimum time required to develop substantially standard mechanical properties. There are

Table 13.4 Recommended heat treatments for wrought-aluminum alloys

	Annealing Temperature			Solution Treatment			Precipitation Treatment		
Alloy	Temp., °C	Time, h	Temper	Temp., °C	Time, h	Temper	Temp., °C	Time, h	Temper
1100	340	*	O						
3003	410	*	O						
3004	400	*	O						
5052	340	*	O						
2014	410	3*	O	505	‡	T4	170	10	T6
2017	410	3†	O	505	‡		190	9	T84
2024	410	3†	O	490	‡	T4	190	12	T81
7075	410	3*	O	465	‡		120	24	T6
2025	410	3*	O	515	‡	T4	170	10	T6
4032	410	3*	O	510	‡	T4	170	10	T6
6061	410	3*	O	520	‡	T4	160 or	18	T6
							175	8	T6

*Time in furnace need only be long enough to bring all parts to the annealing temperature; cooling rate is unimportant.

†To obtain full softening, cooling should be at a rate no faster than 30°C/h down to 260°C; the subsequent rate is unimportant.

‡Time may vary from 10 min for sheet in salt bath, 60 min for plate in air to at least 4 h in air for average forgings. Time for clad products should be minimized to prevent diffusion of alloying elements from core to surface. Rapid transfer from the furnace to a cold-water quench is recommended except for large forgings, which may be quenched in hot water.

chemical and metallographic methods for determining whether or not this time permitted copper to diffuse to the surface of the clad material to an unfavorable extent.

Table 13.4 illustrates that the 7075 alloy shows appreciably higher strengths than 2024. Its yield strength is frequently higher than the tensile strength of 2024. The alloy is not so readily formed as 2024, and it shows more sensitivity to cracking under the combined influence of a tensile stress and corrosive environment. The latter tendency can be minimized or eliminated by surface peening, usually with steel shot. This treatment induces a residual compressive stress, as shown on page 297. Thus, it reduces the actual surface tensile stress that may result from a given bending force.

13.3 CAST ALUMINUM ALLOYS

Aluminum alloys may be cast in sand molds, in gravity-fed iron molds (permanent molds), or in steel dies under pressure. Solidification is much more rapid when the alloys are cast in metal molds. Therefore, a given alloy will be finer-grained and stronger as a permanent mold casting (Table 13.5).

Table 13.5 Effect of solidification rate, slowest for sand castings and fastest for pressure die castings, on the properties of three common aluminum casting alloys

		Sand Cast		Permanent Mold Cast		Pressure Die Cast	
Alloy	Composition, %	Tensile Strength, MPa	Elongation in 50 mm, %	Tensile Strength, MPa	Elongation in 50 mm, %	Tensile Strength, MPa	Elongation in 50 mm, %
43F	5 Si	135	8	160	10	230	9
214F	4 Mg, 1.8 Zn	175	9	190	7	280	10
356 T61	7 Si, 0.3 Mg	230	4	265	5		

The alloys are usually obtained, made up to proper composition, in the form of pigs, although virgin aluminum may be alloyed with "master" alloys (for example, Al + 33% Cu) to obtain the specified composition. In addition, there is usually 25 to 75% secondary metal or scrap to be used up. Melting is customarily done in iron pots with the iron protected by a lime or similar ceramic coating to prevent the pickup of iron by liquid metal. The deleterious effect of iron on the properties of a cast 10% Si alloy is shown by the data of Table 13.6. The reduction in both strength and ductility is caused by an increasing amount of coarse Al-Fe-Si crystallites in place of fine eutectiferous silicon. The data of Table 6.2 indicate that iron impurity combines with copper and aluminum, forming an insoluble Al-Cu-Fe compound. This reduces the amount of copper that can dissolve in aluminum and therefore has an adverse effect on the properties of heat-treated Al–4.5% Cu alloys. Micrographs 13.1 to 13.14 illustrate microstructural details of several cast alloys.

In melting aluminum alloys, it is necessary to avoid overheating for two reasons: (1) As discussed in Chap. 3, overheated metal, when cast into a mold, solidifies more slowly because of the additional heat to be removed and, therefore, a coarser grain and weaker structure will be obtained. (2) Aluminum at high

Table 13.6 Effect of iron impurity on the properties of modified chill-cast Al-10% Si alloy

% Si	% Fe	Tensile Strength, MPa	Elongation, in 50 mm, %	Reduction in Area, %	BHN
10.8	0.29	217	14.0	15.3	62
10.8	0.79	216	9.8	11.6	65
10.3	0.90	210	6.0	6.2	65
10.1	1.13	171	2.5	2.2	66
10.4	1.60	126	1.5	1.0	68
10.2	2.08	78	1.0	0.2	70

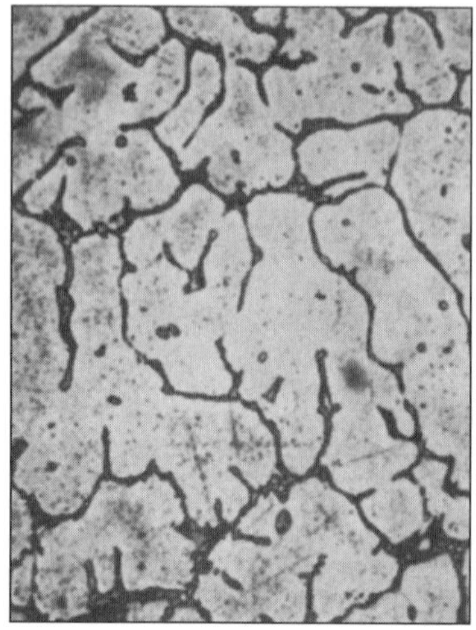

Micrograph 13.1 Al + 8% Cu + about 1 to 1½% Fe and Si as impurities; ×50; 0.5% HF, then 20% hot H_2SO_4 etch. This is a hypoeutectic structure consisting of cored primary α_{Al} dendrites (the dendritic characteristic is not very evident) surrounded by the eutectic of α_{Al} and θ (or $CuAl_2$).

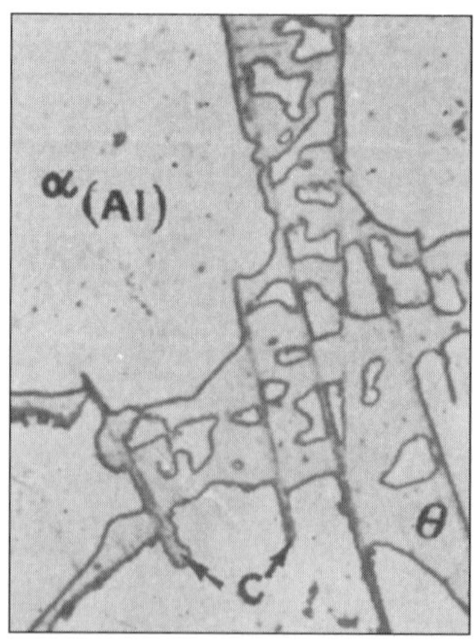

Micrograph 13.2 Same alloy (8% Cu) ×1000; 0.5% HF, then hot 20% H_2SO_4 etch. A greatly magnified view of the α_{Al} and θ_{CuAl_2} eutectic shows the brittle θ phase is continuous, probably because this eutectic consists of 58% θ and 42% α_{Al} (proportions calculated on a weight basis). The needles (marked *C*) extending through the eutectic are an Al-Cu-Fe compound originating from the iron impurity.

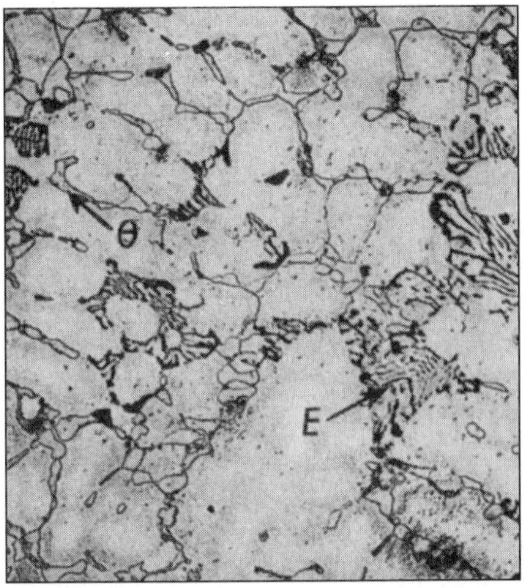

Micrograph 13.3 Al + 4% Cu with controlled impurities of Fe and Si (alloy 195); ×75; 0.5% HF etch, then hot 20% H_2SO_4. The as-cast structure reproduced here shows cored dendrites of α_{Al} surrounded by the $\alpha_{Al} + \theta$ eutectic and a eutectic structure of α_{Al} and Al-Fe-Si compound (*E*) originating from the impurities. Although the copper content of this alloy lies to the left of the $\alpha_{Al} + \theta$ eutectic horizontal, some eutectic is found in the cast structure because of the metastable position of the solidus on rapid cooling of the casting.

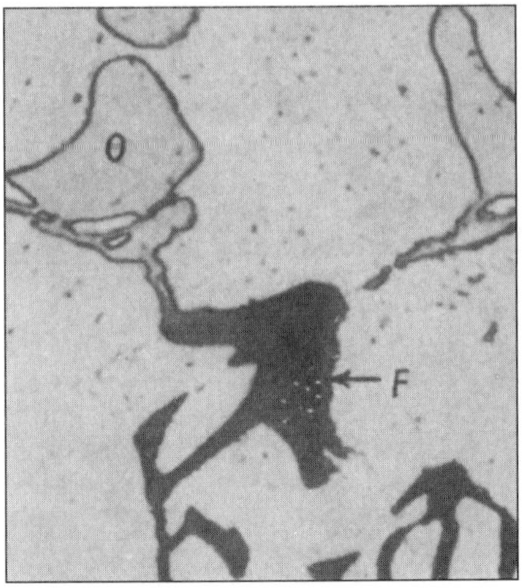

Micrograph 13.4 Same alloy (4½% Cu) ×1000; 0.5% HF etch, then hot 20% H_2SO_4. Again, the *q* present does not appear to be part of a eutectic structure, since the α_{Al} of the eutectic is not inside the *q* but outside, in contact with and indistinguishable from the primary α_{Al}. The θ and Al-Fe-Si compound appear to be isomorphous, since there is a continuity from one structure to the other, with a gradation in the coloring or degree of attack by the etchant.

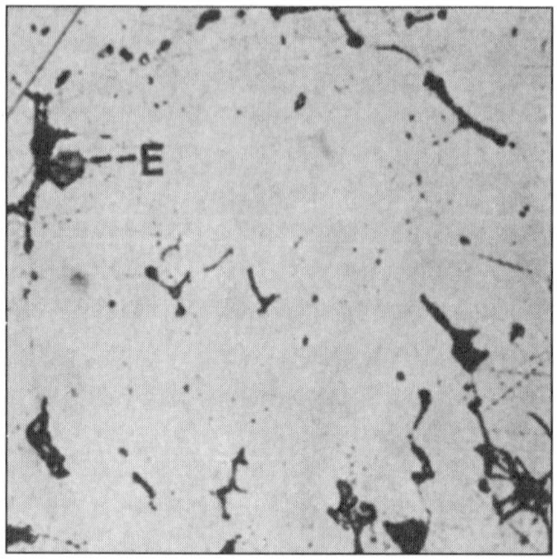

Micrograph 13.5 Same alloy (4.5% Cu) ×75; 0.5% HF, then hot 20% H_2SO_4; structure after heat treatment as follows: 15 h at 510°C, followed by a water quench; reheated 15 min in high-pressure steam. This structure should be compared with that of Micro. 13.3. It is evident that the heat treatment at 510°C has caused the $CuAl_2$ to dissolve in the α_{Al} matrix (thus the high-temperature treatment is called a *solution anneal* or *solution heat treatment*). Quenching in water after the high-temperature soak prevented precipitation of the dissolved θ upon cooling the alloy to temperatures at which two phases exist according to the diagram. Reheating the metastable or supersaturated solid solution has not caused particles of $CuAl_2$ to form in a size visible at this magnification. Particles of the Al-Fe-Si compound (*E*) are still present, since this phase has no measurable solubility in α_{Al} and remains substantially unchanged through the heat treatment.

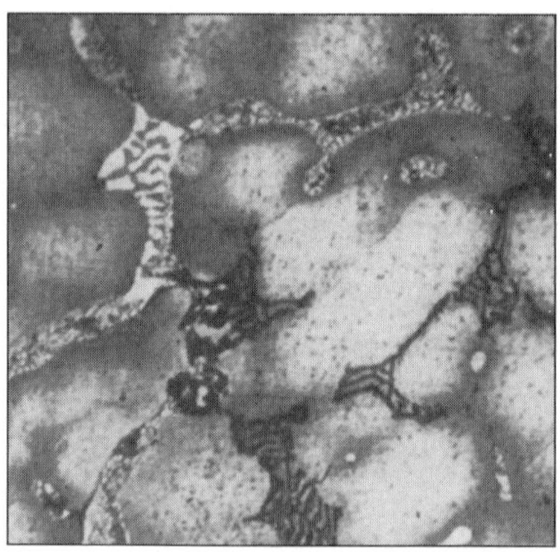

Micrograph 13.6 Same alloy (4.5% Cu) ×500; etched with HCl, HNO_3, and HF in water (Keller's reagent); structure after heating to 575°C and quenching. According to the Al-Cu phase diagram, when this alloy is heated above about 565°C, it is in a two-phase field, α_{Al} + liquid, and the liquid has an almost eutectic concentration of copper. The liquid phase forms at the α_{Al} grain boundaries and upon quenching must solidify there, in large part as a eutectic, which in this system is brittle. With a nearly continuous brittle structure enveloping each grain, the total structure is now brittle and weak. The etch used for this micrograph has brought out coring in the aluminum matrix, i.e., variations in the amount of dissolved copper. Although the binary Al-Cu diagram shows the safe solution annealing temperature range to be 510 to 565°C, the presence of a eutectic network in the original casting of this alloy would set the upper limit at 545°C. More importantly, the presence of iron and silicon impurities means that a ternary or quaternary eutectic exists, with a melting point of about 525°C, so that the safe heat-treatment temperature range is quite narrow, 505 to 520°C. The lower limit is set by the necessity for dissolving all the copper in order to obtain optimum properties, while the upper limit is set by the melting point of any eutectic present in the structure.

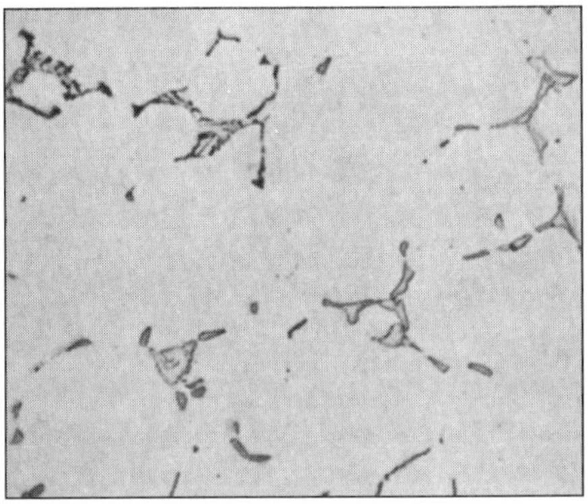

Micrograph 13.7 Same 195 alloy (4.5% Cu; Fe and Si impurities); heated only 6 h at 510°C and quenched; ×100; 0.5% HF, then 20% H_2SO_4 etchants. The θ compound, here a light clear gray, has not been completely dissolved because of too short a solution treatment. The amount present has decreased and the interdendritic continuity broken up. However, less than optimum properties would be shown by this structure.

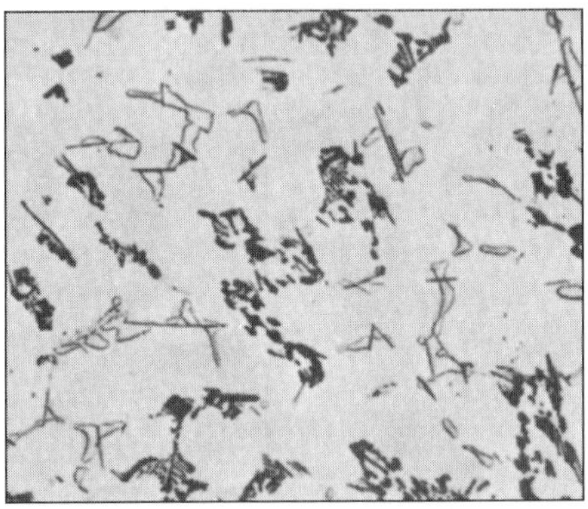

Micrograph 13.8 Alloy 122 (10% Cu, 0.2% Mg; Fe and Si impurities); chill cast, heated 8 h at 510°C and quenched; ×100; 0.5% HF, then 20% H_2SO_4 etchants. The solution treatment of this alloy cannot, of course, dissolve all the θ phase, but the alloy becomes somewhat tougher when the continuity of the eutectiferous θ is broken up. Furthermore, the hardness of the alloy is increased and dimensional stability gained by a subsequent precipitation treatment.

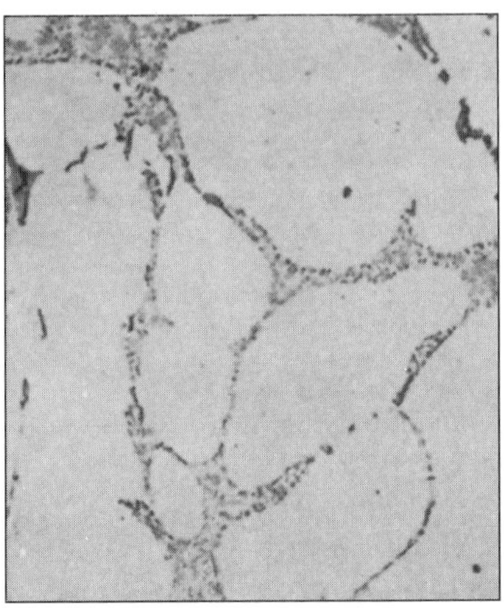

Micrograph 13.9 Al + 3% Si + 0.3% Mg, sodium-modified sand casting, as cast; ×100; 20% H_2SO_4 etch. This low-silicon hypoeutectic alloy shows primary aluminum and a very fine interdendritic eutectic network containing silicon crystallites and also Mg_2Si (black).

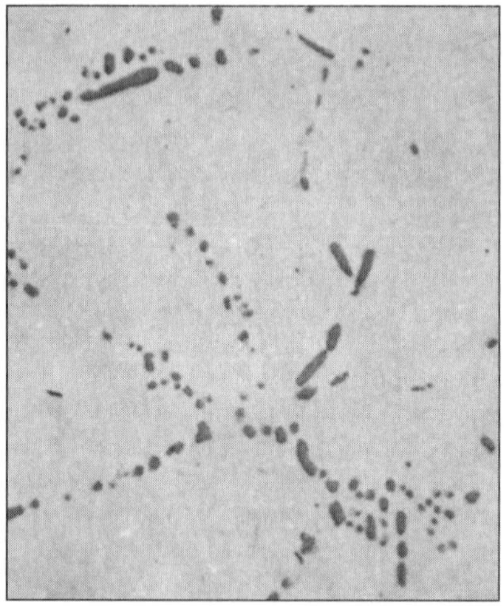

Micrograph 13.10 Alloy of Micro. 13.9, heat-treated 20 h at 540°C; ×100, 0.5% HF etch. The long-time solution heat treatment has dissolved some silicon crystallites and the Mg_2Si. Its most marked effect, however, has been to agglomerate the undissolved silicon crystallites. After the α phase is saturated, remaining silicon particles continue to grow in a spheroidal manner. This occurs by smaller ones dissolving and an equal amount of silicon precipitating from the now supersaturated α onto other silicon particles that grow.

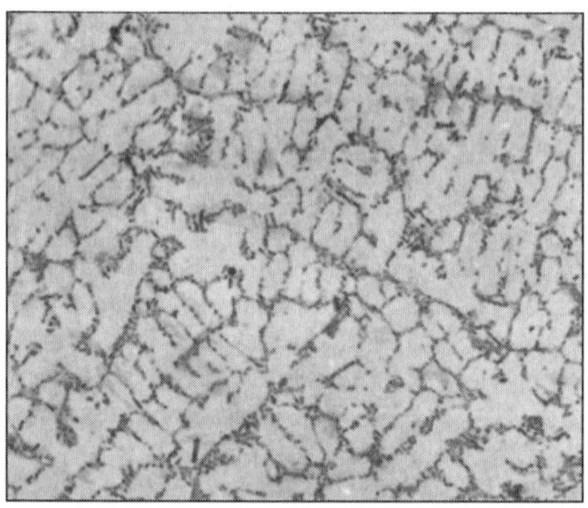

Micrograph 13.11 Al + 6% Si, chill cast; ×100; 0.5% HF etch. In this chill casting, the amount of eutectic is greater, naturally, than in the 3% Si alloy (Micro. 7.13). Note that chill casting has given fine, eutectiferous silicon particles, but they are not so fine and are more platelike than those in the modified alloy.

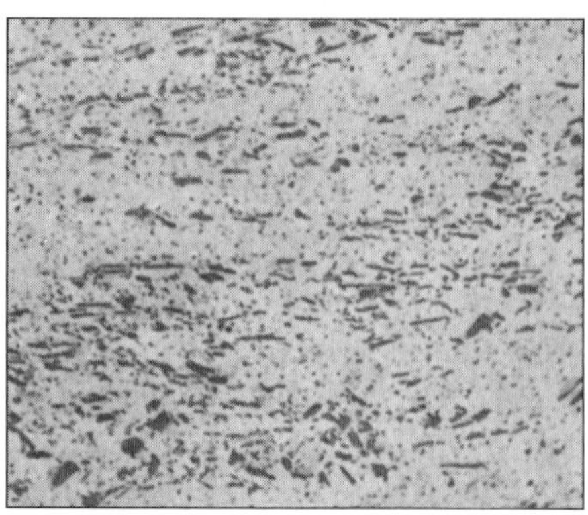

Micrograph 13.12 Al + 6% Si, hot forged at about 500°C; ×100; 0.5% HF etch. Forging has broken up the eutectic structure and aligned silicon crystallites in the flow direction. The temperature of hot work has also induced growth of the eutectiferous crystallites. Slow cooling from the forging temperature has caused precipitation of silicon from solid solution, which gives the "dirty" background.

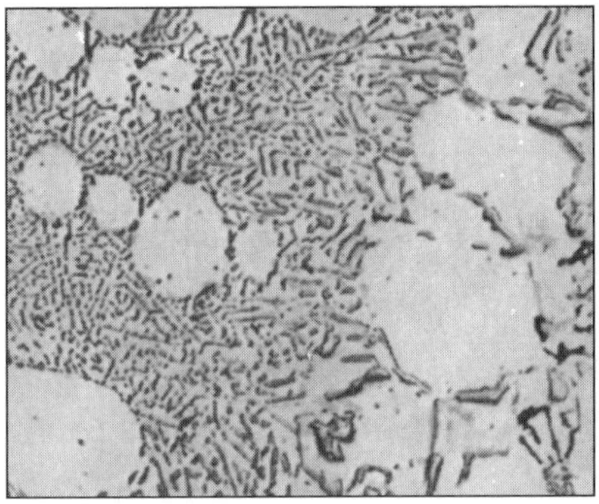

Micrograph 13.13 Al + 11% Si + 0.3% Mg, chill cast; ×100; 0.5% HF etch. There is less primary α now and the eutectiferous silicon is finer in some places than in others because heat released by the eutectic solidification causes slower freezing of the final liquid.

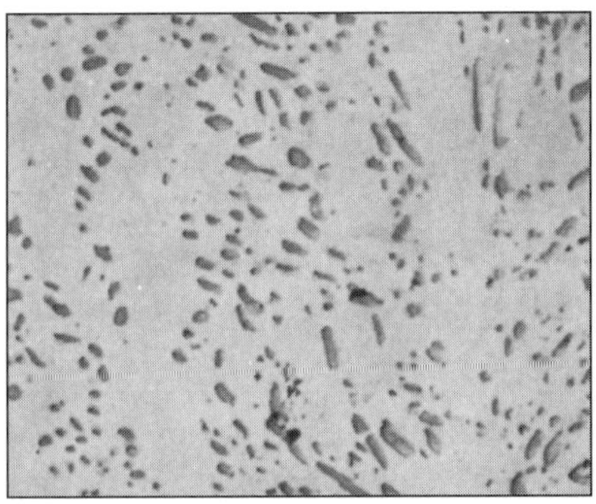

Micrograph 13.14 Same specimen as Micro. 13.13, heated 20 h at 540°C and quenched; ×100; 0.5% HF etch. As in Micro. 13.10, the solution heat treatment has resulted in a pronounced agglomeration of undissolved silicon crystallites and, relatedly, a diminution in the number of particles.

temperatures reacts readily with water vapor to form aluminum oxide and hydrogen. The liquid metal aggressively dissolves this gas. As the temperature decreases, the hydrogen becomes less soluble, and during freezing the solubility drops very sharply. At the same time, there are physical factors such as dendrites to encourage hydrogen bubble formation and, unfortunately, simultaneously to prevent the escape of the bubbles. Therefore, aluminum alloys melted on humid days or in furnaces fired with city gas (yielding considerable water vapor in the products of combustion) are always more porous than castings made on dry days or melted in electric furnaces.

It would be too expensive to melt with electricity in air-conditioned foundries, so hydrogen pickup is minimized by avoiding overheating and then removed, if necessary, by flushing with neutral gas. Dry nitrogen bubbled through the liquid before pouring the castings will remove hydrogen by physical means. Chlorine gas is more expensive but when bubbled through the liquid not only removes hydrogen but also chemically purifies the bath by removing oxide or dross films that are entrapped during melting.

The casting of the alloys follows normal practice. Although overheating is undesirable, the liquid must be sufficiently hot to fill the mold before solidification begins, to avoid *cold shuts* or *misruns*. These are discontinuities arising from streams of metal coming from different directions that make physical contact without completely fusing because of heavy oxide films and lack of fluidity.

The casting characteristics of pure aluminum are improved by alloying even more than the mechanical properties. The higher-silicon-content alloys listed in Table 13.7 all exhibit very good fluidity in the liquid state, are not hot-short or susceptible to hot cracks, and are also less susceptible to shrinkage cavities. Since these cavities tend to be interdendritic and structurally connected, this means that the Al-Si alloys such as 43, 108, 319, and 356 are all well suited for pressure-tight parts. The choice between them might depend on the mechanical properties required and the cost, these usually being directly proportional. Relative properties are given by the data of Table 13.8.

Table 13.7 Composition of some common aluminum casting alloys

	Composition, %					
Alloy	**Cu**	**Si**	**Mg**	**Other**	**Heat-Treatable**	**Characteristics**
112	7.0			1.7 Zn	*	General-purpose castings
212	8.0	1.2		1.0 Fe	*	Better castability than 112
195	4.5	0.8			Yes	High strength
43		5.0			No	Excellent castability
108	4.0	3.0			*	Strength and castability
319	3.5	6.3			Yes	Strength and castability
356		7.0	0.3		Yes	Strength and castability
220			10.0		Yes	Highest strength
142	4.0		1.5	2.0 Ni	Yes	Good hot strength
A132	1.0	12.0	1.2	0.8 Fe, 2.5 Ni	Yes	Hot strength; low thermal expansion

*Alloy not usually heat-treated, although some improvement in properties is possible.

Table 13.8 Typical properties of some common aluminum casting alloys in sand-cast form

Alloy	Condition	Tensile Strength, MPa	Yield Strength, MPa	Elongation in 50 mm, %	BHN	Endurance Limit, MPa	Thermal Expansion, 30–100°C
112-F	As cast	170	105	1.5	70	65	12.2
212-F	As cast	160	100	2.0	65	55	12.2
195-T4	12 h, 515°C, quench	225	110	8.5	60	40	12.2
195-T6	12 h, 515°C 4 h, 155°C	250	170	5.0	75	45	12.2
195-T62	12 h, 515°C 14 h, 155°C	280	210	2.0	95	50	
43-F	As cast	135	65	6.0	40	45	12.2
108-F	As cast	150	100	2.5	55	55	12.2
319-F	As cast	190	125	2.0	70	70	
319-T6	12 h, 515°C	250	170	2.0	80	70	
319-T6*	4 h, 155°C	280	190	3.0	95		
356-T51	8 h, 228°C	175	140	2.0	60	50	11.9
356-T6	12 h, 540°C 4 h, 155°C	230	240	4.0	70	55	11.9
356-T7	12 h, 540°C 4 h, 155°C	240	210	2.0	75		
356-T71	12 h, 540°C 3 h, 245°C	195	150	4.5	60		
356-T7*	12 h, 540°C 4 h, 155°C	280	190	5.0	90		
220-T4	Solution and quench	320	175	14.0	75	50	13.6
142-T21	3 h, 345°C	190	125	1.0	70	50	12.5
142-T77	6 h, 520°C 2 h, 345°C	105 190	195	0.5	85	55	
132-T551*	16 h, 170°C	265	195	0.5	105		10.5
132-T65*	8 h, 515°C 14 h, 170°C	330	300	0.5	125		

*Alloy cast in permanent mold.

The higher-copper-content alloys, for example, 112 and 212 in Table 13.7, are characterized by good casting characteristics and also better machinability than the silicon alloys. They are not so suitable, however, for pressure-tight castings or where both thick and thin sections adjoin in the same casting. These alloys are not heat-treated, although they would respond favorably, as could be anticipated from the phase diagram. The discussion of Micro. 13.8 would lead one to guess that, although the embrittling interdendritic eutectic network of α $CuAl_2$ could not be completely eliminated, the continuity of the $CuAl_2$ phase could be disrupted by spheroidizing effects with an improvement in ductility.

The 4.5% Cu alloy, 195, is more difficult to cast but, when heat-treated, has the best mechanical properties of any of the common sand-casting alloys. As Table 13.8 illustrates, its mechanical properties are poorer than those of the 10% Mg alloy, 220. In addition, the latter alloy is lighter, more resistant to corrosion, and more machinable. Unfortunately, 220, like Mg-base alloys, requires special inhibitors in the sand to prevent deep surface defects and in general shows poorer castability. These difficulties plus the high cost have prevented the 10% Mg alloy from being used extensively.

To return to the effects of impurities, the data of Table 13.9 in comparison with those of Table 13.8 show how noticeably superior are the properties of the 4.5% Cu alloy when made from high-purity aluminum. However, since the higher-purity base metal costs about twice as much per pound as ordinary primary aluminum, there is little industrial interest in the high-purity alloys.

The effect of magnesium variations, at three different silicon contents, on alloys of the 356 type, is shown by the graphs of Fig. 13.1. It is apparent that magnesium is more potent as a strengthener than silicon. In these alloys, the response to heat treatment is associated with the solution of the compound Mg_2Si and its precipitation during aging heat treatment.

Table 13.9 Effect of heat treatment on the properties of sand-cast Al–4.5% Cu alloy made of high-purity (99.93%) aluminum

Condition of Alloy	Tensile Strength, MPa	Yield Strength, MPa	Elongation in 50 mm, %	BHN
As cast, aged room temp.	140	60	7.5	45
1 h at 540°C, quenched, aged room temp. 2 days	225	160	5.5	76
8 h at 540°C, quenched, aged room temp. 2 days	280	160	14.6	74
40 h at 540°C, quenched, aged room temp. 2 days	295	170	19.0	83
40 h at 540°C, quenched, tested immediately	250	120	20.7	62

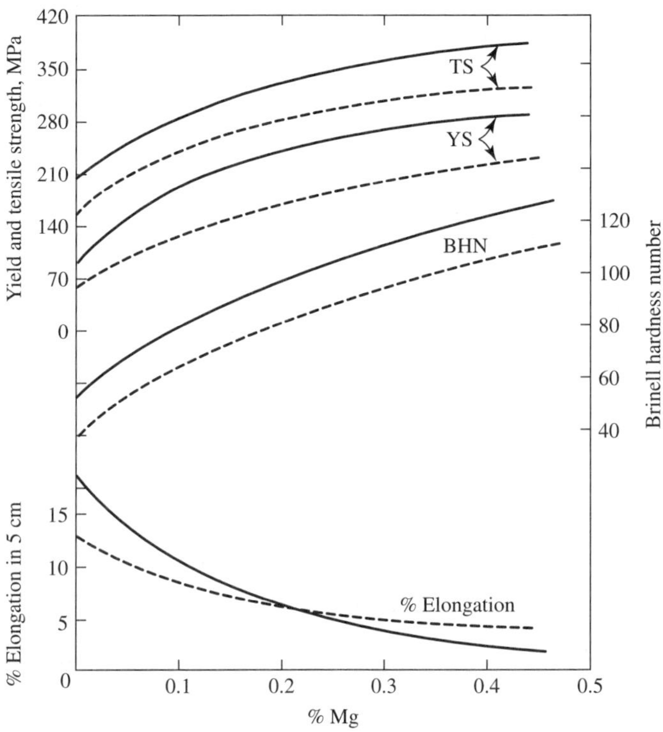

Figure 13.1 Effect of variation of magnesium content upon the mechanical properties of aluminum +5% Si alloy (dashed lines) and Al + 13% Si (solid lines). These data are based on tests of chill-cast test bars, solution-treated 2 h at 550°C, quenched and aged 20 h at 155°C. The alloys here are of the alloy 356 (7% Si, 0.5% Mg).

13.4 RESIDUAL STRESSES IN ALUMINUM ALLOYS

Aluminum alloys, like all other metals and alloys, are subject to distortion and residual stresses when quenched in cold water, as a result of the temperature gradients developed during cooling. Contraction of the more rapidly cooling surface compresses the interior, which plastically deforms to conform to the shrunken interior. As the center cools to the same temperature as the surface, it attempts to contract, but, by this time, the surface metal is cold and not very plastic. As a result, it is under an elastic compressive stress, and since the rigidity of the surface prevents the interior from attaining its stable dimensions, there is a balancing tensile stress in the central region. The development of residual or macrostresses, resulting from differential plastic deformation induced by thermal gradients of quenching, is illustrated for cylindrical shapes by Fig. 13.2. The intensity of the stresses that may be developed in some common alloys is given by the data of Table 13.10. These data were obtained

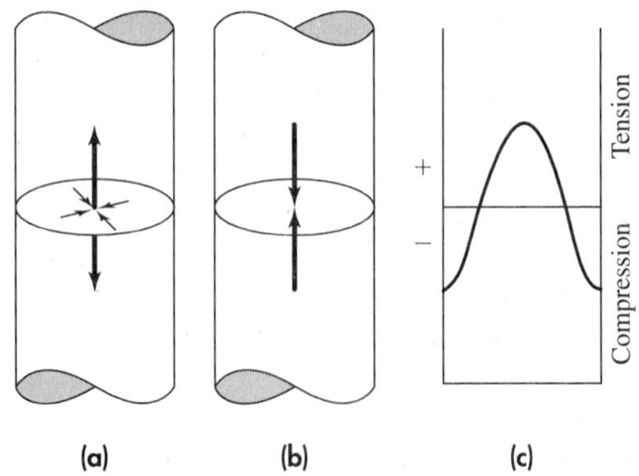

Figure 13.2 Origin of residual stresses upon quenching. (*a*) Cylinder immediately after immersion in quenching fluid; surface contracts and, being colder and stronger than the hot center, squeezes this section plastically along axis. (*b*) Later stage of quench when center cools and contracts, a movement resisted by the now cold and relatively stiff surface. (*c*) Longitudinal residual stress distribution in a radial direction across the cylinder; compression at the surface, changing to tension at the center.

Table 13.10 Length changes upon boring and residual stresses of aluminum alloy 122 cylinders (6.4 cm in diameter by 25 cm long), quenched from a solution treatment at 480°C

Cooling Conditions	Contraction in Total Length, micron	Surface Stress, MPa	Center stress, MPa
Annealed alloy	10	−4.1	4.1
Quenched, boiling water	25	−15.9	9.0
Quenched, ice water	30	−134.5	111.0
Annealed at 225°C	17	−39.3	58.6

by machining away part of the quenched cylinders, measuring the deformation that occurs when the stress system is thus unbalanced, and calculating the stress associated with the deformation by the equations of Sachs.

Comparable residual stresses may result from forming operations that involve differential plastic deformation, e.g., bending. The schematic drawing of Fig. 13.3 shows the surface that was deformed in *compression* has a residual *tensile* stress upon bending, and vice versa.

A third source of residual stresses is differential plastic deformation deliberately imposed by surface rolling or peening, which deforms the surface into tension as in Fig. 13.4 and leaves it in a state of elastic compression. This has already been referred to in a previous section as a means of preventing tensile stress-corrosion cracking in certain susceptible alloys such as 7075. The introduction of favorable compressive stresses by this means is also common practice in highly stressed steels and other alloys.

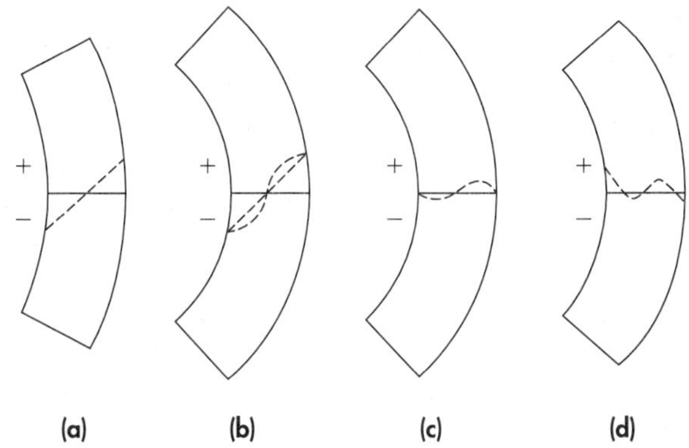

Figure 13.3 Origin of residual stresses upon cold bending. (*a*) Strip bent so that surface stresses (dotted line) just reach elastic limit of the metal at the surfaces; (*b*) continued bending causes plastic flow at surfaces, raises stress at inner zones nearer to the elastic limit; (*c*) release of the bending force causes elastic springback, here shown carried to point where surface stresses are zero but stress distribution makes this position unstable; (*d*) further springback to balance the stresses on either side of the neutral axis results in residual, and vice versa.

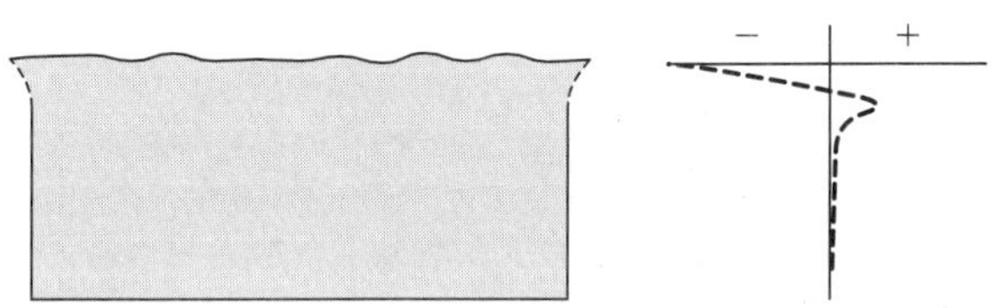

Figure 13.4 Shot peening of the immediate top surface of the block (left side) tends to spread it. Rigidity of the metal underneath prevents bending or spreading; thus the surface is in compression (right-hand schematic stress plot), with a balancing layer underneath in tension.

Residual stresses, whatever their source, can be reduced or virtually eliminated by a slight reheating, which, as described in Chap. 4, lowers the elastic limit and permits elastic stress to be reduced by a differential plastic flow. The data of Fig. 13.5 show a typical stress-relief, or relaxation, curve showing the effects of time at constant temperature on residual stress, Brinell hardness, and growth of aluminum alloy 122.

To minimize the distortion and residual stresses resulting from drastic quenches, many alloys are quenched in hot water, which cools the metal more slowly and, relatedly, minimizes the temperature gradients that cause residual stresses. The slower rate of cooling is still rapid enough to retain the solute in a supersaturated matrix solution *except at grain boundaries*.

Grain-boundary precipitation during cooling (or aging) may not adversely affect the mechanical properties attained by the alloy but may have serious consequences for its resistance to corrosion and for the physical properties shown by corroded specimens. A uniform chemical composition throughout the structure

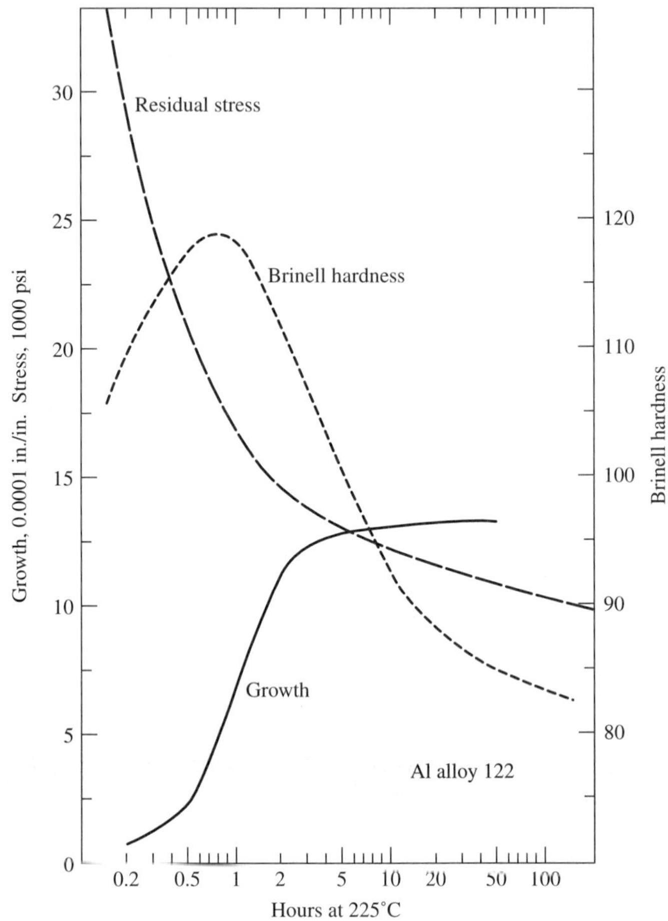

Figure 13.5 Rate of stress relief at 225°C of aluminum alloy 122 as quenched from a solution treatment. The accompanying precipitation hardening and later overaging, or softening, is shown, and also the dimensional growth which accompanies precipitation.

generally assures a uniform rate of corrosion, which in aluminum alloys is quite slow. Alloys of aluminum with 4 to 5% Cu, as quenched in cold water, are substantially in this condition, for, if precipitation occurs, it takes place throughout the structure upon planes of plastic movement during the quenching and, as a result, corrosion is uniform. When these alloys are quenched in hot water to avoid distortion and stresses, the subsequent precipitation is concentrated at grain boundaries, with a related concentration gradient of copper from the boundary regions to the remainder of the grains. The boundary area, of purer aluminum (since precipitation of copper is more complete), acts as an anode in an electrolyte, with the large area of the grain acting as a cathode when this alloy is subjected to corrosion. As a result, the attack is chiefly at grain boundaries, and if the alloy is in relatively thin sheet as in aircraft structures, it can be seriously embrittled. The difficulty can be overcome by having the alloy coated with pure aluminum. By the electrolytic effect mentioned above, this tends to protect the alloy even at the edges of the sheet or in regions of cracks or other surface damage.

13.5 ALUMINUM-LITHIUM ALLOYS

Research on the alloying of lithium with aluminum began in Germany in the 1920s. The first commercially available aluminum-lithium (Al-Li) alloy, ×2020, was used in 1957 in the wings and horizontal stabilizers of an experimental Navy aircraft. It contained about 1.1 wt % lithium. In 1969, the alloy was removed from production because of segregation effects, low toughness and ductility, and problems during casting because of the high reactivity of lithium. The stimulus to again produce aluminum-lithium alloys was the emergence of fiber-reinforced composites as possible substitutes for conventional aluminum in aircraft.

Besides beryllium, which forms an extremely toxic oxide, lithium is the only alloying element that both decreases the density and increases the elastic modulus of aluminum. The low solubility of beryllium in aluminum, less than 0.03 wt %, however, leads to massive segregation of the Be-rich phase upon solidification; poor mechanical properties result.

For each 1 wt % lithium addition, the density decreases by about 3% and Young's modulus increases by about 5% for lithium additions up to 4 wt %. It is not possible to continue to simply increase the lithium content beyond a level of about 3 wt % in production since deleterious effects are observed on the toughness and ductility.

Some of the original industrial and military goals of aluminum-lithium alloy research have included:

1. An alloy to replace fracture-tough 2024-T3x
2. An alloy to replace high-strength 7075-T6x.
3. An alloy to replace moderate-strength, stress-corrosion-resistant 7075-T73x.

The reduction in density of commercially available Al-Li alloys is not as great as researchers originally anticipated. Current alloys have 7 to 9% lower densities than 7075. The promotional claim of the early 1970s that Al-Li alloys may offer a 10% reduction is now almost forgotten. Several newly developed Al-Li alloys have been produced with better toughness and stress-corrosion-cracking (SCC) resistance than 7075 but with equivalent mechanical properties.

There are many reasons for the developmental problems of aluminum-lithium alloys. They are harder to cast without experiencing cracking because of the segregation and porosity, which are difficult to control. There is also a decrease in toughness and SCC resistance with increasing lithium content. Al-Li usually delaminates along the normal short-transverse (S-T) direction (Fig. 13.6) during fracture. As toughness is increased by aging, the SCC resistance decreases, particularly in the S-T direction.

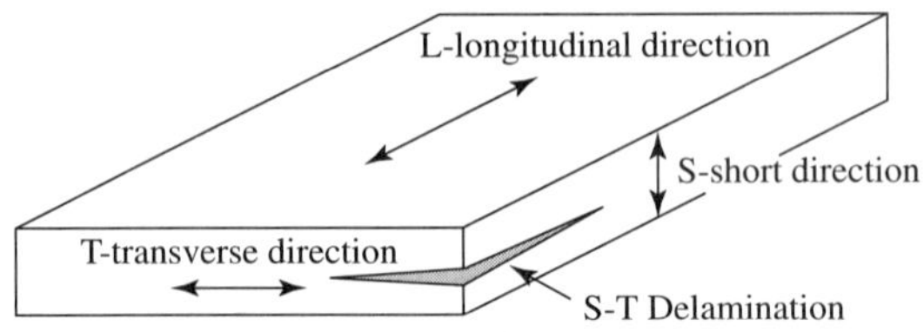

Figure 13.6 Definition of short transverse (S-T) direction.

The commercial Al-Li alloy that was selected for testing contains major constituents of copper and zirconium in addition to the lithium. The following review of its physical metallurgy explains the purpose of each constituent, some basics of how the commercial alloys achieve their desired properties, and an introduction to the microstructural development in cast Al-Li alloys.

Physical Metallurgy of Al-Li Alloys

Precipitation-hardened aluminum-lithium alloys obtain their strength from the formation of a large volume fraction of a metastable, ordered, and coherent δ' (Al_3Li) phase having a $L1_2$-type superlattice structure. The $L1_2$ lattice structure consists of eight shared corner sites occupied by lithium and six shared faces occupied by aluminum similar to the Cu_3Au structure. In Al-Cu-Li alloys, additional strength is achieved by the coprecipitation of copper-rich phases independent of δ' precipitation. For example, in high-lithium–low-copper alloys ($>$ 2 wt % Li, $<$ 2 wt % Cu) the decomposition of the supersaturated solid solution (α_{ss}) occurs by the reaction

$$\alpha_{ss} \begin{cases} \rightarrow \delta'\ (Al_3Li) \rightarrow \gamma\ (AlLi), \text{ cubic B32 (NaT1)} \\ \rightarrow T_1\ (Al_2CuLi), \text{ hexagonal} \end{cases}$$

The less than optimal deformation and fracture behavior of Al-Li–type alloys has been attributed to one or several of the following:

1. Strain localization that results from planar slip of shearable precipitates
2. Formation of coarse intermetallics during solidification
3. High levels of trapped hydrogen
4. Grain-boundary segregation of tramp elements such as sodium, potassium, and sulfur

Unlike lithium, these tramp elements have virtually no solid solubility in aluminum and may lead to grain-boundary segregation.

During plastic deformation of Al-Li alloys, the coherent δ' precipitates are sheared by dislocations. The planar slip associated with δ' shearing leads to heavily localized dislocation pileups at grain boundaries which, in turn, generate stress concentrations across grain boundaries.

Al-Li alloys have been reported to contain over ten times the amount of hydrogen normally found in high-strength, nonlithium aluminum alloys. This has been attributed to a large increase in the solubility of hydrogen in the lithium-alloyed matrix and to the formation of a hydrogen-rich phase during solidification. Investigators have suggested the poor ductility of Al-Li alloys might be caused by the formation of a stable hydride of either lithium as LiH or aluminum and lithium as Li_3AlH.

Some of the approaches for improving toughness in Al-Li-X alloys involve:

1. Encouraging dislocation cross slip or precipitate bypassing in the alloy by modification of the lattice parameters to increase mismatch in the Al-Al_3Li system
2. Introducing secondary precipitation systems
3. Using a dispersion-hardening system in addition to the precipitation system

Other approaches include grain refinement via additions of Mn, Zr, Cr, or Co and minimizing the presence of tramp elements (potassium, sodium, and sulfur).

Zirconium has been found to be the most effective element for inhibiting recrystallization in Al-Li alloys. An addition of 0.2 wt % Zr is sufficient to completely prevent recrystallization. The benefit is that unrecrystallized alloys have much higher strengths than either partially or fully recrystallized alloys.

The close match between the lattice parameters of the precipitate and matrix in Al-Li alloys results in a small misfit strain and leads to a homogeneous distribution of coherent, spherical δ' precipitates. This particular precipitate mechanism is analogous to the δ-δ' (Ni-Ni_3Al) system that gives Ni-base superalloys their high-temperature ductility.

The improvements in elastic modulus and density in the alloy occur whether the lithium is present in solid solution or as δ' precipitates. However, improvements in strength accompany the nucleation and growth of δ', and the magnitude of the effect increases with the volume fraction and the size of δ'. Copper is added to increase the strength and aid in homogenizing deformation, since it coprecipitates to form T_1 with δ'. Table 13.11 illustrates changes in strength and ductility accompanying these alloy additions. The last alloy listed in the table is similar to the commercial alloy used in the present experiments.

Table 13.11 Effect of alloy additions on mechanical properties of Al-Li alloys

Alloy, wt %	Elastic Modulus, GPa	Yield Strength, MPa	Tensile Strength, MPa	Elongation, %
Pure Al	68.0	12	47	60
Al-1.5Li	76.5	46	94	30
Al-2.3Li	80.5	195	278	7
Al-2.5Li-2.5Cu-0.1Zr	79.4	455	504	4

PROBLEMS

1. Why is the corrosion resistance of all Al-Cu alloys, as-cast or heat-treated, poorer than that of pure aluminum or the Al-Si alloys?
2. Silicon is typically added to steel to make it more machinable, so why are the Al-Cu alloys more machinable than the Al-Si alloys?
3. If a complex shape is to be formed at room temperature from alloy 2024 and high strength in the final shape is required, what temper should the alloy be in if (*a*) the shape can be heat-treated subsequently and (*b*) no heat treatment (above room temperature) is possible after forming?
4. If an alloy is "burnt" during heat treatment, why are some voids usually found in the zones that were liquid at the high temperature?
5. Assume that alloy 1100-O sheet, upon cold drawing into a deep rectangular box, shows cracking at the corners. How would you change the box design or the alloy condition to avoid the cracking?
6. Alloy 2024 can be formed more easily if the deformation is performed at 200 to 230°C and if the time at this temperature is short enough to prevent overaging. Why is forming easier at this temperature? Would the deformation be considered hot working or cold working?
7. Why are the solution heat-treatment times greater for castings than for equivalent wrought alloys and, among wrought alloys, greater for forgings than for sheet?
8. What elements may be used as grain-refining agents in aluminum, and in what applications would this be beneficial?

REFERENCES

Hatch, J. E.: *Aluminum: Properties and Physical Metallurgy,* ASM Int., Materials Park, OH, 1984.

Brooks, C. R.: *Heat Treatment, Structure and Properties of Nonferrous Alloys,* ASM Int., Materials Park, OH, 1990.

ASM Handbook, vol. 2, *Properties and Selection: Nonferrous Alloys and Special Purpose Materials,* ASM Int., Materials Park, OH, 1990.

VanderVoort, G. F. (Ed.): *Atlas of Time-Temperature Diagrams for Nonferrous Alloys,* ASM Int., Materials Park, OH, 1991.

Reed-Hill, R. E., and R. Abbaschian: *Physical Metallurgy Principles,* 3d ed., PWS Publishing Co., Boston, MA, 1994.

Christian, J. W.: *The Theory of Transformations in Metals and Alloys,* Pergamon Press, Oxford, England, 1965.

Porter, D. A., and K. E. Easterling: *Phase Transformations in Metals and Alloys,* Van Nostrand Reinhold, Workingham, Berkshire, England, 1983.

chapter

14

Copper and Copper Alloys

Copper is one of the oldest known metals and is today a metal valuable to our industrial and technological economy. It has many useful alloys, some of which have been introduced in previous chapters as examples of specific strengthening mechanisms, namely:

Cu-Ni alloys as examples of solid-solution hardening (Chap. 3)

Cu-Zn solid solutions or brasses for work hardening (Chap. 4)

Cu-Be alloy as a second example of precipitation hardening (Chap. 6)

14.1 COPPER ALLOY DESIGNATIONS

The Copper Development Association (CDA) administers the copper designation system that is most commonly used in the United States. Unified Numbering System (UNS) classifications follow virtually the same five-digit structure. The prefix C indicates a copper alloy, C10100 to C79900 represent the wrought alloys, and C80000 to C99900 are reserved for casting alloys. Some of the major alloy systems include:

Wrought alloys:

C10000 to C19900	Coppers (>99.3%), high-copper alloys (>96%)
C20000 to C29900	Common brasses (Cu-Zn alloys)
C30000 to C39900	Leaded brasses (Cu-Zn-Pb alloys)
C40000 to C49900	Tin brasses (Cu-Zn-Sn alloys)
C50000 to C59900	Tin bronzes (Cu-Sn alloys)
C60000 to C69900	Aluminum bronzes, silicon bronzes, manganese bronzes
C70000 to C79900	Copper nickels, nickel silvers (Cu-Ni-Zn alloys)

Casting alloys:

C80000 to C89900	Cast coppers, cast high-copper alloys, cast leaded brasses, cast tin brasses, cast manganese bronzes, cast Cu-Zn-Si alloys.
C90000 to C99900	Cast tin bronzes, cast leaded tin bronzes, cast copper nickels, cast aluminum bronzes, cast silicon-aluminum bronzes

14.2 UNALLOYED COPPERS

The major uses of unalloyed copper are based principally on its high electrical conductivity. Since this property is adversely affected by almost all other elements that go into solid solution in copper, metal purity becomes a major concern.

Copper ores are generally sulfides with zinc, lead, and other sulfides present. Smelting produces a very impure metal that is electrolytically refined, resulting in high purity. However, the electrorefined metal must be melted, and in commercial practice, it picks up oxygen from gases in contact with the liquid metal. Oxygen atoms are too small to go into substitutional solid solution and too large to be dissolved interstitially to any degree. Thus, it exists in solidified copper as Cu_2O, cuprous oxide. The major differentiation between the several grades of unalloyed copper is the amount of oxygen present and/or the amount of residual deoxidizer element. The grades of unalloyed copper are as follows:

Electrolytic tough pitch (ETP) copper: 99.95% Cu 0.04% oxygen, conductivity 101% IACS. This is the standard electrical-wire grade of copper identified as CDA 110. The oxygen is present as Cu_2O, which forms a eutectic with Cu (Micros. 14.1 and 14.2). Hot rolling breaks up the eutectic structure so that the Cu_2O ends up as discrete, rounded particles (Micros. 14.3 and 14.4). ETP copper should not be subjected to high temperatures, e.g., above 500°C, in the presence of reducing gases containing hydrogen. Hydrogen atoms readily diffuse into the solid metal and react with the Cu_2O to form H_2O. Sufficient local pressure develops from the steam that is formed to create internal holes, which can lead to severe embrittlement (Micro. 14.5).

Oxygen-free, high-conductivity (OFHC) copper: 99.95% Cu, 95% IACS conductivity, CDA 102. By melting and pouring electrorefined copper under a carbon monoxide–nitrogen atmosphere, the presence of oxygen in the copper can be eliminated. However, oxygen tends to remove certain impurities that affect conductivity so that the conductivity of this copper, though good, is definitely poorer than that of ETP copper. It is, of course, normally immune to hydrogen embrittlement, but oxygen can be diffused into the solid OFHC copper at elevated temperatures and it then becomes subject to embrittlement (Micro. 14.5).

Deoxidized low-phosphorus (DLP) copper: 99.92% Cu + 0.009% P, up to 100% IACS conductivity. Oxygen can be removed from liquid copper with elemental phosphorus, using a basic slag to remove the deoxidation product, P_2O_5. It is not

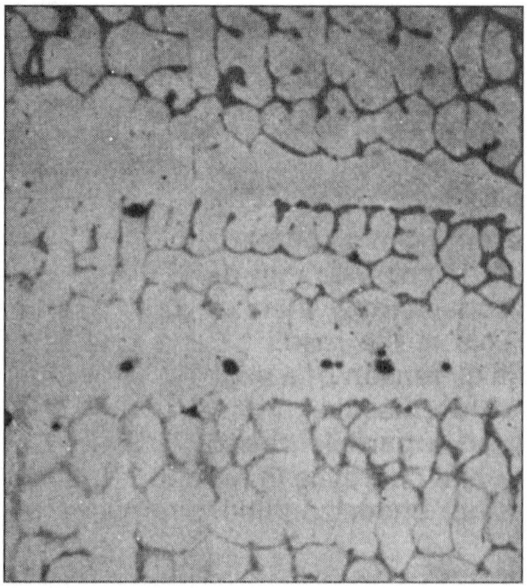

Micrograph 14.1 Cast electrolytic tough pitch copper (99.95% Cu, 0.03% O_2); ×50; NH_4OH-H_2O_2 etch. All metals solidify from the liquid state by the growth of crystals, which, because of preferred growth in certain directions, form as open, treelike structures called dendrites (see Chap. 3). At a later stage in growth, when different dendrites are in contact, the open spaces between them are filled in with more of the crystalline element. If impurities are present, they will usually be of lower melting point or will form a structure of lower melting point, which means that they will be concentrated in the parts last to freeze, i.e., the open spaces between dendrites. This structure of a cast wire bar shows nearly pure copper in the form of cells, actually the intersections of dendritic arms with the surface of polish. The oxygen impurity, present as Cu_2O (cuprous oxide) particles, forms with copper the dark structure of lower melting point, outlining the dendritic cells. The black spots are pores or holes in the cast metal.

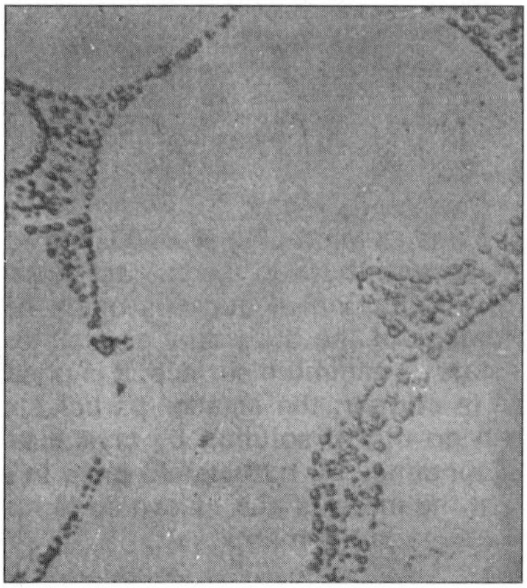

Micrograph 14.2 Same as Micro. 14.1 at ×500. This shows in detail the dark structure outlining the copper dendrites. The cuprous oxide particles or crystallites are seen to be globular bodies dispersed in a copper background. Note that copper is continuous and that the brittle oxide, while forming a network, actually consists of separate, discrete crystalline particles. (This alloy structure is that of a hypoeutectic; see Chap. 5.)

Micrograph 14.3 Same as Micro. 14.1 after hot rolling; $\times 50$; NH_4OH-H_2O_2 etch. The dendritic structure of Micro. 14.1, showing probably only two separate crystal (or dendrite) orientations, has been completely obliterated by the hot deformation. Now the specimen shows hundreds of very small, individual crystals. Parallel straight lines extending across many of the crystals outline annealing twins that appear after a metal has been deformed and annealed or equivalently deformed at a high temperature. In addition to changing the grain size of the copper, the hot rolling has destroyed the interdendritic network of cuprous oxide particles and caused the oxide to be aligned as stringers of particles in the direction of hot working.

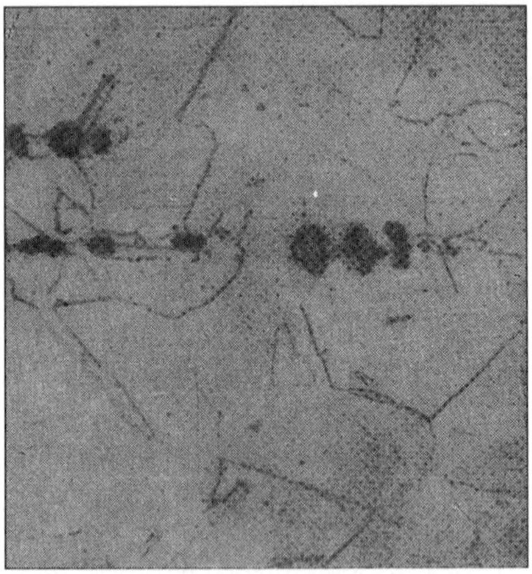

Micrograph 14.4 Same as Micro. 14.3 at $\times 500$. It is apparent from a comparison of the size of these oxide particles with those in the as-cast structure (Micro. 14.2) that the hot rolling not only changed the distribution of cuprous oxide but also considerably increased the size and decreased the number of the individual crystallites. This is the result of attempts by the oxide to reach a state of minimum surface. It is possible by reason of a slight solid solubility of cuprous oxide in copper; the smaller particles of oxide dissolve, and a corresponding amount must then go out of solution by crystallizing on a particle already present. Thus there is a general tendency for particles to grow in size and decrease in number, but this is possible only when the metal is at an elevated temperature under conditions of slight solid solubility of the particle in the matrix.

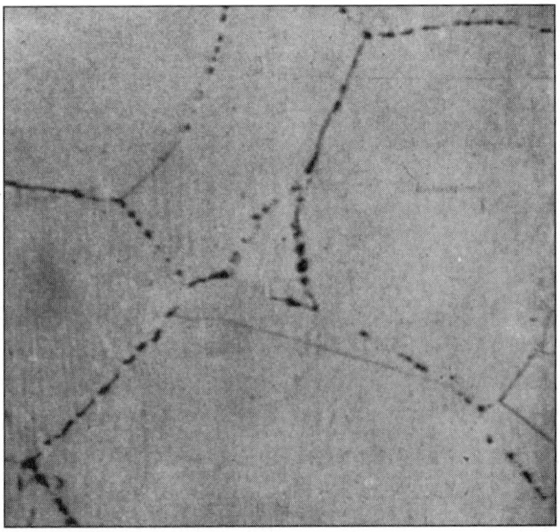

Micrograph 14.5 Oxygen-free high-conductivity copper, heated in an oxygen-bearing atmosphere for 2 h at 900°C; reheated 2 h at 900°C in hydrogen, ×200; potassium bichromate etch. Although the original copper was free of oxygen, some entered the metal during the first heat treatment, to a content of about 0.008% O_2. On reheating in hydrogen, atoms of hydrogen diffused into the copper. They reacted with cuprous oxide at the grain boundaries to form steam ($H_2 + Cu_2O \quad 2Cu + H_2O$) at a high pressure and temperature, and the steam created a network of fine holes along the grain boundaries. (Note that the twin bands do not contain any pores. This is further evidence of the continuity of the atomic lattice at twin bands and discontinuity at grain boundaries.) Copper in this condition may show a strength of only 35 MPa with zero elongation as compared with a normal 220 MPa and 40% elongation.

easy to control the chemistry of this reaction, and a small amount of residual phosphorus is very detrimental to good conductivity.

Deoxidized high-phosphorus (DHP) copper: 99.95% Cu + 0.02% P, 85% IACS conductivity.

Silver-bearing copper: 99.9% Cu + 0.03% Ag. This is "Lake" copper, not electrolytically refined. Silver is completely soluble in copper, at least up to 7+%, and has no reaction with, or effect upon, the oxygen present. It is the one element that has no adverse effect on conductivity. Its major beneficial effect is to raise the softening temperature or creep strength of the copper, i.e., a solid-solution strengthening effect.

14.3 BRASSES: Cu-Zn ALLOYS

The Cu-Zn Phase Diagram

The Cu-Zn (brass) phase diagram (Fig. 14.1) is of interest because it displays a peritectic reaction and a number of intermediate phases. The β phase is an fcc solid solution of Zn in Cu; the α phase has a bcc structure and, as pointed out earlier,

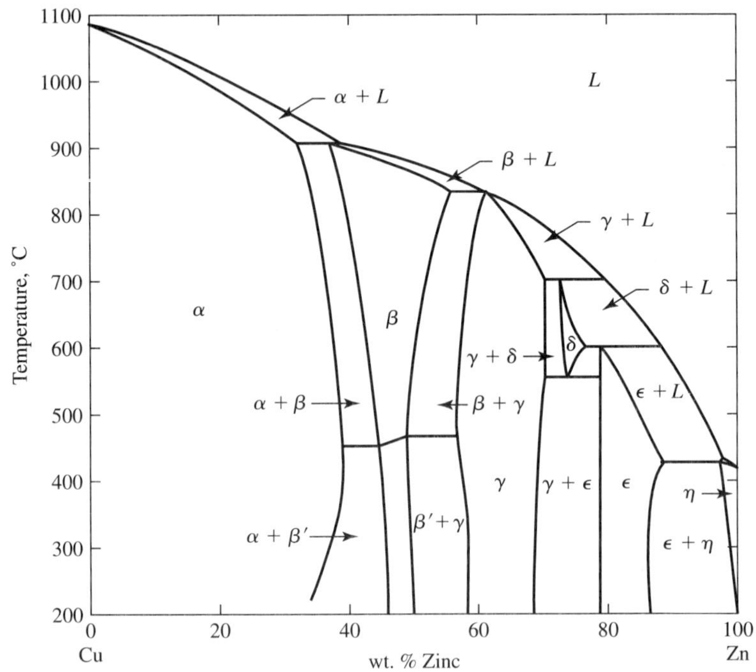

Figure 14.1 The Cu-Zn phase diagram, a series of peritectic reactions plus a eutectoid where the delta phase transforms on cooling to gamma plus epsilon. There is also an ordering reaction of beta to beta prime.

transforms at about 460°C to the β' phase, an ordered solid solution. The γ and δ phases have complex crystal structures and are so brittle that alloys containing them are of no commercial importance.

In liquid alloys containing 32.5 to 38.5% Zn, the β phase originates at 905°C, with a composition of 37% Zn, as a result of a reaction between α (32.5% Zn) and liquid (38.5% Zn), written in the form α + liquid $\rightleftharpoons \beta$. This is called a peritectic, from the Greek meaning "around," since, during the reaction, α crystallites will be surrounded by the reaction product β, which is surrounded by liquid. It is very unusual, however, for a peritectic structure to be visible in the microstructure. Because of structural considerations, the peritectic reaction does not rank in industrial importance with the eutectic.

An alloy of 65% Cu–35% Zn, under equilibrium conditions, consists of homogeneous grains of α, up to about 780°C. Upon heating above this temperature, the alloy enters the two-phase field $\alpha + \beta$, which means that β crystals of a higher zinc content (about 39% Zn) form, with a corresponding decrease in zinc content of the remaining α. On heating to increasingly higher temperatures, the amount of α decreases and that of β increases, as may be quantitatively estimated by use of the lever rule. At the same time, the concentration of zinc in both phases changes in accordance with the slope of the phase boundary limits. At 905°C, some α remains,

$$\%\alpha = \frac{37 - 35}{37 - 32.5}(100) = 44\%\ \alpha + 56\%\ \beta$$

and, by supplying additional heat, the peritectic decomposition of the β forms more α and some liquid,

$$\%\beta = \frac{38.5 - 35}{38.5 - 32.5}(100) = 58\%\ \alpha + 42\%\ \text{liquid}$$

All these changes are reversible under equilibrium conditions as approached by very slow cooling.

The 60% Cu–40% Zn alloy contains some β (or β') at all temperatures; at room temperatures, this amounts to about [(40 − 38)/(45.5 − 38)] (100), or 26%, but on heating above 453°C the proportion of β increases, while that of α decreases, and the zinc concentration in both phases decreases. At about 780°C, the alloy enters the single β-phase field, and it will remain a completely β structure up to the temperature of the solidus line, where melting begins. Again, these changes are reversible under equilibrium conditions of cooling or the nearest approach to that almost unattainable state. It is unattainable during a cooling cycle because the diagram requires that, under equilibrium conditions, both the relative amounts of α and β and, simultaneously, their composition (or zinc contents) shall change. This requirement means that zinc and copper atoms must travel continually through the lattice (in opposite directions), not merely at the boundaries of the two phases but through each large crystallite in order to maintain homogeneity of the phases. The metastable position of the solidus line encountered in the simple solid-solution diagram has already been described as originating from incomplete diffusion, or atomic interchange, between two elements in a single phase. In a similar manner, the boundaries of the $\alpha + \beta$ field are subject to displacement to the left under ordinary cooling conditions. Thus the 60-40 alloy, after cooling in air to room temperature, may show considerably more than the calculated 26% of the β phase.

Since even normal cooling rates tend to result in metastable positions of the phase-field boundaries of the diagram, very fast cooling rates, by preventing diffusion, may alter sloping boundaries so greatly as to make them appear vertical; i.e., the equilibrium structural condition for a high temperature may be at least partially preserved, by quenching, for observation at room temperature.

An alloy of 62.5% Cu–37.5% Zn is of particular interest in that under equilibrium conditions it has a completely β structure at and a uniform α structure below about 500°C. Upon extremely slow cooling from 900°C, it will gradually transform to α on passing through the $\alpha + \beta$ field, and this transformation will be of a diffusion type, i.e., accompanied by continual changes in the zinc concentration of the α and β phases. On more rapid cooling to room temperature, there will be some residual β (or β') in a metastable condition, and the amount of β will increase with increase in cooling rate. However, an extremely drastic quench from 900°C into an iced brine solution results in an entirely different structure. There is no opportunity for the

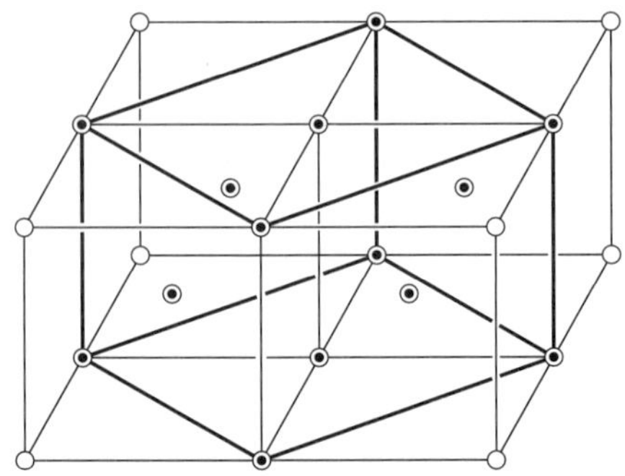

Figure 14.2 Four unit cells of the bcc structure β; specific atoms marked with black centers indicate that the structure might also be considered as fct.

time-consuming diffusion type of α formation in the temperature range 900 to 500°C, but the instability of β at the low temperature causes it to transform at about -14°C to a face-centered structure similar to α but differing in that one edge of the cube is longer than the other two. It is then called a face-centered tetragonal (fct) structure. The high-temperature β and the room-temperature α are of the same composition, and the change from bcc β to fcc α requires only a slight contraction of the lattice in two directions and an expansion in the third (Fig. 14.2). If the dimensional readjustment is incomplete (because of lattice rigidity at low temperatures), the intermediate, unstable tetragonal lattice is found.

The diffusionless transformation of the 62.5-37.5 brass is of no commercial importance, since the requirement of heating to within a few degrees of the melting point gives very coarse grains, the drastic quench required can be obtained only for relatively thin metal sections, and probably the properties are undesirable. However, the structure and its origin are quite similar to the martensitic structure of steels, which is of paramount industrial importance.

Microstructures of Two-Phase Brasses

Several of the various alpha-beta structures obtainable by heat treatments as described previously are illustrated by Micros. 14.6 to 14.11. Although not shown here, most copper-alloy castings are at least two-phase in structure and furthermore contain lead. Lead is frequently added to copper alloys in order to improve their machinability. Although lead dissolves in liquid copper above 1083°C, it is insoluble in solid copper. Copper crystals start to freeze out of a Cu-Pb liquid alloy at about 1083°C and continue to form, enriching the liquid in Pb, until a temperature of 954°C is reached. Here the remaining liquid goes through an invariant monotectic reaction:

$$\text{Liquid (36\% Pb)} \rightarrow \text{Cu (0\% Pb)} + \text{liquid (87\% Pb)}$$

Upon final freezing of practically pure lead at 326°C, the Pb is distributed in the copper interdendritically as small globules. This distribution, as in most second-phase

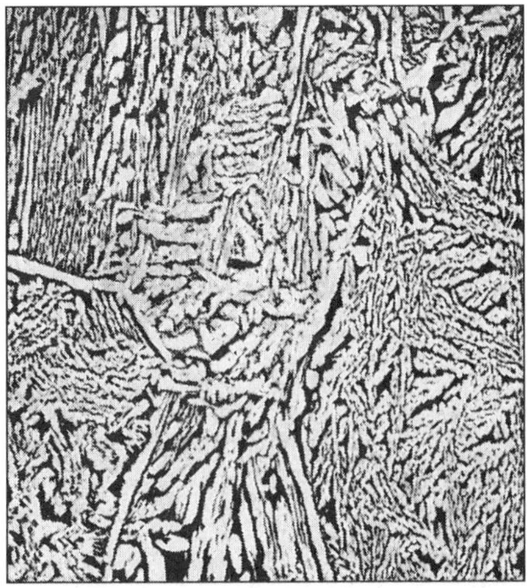

Micrograph 14.6 Extruded and air-cooled section of 60% Cu–40% Zn alloy (Muntz metal); ×50; $FeCl_3$ etch (black phase is β'). At the extrusion temperature, the alloy was completely β, and upon deformation and cooling α (white) formed in the β structure, first at the β grain boundaries and then inside the β grains. The α formation at the β boundaries indicates the size of the 5 grains at the high temperature; parts of about six of the former β grains are visible. The α crystals, during initial formation in the β structure, must have their atoms in conformity with atoms of the β crystals. This is a general rule for the formation of any new solid phase in a solid matrix of different crystal structure (compare the precipitation of θ_{CuAl_2} from α_{Al} during age hardening, page 132).

Usually, only one type of plane of the matrix and one type of plane of the new phase has atoms in a similar pattern when compared in a specific direction. For example, Fig. 14.2 shows that the base plane of the fct (cube plane when it shifts to fcc α) matches the cube plane of the bcc β when the edge of the base of the tetragon is in the direction of the face diagonal of the body-centered cube. This conformity results in an alignment of the new phase α in only certain planar directions of the β phase. Sometimes these planes are clearly outlined by the new phase, and the structure may then be said to show a Widmanstätten pattern, as is evident here to some extent. Frequently, later growth stages of the new phase may form an equiaxed shape, which obscures its crystallographic relationship to the matrix.

Micrograph 14.7 The same Muntz metal quenched (in a 0.9-mm-thick section) in water from 825°C; ×50; etched with NH_4OH-H_2O_2 so that α is dark and β' light (colors reversed from previous specimen). At the high temperature, this structure was completely β (see phase diagram). The quench preserved most of the β but did not suppress completely the formation of α, particularly at the β grain boundaries. Note the directional character of the α forming as plates extending from the boundary into the β' grains (a Widmanstätten characteristic). The relationship is also shown by a few very small isolated platelets of α within the β' grains. The β grain size is very coarse, as a result of grain growth in the single-phase range, 780 to 825°C. It was difficult to find sections showing as many as three grains for this micrograph.

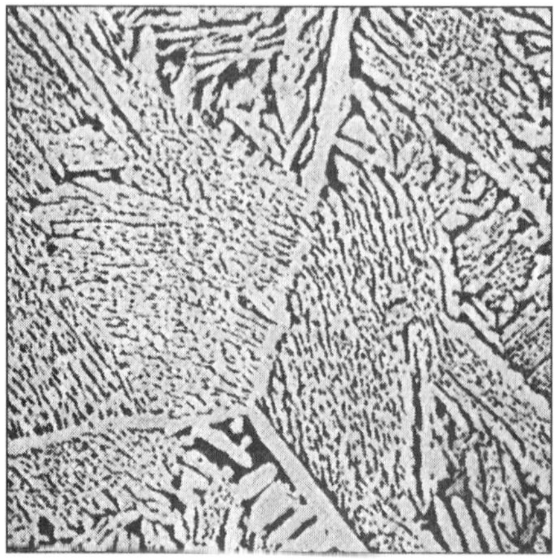

Micrograph 14.8 Same as Micro. 14.7 (quenched from 825°C) after reheating 1 h at 450°C; ×50; etched with $FeCl_3$ (colors again reversed so that α is light and β' dark). The unstable β' of the quenched alloy changed over to α upon heating in the low-temperature range, resulting in the attainment of approximately equilibrium proportions of the two phases. The initial platelets of α at the grain boundaries have grown farther into the former β grains, and their shape, as plates, is still evident by the shape of residual β crystallites between the α plates. The very small plates or needles of α visible in Micro. 14.7 have grown in a similar manner so that the quenched and annealed structure resembles that of the extruded stock (Micro. 14.6).

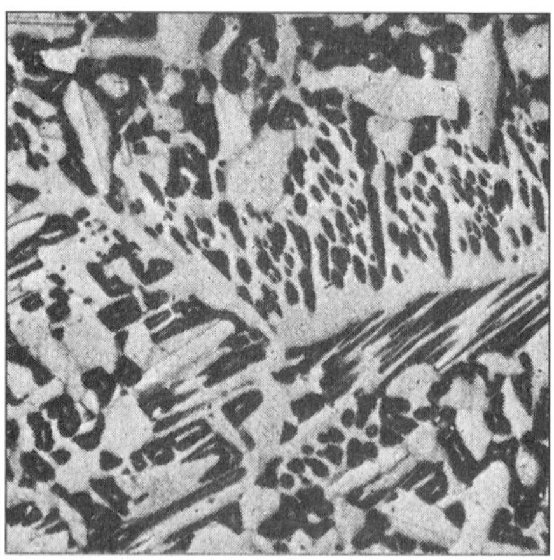

Micrograph 14.9 A different 60% Cu–40% Zn alloy quenched from 825°C and reheated at 500°C; ×200; $FeCl_3$ etch. This structure differs from Micro. 14.8, in that, while α crystals initially grew in a needlelike structure from the β grain boundaries, at a later stage the α formed in more or less equiaxed shapes and the residual β' crystallites do not show any crystallographic pattern or relationship of origin. This is frequently, although not universally, true: that the new phase formed at a high temperature under equilibrium cooling conditions will show a crystallographic pattern while that formed by annealing a quenched metastable structure is equiaxed (compare carbide structures from various heat treatment of steels, Chap. 8). Almost always, however, it is fairly easy to distinguish between structures formed at high or low temperatures. Note the twin bands, faintly visible, in the α crystals of this structure, formed as a result of strain accompanying the transformation.

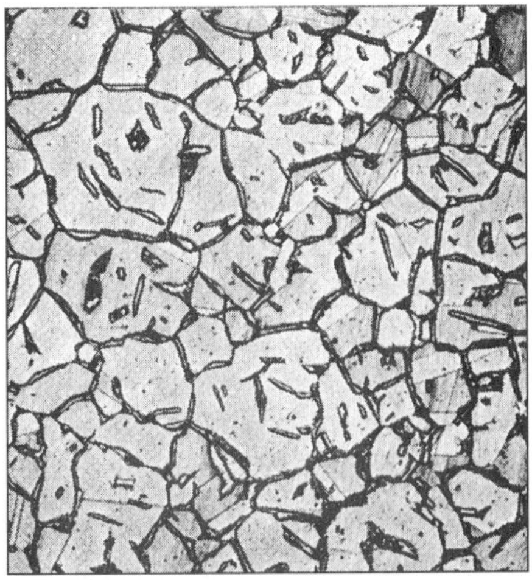

Micrograph 14.10 Worked and annealed 65% Cu–35% Zn alloy (common high brass) quenched from 825°C; ×50; ammonium persulfate etch. As ordinarily worked and annealed, this brass will have a uniform α structure showing annealing twins and a grain size characteristic of the specific rolling and annealing schedule. However, upon heating to 825°C, it enters the $\alpha + \beta$ field (see phase diagram). The new phase, β, forms predominantly at the α grain boundaries and, to a somewhat lesser extent, within the α grains. That forming inside the grains shows the crystallographic relationship required of β forming in α; the β is in lens-shaped plates on specific crystal planes of the α. Quenching from the high temperature has preserved most of the β as β'.

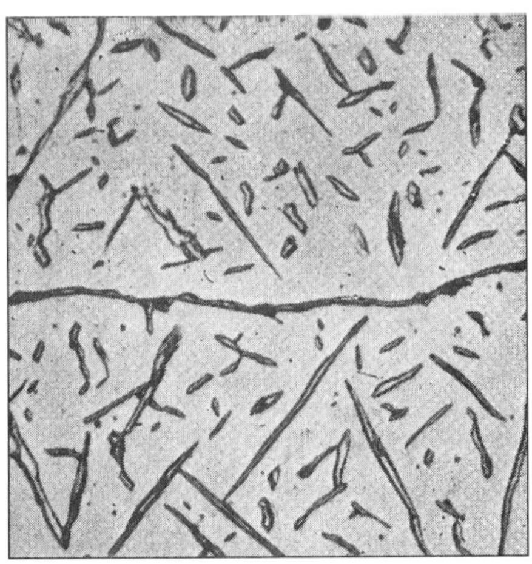

Micrograph 14.11 A cast 65–35 brass quenched from 825°C; ×50; ammonia peroxide etch. The very coarse-grained cast structure shows no twins, but the Widmanstätten pattern of β' platelets in two grains of α is beautifully illustrated.

structures, is to a great degree predictable theoretically. Furthermore, the reasoning followed here for just copper is equally applicable to brasses and other copper alloys with complex microstructures.

Theory of Two-Phase Microstructures

When equilibrium conditions obtain, as in materials heat-treated at high temperature and slowly cooled, the interfacial surface energies of the various phases present determine the structure. Consider, for example, a particle of β phase within, and therefore completely surrounded by, a grain of a phase. There is a certain energy per unit area $\gamma_{\alpha\beta}$ associated with the α-β interface, and the total interfacial energy will be least when the β particle has a spherical shape, since a sphere has the smallest surface area per unit volume of any solid figure.

In a matrix of polycrystalline α phase in which some β has formed, the β phase is most likely to appear in the grain boundaries of the α phase or the points where three α grain boundaries intersect. The reason for this is that in any case the β phase will be surrounded by an interface, but by occupying an α grain boundary, part of the α grain-boundary interface is eliminated and the total energy of the system is lowered (Fig. 14.3). To find the shape taken by the particles of β phase, consider the following situation: two grains of a phase meet with one of β phase as in Fig. 14.4.

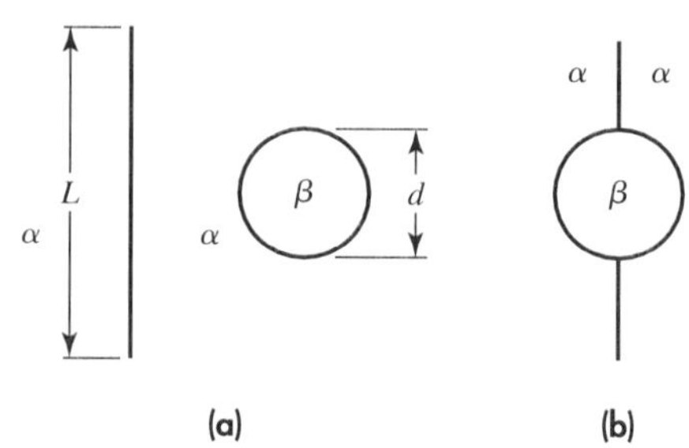

Figure 14.3 The decreased surface energy which favors the formation of the precipitate of β phase in the α-phase boundary. In (a) total surface energy in the plane of the drawing is $L\gamma_{\alpha\alpha} + \pi d_{\gamma\alpha\beta}$, while in (b) it is only $(L - d)_{\gamma_{\alpha\alpha}} + \pi d_{\alpha\beta}$.

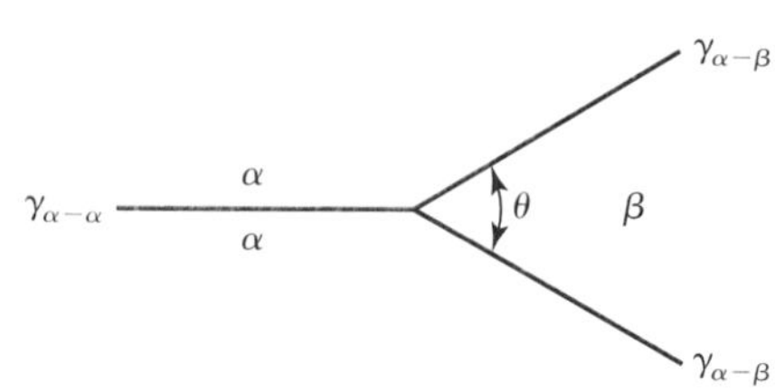

Figure 14.4 The dihedral angle θ is the angle between the two α-β interfaces.

Upon remembering (Chap. 1) that surface energy per unit area is equivalent to force per unit length, we see that a triangle of forces is established at the meeting point of the three phases and these forces will be in equilibrium if

$$\gamma_{\alpha\alpha} = 2\gamma_{\alpha\beta} \cos(\theta/2)$$

The angle θ is called the dihedral angle for a β and an α grain boundary. If $\gamma_{\alpha\beta} > \frac{1}{2}\gamma_{\alpha\alpha}$, θ will be a finite angle; if $\gamma_{\alpha\beta} = \gamma_{\alpha\alpha}$, $\theta = 120°$; and if $\gamma_{\alpha\beta} > \gamma_{\alpha\alpha}$, $\theta > 120°$. If, however, $\gamma_{\alpha\alpha} > 2\gamma_{\alpha\beta}$, the above equation cannot be satisfied and no equilibrium will exist. Instead, the β phase will penetrate or spread along the α grain boundary as a thin film. In this case, if the β phase is brittle or has a low melting point, i.e., below the intended service temperature, the mechanical properties of the alloy may be seriously impaired even though the α matrix is both ductile and strong.

The actual measurement of the dihedral angle, shown in Fig. 14.4, in a microstructure is complicated by the fact that the plane of the specimen surface in which the angle is measured will not, in general, be perpendicular to planes of the grain boundaries. Depending on how the surface of the specimen cuts through the dihedral angle, apparent dihedral angles ranging from 0 to 180° may be observed for any value of θ. It can be shown, however, that, in a surface cutting through many equal dihedral angles at random, the most probable apparent angle is the true angle. Therefore, the most frequently observed dihedral angle in a microstructure can be taken as approximating the true value of the angle θ.

When a second phase β is located at the line juncture (point in the micrographs) of intersection of three α grain boundaries, the shape of the β-phase particle is again determined by interfacial energies as indicated by the dihedral angle θ, but in a more complicated way, as is illustrated in Fig. 14.5. Two different types of β-phase distribution for two different dihedral angles are illustrated by the sketches of Fig. 14.6. In a sample of alloy, the actual shapes of the particles of second phase are, of course, a matter of great variety and complexity, as illustrated by the right-hand sketch in Fig. 14.6; the dihedral angle, however, remains the same in this illustration.

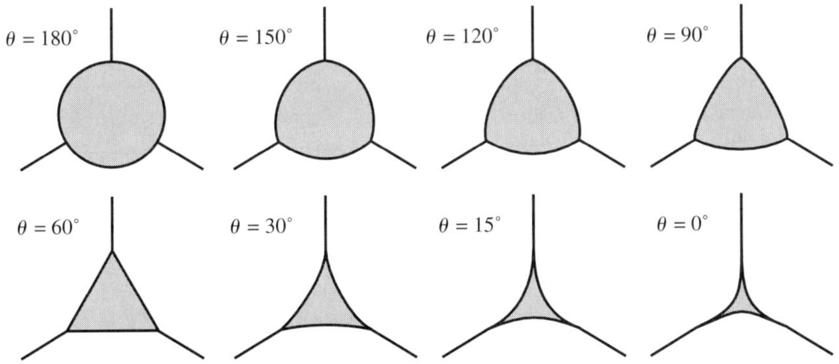

Figure 14.5 As the dihedral angle changes, the shape assumed by a second phase located at the line of intersection of three matrix grains varies in the manner shown.

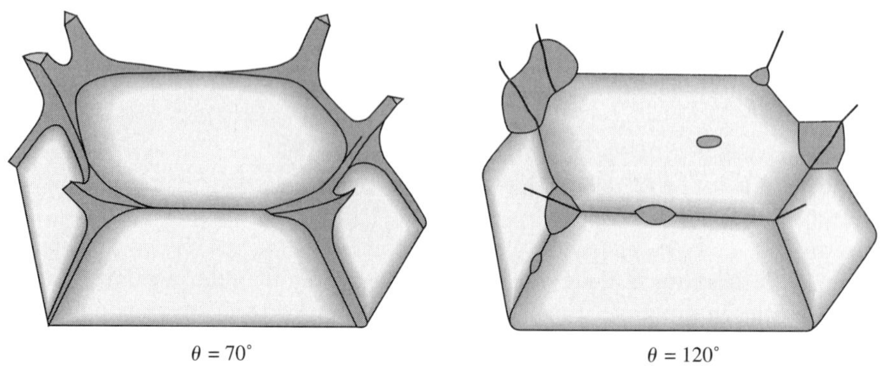

Figure 14.6 Distribution of a second phase in the grain boundaries of the matrix phase. For the case on the left $\theta = 70°$; on the right $\theta \sim 120°$.

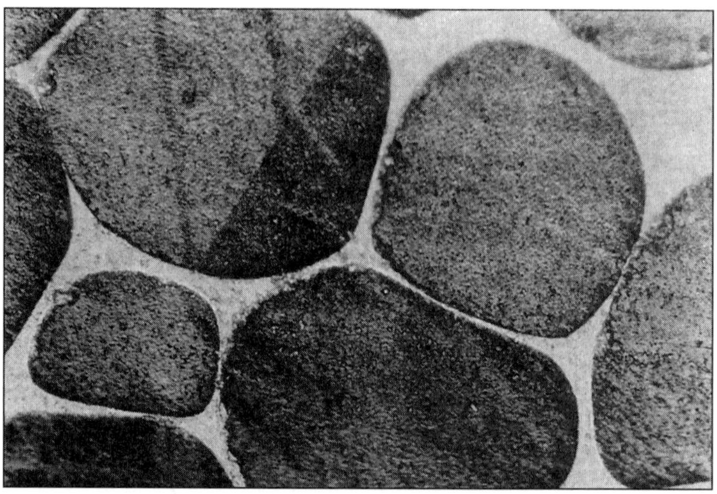

Micrograph 14.12 85% Cu–15% Ag; ×500; held at 850°C for 1 h and then quenched in water; dichromate etch followed by ferric chloride contrast etch. This is a simple eutectic system with a eutectic temperature of 779°C. At 850°C, the structure was roughly 15% liquid and 80% Cu-rich solid. Complete penetration of liquid between solid grains is evident. The dihedral angle θ is zero in this case.

Samples illustrating the above effects can be made by heating alloys into a two-phase region, where they are partly liquid, and then quenching. Micrograph 14.12, for example, shows the structure of a partially melted Ag-Cu alloy quenched from 850°C. The β phase, liquid in this case, has a dihedral angle $\theta = 0°$ and so has penetrated along all grain boundaries, separating the individual grains. At high temperature the alloy in this condition would have no strength whatever, and it is clear that it could not be hot-worked at any temperature above the solidus without falling

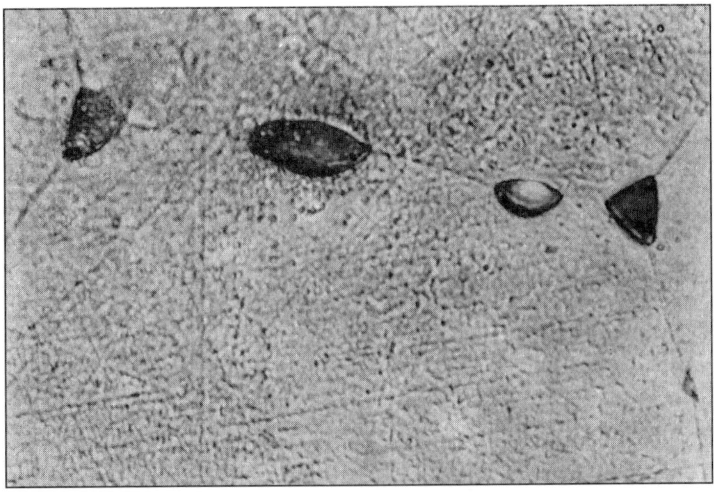

Micrograph 14.13 97% Cu–3% Pb; ×1000; air-cooled from molten state; dichromate etch followed by ferric chloride contrast etch. θ is 70°.

apart. A dihedral angle of 0° between liquid and solid can also lead to the rapid failure of a solid metal used to contain a liquid metal. If the metal of the container is even slightly soluble in the liquid, solution of the solid will begin at the grain boundaries, since these are regions of relatively high energy. If $\theta > 0°$, solution of the boundaries will stop as soon as the equilibrium dihedral angle is reached and there will be only a slow wasting away of the metal surface. If $\theta < 0°$, however, the liquid will continue to penetrate the grain boundaries until the metal disintegrates. The attack on brass by mercury is a familiar example of this phenomenon. The presence of tensile stress in the solid metal accelerates the effect, leading to catastrophic failures.

In cases where θ is large enough, the occurrence of a liquid phase may do no particular harm to the solid. The presence of lead in copper and brass is particularly interesting in this connection. In copper, lead forms a dihedral angle of $\theta = 70°$ (Micro. 14.13), and there is some tendency for the lead to cause fractures of copper at the grain boundaries during hot rolling. Increasing the content of zinc causes θ to increase until, when the $\alpha + \beta$ brass phase field is reached, θ for molten lead is 110° and there is no tendency whatever for failure to occur at grain boundaries at high temperature. The small amounts of lead added to β brass improve its machinability. The lead particles aid in breaking up the chips formed by cutting tools, resulting in a much improved surface finish.

Wrought Brasses

Alloys with up to 35% Zn are, as already evidenced, a solid solution, whereas the 40% Zu alloy, Muntz metal, has an $\alpha + \beta$ structure. The α alloys are differentiated

Table 14.1 Tensile properties of brasses in the form of 1-mm strip

			Properties as Annealed to 0.050-mm Grain Size		Properties as Cold-Rolled to Half Hard Temper	
Zinc %	CDA Alloy No.*	Structure	Tensile Strength, MPa	Elongation in 50 mm, %	Tensile Strength, MPa	Elongation in 50 mm, %
0	110	α	225	45	295	14
10	220	α	260	45	365	11
20	240	α	310	50	425	18
30	260	α	330	62	435	23
40	280	$\alpha + \beta$	380	45	490	15

*Copper Development Association number.

by gradual changes in color, strength, ductility, corrosion resistance, and cost. Changes in tensile properties as related to zinc content are shown by the data of Table 14.1. It is noteworthy that up to 35% Zn, increase in Zn content increases both strength and ductility. Specific alloys in the α brasses are:

1. The 5% Zn alloy, known as gilding metal, has a golden color and is used for cheap jewelry or gilding purposes.
2. The 10% Zn alloy, known as commercial bronze, has a bronze color and is cheaper than the tin bronzes.
3. The 15% Zn alloy has a red tint somewhat like copper and is called red brass. Like copper and the preceding alloys, it is resistant to cracking under the combined action of elastic stresses and corrosive agents such as ammoniacal solutions or vapors and mercury salts.
4. The 30% Zn alloy has higher strength and ductility than any of these alloys, and this combination makes the 70-30 solid solution particularly well suited to severe cold-forming operations. One of the most severe operations is drawing a deep cup shape, e.g., a cartridge case, from a flat circular blank. The ease of fabrication by cold drawing to such shapes, plus cold-worked strength and the general resistance to corrosion, has caused this alloy to be extensively used for cartridge cases.

14.4 TIN BRONZES: Cu-Sn ALLOYS

Tin bronze is one of the oldest alloys. Many commercial wrought bronzes are essentially single-phase alloys containing about 5% Sn in solid solution. A study of the phase diagram shown in Fig. 14.7 indicates that even these wrought 5% Sn

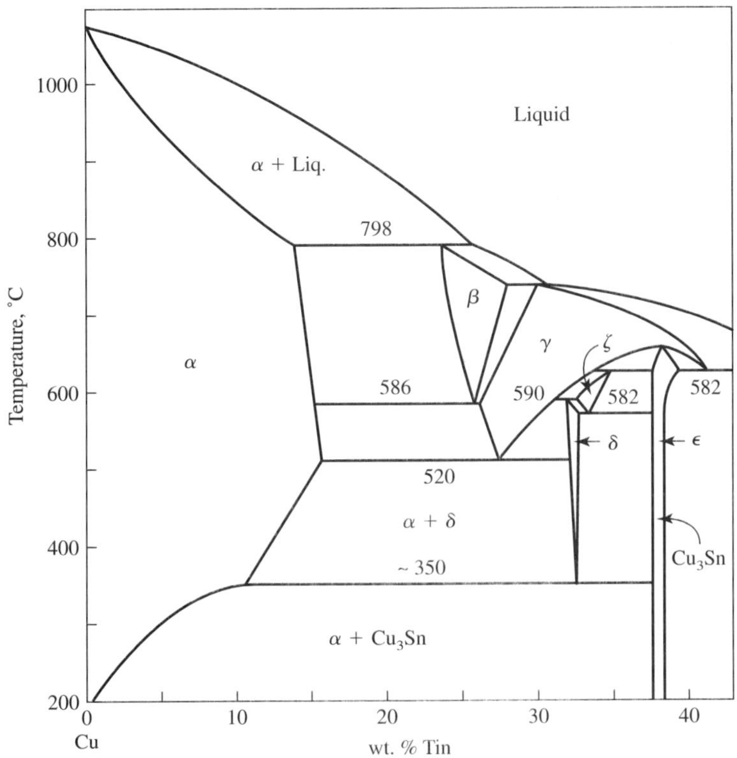

Figure 14.7 Phase diagram of the Cu-Sn system up to 40% Sn. The ε phase is also called Cu_3Sn, with an electron-atom ratio of 7:4.

bronzes should show two phases at ordinary temperatures, α plus precipitated ε or Cu_3Sn. Presumably, the temperature at which the α becomes unstable and tends to precipitate Cu_3Sn, 300°C for the 5% alloy, is so low that the precipitate does not form, at least in visible size. Deformation would presumably accelerate precipitation in the metastable, supersaturated α matrix.

Most cast binary tin bronzes contain approximately 10% Sn. According to the phase diagram, a 90% Cu–10% Sn alloy should solidify as a single-phase alloy and remain unchanged to just below 360°C, where, again, a supersaturated condition would be developed, with the tendency for precipitation of Cu_3Sn limited or suppressed by the low diffusion rate at these temperatures. This structure, theoretically all α for the 10% Sn alloy, is practically never observed because the alloy almost never solidifies under equilibrium conditions. Coring during dendritic solidification of α results in a metastable solidus placed to the left and under the equilibrium solidus. Therefore, at 798°C some interdendritic liquid remains, and this peritectically reacts with α to form interdendritic β, namely, α + liquid $\rightarrow \beta$.

The phase diagram for the higher-tin-content alloys, 15 to 37% Sn, is quite complicated, but an understanding of what happens to the β on cooling below 798°C requires a brief discussion of this part of the diagram. During cooling from 798 to 520°C, the changes are similar to those for high brasses: β diminishes in quantity as part of it changes to α; simultaneously the β becomes richer in tin. These changes are predicted by the slopes of the lines separating the α-, $(\alpha + \beta)$-, and β-phase fields. Thus far, the changes are exactly analogous to those described earlier for the Cu-Zn system.

At 586°C, the β phase disappears completely, undergoing the reaction $\beta \rightarrow \alpha + \gamma$. The two sloping lines from above converge to meet the horizontal line at a composition of 26.8% Sn. This construction is identical to that of the eutectic, the only difference being that there is a solid phase above the horizontal, rather than a liquid. Transformations of this type, called eutectoids, were covered in detail in Chap. 5 for steels. At the moment, discussion will be limited to pointing out that cooling through 520°C results in the disappearance of γ with an increase in the amount of α already present in the 10% Sn alloy postulated at the start, and the appearance of a new phase of higher tin content called δ, which is a brittle compound. This compound is still mechanically distributed in the interdendritic spaces where β was originally formed, peritectically.

As the alloy cools from 520 to 360°C, the α should change in composition, becoming lower in tin content, while the δ phase becomes richer in tin. These changes, predicted by the slopes of the boundaries of the $\alpha + \delta$ field, would not ordinarily be realized because of the slowness of the diffusion required to permit these compositional adjustments.

The boundaries of the δ-phase field converge at a horizontal line at 360°C. The construction of the diagram and the reaction at this temperature are nearly identical to those just discussed; here the eutectoidal reaction on cooling is

$$\delta \rightarrow \alpha + \varepsilon_{Cu_3Sn}$$

Whereas the γ eutectoid reaction is difficult to suppress, the δ eutectoid reaction is sluggish and difficult to achieve completely because of low atomic mobility at this relatively low temperature and the greater compositional changes involved. The new phase ε is hcp, brittle, and of a composition equivalent to the compound Cu_3Sn.

14.5 SILICON AND ALUMINUM BRONZES

Over 3000 years ago, the Chinese were producing tin bronzes and, because of their properties, the word bronze has had a connotation of a superior material. In this century, when metallic silicon and aluminum were first used, the alloys were called silicon bronzes or aluminum bronzes when combined with copper; their colors were similar to tin bronze. There are peritectic and eutectoid reactions and comparable successive phases, i.e., α, β, γ, δ, and ε. However, commercial wrought alloys are basically α structures. Typical strength properties are given in Table 14.2.

Table 14.2 Tensile properties of Sn, Si, and Al bronzes in the form of rolled strip

		Annealed to 0.035-mm Grain Size		Cold-Rolled to Hard Temper	
Bronze Type	**CDA Alloy No.***	**Tensile Strength, MPa**	**Elongation in 50 mm, %**	**Tensile Strength, MPa**	**Elongation in 50 mm, %**
Tin:					
5% Sn	510	340	57	575	8
10% Sn	521	435	68	690	16
Silicon:					
1.5% 420	651	280	30	455	10
3.0% Si	655	420	60	660	8
Aluminum:					
5% Al	608	420	65	700	8

*Copper Development Association number.

14.6 CAST COPPER-BASE ALLOYS

Copper alloys used for castings differ in some ways from their nominally equivalent wrought alloys. The cast alloys have more compositional latitude since there is no concern with hot or cold workability. They can be more complex structurally in order to achieve the optimum compromise between various concerns, namely, strength, solidification shrinkage and other castability factors, grain control, machinability, etc. Table 14.3 shows that many alloys have three or four elements added to the basic copper. This listing is far from complete but is typical of the many choices available.

Table 14.3 Composition and strength properties of some copper casting alloys

			Alloy Composition, %						
Common Name	**CDA No.**	**ASTM Alloy**	**Sn**	**Pb**	**Zn**	**Fe**	**Other**	**Tensile Strength, MPa**	**Elongation in 50 mm, %**
Leaded red brass	836	B62	5	5	5	...	...	245	32
Tin bronze	905	B143-1A	10	...	2	...	...	315	33
High-leaded tin bronze	937	B144-3A	10	10	...	...	...	270	30
High-strength Mn-bronze		B147-8C	...	...	39	1	1.25Al, 0.25Mn	495	40
Al-bronze, as cast	955	B148-9D	...	...	...	4	4Ni, 11Al	665	7–20
Ni-Sn bronze, as cast		B292-A	5	...	2	...	Ni	315	25

PROBLEMS

1. What would be the differences, if any, in microstructure between the following: (*a*) 80% Cu–20% Zn, quenched from 850°C; (*b*) 80% Cu–20% Zn, slow-cooled from 850°C; (*c*) 60% Cu–40% Zn, quenched from 850°C; and (*d*) 60% Cu–40% Zn, slow-cooled from 850°C?
2. What effect on the mechanical properties of an alloy might be anticipated if it has two phases and the second phase forms the following dihedral angles: $\theta = 30°$, $\theta = 70°$, $\theta = 120°$? Will the properties of the primary and second phases also influence your answer?
3. A cast ETP copper alloy is found to have low-tensile ductility. Metallographic examination shows the structure to contain microcavities. Indicate three possible causes for the formation of cavities in this alloy. How would you distinguish between these?
4. Micrograph 14.7 shows a Cu-Ag alloy heated into the two-phase, $\alpha + L$, region and quenched. Indicate how, from this photomicrograph, you can tell that this structure was a reheated rather than an as-cast alloy.
5. Micrograph 14.2 shows a Muntz metal structure that is nearly all β phase but with extremely coarse β grain size. How could this alloy be treated to produce nearly all β phase but with a finer grain size?

REFERENCES

ASM Handbook, vol. 2, *Properties and Selection: Nonferrous Alloys and Special Purpose Materials,* ASM Int., Materials Park, OH, 1990.

Brooks, C. R.: *Heat Treatment, Structure and Properties of Nonferrous Alloys,* ASM Int., Materials Park, OH, 1990.

Smith, W. F.: *Principles of Materials Science and Engineering,* McGraw-Hill, New York, 1986.

2000 Annual Book of ASTM Standards, vol. 02.01, *Copper and Copper Alloys,* ASTM, West Conshohocken, PA, 2000.

chapter

15

Magnesium Alloys

The ready availability of an important "ore" of magnesium, i.e., seawater, the light weight of the metal, the excellent machinability, and the good strength properties of its alloys ensure an ever-increasing utilization. Two disadvantages to its more general use have been its relatively high cost and chemical reactivity, which are being overcome by increased production and research, respectively. The tremendous demand for this metal during World War II was responsible for a great expansion in primary production facilities and a widespread dissemination of knowledge regarding the fabrication of its alloys by casting, welding, forming, etc. Now, the metal is no longer regarded as a hazardous and unreliable material but as a common engineering metal that, in alloyed form, is competitive for many applications with aluminum, copper, and ferrous alloys.

15.1 MAGNESIUM ALLOY DESIGNATIONS

Pertinent Phase Diagrams

The most important alloys of magnesium contain aluminum. Therefore, the magnesium end of an Al-Mg binary phase diagram is shown in Fig. 15.1. The aluminum-rich end of this diagram was presented in Fig. 5.3.

At each end of the diagram, there is a eutectic between a brittle compound and a terminal solid solution. The compound in each case is of relatively low stability, having a relatively low melting point: necessarily, therefore, the eutectic melting points are moderately low. The eutectic structures, as shown by the lever rule, contain more compound than α solid solution, and since the compounds are brittle, alloys with a eutectic network are brittle. Also, in each case, the solid solution has a relatively high solute content at the eutectic temperature and a decreasing solubility with decreasing temperature. Consequently, age-hardening possibilities exist.

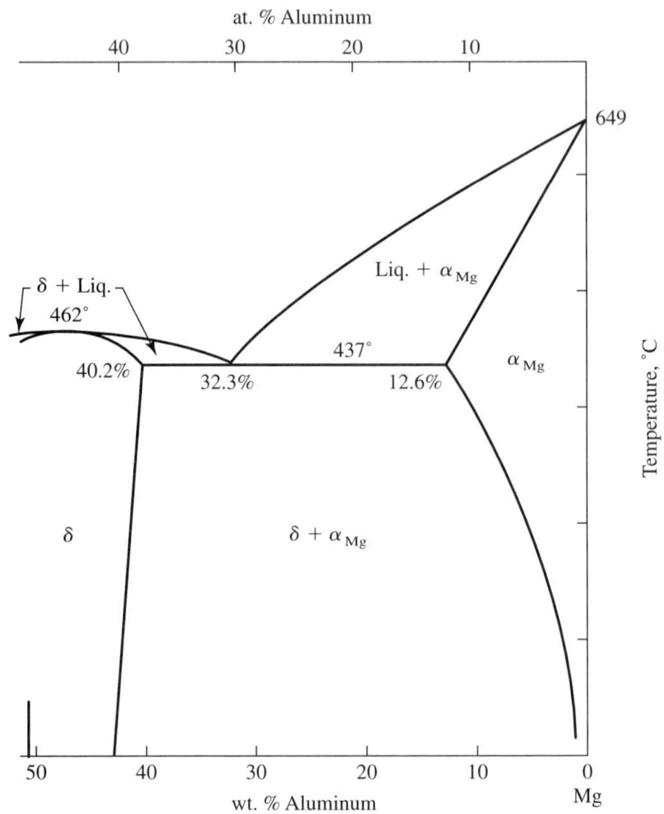

Figure 15.1 Magnesium end of the Al-Mg phase diagram. The δ phase is also known as the compound $Mg_{17}Al_{12}$, a designation based on crystallographic evidence rather than composition, since Mg_3Al_{12} would be simpler and within the δ homogeneity field.

Zinc and manganese are also frequently alloyed with magnesium. In the binary Mg-Zn phase diagram a eutectic is found between a terminal solid solution and a brittle phase Mg-Zn. Both the Mg-Zn and Mg-Mn systems show solid solubility decreasing with decrease in temperature.

Since commercially useful alloys contain appreciable concentrations of both aluminum and zinc, a ternary diagram is required to show the phase conditions of these complex alloys. Not all data for the magnesium-rich alloys of Mg-Al-Zn can be shown in a single diagram, but the change of solid solubility with temperature and the identity of the second phase in cast alloys, data of most value in the discussion of heat treatment and structures, are shown in Fig. 15.2 together with the compositions of two commercial sand-casting alloys. Alloy AZ92 shows only the binary compound $Mg_{17}Al_{12}$, while alloy AZ63 shows in addition to this some ternary compound $Mg_3Zn_2Al_3$. The composition of the ternary compound is such that five Mg atoms in $Mg_{17}Al_{12}$ have been replaced by eight Zn atoms.

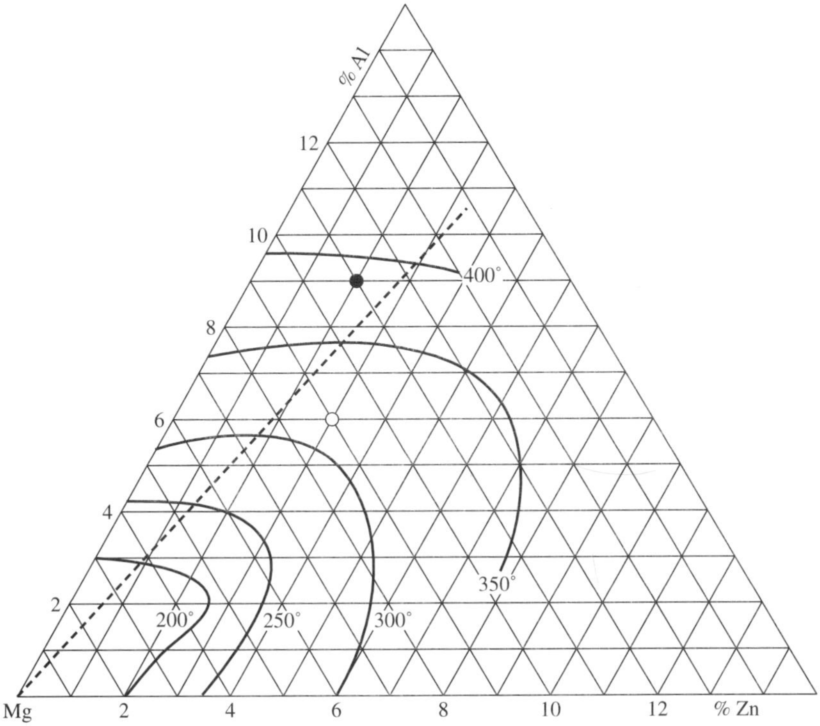

Figure 15.2 Ternary section at the magnesium corner of the Mg-Al-Zn system with isothermals showing the solid solubility limits at the indicated temperatures. The dotted line separates structural fields in as-cast alloys. To the left, structures consist of the solid solution and massive $Mg_{17}Al_{12}$; this area includes alloy AZ92 at the solid black circle. To the right of the dotted line, as-cast structures consist of the solid solution, massive $Mg_{17}Al_{12}$ and the ternary compound $Mg_3Zn_2Al_3$ as, for example, in alloy AZ63 at the open-circle composition.

Magnesium Alloy Identification

There is a standard four-part system for identifying magnesium alloys, like the AZ63 just mentioned.

1. Two capital letters indicate the two principal alloy elements, put in order of decreasing percentage. The letters employed are:

A	Aluminum	H	Thorium	Q	Silver
B	Bismuth	K	Zirconium	R	Chromium
C	Copper	L	Beryllium	S	Silicon
D	Cadmium	M	Manganese	T	Tin
E	Rare earth	N	Nickel	Z	Zinc
F	Iron	P	Lead		

2. Two digits indicate the rounded-off percentages of the two main elements in the same order as the letters. Thus AZ63 consists mainly of Mg + 6% Al + 3% Zn.

3. A capital letter distinguishes between alloys of the same major composition but differing in minor elements. The letters represent chronological sequence of development.

4. A letter and number indicate the condition and properties. These include:

F	As fabricated
O	Annealed
H10, H11	Slightly strain-hardened
H23, H24, H26	Strain-hardened and partially annealed
T4	Solution heat-treated
T5	Artificially aged only
T6	Heat-treated and artificially aged

15.2 THE NATURE OF MAGNESIUM ALLOYING

Crystallography Magnesium has an hcp structure and can be expected to form complete solid solutions only with metals having similar structures of nearly the same atom size and electrochemical characteristics. Actually, of the other common hcp metals, zinc and beryllium do not meet both requirements and therefore do not form continuous solid solutions with magnesium. Cadmium does meet the requirements and shows unlimited solid solubility in magnesium.

Relative Atomic Sizes It was pointed out earlier that alloying elements must have an atomic size within 15% of that of the solvent metal for extensive solid solutions to form. About half the potential metallic alloying elements for magnesium is within the favorable 15% limit; about one-tenth is on the borderline; and the remainder are outside. The size factor thus initially limits the solid-solution possibilities. Its effect on solid solubility where electrochemical and valency factors are constant is shown in Fig. 15.3.

Valency Factor It has already been pointed out that as valences of solvent and solute become more unlike, solid solubility is more restricted.[1] This is shown by Fig. 15.4, phase diagrams of magnesium-rich alloys with elements in the long-period VB of

[1]W. Hume-Rothery, R. E. Smallman, and C. W. Haworth, *The Structure of Metals and Alloys*, Institute of Metals and Institution of Metallurgists, London, 1969.

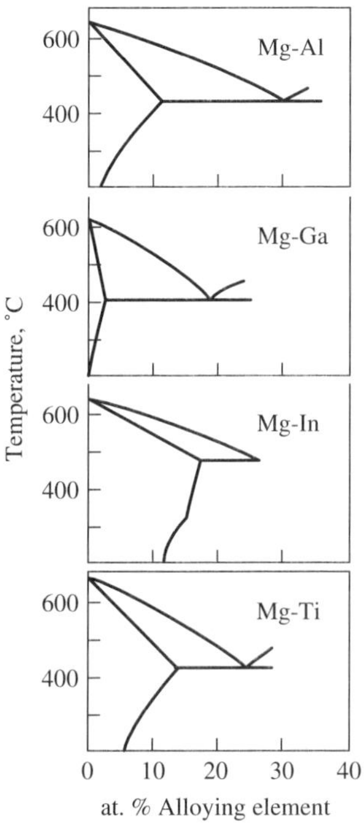

	Group	Lattice	Valence	Size Factor, %
Silver	IB	fcc	1	10
Cadmium	IIB	hcp	2	7
Indium	IIIB	fct	3	6
Tin	IVB	bct	4	12
Antimony	VB	rh	5	10

Figure 15.3 Influence of size factor on the solid solubility of periodic table group IIIB elements in magnesium, as shown by the binary phase diagrams.

the periodic table. All are within the favorable zone with regard to atomic size. The figure also illustrates the relative valence effect, i.e., that an element of higher valence is more soluble in a metal than one of comparable lower valence. Thus, although the valence difference between Mg-Ag and Mg-In is the same, univalent silver is less soluble in divalent magnesium than is trivalent indium.

Electrochemical Factor Magnesium is a strongly electropositive element, and when it is alloyed with electronegative elements, compounds are almost invariably formed, despite occasional factors favorable for solid-solution formation. These compounds are practically always of the ionic or NaCl type or of the Laves type of structure and have compositions in accordance with normal chemical-valence rules.[2]

Compounds may also be formed when the metal is alloyed with weakly electropositive or weakly electronegative elements. The compositions of these compounds do not correspond to normal valences: they are less stable and have lower melting points. However, the elements involved are the more important alloying agents. There is a fairly good relationship between the stability of the compound as

[2]It should be emphasized that, in intermetallic compounds, formulas corresponding to the normal valences of the elements are comparatively rare, except for compounds of strongly electropositive metals with elements from groups IVB, VB, VIB, and VIIB.

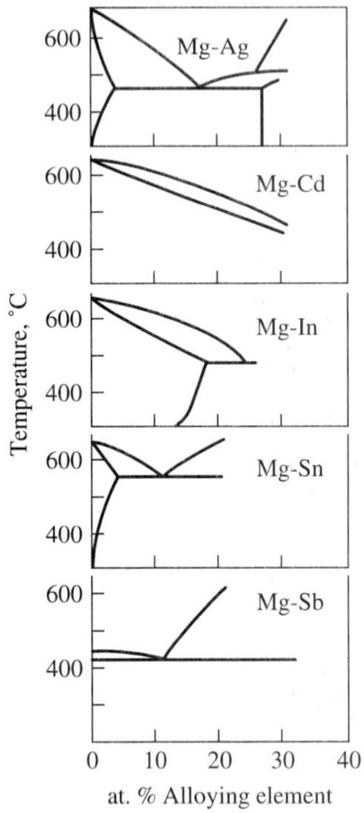

	Period	Lattice	Valence	Size Factor, %
Aluminum	3	fcc	3	10
Gallium	4	orth rh	3	13
Indium	5	fct	3	6
Thallium	6	hcp	3	10

Figure 15.4 Influence of valence on the solid solubility of long-period V elements in magnesium as shown by the binary phase diagrams.

shown by its melting point and its influence on the metallurgically important phase diagrams. Since most magnesium-alloy systems show eutectics between the compounds and the terminal, or magnesium-rich, solid solution, the effects discussed are best summarized by the following:

1. The higher the melting point of the compound the less its solubility in magnesium.
2. The higher the melting point of the compound the higher the melting point of the eutectic.
3. The higher the eutectic melting point the lower the solute element content of the eutectic (or the nearer the eutectic composition to the magnesium end of the diagram).

15.3 CAST MAGNESIUM ALLOYS

Magnesium is a highly reactive metal under certain conditions; yet it can safely be melted in iron or graphite crucibles. Clays are unsatisfactory, since the SiO_2 present is readily reduced, with the silicon forming Mg_2Si as $4Mg + SiO_2 \rightarrow 2MgO + Mg_2Si$. Usually the primary producer markets alloys in the forms of ingots and "hardeners" made up to chemical specifications, although the principal alloying elements can be conveniently added in the foundry in the amounts needed. Since magnesium and its alloys react readily with air in the molten state, it is necessary to use a protective flux during all the processing cycle up to the time of pouring.

The fluxes used are mixtures of various proportions of $MgCl_2$, KCl, CaF_2, MgO, and $BaCl_2$. The specific composition of the flux depends on whether it is a crucible flux or an open-pot flux and on the alloy being processed. These fluxes have roughly the same density when melted as the liquid alloy. The crucible flux fuses to form a liquid cover when first placed on the melt but dries out to a scaly crust on standing at about 760 to 790°C or during superheating. The open-pot flux fuses to a liquid and remains a liquid during its useful life. This flux is used for metal protection in the premelting unit and for protection and cleaning in open-pot practice.

The metal is protected just before pouring by a mixture of equal parts by volume of coarse sulfur and fine boric acid or by any of several proprietary materials.

The reactivity of the metal causes reactions between the metal and water in the sand of "green" sand molds or oxygen in dry sand cores. These reactions result in blackening of the skin of the casting to an appreciable depth with local porosity and gray oxide powder effects, called burning. To avoid these defects, which markedly reduce strength properties, the sands are mixed with addition agents such as sulfur, boric acid, KBF_4, and ammonium silicofluoride, specifically known as inhibitors. For example, 0.5% of boric acid and an equal amount of sulfur in dry sand cores give castings that are clean and bright on the surface and on fractured sections right up to the surface.

Micrographs 15.1 to 15.7 illustrate the most common structural features in cast and wrought magnesium alloys. Nominal compositions and typical properties are

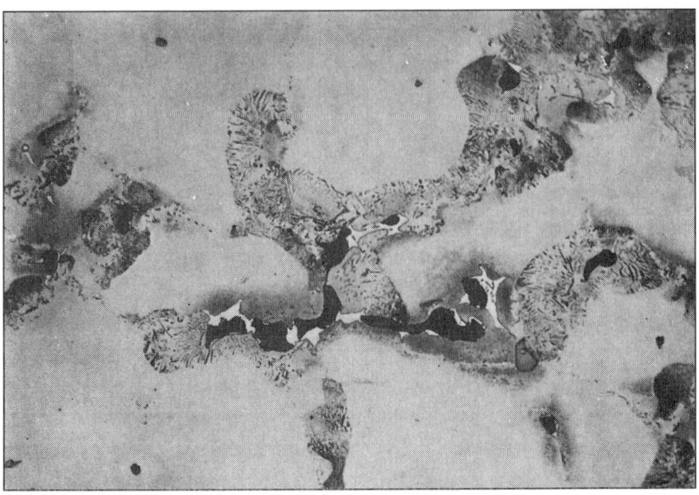

Micrograph 15.1 Alloy AZ63A-T6 (6% Al, 3% Zn, 0.2% Mn) in the solution-treated and aged condition; etched with glycol reagent; ×500. After the solution heat treatment, the structure consisted of polygonal grains of the supersaturated magnesium solid solution, with a few particles of rounded manganese compound (gray). Aging induced precipitation of two types: general, continuous precipitation of particles here not clearly resolved; the lamellar or pearlitic (see Chap. 7) discontinuous precipitate of $Mg_{17}Al_{12}$, which started at the solvent grain boundaries and progressed only a short distance.

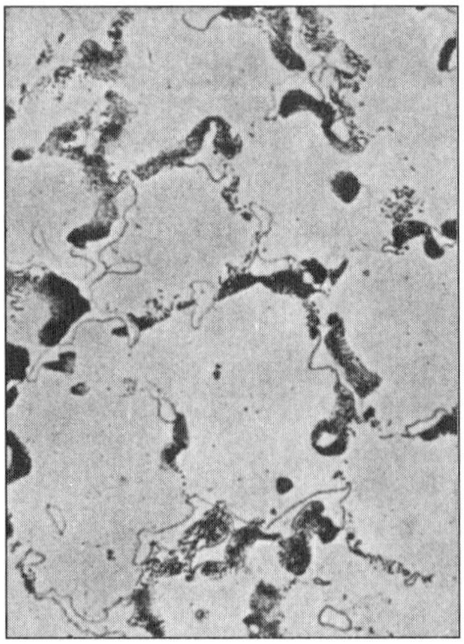

Micrograph 15.2 Alloy AZ92A-F (9% Al, 2% Zn, 0.1% Mn) in the as-cast condition, glycol etch; ×250. The structure shows massive eutectiferous $Mg_{17}Al_{12}$ as the clearly outlined and more or less continuous network. The eutectic has "separated"; i.e., the α Mg phase of the eutectic is connected to, and indistinguishable from, the primary α Mg dendrites, leaving massive compound. Slow cooling after solidification permitted precipitation of fine $Mg_{17}Al_{12}$ in lamellar form from the aluminum-rich cored α Mg near the compound. Note the absence of the ternary compound $Mg_3Al_2Zn_3$.

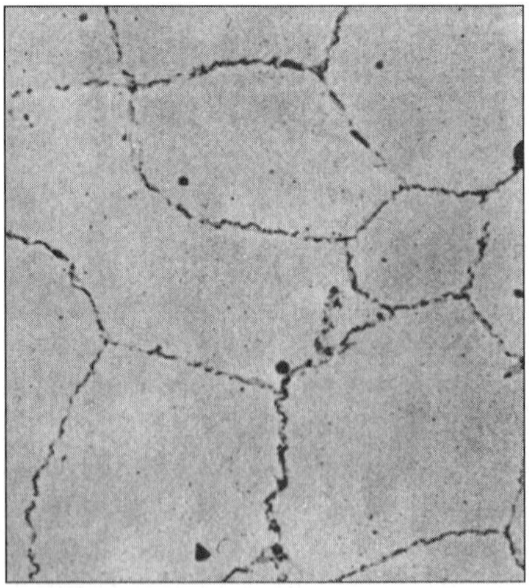

Micrograph 15.3 Alloy AZ92A-T4 (9% Al, 2% Zn, 0.1% Mn), cast and solution heat-treated; acetic glycol etch; ×250. This is a typical solution-treated structure, with all $Mg_{17}Al_{12}$ from the cast structure in solution. The casting was air-cooled from the solution temperature, and the wavy irregular grain boundaries are characteristic of this cooling rate. Probably localized precipitation occurs during cooling.

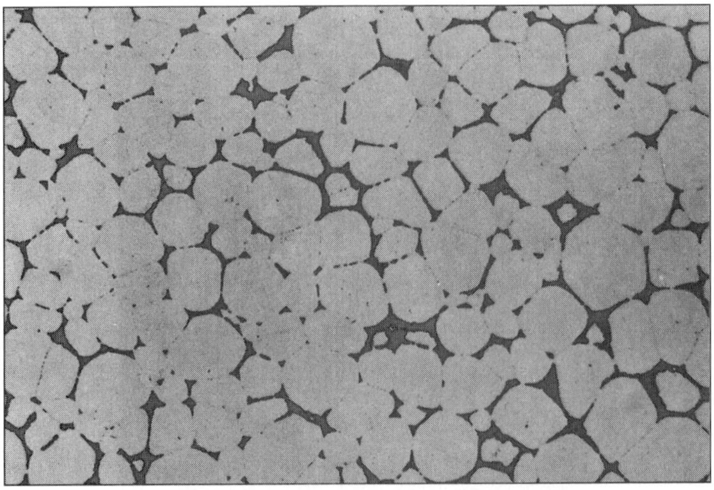

Micrograph 15.4 Alloy EZ33A-T5 (3.2% mischmetal, a rare-earth alloy containing cerium as a major component, 2.8% Zn, 0.6% Zr) in the cast and aged condition, etched with glycol reagent; ×250. The massive compound localized between the grains is usually stubby and thick. Mechanical properties are improved when the compound particles are coalesced into short particles and the continuity of the matrix grains is high.

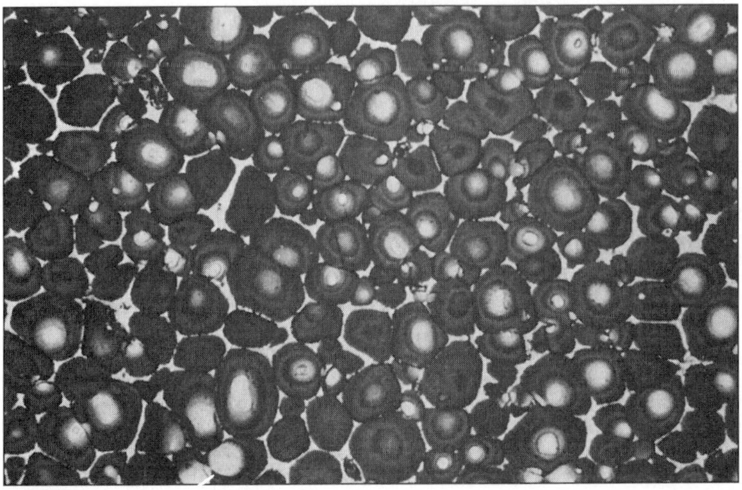

Micrograph 15.5 In the structure of Micro. 15.4, when further etched alternately with glycol and the phosphopicral etchant, the matrix grains show a symmetrical stain pattern associated with the coring effect of Zr.

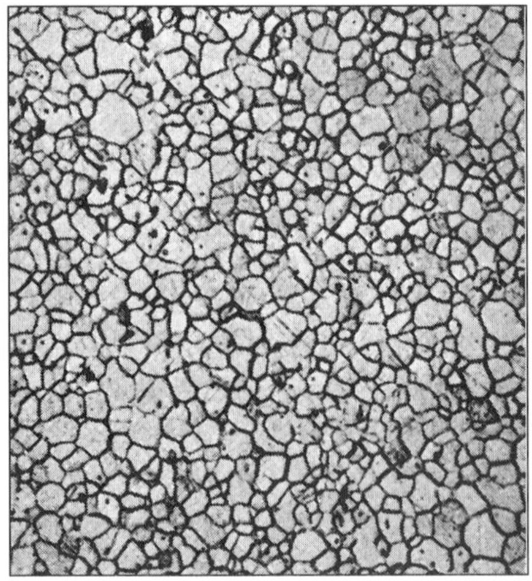

Micrograph 15.6 Alloy AZ31B as annealed sheet; etched with 5% picral (100 ml), H_2O (10 ml), and glacial acetic acid (5 ml); ×250. A slight color contrast between grains of the uniform solid solution is caused by orientation differences. The absence of mechanical twins is typical of wrought hcp metals after annealing.

Micrograph 15.7 Structure of a rod extruded from alloy (P) ZK60B pellets (6.2% Zn, 0.6% Zr); etched with acetic picral reagent [6% picral (70 ml), H_2O (10 ml); 10 ml acetic acid]; ×250. Extrusions made with ZK60B pellets possess a fine and uniform grain size that is essentially independent of section size. Mechanical properties are greatly improved when grain size can be maintained in the vicinity of 8 micrometers. Pellets have the capacity to develop fine grain, since each pellet is essentially a cast ingot with fine grain. Alloy microsegregation is both sharp and on a fine scale, and grain growth is inhibited by the interference offered by alloy gradients.

Table 15.1 Composition and properties of cast alloys

	Alloy (ASTM)			
	AZ63A	**AZ92A**	**ZE41 A-T5**	**ZK51 AT5**
Al,%	5.3–6.7	8.3–9.7		
Zn,%	2.5–3.5	1.6–2.4	3.5–5.0	3.6–5.5
Mn,%	0.15 min	0.10 min		
Zr,%			0.4–1.0	0.55–1.0
Rare earths, %			0.75–1.75	
Tensile modulus, GPa	45	45	45	45
Tensile strength as cast, MPa	205	170		
Solution-treated	280	280	210	280
Yield strength, MPa:				
As cast	100	100		
Solution-treated	90	100	140	170
Elongation, % in 50 mm:				
As cast	6	2		
Solution-treated	12	9	3.5	8
Solution treatment temp., °C	390	410		

given in Table 15.1, together with the heat treatments ordinarily employed to obtain these properties. The more common and older alloys are those of the Mg-Al-Zn system such as AZ63A and AZ92A. Other systems, Mg-Zn-Zr, Mg-RE-Zr, and Mg-Th-Zr, also are being used commercially. Alloys from these systems are heat-treatable. The slightly lower solution treatment for alloy AZ63A compared with AZ92A results from the increased zinc content, which lowers the eutectic melting temperature.

Burning during heat treatment of magnesium alloys may occasionally occur. It manifests itself in three ways: (1) surface exudations, (2) a gray-black powder on the surface, and (3) voids on the surface and in the interior.

The cause of burning may be too high a solution-treatment temperature, an effect comparable with burning of aluminum alloys. It may also be caused by heating too rapidly to the proper temperature: the microsegregation typical of the cast structures is then responsible. A third cause of burning may be the presence of water vapor in the surrounding atmosphere and the absence of sulfur dioxide. Even 1% SO_2 has an inhibiting effect on the H_2O-Mg or H_2O-$Mg_3Al_2Zn_3$ reaction (in alloy AZ63). For this reason, sulfur dioxide is customarily added to the furnace atmosphere during solution heat treatment of magnesium alloys.

From the phase diagrams and previous microstructures, it is evident that these alloys have brittle, interdendritic, eutectiferous networks in the as-cast structure, which are reduced or eliminated by the solution treatment. Thus the solution-treated alloys have far better ductility than the as-cast alloys, and at the same time the tensile strength is considerably increased. The aging treatment does not affect tensile strength but notably increases yield strength, which, as earlier remarked, is the more important property in engineering design. The decrease in ductility that occurs on aging is related to heavy precipitation, but this treatment leaves the casting with sufficient ductility for many uses.

The comparative properties of alloys AZ92A and AZ63A show that the former has somewhat better yield strength and less ductility. However, the relative casting characteristics are also important in choosing between the two alloys. Since alloy AZ92A is less subject to microporosity (very fine pores rather uniformly distributed through the structure) and has generally a better castability, it is the more generally used alloy.

The solid solution obtained by the solution heat treatment is generally air-cooled to avoid distortion and cracking. However, it has been found that, as in the case of aluminum alloy 2024, more rapid cooling eliminates the tendency for grain-boundary precipitation, which occurs to some extent upon air cooling. The subsequent aging results in more uniform precipitation, less boundary concentration of the precipitant, and correspondingly better strength and ductility. These effects are shown by the data of Table 15.2.

Water quenching of the cast magnesium alloys may cause cracking if the alloy is quenched from within the hot-short range, i.e., from a temperature that is too high or water that is too cold. The resulting thermal gradients give rise to high cooling stresses. Figure 15.5 shows the limits of the relationship of metal and water temperatures within which cracking may or may not occur. Quenched within these limits, the alloy will show residual stresses, but the subsequent aging treatment will reduce the macrostresses to a negligible value. After aging, as shown in Table 15.2, both strength and ductility are usually at least 10% better than those obtained with the usual air-cool and aging treatment.

Table 15.2 Effect of cooling rate from the solution treatment on properties of subsequently aged sand-cast magnesium alloys

	Alloy AZ92A			Alloy AZ63A		
Cooling Time at 410–190°C, s	Tensile Strength, MPa	Yield Strength, MPa	% Elongation in 50 mm	Tensile Strength, MPa	Yield Strength, MPa	% Elongation in 50 mm
190 (air cool)	280	170	2.0	280	135	5.0
65 (oil quench)	295	175	1.8	280	165	3.4
5.5 (water 90°C)	320	210	2.2	300	150	5.7
0.5 (cold-water spray)	340	210	3.5	315	150	7.3

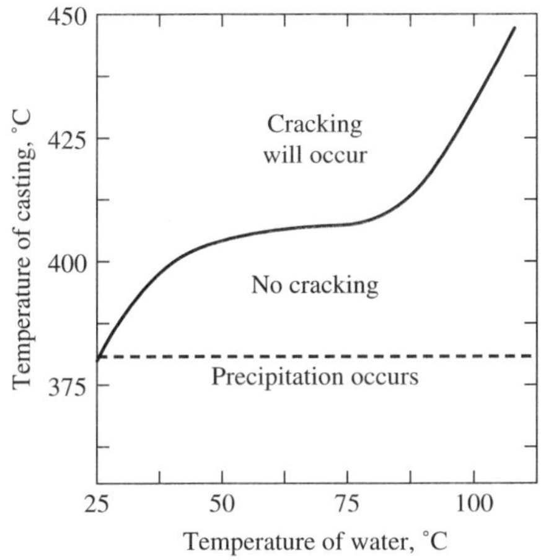

Figure 15.5 The cracking tendency of casting alloy AZ92A upon water quenching from the solution heat treatment as related to the temperature of the casting and of the quenching water.

Alloys in the Mg-Zn-Zr system develop high yield strengths with reasonably good ductility. However, the somewhat higher cost of the alloys as compared with the Mg-Al-Zn alloys and foundry castability problems associated with shrinkage, incidence of microporosity, and cracking are limiting their application. Also, these alloys cannot be repair-welded very readily. Generally, they are suggested for simple, highly stressed parts of uniform cross section, although as experience with the alloys is increased, it can be expected that more complex parts can be produced. Additions of rare-earth (RE) metals and thorium in this system result in some reduction in room-temperature-strength properties but somewhat improve castability and weldability.

It may be noted that aluminum and zinc are the alloying elements that are most effective in promoting room-temperature strength and ductility in magnesium alloys, while thorium is necessary for the high-temperature properties. The Mg-RE-Zr systems of alloys are found to be useful, depending upon the operating conditions, up to 320°C. These alloys are relatively free of microporosity but are more prone to superficial shrinkage defects and to dross inclusions than the Mg-Al-Zn alloys. For this reason they may be more difficult to cast in some designs than the Mg-Al-Zn alloys. EZ33A alloy has a very low tendency toward microporosity and is useful in applications requiring pressure tightness.

The Mg-Th-Zr alloys are intended primarily for elevated-temperature applications at 200°C and above when properties superior to those of the Mg-RE-Zr system are required. These alloys are less castable than Mg-RE-Zr alloys because oxide inclusions and defects attributable to gating turbulence are harder to control. The tendency toward inclusions observed in the Mg-Th-Zr alloys is particularly marked in thin-walled parts that require a rapid pouring rate. These alloys are found to have adequate castability for the production of complex parts containing sections of moderate to heavy wall thickness.

Grain-Size Control of Cast Alloys

Magnesium alloys containing appreciable amounts of aluminum are customarily superheated in the liquid state before casting in order to obtain a very fine as-cast grain size. This is in direct contradiction to the generalization in Chap. 3 that overheating a liquid metal before casting results in a coarser grain, since all potential nuclei in the liquid are dissolved (on the assumption of subsequent cooling to the same pouring temperature). In Chap. 13, overheating of aluminum alloys was shown to be further undesirable because of hydrogen absorption by the liquid and the resulting gas porosity of the casting. Yet Mg-Al casting alloys, after being heated 100 or 200°C above their melting points and then cooled to their proper pouring temperatures, show a much finer as-cast grain size than when just barely melted and cast.[3]

The usual procedure is to melt the alloy and, at about 730°C, refine by treating with a flux that is stirred into the metal to remove oxide dross films. Then a covering flux is added and the temperature is raised between 870 and 930°C. No holding time at this temperature is necessary beyond that required to equalize the temperature. The liquid metal is cooled in the furnace pot to the proper pouring temperature of about 730°C, again held only to equalize the temperature, and the metal poured into the mold. The process of superheating is obviously expensive by reason of fuel costs, the long nonproductive time required for heating and cooling, and the decreased life of furnace refractories and melting pots.

Many researchers responding to the need for increased and more rapid production have shown that other methods may be equally effective in achieving grain refinement of magnesium-alloy castings. Some of these methods are as follows:

1. Vigorous stirring at 760°C (applicable only to small melts)
2. Bubbling acetylene, methane, propane, or carbon tetrachloride through the liquid at about 760°C
3. Stirring into the melt up to 1.0% C as powdered graphite, lampblack, or aluminum carbide, Al_4C_3

Metal refined by superheating loses most of its fine-grained characteristics upon remelting but, when refined by carbon additions, a good part of the effect is retained upon remelting scrapped parts of castings, etc. The fact that grain refinement by carbon additions is possible only in aluminum bearing alloys such as AZ92A or AZ63A makes it seem probable that Al_4C_3 is the responsible nucleating agent in heats where carbon is present.

Pressure Die-Cast Magnesium Alloys

Magnesium-alloy castings considered so far generally are produced in specially treated sand molds. However, the relatively low melting points of these alloys and

[3]Pure magnesium and the alloy containing only manganese are exceptions and do not show the superheating grain refinement.

Table 15.3 Composition and room-temperature properties of three magnesium die-cast alloys and one equivalent alloy, sand-cast and heat-treated

Alloy	Composition, % Al	Mn	Si	Zn	Tensile Strength, MPa	Yield Strength, MPa	Elongation in 50 mm, %	Typical Use
Die-cast:								
AM60A	6	0.2	...	...	210	120	65	Die-cast automotive wheels
AS41A	4	0.3	1.0	...	225	155	4	Crankcase of air-cooled automotive engine*
AZ91B	9	0.2	...	0.8	240	160	4	Most commonly used die-casting alloy
Sand-cast:								
AZ92-T6	9	...	...	2.0	280	150	2	

*Chosen for improved creep strength.

their nonreactivity with iron or steel make it possible to produce pressure die castings of Mg alloys whenever a relatively large number of units are to be produced. An example is die-cast automobile wheels, frequently used for sports cars and racing cars. The most frequently used Mg die-casting alloys are listed in Table 15.3, which includes their composition and properties and, for comparison, an equivalent sand-cast alloy.

15.4 PROPERTIES OF MAGNESIUM ALLOYS

The five alloys of magnesium that are produced in wrought forms are given in Table 15.4. Typical mechanical properties of the last four alloys, extruded in rod form, are given in Table 15.5. Of these alloys, only AZ80A, with the highest aluminum content, and ZK60A show an age-hardening response upon heat treatment. Since the extrusion process is carried out at approximately the solution-heat-treatment temperature and the extruded shape cools in air fairly quickly, it is usually necessary only to age the alloy after extrusion. The same is generally true of forgings and other wrought forms, although, if hot working is carried out at too low a temperature, the regular solution heat treatment may be employed prior to aging. The aging treatment does not markedly affect the properties listed in Table 15.5, but it is effective in increasing the creep strength of the alloy at elevated temperatures.

In the discussion of wrought alloys of magnesium, the hcp structure of the metal becomes significant. During hot working the individual crystals or grains deform

Table 15.4 Compositions of the principal wrought Mg alloys

Alloy	Analysis, % Al	Zn	Mn	Forms Available
AZ31B	3.0	1.0	0.3	Extrusions, plate, sheet
AZ61A	6.5	1.0	0.2	Extrusions, forgings, sheet
M1A	. . .	1.5	. . .	Extrusions, forgings
AZ80A	8.5	0.5	0.2	Extrusions, forgings
ZK60A	. . .	5.5	(0.7 Zr)	Extrusions, forgings

Table 15.5 Composition and properties of wrought alloys of magnesium

	ASTM M 1A-F	AZ61A-F	AZ80A-T5	ZK60A-T5
Al,%		5.8–7.2	7.8–9.2	
Zn,%		0.4–1.5	0–0.2	4.8–6.2
Mn,%	1.2 min	0.15 min	0.15 min	
Zr,%				0.45 min
Density, g/cm^3	1.77	1.80	1.80	1.83
Melting range, °C	648–649	510–615	480–600	520–635
Tensile modulus, GPa	45	45	45	45
Tensile strength, MPa	250	300	350	340
Tensile yield strength, MPa	160	180	240	265
Compressive yield strength, MPa	70	120	195	195
Elongation in 50 mm, %	5–12	14–17	6–8	11–14

primarily by basal slip. As a result, the basal planes rotate so that they tend to become parallel to the working surface of the wrought material. A magnesium crystal ($c/a = 1.624$) deforms by mechanical twinning when compressed along the a axis: in a wrought magnesium alloy, therefore, the grains are oriented in such a way that they twin during a compression test, and it is for this reason that the compressive yield strength is generally lower than the tensile yield strength.

Cold working of magnesium alloys is possible only to a limited extent. At elevated temperatures, slip planes other than the basal become more active, and the sheet alloys can readily be drawn into complex or deep shapes if the metal is deformed above 200°C. Forging is performed at a higher temperature, about 300 to 400°C; it is by hydraulic presses rather than by hammers.

Included in Table 15.5 are data on a newer alloy of magnesium containing no aluminum but about 5% Zn and 0.7% Zr. This alloy may supplant AZ80A as the high-strength alloy since it has a higher yield strength plus a much greater resistance to notch brittleness, with toughness equal to that of the lower-strength alloys. The superior properties of the new alloy are accompanied by a greater cost, which somewhat limits its application.

The cerium alloy EM62 (6% Ce + 2% Mn) can also be produced in wrought form with the cerium content reduced. Again, the advantage is in high-temperature strength. The creep rates for this type of alloy are of a new order of magnitude. For example, the regular high-strength alloy AZ80A creeps at a rate of 0.7% in 1000 h under 30 MPa stress at 150°C, while forged alloy EM22 (2% Ce + 2% Mn) creeps only 0.3% in 1000 h at 140 MPa stress—less than half the rate at five times the stress.

Corrosion Resistance The first important uses of metallic magnesium were in the field of pyrotechnics and for chemical reagents, such as the Grignard reagent, used in the synthesis of organic compounds. These applications take advantage of the chemical activity of the finely divided metal. With the introduction of alloys having favorable weight-strength ratios, serious problems regarding the chemical stability of the material arose. From the standpoint of corrosion resistance to ordinary outdoor atmospheres, substantial improvement over the earlier alloys has been attained by the use of higher-purity magnesium and by modifications of foundry practice, particularly with respect to fluxing treatments. Although commercial alloys are reasonably stable under inland atmospheres, it is desirable to paint the metal unless it is definitely known that the exposure conditions are not unfavorable. Seacoast locations involving direct contact with salt products are definitely corrosive to magnesium alloys.

Formability The crystal structure that limits the cold forming of magnesium allows the attainment of excellent plasticity at slightly elevated temperatures. Some shapes which are made in one draw from warmed magnesium-alloy sheet require two draws if made of steel or aluminum.

Notch Sensitivity The strength properties of magnesium alloys, both wrought and cast forms, may be impaired by notches or sharp discontinuities that serve as stress raisers. The marked notch sensitivity of magnesium alloys requires particular care in the design of stressed parts to avoid abrupt changes in cross section and to provide ample fillets of smooth curvatures. Stressed magnesium parts may be also dangerously weakened by faulty machining operations that produce notch effects.

Modulus of Elasticity The tensile elastic modulus for magnesium is 45 GPa compared with 70 GPa for aluminum alloys and 200 GPa for steel. Thus, for the same dimensions, an elastically stressed magnesium alloy will deflect over four times as much as steel and 50% more than aluminum. Since the relative densities are 1.8, 2.8, and 7.9, equal weights of the three materials would show magnesium to be the stiffest, while steel would show the greatest deflection. The extreme lightness of the metal, therefore, more than compensates for its low modulus. Many structures of magnesium, if made to the same weight as a comparable aluminum alloy, may be sufficiently bulkier or thicker to eliminate the need for stiffening ribs or members, thus simplifying design and lowering fabrication or assembly costs.

Machinability The low plasticity, particularly local, of the metal at room temperatures makes this metal the most machinable of all. It can be cut at high speeds to a beautiful finish, with little tool heating or wear. One large firm lists machining costs of aluminum castings as 25% higher, bronze castings 35% higher, and iron castings 50% higher than magnesium castings. The combustibility of magnesium dust and chips constitutes a hazard that can be eliminated by proper safety precautions: washing fine cuttings away continually, collecting and disposing of dust, etc.

Weldability Magnesium alloys are most commonly welded by the gas shielded-arc welding processes. The gas tungsten-arc process utilizes a tungsten electrode, magnesium-alloy filler rod, and an inert gas such as argon or helium as arc shield. In the gas metal-arc process, a continuously fed magnesium-alloy wire acts as the electrode for maintaining the arc while an argon-gas shield prevents oxidation of the weld puddle. No flux is required, and welding operations similar to those performed on steel and aluminum can be performed with these processes. With the advent of the gas shielded-arc processes, gas welding of magnesium alloys has decreased because of the corrosion problems associated with the fluxes used. Currently, gas welding is used mainly for emergency field repair until a more permanent repair can be made or a replacement part obtained. Gas welding involves the use of a gas torch, magnesium-alloy filler rod, and a chloride-base flux. Welding is confined to joint designs that can be thoroughly cleaned of flux afterward. The usual welding problems of residual stresses and the tendency for certain alloys to be crack-sensitive can be overcome by care, preheating, and postheating for stress relief. Torch, furnace, and flux-dip brazing methods have also been applied to magnesium alloys. Currently, flux-dip brazing is in greatest use because of its greater speed and applicability to more difficult joint designs than possible with furnace or torch brazing. Magnesium alloys are also readily weldable by resistance-welding methods such as spot, seam, and flash welding utilizing equipment commonly employed for aluminum welding.

Cost Magnesium in alloyed form costs more per pound than its chief industrial competitors, steel, aluminum, and copper. It best competes with these metals when one or more of its most important characteristics can be employed to advantage, and under such conditions, it may be cheaper than a competitive metal. Specifically:

1. It is cheaper as a photoengraving material than copper.
2. It is cheaper for cathodic protection purposes than zinc. In this case, the principal competitor is generated electricity.
3. It is cheaper in the form of complex die castings than any other material. This is because steel cannot be die-cast and machining is expensive; also, because aluminum has a higher heat content than magnesium and cannot be cast as rapidly.

PROBLEMS

1. Calculate the maximum tensile and yield loads of a round test bar of magnesium alloy AZ92A-T4 whose cross-sectional area was such that its weight per unit length was identical to that of the comparable cast-aluminum alloy 220-T4. Which alloy rod is stronger?
2. Make the same calculation and comparison as in Prob. 1 for the strongest wrought alloy of (*a*) aluminum, 7075T-6, and (*b*) magnesium, ZK60A-T5.
3. Why is it that wrought magnesium alloys, particularly sheet, are produced from high-purity magnesium, whereas cast alloys are made up with ordinary electrolytic metal?

REFERENCES

Avedesian, M., and H. Baker (Eds.): *ASM Specialty Handbook: Magnesium and Magnesium Alloys*, ASM Int., Materials Park, OH, 1998.

Heat Treater's Guide: Practices and Procedures for Nonferrous Alloys, ASM Int., Materials Park, OH, 1996.

Fatigue Databook: Light Structural Alloys, ASM Int., Materials Park, OH, 1995.

ASM Handbook, vol. 2, *Properties and Selection: Nonferrous Alloys and Special Purpose Alloys*, ASM Int., Materials Park, OH, 1991.

Brooks, C. R.: *Heat Treatment, Structure and Properties of Nonferrous Alloys,* ASM Int., Materials Park, OH, 1990.

chapter

16

Titanium Alloys

Titanium and zirconium both have unique properties, which fit them for certain vital applications. Their chemical and physical characteristics are closely related. Titanium ores are abundant. The metal is characterized by a high strength-to-weight ratio, a high melting point, and excellent corrosion resistance. These properties appeared so useful for structures of very high-speed aircraft that large amounts of money were committed in the 1950s to research and the development of titanium production facilities. As a result, this metal was changed from a laboratory curiosity to a major article of commerce. Zirconium, not so abundant as titanium, finds important applications in nuclear technology because of its unusually low absorption cross section for thermal neutrons. Both metals are expensive to win from their ores, so that their use in a particular application must be justified in terms of their unique properties. In this chapter, only the technology of titanium will be covered, although most of it also applies to zirconium.

16.1 UNALLOYED TITANIUM

At temperatures below 882°C, the crystal structure of titanium is hcp, with a c/a ratio about 2% less than the ideal value of 1.633, as shown in Table 16.1. Plastic deformation of hcp α titanium occurs by both slip and twinning. Slip occurs in the $[11\bar{2}0]$ direction, the close-packed direction, as in the other hcp metals. However, in contrast to zinc and cadmium where only the basal plane is a slip plane, slip in

Table 16.1 Axial ratios of hexagonal metals

Metal	c/a	Metal	c/a
Cadmium	1.886	Zirconium	1.589
Zinc	1.856	Titanium	1.587
Magnesium	1.624	Beryllium	1.568

titanium occurs on the $(10\bar{1}0)$ and $(10\bar{1}1)$ planes as well as on the basal plane. Of these, the $(10\bar{1}0)$ planes predominate and the others are active only for special orientations. This is in accordance with the rule, which states that slip usually occurs on the most widely spaced planes and in close-packed directions. In zinc and cadmium the *c*/*a* ratio is much larger than ideal, and the basal planes are consequently widely spaced. In magnesium, slip still occurs predominantly on the basal planes, $(10\bar{1}1)$ slip becoming active only at elevated temperatures. In titanium, with *c*/*a* less than ideal, the basal planes are closely spaced and are less favored for slip.

Titanium also deforms readily by twinning; at room temperature $(10\bar{1}2)$, $(11\bar{2}1)$ $(11\bar{2}2)$, $(11\bar{2}4)$, and $(11\bar{2}3)$ can all serve as twin planes. At low temperature (−196°C) slip occurs only on $(10\bar{1}0)$ planes, and twinning is the primary mode of deformation. Aluminum as an alloying element tends to reduce the amount of $(10\bar{1}2)$ twinning but does not eliminate it. At high temperatures, slip is dominant, although $(10\bar{1}2)$ twinning may persist up to about 700°C. The large number of slip and twinning systems available in titanium means that this metal can be much more extensively cold-worked than either zinc or magnesium. The contrast with the cold rolling of magnesium is particularly marked: under the stresses developed in rolling, neither titanium nor magnesium can twin on $(10\bar{1}2)$: but in titanium, action of the other twin planes permits extensive cold deformation.

Titanium has a very great affinity for the gaseous elements hydrogen, carbon, nitrogen, and oxygen, all of which form interstitial solid solutions. Even the purest metal available is not free of these elements and commercial grades contain them in significant amounts. They all tend to increase strength and hardness. Since oxygen is usually present in the greatest amount, the total effect of O, N, and C, expressed in terms of percent oxygen equivalent, is plotted vs. hardness, strength, and ductility (reduction of area in tensile tests) in Fig. 16.1.

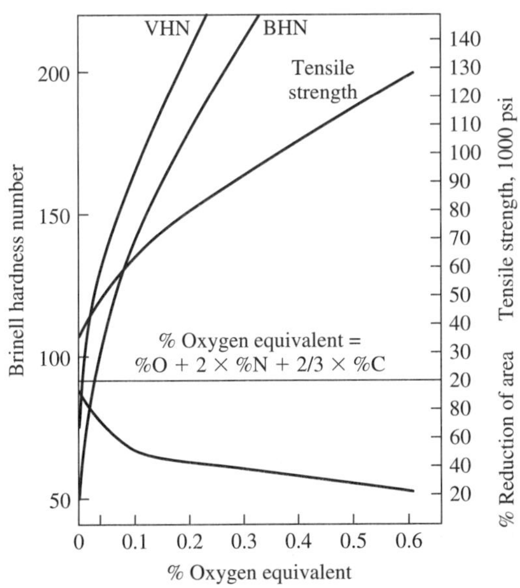

Figure 16.1 Hardness and tensile properties of titanium vs. the oxygen equivalent, i.e., the summation of the relative effects of oxygen, nitrogen, and carbon impurities in the titanium.

Table 16.2

Grade	Yield Strength, MPa (10^3 lb/in^2)
A40	280–420 (40–60)
A55	385–560 (55–80)
A70	490–665 (70–95)

Since chemical analysis for small amounts of interstitial impurities is difficult, the purity of commercially pure titanium is specified in terms of its yield strength at 0.2% offset, as shown in Table 16.2. The lowest-strength grade (highest purity) is used when a maximum of ductility and formability is required; the others provide a combination of moderate strength and outstanding corrosion resistance.

Allotropic Forms Both titanium and zirconium transform to the bcc structure (β) on heating to a high temperature (882.5°C for titanium and 865°C for zirconium), and there is every reason to believe that the mechanism of transformation is the same for the two metals. If on cooling, the bcc form is held just below the transformation temperature, hcp α forms from the β very slowly. At lower temperatures the rate of transformation is more rapid. It is not possible to retain the β phase by sudden cooling no matter how rapid the quench. This observation is readily explained if it is assumed that the transformation mechanism does not require diffusion. From a study of the orientation relationships between transformed and parent material in large grains of zirconium, W. G. Burgers deduced the mechanism of transformation illustrated in Fig. 16.2. The actual transformation mechanism may not follow the exact sequence of movements shown in Fig. 16.2, but it is clear that the bcc to hcp transformation can occur without the aid of diffusion-controlled atom

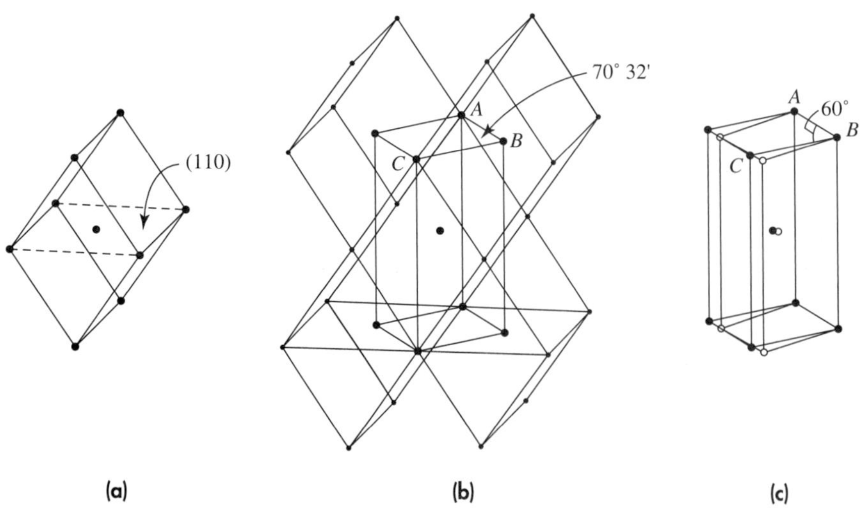

Figure 16.2 The diffusionless transformation of α titanium to β titanium. (*a*) A bcc unit cell of β titanium; (*b*) an array of the cells shown in (*a*), with a potential hcp cell outlined; (*c*) shear of the cell in (*b*) into an hcp cell α titanium.

movements. It satisfies the experimentally observed condition that the {110} planes of the β phase are parallel to the (0001) plane of the α phase, which is formed.

16.2 PHASE DIAGRAMS OF TITANIUM ALLOYS

The phase diagrams of binary alloys containing titanium display a rich variety of complex phase relations. The characteristics of these diagrams, which are significant for the industrially important alloys, can be classified quite simply. The most important question is whether a given alloying element stabilizes the α phase or the β phase. Stabilizing the α phase means that as solute is added, the α-β phase transition temperature is raised: β stabilization implies the converse. Aluminum is an α stabilizer, as is seen from the Al-Ti phase diagram (Fig. 16.3). Other important α stabilizers are O, C, and N.

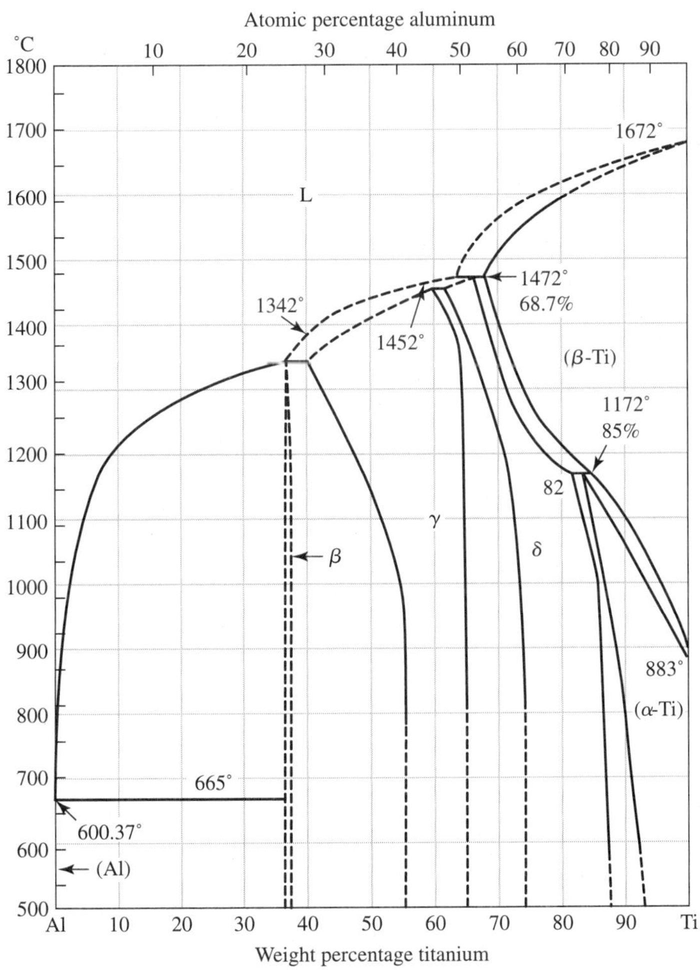

Figure 16.3 The Al-Ti phase diagram. Note that in addition to the high-temperature β-Ti phase on the right-hand side, there is an intermetallic compound Al_3Ti, also designated β, in the Al-base alloys.

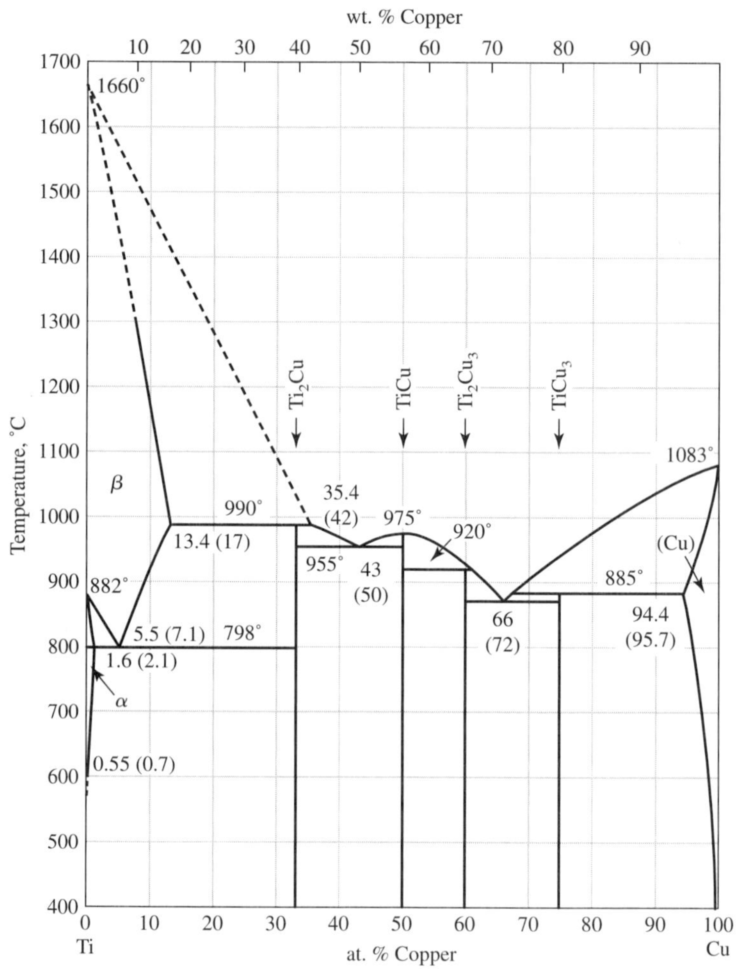

Figure 16.4 Phase diagram for the Ti-Cu system.

With the exception of copper, the important β stabilizers are all transition elements (molybdenum, vanadium, manganese, chromium, and iron). The Ti-Cu diagram (Fig. 16.4) shows how strongly copper suppresses the α-phase field and also the occurrence of the large number of intermetallic compounds characteristic of this type of alloy.

The characteristic phase diagrams of titanium alloys can be divided into three groups: the β isomorphous, eutectoid, and peritectoid types. The first two types include the β stabilizers, the last one the α stabilizers. A typical β-isomorphous diagram is that of the Ti-V system (Fig. 16.5), in which there is a continuous field of β solid solution ranging from pure Ti to pure V. The Ti-Cu diagram of Fig. 16.4

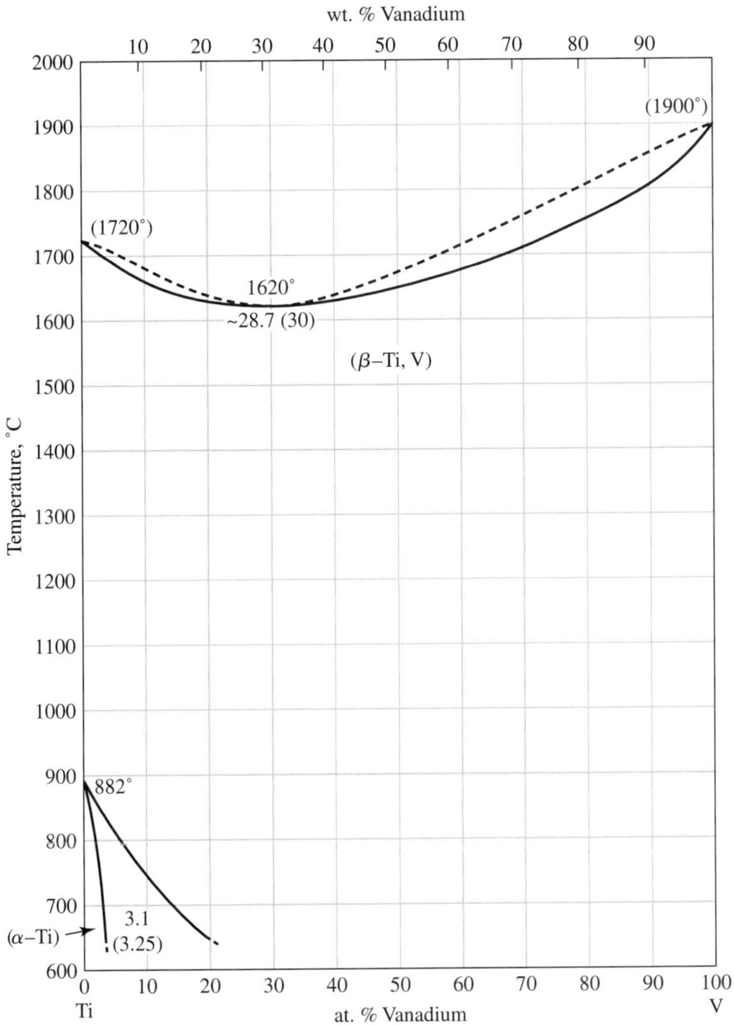

Figure 16.5 Phase diagram for the Ti-V system.

illustrates the eutectoid type. In the eutectoid systems, intermetallic compounds always occur, in contrast with the β-isomorphous systems. The Al-Ti diagram (Fig. 16.3) is of the peritectoid type.

Because of weight and other considerations, Al is the most widely used alloying element in titanium. Inspection of Fig. 16.3 indicates that an alloy of Ti-17 wt % Al will pass through the following phase fields upon equilibrium cooling from the β field: $\beta \rightarrow \beta + \delta\ (Ti_3Al) \rightarrow \alpha + \delta \rightarrow \delta\ (Ti_3Al)$. The Ti_3Al phase, which embrittles α Ti when it is present in sufficient amounts, is an ordered α structure, which is sometimes referred to as α_2. It can be considered to consist of four α cells arranged in space so that the c_0 parameter is about the same length as that of α and the a_0 parameter is about twice that of α. The stoichiometric composition for Ti_3Al exists within the solid-solution region of the δ phase.

The β phase in Ti alloys can be retained upon quenching from either the all-β field or from the $\alpha + \beta$ field, provided sufficient β-stabilizing alloying elements such as Mo, V, Mn, or Cr are present. The decision must be made whether an all-α, all-β, or mixed $\alpha + \beta$ structure is desired for any specific service.

The important characteristics of a number of titanium alloy systems are summarized in Table 16.3.

Table 16.3 Characteristics of binary titanium phase diagrams

β Isomorphous Systems

Element	Max. Extent of α-Phase Field, wt %
Mo	0.8
V	3.5
Ta	12.0
Cb	4.0

Eutectoid Systems

Element	Compounds Formed	Eutectoid Composition, wt %	Eutectoid Temperature, °C	Max. Extent of β-Phase Field, wt %
Mn	TiMn, $TiMn_2$	20	550	33
Fe	TiFe, $TiFe_2$	15	595	25
Cr	$TiCr_2$	15	670	100
Cu	$TiCu_2$, TiCu, etc.	7.1	789	17
Ni	Ti_2Ni, TiNi, $TiNi_3$	5.5	770	13

Peritectoid Systems

Element	Solid-Solution Type	Max. Extent of α-Phase Field, wt %	Max. Temperature of α-phase Field, °C	Max. Extent of β-Phase Field, wt %
Al	Substitutional	17.5	1172	35
C	Interstitial	0.48	920	0.75
B	Interstitial	....	885	
O	Interstitial	15.5	1900	1.8
N	Interstitial	7.0	2350	1.9

The all-α alloys develop good strength and toughness and have superior resistance to oxygen contamination at elevated temperatures but have relatively poor forming characteristics. All-β alloys, having a bcc structure, display much better formability, have good strength both hot and cold, but are more vulnerable to contamination from the atmosphere. They also have a high density (since the β stabilizers are transition elements) so that, for a given strength, they have a lower strength-to-weight ratio. The $\alpha + \beta$ alloys represent a compromise between these characteristics: they have good formability and cold strength but are weak when hot. Many titanium alloys can be strengthened by heat treatment as well as by solid-solution hardening and cold working.

The simplest type of titanium alloy is the all-α type formed by the addition of an α stabilizer, usually aluminum. Aluminum has a strong solid-solution hardening effect on titanium, as shown in Fig. 16.6. Additions of tin further strengthen the α phase without a significant loss of ductility: so a composition frequently chosen for good strength properties is Ti + 5% Al + 2.5% Sn.

The $\alpha + \beta$ alloys with small β-stabilizer content, such as Ti + 6% Al + 4% V, or $\alpha + \beta$ alloys with heavily alloyed 6% and with α containing Mo, such as Ti + 6% Al + 2% Sn + 4% Zr + 6% Mo, are also used at elevated temperatures. The strength-weight ratio properties of these alloys as a function of temperature are compared with aluminum and steel in Fig. 16.7. Of the three titanium alloys, the all-α Ti + 5% Al + 2.5% Sn not only has the lowest strength-to-density ratio but also the lowest strength as well because it cannot be heat-treated. To be heat-treatable (as

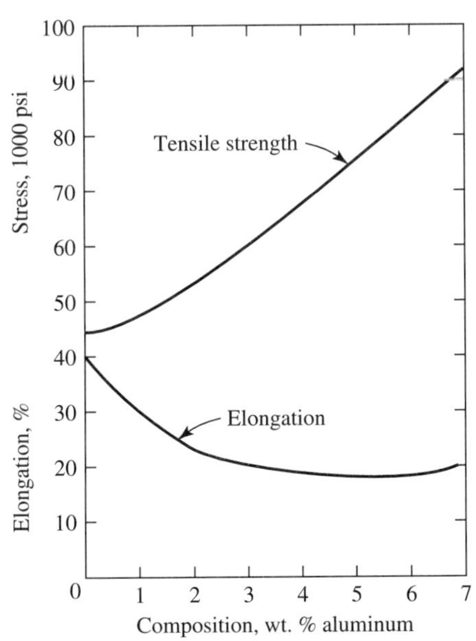

Figure 16.6 The hardening of titanium by aluminum in solid solution.

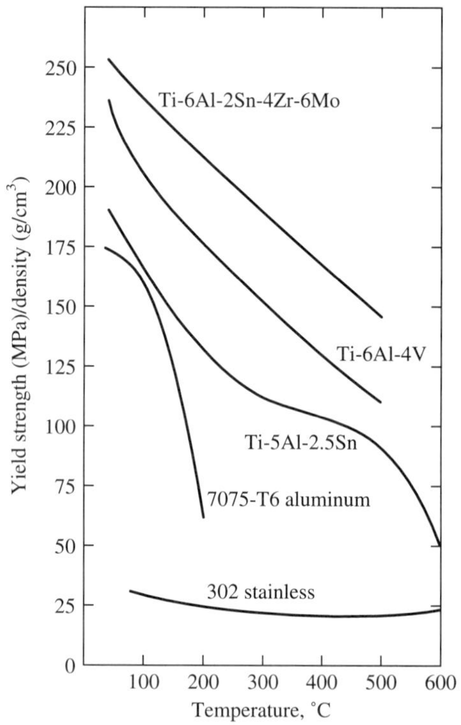

Figure 16.7 The yield-strength-density ratio for three commercial titanium alloys compared with those for a high-strength aluminum alloy and type 302 stainless steel.

discussed shortly) an alloy must have appreciable amounts of β present. The Ti + 6% Al + 2% Sn + 4% Zr + 6% Mo, with the most α, has been strengthened the most by heat treatment. After the strengthening treatment, an alloy should be employed at a temperature below the lowest heat-treatment temperature. At an elevated temperature of service, the amount of β remaining from the strengthening treatment would tend to decrease as a result of phase-diagram requirements. During the period of useful life, only minor changes in mechanical properties and in microstructure can be tolerated. Some indication of the relative stability of the Ti + 6% Al + 2% Sn + 4% Zr + 6% Mo alloy is its relatively slow decay in strength with increasing temperature.

All three titanium alloys of Fig. 16.7 are used in the form of forgings, bars, and rings, which find extensive applications in jet engines. The Ti + 6% Al + 4% V and Ti + 5% Al + 2.5% Sn alloys are also produced in the form of plate, sheet, and strip.

Most titanium alloys used commercially consist of $\alpha + \beta$ structures, and the Ti + 6% Al + 4% V composition is the most widely used alloy. However, several all-β alloys have been developed because of the inherent superiority of the bcc structure over the hcp structure with respect to bending and cold-forming operations. After forming in the soft condition, the alloys may be heat-treated to high-strength levels. Thus, they are particularly suited for applications such as pressure vessels requiring maximum strength-to-weight ratio, honeycomb panels, and missile casings. Three current β alloys are Ti + 13% V + 11% Cr + 3% Al, Ti + 11.5% Mo + 6% Zr + 4.5% Sn, and Ti + 8% Mo + 8% V + 2% Fe + 3% Al.

The all-β alloys and many $\alpha + \beta$ alloys whose compositions are not given here have multicomponent additions that serve to enhance solid-solution strengthening of both phases or to enlarge the strengthening potential of β on aging. Often alloying elements are added on the synergistic principle that multiple alloying can produce greater strengthening than would be expected on the basis of the sum of effects produced by each individual addition. Low β-containing $\alpha + \beta$ alloys sometimes have small silicon additions. Silicon forms an insoluble silicide, which serves to increase elevated-temperature creep resistance.

16.3 HEAT TREATMENT OF TITANIUM ALLOYS

As pointed out earlier, titanium alloys containing β-stabilizer additions can retain the β with appropriately rapid cooling from either the β or $\alpha + \beta$ phase fields if the β contains sufficient stabilizer. The ability to retain the β phase is not the same for each alloying element, as indicated in Table 16.4. If insufficient β stabilizer is present during cooling to retain the β, it may decompose partially by several mechanisms. Depending on alloy composition, some athermal ω whose structure is not fully understood or martensite may form. The composition range over which these decomposition products exist overlaps. At low alloy compositions only martensite is formed, and at high-β-stabilizer contents only some athermal ω is found.

Table 16.4

Alloy System	Approximate Minimum Alloy Composition to Retain β in Thin Section Sizes upon Water Quenching, wt %
Ti-Fe	4–6
Ti-Mn	4–6.5
Ti-Cr	6–7
Ti-Mo	11
Ti-Cu	13
Ti-V	16

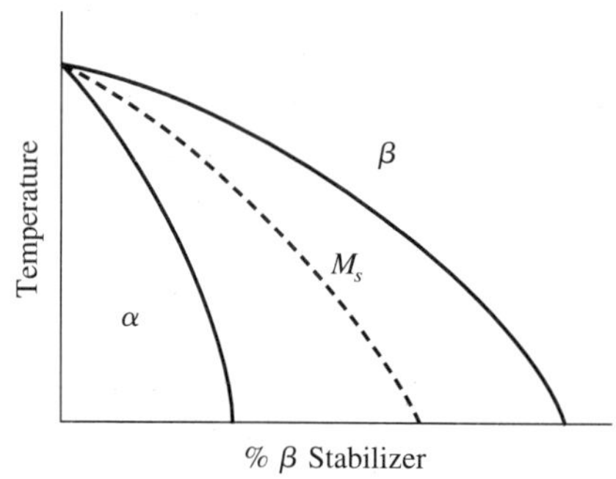

Figure 16.8 Schematic phase diagram showing the dependence on composition of the martensitic-start M_s temperature of β-stabilized types of titanium alloys.

The athermal ω phase forms as a result of very small atomic displacements and hence cannot be suppressed during quenching in the composition range where it can form. The term martensite (Chap. 7) refers to a diffusionless shear transformation, which occurs during quenching. The temperature at which martensite will start to form during quenching, the M_s temperature, depends on the alloy content, as shown schematically in Fig. 16.8. The temperature at which the alloy is completely transformed to martensite, the M_f temperature, is not well defined. As in Fe-C alloys, martensite may form as a massive structure or as a needlelike structure. The latter becomes more evident as the alloy content increases in the composition range where martensite forms.

Two types of martensite, α' and α'', have been found to form. The α' structure is hcp, while the α'' is orthorhombic. The decomposition of martensite, which is supersaturated with respect to β, is the basis for strengthening the Ti + 6% Al + 4% V alloy. Martensite has been reported to decompose by the precipitation of small particles of α or β. The strengthening which generally accompanies this precipitation is much smaller than the strengthening produced by martensite in steel. In the case of the Ti + 6% Al + 2% Sn + 4% Zr + 6% Mo alloy, however, the fine precipitate of β markedly strengthens the alloy and drastically reduces the ductility, which cannot be recovered even with very long overaging times. As a result, martensite formation in this alloy is avoided by heat treating low enough in the $\alpha + \beta$ field when the alloy content is sufficient to depress the M_s below room temperature. Strengthening of this alloy, as will be discussed, occurs by the precipitation of α.

Retained β may decompose on aging to form a complex hexagonal precipitate, also known as ω, at temperatures below 550°C, or β', which is a bcc solid solution lower in β-stabilizer content than the original alloy content. Although ω precipitation greatly strengthens the β, in sufficient amounts it reduces the ductility essentially to zero. In some instances it has been found that slip, cutting the ω particles, causes them to dissolve, thus causing a local softening. This softening permits further deformation to continue locally until voids form, then link up, and cause fracture. On continued aging at a high enough temperature, both ω and β decompose to form α. At still higher temperatures α will form directly from the β without any of the intermediate structure.

The α form is lower in density than β, and when it precipitates, it is deformed by the β matrix and in turn causes deformation of the α. The highly deformed α is strengthened sufficiently by this deformation to serve as an obstacle to dislocation movement in β, thus strengthening the β. The increased dislocation density in β developed by α precipitation serves the same function. It is this process that strengthens the Ti + 6% Al + 2% Sn + 4% Zr + 6% Mo alloy. Heat treatment of $\alpha + \beta$ alloys that are to be strengthened by aging is a two-step process. The first step, known as a solution treatment, is carried out high in the $\alpha + \beta$ field and provides sufficient β to produce the desired strengthening and to fix the morphology of α. The α formed during this heat treatment is known as primary α. After quenching from the solution treatment, the alloy is reheated to a temperature usually below 650°C for aging.

16.4 PROPERTIES OF TITANIUM ALLOYS

Effect of α Morphology

The morphology, i.e., shape, of α in an $\alpha + \beta$ alloy plays an important role in determining tensile ductility, resistance to fracture in the presence of a preexisting crack, and development of fatigue cracks. Micrographs 16.1 to 16.7 give details of

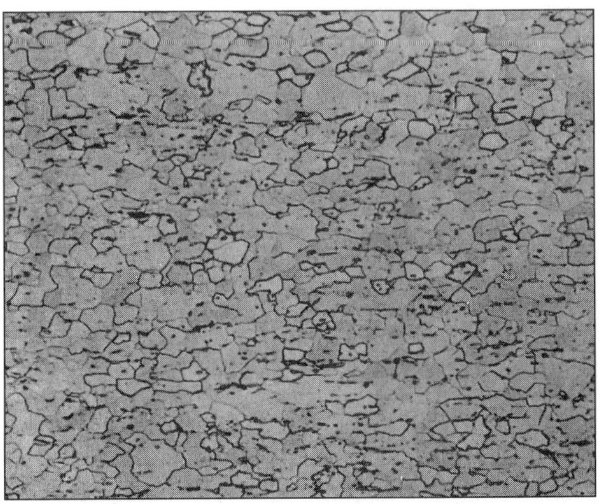

Micrograph 16.1 Commercial, unalloyed titanium, grade A70; HF + $2HNO_3$ etch; ×300. This sample was taken from an annealed sheet and shows a typical α structure. The dark particles are retained β, resulting from the small amount of residual iron impurity.

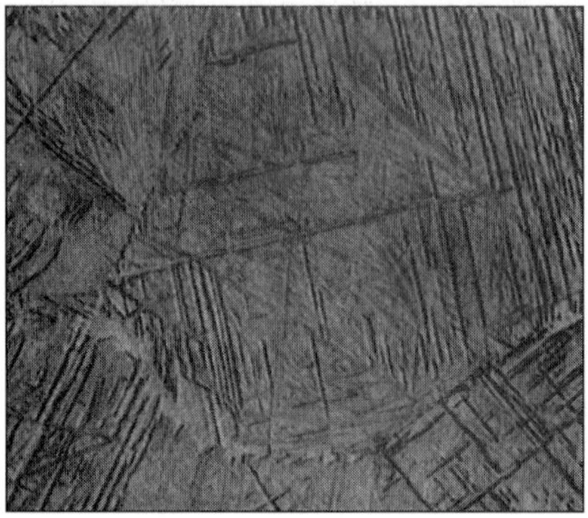

Micrograph 16.2 Ti + 6% Al + 4% V, quenched from the β field, etched with benzol chloride + HF; ×500. Quenching produced martensite. However, the center grain indicates that quenching was not rapid enough to prevent some α formation at β grain boundaries.

Micrograph 16.3 Ti + 6% Al + 4% V, furnace-cooled from the β field into the $\alpha + \beta$ field at 930°C, held 46 h to coarsen the structure, water-quenched, reheated to 770°C, and again water-quenched; benzol chloride + HF etch; ×500. The structure shows that α formed first at the original β grain boundaries and Widmanstätten α within the former β grains. The β remaining after the 930°C hold transformed to martensite upon quenching. This martensite, although decomposed at 770°C, still reveals the characteristic martensite appearance.

Micrograph 16.4 Ti + 6% Al + 4% V, held 72 h at 930°C and water-quenched, reheated 2 h at 740°C, and quenched; benzol chloride + HF etch; ×500. Equiaxed α (dark) formed at 930°C in a matrix of decomposed β martensite. This micrograph, together with Micros. 16.2 and 16.3, shows that β will transform to martensite upon quenching, whether or not α is present, if the β is of proper composition.

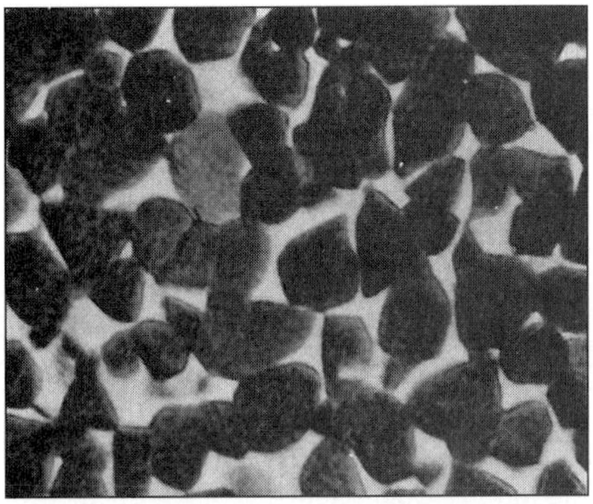

Micrograph 16.5 Ti + 6% Al + 4% V, held 20 h at 910°C, furnace-cooled to 790°C, held 1 h and quenched; benzol chloride + HF etch; ×1000. The structure consists of equiaxed α (dark) and retained β (light).

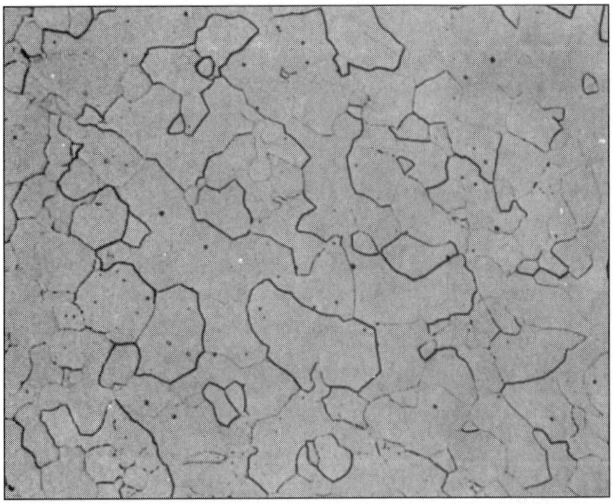

Micrograph 16.6 Ti + 13% V + 11% Cr + 3% Al; HF + $2HNO_3$ etch; ×300. This is from a sample of sheet solution treated at 820°C and water-quenched. The structure is entirely β phase.

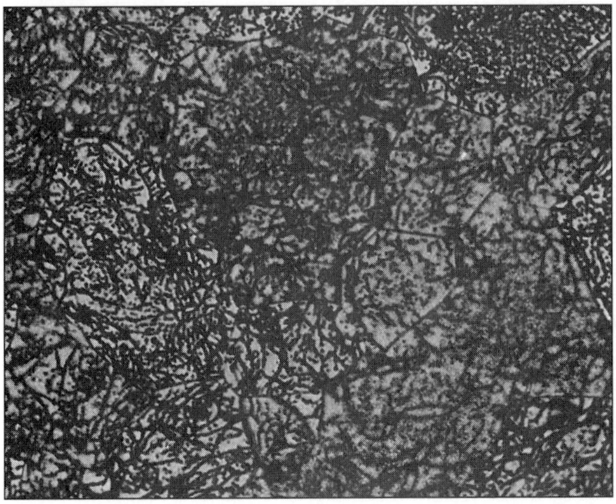

Micrograph 16.7 The sample of Micro. 16.6 after aging for 24 h at 485°C. The structure consists of β plus a darkly etched network of α. Traces of the original α grain boundaries can be seen, as well as the different appearance of the α in different β grains.

the microstructure of unalloyed titanium and several titanium alloys. Primary α exists in two forms, Widmanstätten plates plus grain-boundary α (Micro. 11.3) or equiaxed α (Micro. 11.4). The first structure forms on cooling either from the β into the $\alpha + \beta$ field or from high in the $\alpha + \beta$ field to some lower temperature in the same two-phase field. An orientation relationship exists between the Widmanstätten α plates, which form within the grains and the β grains in which they form. The same orientation relationship exists between the α at grain boundaries and one of the two grains sharing the boundary. This orientation relationship has special significance in plastic deformation of $\alpha + \beta$ structures.

Equiaxed α will develop from the platelet form by hot working in the $\alpha + \beta$ field followed by recrystallization in the same two-phase field, but not necessarily at the same temperature used for hot working. The α created as a result of this process exists, in a β-matrix alloy, either at edges shared by three β grains or along the faces shared by two β grains. The $\alpha + \beta$ interfaces in both structures may serve as sites for void formation during tensile testing or during tensile fracture in the region ahead of a preexisting crack. Both Widmanstätten and grain-boundary α particles having relatively long interfaces with β provide a ready path for void growth. Voids in the platelet structure reach a critical size for fracture at lower strains than they do in equiaxed structures, where the α particles also serve as crack stoppers. Thus the platelet structures usually have lower ductility than equiaxed ones. For example, the Ti + 5% Al + 5% V + 1% Fe + 0.5% Cu alloy, heat-treated to a yield strength of 163,000 lb/in^2, shows a tensile reduction in area of 59% for an equiaxed structure but only 29% for a platelet structure of the same yield strength.

The orientation relationship for the platelet structure is the Burgers orientation relationship described earlier in connection with Fig. 16.2. The crystallographic relationship is designated:

$$(0001)\alpha \parallel (110)\beta$$

$$<11\bar{2}0>\alpha \parallel <111>\beta$$

This means that not only the (0001) plane of α is parallel to a (110) plane of β but also a $<11\bar{2}0>$ direction in the basal α plane is parallel to a $<111>$ direction in the corresponding $<110>\beta$ plane. This orientation relationship also means that a slip system in α, that is, $(0001)<11\bar{2}0>$, is parallel to a slip system in β, that is, $(110)<111>$. As a result slip starting in α can readily be transmitted to β. Therefore, fatigue cracks, which require slight plastic deformation to form, can develop more readily in the platelet structure than in the equiaxed α morphology.

It may be concluded that if for a particular application one mode of failure is more predominant than any other, a particular morphology may be selected to provide maximum resistance to this mode of failure. It is not uncommon practice in industry to develop the knowledge of how to produce a specific desired microstructure and to require the producers to provide material whose microstructure the purchaser has preselected. Table 16.5 gives typical properties of the three major types of titanium-alloy structures in both annealed and heat-treated forms.

Table 16.5 Properties of titanium alloys with typical structures

		Annealed			Heat-Treated		
Type	Composition, %	Tensile Strength, MPa	Yield Strength, MPa	Reduction in Area, %	Tensile Strength, MPa	Yield Strength, MPa	Reduction in Area, %
All α	Ti-5Al-2.5Sn	800	770	25			
$\alpha + \beta$	Ti-6Al-4V*	910	840	25	1085	1010	15
	Ti-6Al-2Sn-4Zn-6Mo†	1120	1050	20	1190	1120	20
All β	Ti-13V-11Cr-3Al	875	840	25	1190	1120	10

*Annealing treatment: heat from 700 to 870°C, 1 to 8 h, air-cool; hardening heat treatment 940°C, 1 h, water-quench; age at 485°C for 8 h, air-cool.

†Annealing treatment: 840°C for 1 h, air-cool plus 650°C for 1 to 8 h, air-cool; hardening heat treatment 870°C for 1 h water-quench; age at 600°C for 8 h, air-cool.

Creep Properties of Titanium The somewhat unusual creep behavior displayed by titanium and some of its alloys is illustrated in Fig. 16.9. Under creep conditions the normal type of strain-vs.-time curve is obtained. Figure 16.9 shows the results of creep tests conducted over a range of temperature, with the stress required to maintain a strain rate of 10^{-4} %/h being plotted as a function of test temperature. The yield strength as a function of temperature is shown for comparison. While the yield strength declines steadily with increasing temperature, the stress required to maintain a constant creep rate increases in the range from 100 to 200°C. The origin of this increased creep resistance is not fully understood. It is very likely, however, that it is due to the interaction between interstitial impurities in the titanium and dislocations similar to the strain-aging phenomenon that occurs in steel (see Chap. 8). Below about 300°C the interstitial solute atoms are strongly attracted to

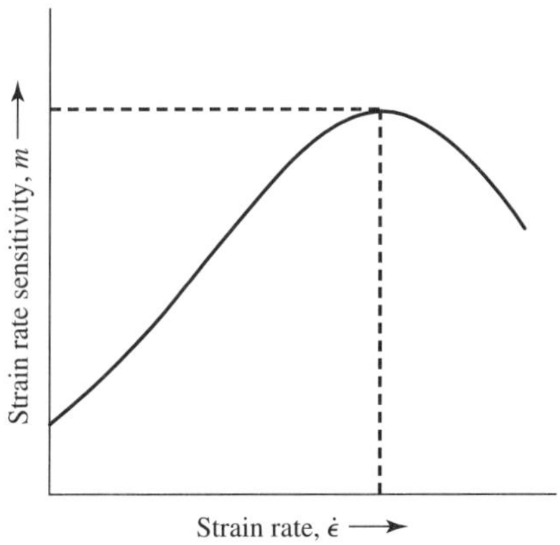

Figure 16.9 Short-time yield strength and creep strength of commercially pure titanium as a function of test temperature.

edge dislocations. This is because an interstitial solute causes considerable dilation in its vicinity; its strain energy is lowered, therefore, when it is in the dilated region under the extra half plane of the dislocation. This means that there is a force of attraction between a dislocation and nearby interstitial solute atoms. To move, either the dislocation must be pulled away from the solutes, which accumulate near it, or the solutes must follow the dislocation by diffusion as it moves slowly along its slip plane during a creep experiment. This added drag, present as long as the mobility of the solute atoms is low, is one possible cause of the enhanced creep resistance observed in impure titanium and in titanium alloys. Cold work improves creep resistance at room temperatures, and annealing several hours at 200°C after cold working further improves creep strength at room temperatures.

16.5 APPLICATIONS OF TITANIUM ALLOYS

Titanium is produced from rutile, TiO_2, by the Kroll process, named after its inventor. The oxide is first converted to $TiCl_4$. Since magnesium has a greater affinity for chlorine than titanium, it is used to reduce the chloride to pure titanium. The metal produced in this manner has a spongelike appearance. To convert the sponge to ingot form, the sponge is pressed into compacts with or without alloying elements as required. A number of compacts are welded together and form the positive electrode in a vacuum-arc furnace, which must be used to avoid pickup of extremely deleterious amounts of oxygen and nitrogen from the air. During melting an arc is established between the water-cooled copper crucible and the titanium anode, and as the metal melts, it drips down into the crucible: 86-cm-diameter, 7-ton ingots are readily produced by this method.

Titanium can be hot-worked by standard steel-mill equipment, and this has greatly facilitated the rapid development of the titanium industry. The reaction of titanium with atmospheric gases, however, is an important factor in hot working. Titanium absorbs hydrogen at temperatures above 150°C, oxygen above 700°C, and nitrogen above 815°C, with a resultant embrittlement. The best hot-working temperature is about 930°C. The usual hot-working procedure is to use an oxidizing atmosphere (to avoid hydrogen contamination) and then to remove the oxygen-embrittled surface layers after working. The absorption of interstitial solutes is also important in welding titanium; shielding of the weld by helium or argon is required. With this protection the all-α alloys are readily welded.

The largest use of titanium by far is for aerospace engines and structures, particularly military and commercial jet aircraft. Approximately 75% of total titanium output is used for aircraft engines and airframes. Industrial uses for titanium in such applications as tubing for heat exchangers, anodes for electrolytic copper production, pumps, valves, and fittings constitute the major portion of the remaining output. Industrial applications for titanium represent the fastest-growing element of expanding titanium use, and this will probably continue with realization of titanium's increasing cost-effectiveness.

PROBLEMS

1. Since copper is a strong β-phase stabilizer, why is it not used extensively in the all-β alloys?
2. Describe the principal reason for use of each of the following alloying elements in titanium alloys: Al, Sn, and Mo.
3. Sketch on a structure cell of titanium the possible slip systems in this metal at room temperature.
4. List the relative advantages of the (1) all-α alloys, (2) all-β alloys, and (3) α-β alloys in titanium applications.
5. Cadmium is a hexagonal metal with $c/a = 1.886$, while beryllium is a hexagonal metal with $c/a = 1.568$. What effect would you expect this to have on the deformation characteristics for these two metals relative to titanium, which has a ratio of 1.587?

REFERENCES

Eylon, D. (Ed.): *Titanium for Energy and Industrial Applications,* The Metallurgical Society of AIME, 1981.

Coolings, E. W.: *Applied Superconductivity, Metallurgy and Physics of Titanium Alloys,* vol. 1, Plenum Press, New York, 1986.

Jaffee, R. I.: "The Physical Metallurgy of Titanium Alloys," *Prog. Met. Phys.,* vol. 7, pp. 65–163, 1958.

Boyer, R., G. Welsch, and E. W. Collings (Eds.): *Materials Properties Handbook: Titanium Alloys,* ASM Int., Materials Park, OH, 1994.

Jaffee, R. I., and H. M. Burte (Eds.): *Titanium Science and Technology* (Proc. Second Int. Conf. on Titanium, Boston), Plenum Press, New York, 1973.

ASM Specialty Handbook: Heat-Resistant Materials, ASM Int., Materials Park, OH, 1997.

Donachie, M. J.: *Titanium, A Technical Guide,* ASM Int., Materials, Park, OH, 1988.

Bloyce, A., P.H. Morton, and T. Bell: "Surface Engineering of Titanium and Titanium Alloys," *Surface Engineering,* vol. 5, pp. 835–851, *ASM Handbook,* ASM Int., Materials Park, OH, 1995.

chapter

17

Metals for High-Temperature Service

The conversion of thermal energy into mechanical or other forms of energy is more efficient the higher the operating temperature of the heat engine used. This statement applies not only to steam turbines, gas combustion turbines, and jet engines, but also to liquid- or solid-fueled missile engines and to nuclear power devices. The factor limiting the temperature actually used is the hot strength of the materials of which the engine is built. These are usually metals, employed because of their high strength, their relative ease of fabrication into complex shapes, and their resistance to brittle fracture under mechanical or thermal stress.

The basic criteria for metals suitable for structural uses at high temperatures are:

1. High melting point or range (the metal must be solid to retain shape)
2. Reasonable strength at temperature of service plus, in some cases, light weight or high stiffness
3. Ability to be fabricated into desired shapes
4. Sufficient ductility at room temperatures and at high operating temperatures to resist brittle fracture
5. Resistance to oxidation, inherent or attained by coatings

Iron and carbon steels have reasonably high melting temperatures but lose strength rapidly at temperatures approaching the A1 transformation temperature. Austenitic stainless steels have appreciably higher strengths than carbon steels at 500 to 650°C but, even modified compositionally, are not suitable for service at 800°C and up. Successive developments of more highly alloyed stainless steels have resulted in the superalloys, one type being iron-nickel-base, another nickel-base, and the third cobalt-base, for service at 750 to 1000°C temperatures, all well above those served by stainless steel, discussed in Chap. 11.

Titanium is a relatively new metal for commercial structural uses, and while it has a high melting point, it is classed as a reactive metal rather than a refractory one. However, alloys of titanium are useful at temperatures up to 480°C. Because of relatively light weight, titanium-alloy structures have a high strength-to-weight ratio and are particularly useful for aircraft and other services where temperatures over a long period of time do not exceed about 500°C. These alloys are discussed in Chap. 16.

Present developments, now mostly in the research stage, are concerned with the very high-melting-point metallic elements classified as refractory metals and defined as those with melting points above 1875°C such as vanadium, niobium, chromium, tantalum, molybdenum, and tungsten or alloys based on these metals. Up to now, there has been no measurable use of these metals for air or space vehicles, but they are needed for very high-speed airplanes, i.e., up to Mach 10 types, for solid-fuel motor nozzles and missile trajectory controls, for advanced power systems, and for similar future services at temperatures above 1100°C. This chapter is concerned with metals for these kinds of high-temperature service.

17.1 HIGH-TEMPERATURE PERFORMANCE OF REFRACTORY METALS

Mechanical Properties

The schematic graph of Fig. 17.1, showing the effects of temperature on strength of the refractory metals, gives three temperature ranges, each a range of fractions of

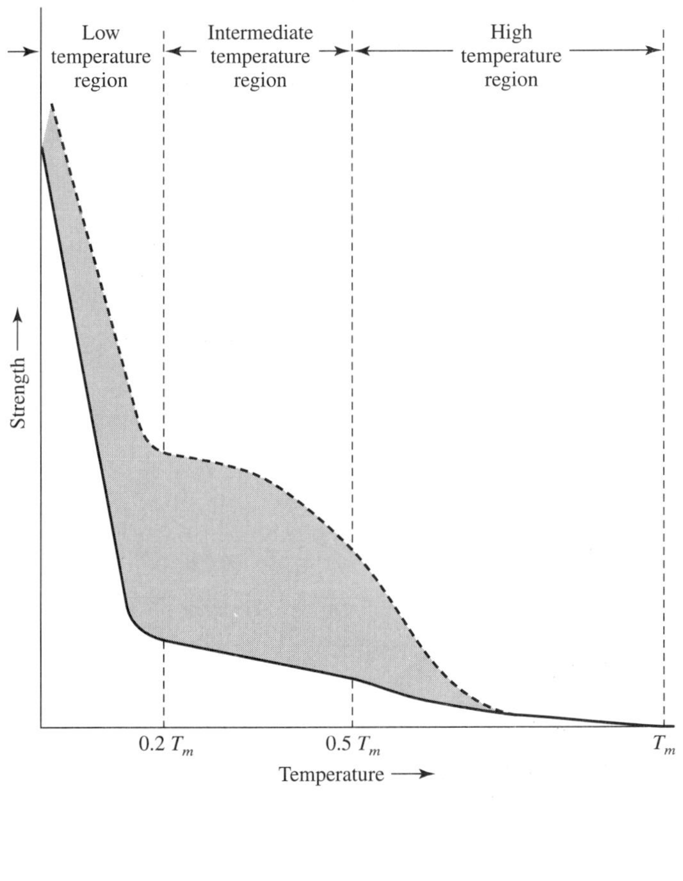

Figure 17.1 Schematic diagram showing effect of temperature on strength of bcc refractory metals (T_m = temperature of melting point). The shaded area represents the range of strengths depending on the purity of the metal, its grain size, and the strain rate of testing.

the melting-point temperature T in kelvin. Figure 17.1 will be used in discussing the properties of metals at elevated temperatures.

0 to 0.2T_m This is the low-temperature range in which strength of bcc metals increases rapidly with decreasing temperature. This appears to be a basic characteristic of the bcc metals in their purest state, although interstitial impurities by locking dislocations add an increment to yield strength. At low temperatures in this range, it becomes easier to initiate and propagate a crack than to initiate plastic deformation; thus brittle behavior will be encountered. In high-melting-point elements, the temperature at which transition from ductile to brittle fracture occurs is generally well above room temperature (except for tantalum, which is ductile down to 4 K). Therefore, brittle failure of structural components may be encountered upon start-up at room temperatures or below, long before the relatively high service temperatures are attained.

Above the ductile-brittle transition temperature range, dislocation pileups activate only slip. Within the transition range, such pileups may initiate either slip or microcracks or both. The individual microcracks may not propagate catastrophically unless several join together and reach a size that would serve as a Griffith crack. In this temperature range, some plastic behavior always precedes brittle fracture. In the nil-ductility region below the transition temperature, dislocation pileups always initiate microcracks, which propagate to produce total failure.

In fcc metals, the temperature dependence of mechanical properties is not nearly as great. Thus, while there is some increase in properties at low temperatures, there is no sharp increase and, more importantly, there is no ductile-to-brittle transition. Thus these metals can be used down to low temperatures without undue concern for brittle fracture.

0.2 to 0.5T_m This is a range where strength changes only slightly with temperature but where impurities, grain size, and other metallurgical variables have very considerable effects on strength and ductility. The purer metals have lower strengths and good ductility; impure specimens show peaks and valleys in both strength and ductility curves vs. temperature.

In this intermediate temperature range, the segregation of impurities at dislocations, i.e., Cottrell "atmospheres," increases strength in bcc metals but in a temperature-dependent manner, since relatively long-range diffusion is involved. In addition to this effect, an applied stress field will cause local ordering of interstitial solute atoms into one of the sites of lower energy corresponding to the three directions of tetragonality of bcc metals. Diffusion is not involved in this process, and the strengthening depends on interstitial impurity concentration but not on temperature.

The second major factor affecting strength in the intermediate temperature range is the degree of strain hardening accompanying deformation vs. the degree of thermal recovery simultaneously occurring. The upper limit of the temperature range is defined as that temperature where these two processes occur at the same rate. Since thermal-recovery processes are strongly time-dependent, the strength of metals in this range is very sensitive to the strain rate. Therefore, short-time tensile tests will give very much higher strength values than creep or 10- to 100-h stress-rupture tests.

Above 0.5T_m This is the high-temperature range where creep processes become the major concern. Creep is a thermally activated process, and in this temperature range the activation energy for creep or stress rupture has been found usually to be the same as for self-diffusion. Figure 17.2 shows the correlation of these activation energies with absolute melting temperature for a large number of different metals of various types.

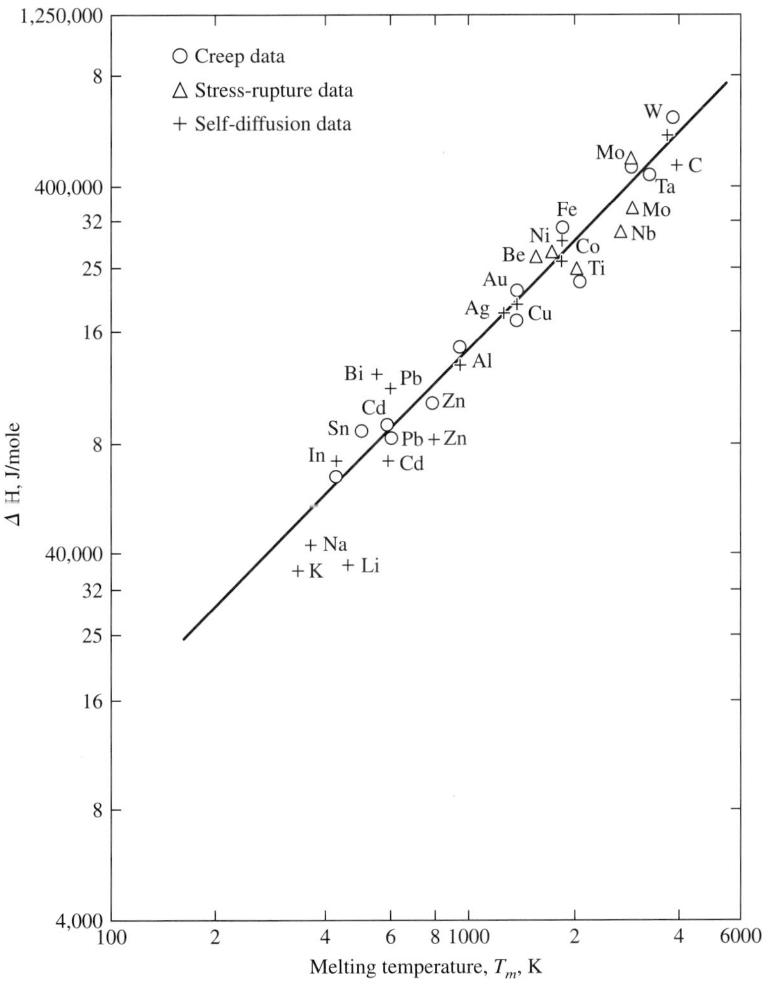

Figure 17.2 Correlation of the activation energies for creep, stress rupture, and self-diffusion with the melting temperature.

Oxidation Behavior

The chemical property of importance in the uses of metals at high temperatures is resistance to oxidation, since service conditions almost always involve exposure to oxygen-bearing gases. Surface oxidation always reduces the effective thickness of the underlying load-bearing metal, and this effect alone can result in serious weakening if the rate of oxidation is high. Subsurface oxidation of alloying elements that form very stable oxides occasionally may strengthen the metal by generating a fine dispersion of such oxide particles within the matrix metal, or it may embrittle the metal if the oxide forms as films along grain boundaries.

Surface oxidation occurs at a linear rate with time when the oxygen-metal reaction at the interface is rate-controlling. This is true when the oxide is porous or cracks or spalls off continuously. A linear rate at high temperatures usually results in excessive oxidation.

When the amount of oxidation is proportional to the square root of the exposure time, the oxidation-time curve is parabolic. In this case, the rate-controlling factor is not reaction at the interface but diffusion of either oxygen or metal ions through the oxide film. In this case, even a small amount of an alloying element, which affects the concentration of lattice vacancies in the oxide, can have major effects on the oxidation rate.

The usual Arrhenius equation, while applicable to cases where a single type of oxide is formed and its growth rate is diffusion-controlled, does not apply in a simple form to the oxidation of refractory metals because two or more oxide layers are formed. In M_xO_y, metal-ion diffusion may be rate-controlling, whereas in $M_{2x}O_y$, the oxygen-ion diffusion rate may be controlling. In general, completely parabolic oxidation is not found for refractory metals at temperatures above 1000°C, and this fact is one of the important limitations to their uses at high temperatures. Nonuniform behavior with increase in time oxidation can result from:

1. Fracture of a nonporous oxide; change from parabolic to linear rate
2. Aging of a nonporous oxide, removing excess vacancies; change from parabolic to cubic rate
3. Sintering of a porous oxide; change from linear toward parabolic rate

17.2 NICKEL- AND IRON-BASE SUPERALLOYS

The British developed the first series of age-hardenable nickel-base superalloys for jet aircraft alloys. These alloys, named Nimonic (Table 17.1), were an outgrowth of the discovery that the addition of a specified amount of titanium to a well-known electric-resistance heating alloy containing 80% Ni and 20% Cr resulted in the formation of an intermetallic compound whose dispersion and the related alloy properties could be controlled by solution and precipitation heat treatments.

Table 17.1 Nominal composition of typical nickel-base superalloys

	Composition, wt %							
Name	**C**	**Cr**	**Al**	**Ti**	**Mo**	**Co**	**B**	**Remainder**
Nimonic 80	0.08	20	1.5	2.4	...	...	...	Ni
Nimonic100	0.20	11	5.0	1.3	5.0	20	...	Ni
Inconel X	0.06	16	0.6	2.5	...	...	...	Ni, 1.5 Cb, 7.0 Fe
Inconel 700	0.10	17	3.3	2.2	3.0	30	0.008	Ni, 0.04 Zr
Astroloy	0.10	15	4.2	4.0	5.0	15.5	...	0.03N
Rene 41	0.09	19	1.5	3.1	10.0	11.0	0.005	
Waspalloy	0.07	19	1.4	3.0	4.3	13.5	0.006	
Rene 85	0.27	9	5.3	3.2	3.2	15	0.015	
IN 100	0.18	10	5.5	3.0	3.0	15	0.014	

Later metallurgical developments in Britain, the United States, and Russia resulted in further improvements in properties, attainable by additions of controlled amounts of other elements such as cobalt, aluminum, molybdenum, and tungsten and minor constituents such as boron, zirconium, etc.

The strength above that of pure nickel of all these alloys depends initially on solid-solution strengthening. Localized segregates of solute atoms result in clusters, "atmospheres," or short-range ordering and a further strengthening. These strained areas where atoms of different size are congregated impede the movement of dislocation lines through the matrix crystals.

The major solid-solution element in these nickel-base alloys is chromium, but molybdenum and tungsten, when present, also give solid-solution strengthening. The Cr-Ni phase diagram of Fig. 17.3 shows the solid solubility of chromium in nickel. Aluminum, niobium, and titanium, while serving other functions in these alloys, also contribute significantly to solid-solution strengthening.

Precipitation hardening is also a major contributor to strength properties of these alloys. At least partial coherency between the matrix and precipitating phase or, equivalently, the local concentrations of solute atoms called Guinier-Preston (GP) zones of the preprecipitate stage, result in local lattice strains that strongly impede the movement of dislocations. In the nickel-base superalloys, the major solution-precipitate phase is $Ni_3(Ti,Al)$. Although both the matrix and this phase, called γ', are fcc, they differ enough in lattice parameter to create coherency precipitate strains and related precipitation hardening.

When boron and zirconium are present, their atoms are respectively smaller and larger in size than the matrix atoms. Small additions of each contribute somewhat to high-temperature strength. It is claimed that the effect is a result of these atoms filling vacancies and lattice imperfections at or near grain boundaries. This would be beneficial, since grain boundaries are excellent sources of the vacancies necessary

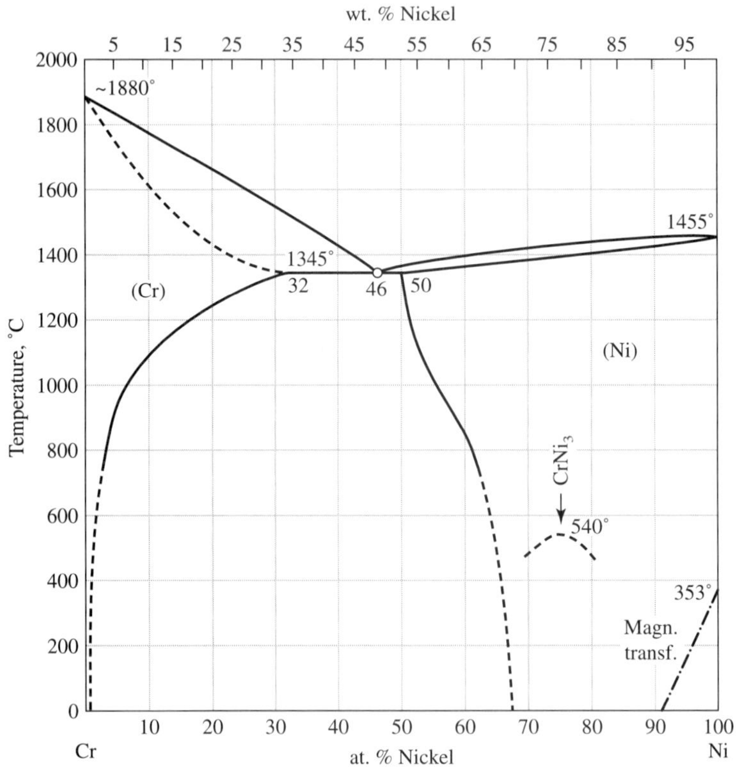

Figure 17.3 Phase diagram for the Cr-Ni alloys.

for dislocation climb, which permits dislocations to move around a barrier, a necessity for plastic deformation at high temperatures.

The other metallurgical structural factor, identified as affecting the properties of these alloys, is the presence and distribution of carbide phases, $M_{23}C_6$ or M_6C, where M can be Cr, Mo, or W. The relative amounts present of the $Ni_3(Ti,Al)$ precipitate and of MC, $M_{23}C_6$, and M_6C as a function of aging temperature and time for a nickel-base superalloy are shown by Fig. 17.4. In these compounds, M can be Ti, Ta, Nb, or V. Grain-boundary carbides must be controlled, since these can increase rupture strength or ductility, depending on their morphology. They also influence the chemical stability of the matrix by immobilizing reacting elements. The $Ni_3(Ti,Al)$ in Fig. 17.4 is a major factor in strengthening the alloy. Sigma phase of Fe-Cr may be present but is deleterious and should be eliminated by composition control. After long-time service at 800 to 850°C, the stable carbides are MC and $M_{23}C_6$.

The typical heat treatments used to obtain optimum properties in these alloys include a 4-h solution treatment at 1065 to 1150°C, followed by air cooling and a precipitation treatment of 16 h at 675 to 815°C. The lower solution- and

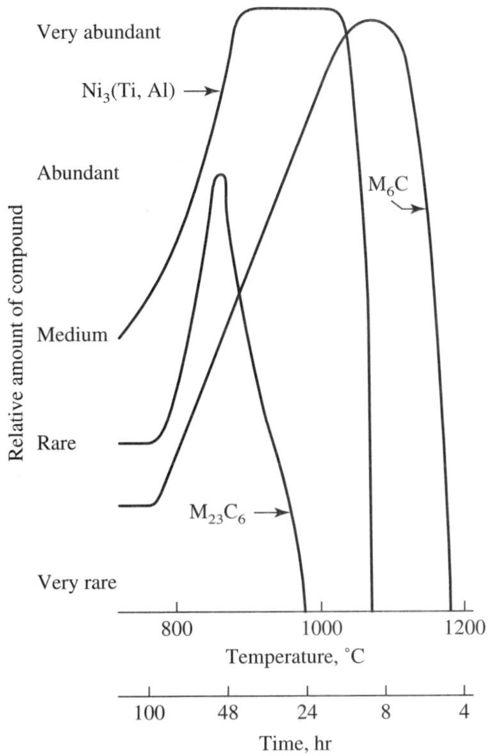

Figure 17.4 Isothermal phase-reaction products in the matrix of two Ni-base superalloys (for compositions, see Table 17.1). For case A, bar stock specimens were water-quenched from 1200°C and then reheated for the indicated times and temperatures. The precipitate phases were extracted and analyzed by x-ray diffraction. For case B, specimens of the as-cast alloy were heated 5000 h at the indicated temperatures before extraction and x-ray analyses.

precipitation-treatment temperatures tend to give finer grain and precipitate size, resulting in optimum short-time tensile properties. Higher temperatures of solution and precipitation give coarser grain sizes and precipitates that lead to better creep properties at high temperatures. Typical properties of several alloys are shown by the data of Table 17.2.

Nickel-base superalloys have inherently good oxidation resistance from the standpoint of general attack as measured by weight loss or gain. As the service temperature approaches 1800°F, however, intergranular oxidation becomes a serious problem. If sections are thin, an intergranular oxidation to a 0.0014-nm depth is significant, and this may occur in 1 h at 1800°F. Both aluminum and titanium in the alloy modify the surface oxide so as to reduce grain-boundary oxidation at the interface and related intergranular oxidation. However, the complexities of behavior here are such that theory is relatively helpless and only careful, controlled experimentation can lead to definitive improvements.

Table 17.2 Typical tensile and rupture strengths of nickel-base superalloys

	Tensile Strengths, MPa			100-h Rupture Strengths, MPa		
Name	**700°C**	**800°C**	**900°C**	**700°C**	**800°C**	**900°C**
Nimonic 80	735	495	240	410	185	
Nimonic 100	1050	735	600	565	300	160
	815°C	**870°C**	**925°C**	**815°C**	**870°C**	**925°C**
Inconel X	455	280	175	170	90	55
Inconel 700	755	630	425	310	195	120
Astroloy	. . .	. . .	. . .	365	260	170
René 41	880	630	405	315	195	125
				650°C	**815°C**	**985°C**
Waspalloy	. . .	. . .	. . .	755	280	50
René 85	. . .	. . .	. . .		500	140
IN 100	. . .	. . .	. . .	. . .	510	175

A related group of superalloys, which are nickel-rich but iron-base, is also widely used. These alloys are hardened by the same basic mechanisms as the nickel-base ones; however, their lower cost makes them a preferred material in many cases. The nickel content for this group of materials varies between 25 and 60% (with iron comprising between 15 and 60%). They have an fcc austenite matrix and are hardened by intermetallic or carbide precipitates. The most common precipitating phases are $\gamma'(Ni_3(Al,Ti))$ and $\gamma''(Ni_3Nb)$, the former being ordered face-centered cubic and the latter being ordered body-centered tetragonal. The composition of the austenite matrix reflects a balance of properties and cost, the lower-nickel alloys being less expensive but having a lower useful temperature range. Since the carbon content is normally low and ferrite stabilizers are present, the minimum nickel content to maintain the austenite matrix is about 25%. Solid-solution strengtheners in this system include chromium, molybdenum, titanium, aluminum, and niobium. Of these elements molybdenum, which expands the austenite lattice, is the most effective. Working in concert with cobalt, which is a weak-solution strengthener but contracts the lattice, the degree of mismatch between the austenite lattice and the γ' precipitate can be controlled by a cobalt-molybdenum balance. Chromium provides the primary oxidation resistance for the alloys. Many of the solid-solution strengthener elements enter the carbides and precipitating phases as well as the matrix. It should be noted that the γ' in these alloys is titanium-rich, rather than aluminum-rich as is the case for the nickel-base superalloys. Other alloying elements, such as boron and zirconium, are added to improve hot ductility.

Carbides are an additional important phase in these alloys, the most important being MC, where M is most often titanium. Other elements entering the carbide phase are molybdenum, niobium, vanadium, zirconium, and tantalum. Second carbide of importance is $M_{23}C_6$, where M is primarily chromium. The carbides may occur in globular forms or as grain-boundary films, the latter being avoided during processing or service since they cause embrittlement. The iron-nickel–base alloys are susceptible to precipitation of one or more secondary intermetallic phases such as η (hexagonal close-packed Ni_3Ti) and δ (orthorhombic Ni_3Nb), which are usually detrimental to properties and are avoided in processing.

Control of the properties of these alloys is usually achieved by using the appropriate solution treatment and aging temperatures for the service intended. The solvus temperatures for the precipitating phases in a given alloy are a function of composition, for example, nickel and titanium levels. At higher levels of these constituents, solution temperatures increase and service temperatures increase as well. Precipitation rates for these superalloys are somewhat slower than those for the nickel-base ones, and thus good weldability and process control are somewhat easier to achieve. Relatively sophisticated thermal and mechanical treatments may therefore be used to achieve good service performance. For example, mechanical working may be combined with thermal treatment to produce recrystallization under such conditions that the amount and morphology of primary and secondary precipitating phases are controlled, and in turn these control resulting grain size. Examples of compositions and properties of some iron-nickel–base superalloys are seen in Tables 17.3 and 17.4, respectively.

Table 17.3 Nominal compositions of typical iron-nickel–base superalloys

	Composition, wt %								
Name	**C**	**Cr**	**Ni**	**Co**	**Al**	**Ti**	**Mo**	**Nb**	**B**	**Remainder**
A-286	0.05	15	26	. . .	0.2	2.2	1.3	. . .	0.003	Fe, 0.03 V
Discalloy	0.04	13.5	26	. . .	0.1	1.7	1.7	. . .	0.01	Fe, 0.02 N
Incoloy 901	0.05	13.5	43	. . .	0.2	2.5	6.2	. . .	. . .	Fe
Incoloy 905	0.05	13.5	38	15	0.7	1.0	. . .	3.0	. . .	Fe
Pyromet 860	0.05	13.5	44	4	1.0	3.0	6.0	. . .	0.01	Fe

Table 17.4 Typical rupture strengths of iron-nickel–base superalloys

	100 h Rupture Strengths, MPa		
Name	**650°C**	**735°C**	**815°C**
A-286	425	245	90
Discalloy	365	210	105
Incoloy 901	560	340	135
Incoloy 903	595		
Pyromet 860	665	420	230

17.3 COBALT-BASE SUPERALLOYS

Cobalt is a transition metal immediately preceding nickel in the fourth period of the periodic table and very similar to nickel in melting point, atom size, and many other physical and chemical properties. Although hcp in structure at room temperatures, it is fcc above 418°C. The allotropic transformation at this relatively low temperature is sluggish, however, because the activation energy is high and the associated change in free energy is small. Therefore, the presence in cobalt-base alloys of some nickel and carbon, both of which lower the transformation temperature, is sufficient not only to assure existence of the fcc structure at service temperatures of 760 to 985°C but also to retain this fcc structure at room temperatures. The latter fact is important in conferring thermal shock resistance, since cobalt alloys are used in jet-engine parts that repeatedly cool down and heat up as such engines are shut off and later restarted.

There are now a number of commercial cobalt-base superalloys with the composition of five typical alloys given in Table 17.5. Chromium, tantalum, tungsten, molybdenum, and nickel all enter the fcc solid-solution matrix and contribute to strengthening by normal solid-solution effects. The relative strengthening effect is dependent on atom sizes; the greater the difference, the less the solubility, but the greater the local lattice distortions and the related strengthening effect. On this basis, the data of Table 17.7 indicate that tantalum and niobium would be the most effective solid-solution hardening element, and this actually is the case. This effect is the basis of alloy SM 302 (Table 17.5). Next in order of decreasing effectiveness (as predicted by the tabulated atomic diameters) are tungsten and molybdenum. Chromium and nickel in solid solution have a relatively lesser strengthening effect. Actually chromium's major usefulness in these alloys is based to a considerable degree on the resistance to oxidation it confers on these alloys. Typical properties of cobalt-base superalloys are given in Table 17.6.

Carbon is an alloying element of major importance since the amount, identity, and distribution of carbides in the structure of these alloys are major factors affecting their high-temperature strengths. The types of carbides found depend on relative

Table 17.5 Nominal compositions of cobalt-base superalloys

		Wt % Composition, Balance Cobalt						
Alloy	**Form**	**Ni**	**Cr**	**Mo**	**W**	**Fe**	**C**	**Other**
HS-25	Wrought	10	20	. . .	15	3	0.1	
HS-21	Cast	2.5	27	5.5	. . .	2*	0.25	0.007 B
HS-31	Cast	10.5	25.5	. . .	7.5	2*	0.5	
HS-151	Cast	1*	20	. . .	12.8	2*	0.05	0.15 Ti
SM 302	Cast	. . .	22	. . .	10	. . .	0.859	Ta, 0.2 Zr, 0.005 B

*Maximum.

Table 17.6 Typical tensile and rupture strengths of cobalt-base superalloys

	Tensile Strength, MPa			Rupture Strength, MPa		
Alloy	**650°C**	**815°C**	**990°C**	**650°C**	**815°C**	**990°C**
HS-25*	720	350	240	. . .	155	50
HS-21†	495	440	225	365	135	65
HS-31†	530	440	200	390	190	80
HS-151†	595	450	. . .	490	265	90
SM 302†	. . .	. . .	. . .	. . .	280	100

*Wrought sheet aged.
†Investment-cast.

concentrations of the various carbide-forming elements and on the heat treatments to which the cast structures may be subjected. The carbides are all more soluble at very high temperatures than at service temperatures, and thus precipitation hardening not only is possible but also appears to be of major importance in these alloys.

A specific example of carbide precipitation is found in alloy HS-31, consisting of 25% Cr, 10% Ni, 8% W, 2% Fe, 0.5% C, with the balance cobalt. In the cast state, its microstructure (Micro. 17.1) consists of the normal cored primary solid-solution dendrites surrounded by interdendritic carbides of eutectiferous origin. Extraction and analysis of these carbides indicate that the major ones are of the

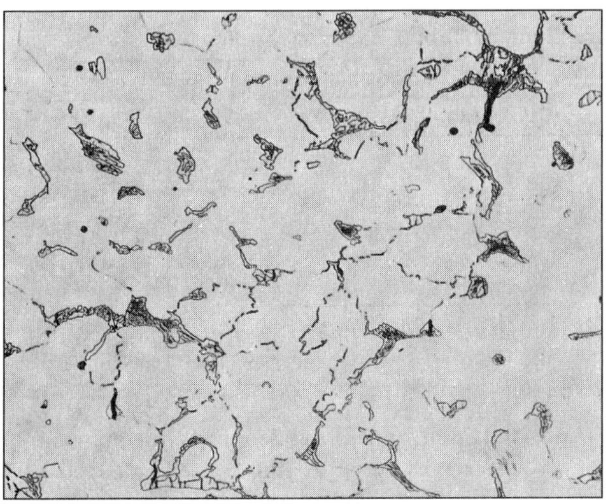

Micrograph 17.1 Cobalt-base superalloy HS-31; ×250; 2% chromic acid etch. As-cast structure; coring of the dendritic matrix is not evident in this etch. Interdendritic carbides, of eutectiferous origin, are Cr_7C_3 and $(CoCrW)_6C$.

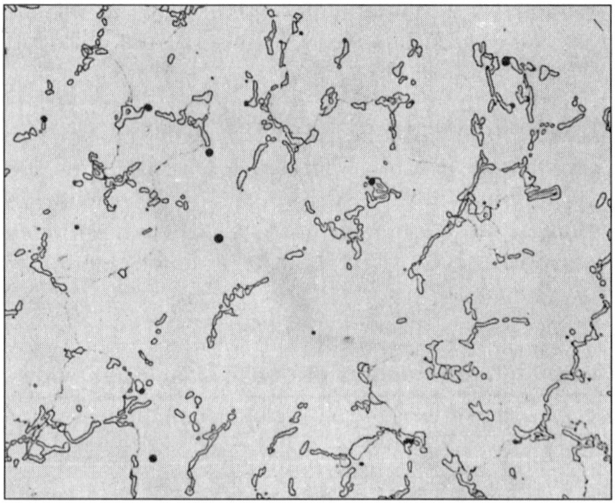

Micrograph 17.2 Cobalt-base superalloy HS-31; ×250; 2% chromic acid etch. Solution-treated at 1200°C; most of the Cr_7C_6 has dissolved, leaving $(CoCrW)_6C$ in a solid-solution matrix.

Cr_7C_3 type, with a small amount of $(CoCrW)_6C$ present. After a solution treatment at 1200°C (Micro 17.2), the total amount of carbides decreases, with that remaining mostly the still undissolved $(CoCrW)_6C$ plus smaller amounts of Cr_7C_3 and $Cr_{23}C_6$. When the solution-treated alloy is subsequently aged at 815°C (Micro 17.3), the $(CoCrW)_6C$ remains unchanged, but a large amount of newly precipitated $Cr_{23}C_6$

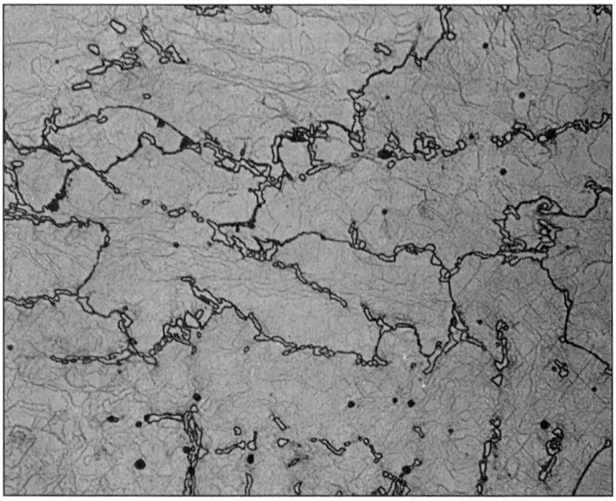

Micrograph 17.3 Cobalt-base superalloy HS-31; ×250; 2% chromic acid etch. After solution treatment, aged 1 h at 815°C; precipitation of $Cr_{23}C_6$ has started on subboundaries in the solid-solution matrix.

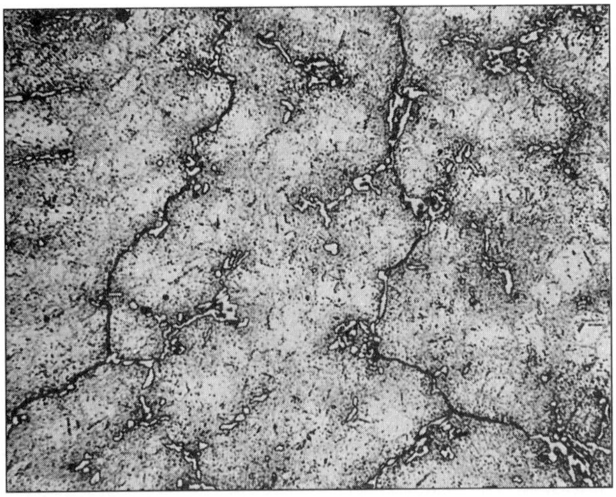

Micrograph 17.4 Cobalt-base superalloy HS-31; ×250; 2% chromic acid etch. Aged 48 h at 815°C; general precipitation of $Cr_{23}C_6$ has occurred throughout the matrix. Note that the prior solution treatment did not homogenize the cored dendritic structure. This now is visually evident in terms of relative amounts of precipitated carbides at the center of dendrites vs. areas near the original interdendritic carbides.

is observed (Micro. 17.4). For a fixed amount of carbon, it can readily be calculated that just changing the initial metastable Cr_7C_3 to stable precipitates of $Cr_{23}C_6$ results in approximately a 50% increase in volume of carbide.

An important member of the family of precision-cast cobalt-base superalloys is SM 302, whose excellent high-temperature strength, e.g., at 985°C, is obtained by solid-solution hardening with the high-melting-point elements tantalum and tungsten plus their carbides. The zirconium and boron present also contribute to high-temperature strength, presumably by reducing the concentration of vacancies or other interactions at grain boundaries. This alloy can be used in the range 815 to 1100°C without requiring a coating to protect it from oxidation.

17.4 VANADIUM, NIOBIUM, AND TANTALUM

The VA refractory metals differ greatly in one respect from their generally similar VIA neighbor transition metals. Vanadium, niobium, and tantalum in the commercially pure state all exhibit good ductility at room temperatures, whereas chromium, molybdenum, and tungsten do not. The reason for this difference cannot be based on melting points or crystal structures (Table 17.7), or even on relative degrees of purity. It is based instead on whether or not the interstitial impurity atoms inevitably present to some degree—hydrogen, carbon, nitrogen, and oxygen—are present in solid solution or in the form of precipitates.

Table 17.7 Property data for refractory metals

Metal	Crystal Structure	Atomic Diameter, Å	Density, g/cm^3	Melting Point, °C	Tensile Modulus, GPa
Ni	fcc	2.491	8.9	1453	210
Co	fcc*	2.497	8.85	1495	210
V	bcc	2.632	6.1	1900	135
Cr	bcc	2.498	7.19	1875	295
Nb	bcc	2.859	8.57	2468	105
Mo	bcc	2.725	10.22	2610	330
Ta	bcc	2.859	16.6	2996	190
W	bcc	2.734	19.3	3410	350
Graphite	hex	1.42	2.25	3727†	40

*At service temperatures above 417°C.
†Sublimes.

Vanadium

Vanadium is a relatively scarce and, relatedly, a costly metal. With a melting point only about 400°C higher than iron, cobalt, and nickel, vanadium is less interesting for high-temperature service than the other refractory metals of generally higher melting points. The pure metal is relatively ductile and does not work-harden appreciably; therefore, it can be worked easily at room temperatures. However, since it oxidizes rapidly at elevated temperatures, it must be protected during hot working. Hot rolling, for example, can be accomplished by sheathing the vanadium ingot in a jacket of steel.

At present, the major use of vanadium metal is in the form of foil used as a bonding agent in producing titanium-clad steel sheet. However, because of its low fission cross section for neutrons, its useful strength at elevated temperatures, and its low density, it has potential uses as a structural member of fast nuclear reactors.

Niobium (Columbium)[1]

Niobium has a high melting point, exceeded only by Mo, Ta, and W in the group of Table 17.5. Like the other VA transition metals, pure niobium has a low brittle-fracture transition temperature and is ductile at room temperatures, permitting normal

[1]This metallic element was first identified in the United States in 1801 and named columbium by its discoverer. Two years later, an eminent Swedish chemist isolated the same element and, not knowing of the American work, named the element niobium. For over 150 years, it was called columbium in the United States and niobium in the rest of the world. However, the American Chemical Society has now officially approved of following the international scientific community in calling the metal niobium.

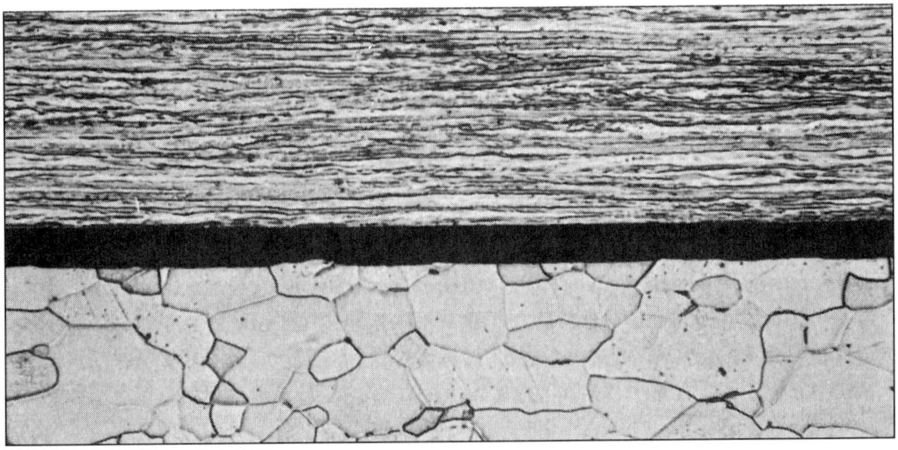

Micrograph 17.5 Niobium sheet rolled 90% at room temperature from 0.5 to 0.05 mm (top sheet) and then annealed 1 h at 1150°C in vacuum (bottom sheet); etch 20 ml HF, 14 ml H_2SO_4, 5 ml HNO_3, 50 ml H_2O; ×500.

cold-working operations to be employed in fabrication processes. Micrograph 17.5 shows that both the cold-rolled and the cold-rolled-recrystallized structures of niobium metal are indistinguishable in appearance from equivalent microstructures of iron or aluminum.

In common with other refractory metals, niobium and its alloys have a tendency to oxidize in air at elevated temperatures, and this had been a serious disadvantage to its use for such service. A complex series of oxides may be formed upon exposure to air at high temperatures, including NbO_x, NbO, NbO_2, and three types of Nb_2O_5. The rate of oxidation keeps changing from linear to parabolic and back to linear, depending on the type and sequential layers of oxides on the metal, through which niobium and oxygen ions must diffuse. But however complex the process, conversion of metal to oxide causes an obvious thinning and related weakening of the structure.

Today, niobium alloys are the only refractory metals that are coated for large-scale oxidation-resistant applications such as jet engine hot sections. The most commonly used coating is Si-20Cr-20Fe for use at temperatures above 425°C. It is applied as approximately 0.08-mm-thick slurry of metal powder in a nitrocellulose suspension with a gelling agent. Reaction bonding is completed by heating to about 1300°C. There are many variations of the silicide coatings including the addition of hafnium silicide for a higher remelt temperature. Other protection concepts for niobium use aluminides, Ni-Cr alloys, zinc, chromium carbide, or noble metals as coating systems applied by flame spray, plasma spray, pack cementation, or hermetic sealing.

Equal in significance to the oxidation process, if not more significant, is the diffusion of oxygen into the metal at a rather rapid rate, producing a hardened but embrittled "oxygen-affected" zone. It has been found that niobium sheet may lose 70% of its ductility when only 15% of its cross section is contaminated by oxygen and that this degree of contamination can occur in 1 min in air at 1095°C.

Table 17.8 Creep rupture strength of niobium alloys

Material	At 1095°C, MPa	At 1200°C, MPa
Pure Nb, recrystallized	35	
D31 (10 Mo, 10 Ti)	80	
D41 (20 Mo, 10 Ti)	105	
FS82 (33 Ta, 0.75 Zr)	135	
F50 (15 W, 5 Mo, 5 Ti, 1 Zr)	140	80
F48 (15 W, 5 Mo, 1 Zr)	245	125

Alloying can greatly improve the resistance of niobium to oxidation and also to oxygen solution, but no single element will improve resistance to both effects. In general Ti, Mo, and W are added, but there are optimum concentrations of each alloying element. Some work indicates the best alloy to be alloy D41 with Nb-10% Ti-6% Mo-20% W. The creep rupture properties of this and other niobium alloys are given in Table 17.8.

Solid-solution strengthening is greatest for solutes having the greatest misfit or difference in atom size, coupled with lowest diffusivity, i.e., highest melting points, to ensure thermal stability. Elements with high melting points meeting these criteria are titanium, tantalum, molybdenum, and tungsten. All these form complete solid solutions with niobium, so the phase diagrams are similar to the Cu-Ni diagram except for Nb-Ti at temperatures where titanium is hexagonal close-packed. As little as 6 at. % W doubles the creep strength of niobium, and even though tungsten is nearly twice as dense as molybdenum, it is a preferred alloying agent at low concentrations.

Second-phase strengthening for high-temperature service is also possible with niobium. There is preference for nonmetal dispersants, e.g., the suboxide ZrO, ThO_2, or carbides, including those of zirconium and hafnium.

Tantalum

Tantalum always occurs in nature associated with niobium in the ore tantalite-columbite, $(FeMn)(TaNb)O_6$. The metal has a melting point of 2996°C, exceeded in the refractory-metal group of Table 17.7 only by tungsten. Although bcc in structure, pure tantalum remains ductile down to liquid-helium temperatures, and even relatively impure metal is ductile at room temperatures.

Arc-melted or electron-beam-melted ingots are readily cold-worked to sheets, which in turn may be fabricated to desired shapes by normal metalworking processes. However, all annealing treatments must be carried out in vacuum or inert gas. Cold-worked metal recrystallizes at temperatures of 990 to 1375°C, depending on the purity of the metal and the amount of prior deformation.

In this connection, it should be recalled that many of the refractory metals are melted by focused electron beams in a high vacuum, which results in high-purity

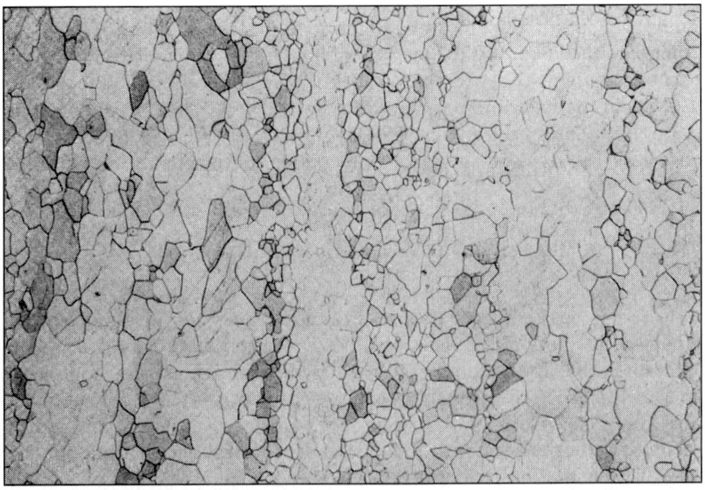

Micrograph 17.6 An electron-beam-melted tantalum billet, 29 × 9 cm cold-upset-forged 72% parallel to axis, then cross-rolled to 82% reduction, and annealed in vacuum for 45 min at 2100°C. The initially very coarse-grained electron-beam-melted billet had considerable differences in stored energy of cold work in the original grains and therefore related differences in recrystallization and grain-growth characteristics upon annealing, which resulted in this nonuniform structure. Etch 20 ml HF, 14 ml H_2SO_4, 5 ml HNO_3, and 50 ml H_2O; ×100.

but coarse grains. Micrograph 17.6 shows the structure of electron-beam-melted tantalum, which was severely cold-worked and then annealed at 2100°C. Although recrystallization was complete upon the subsequent anneal at 2100°C, the coarse-grained ingot structure resulted in differing degrees of stored strain energy in different layers of the metal and, relatedly, differing recrystallization characteristics of these layers upon annealing.

Tantalum is similar to niobium in having poor resistance to oxidation or scaling in air at high temperatures and also in being subject to embrittlement by absorption or solution of oxygen in the metal below the oxide-metal interface. For example, the absorption by tantalum of 1.2% oxygen decreases the tensile reduction in area from 90 to 20%, with a related increase in room-temperature strength from 105 to 690 MPa. Unfortunately, the strengthening effect of dissolved oxygen is not observed at high temperatures.

Strengthening of tantalum for high-temperature service is analogous in principle to that of niobium, to which it is very similar in atom size and electronegativity. Solid-solution strengthening by niobium of lower density (for lighter weight) or by tungsten, precipitation strengthening by carbides, and dispersion hardening by oxides have received only limited study, but all appear to be feasible. Alloys such as Ta + 10% W and Ta + 30% Nb + 5% V have useful strengths in the range 1375 to 1650°C, plus excellent room-temperature fabrication properties. However, on a strength-weight basis, these alloys are less attractive for service below 1375°C than many niobium-base alloys.

17.5 CHROMIUM, MOLYBDENUM, AND TUNGSTEN

Comparison of the high-temperature strength properties of the refractory metals has led to plots like that of Fig. 17.5, where the stress for rupture of these metals in 100 h at a temperature T is plotted against the homologous temperature T/T_m. The data points represent the lowest values reported in the literature for each metal and therefore presumably represent strengths of these metals in the purest states obtainable. It is noticeable that at equivalent fractions of their melting points the VIA metals Cr, Mo, and W are 2 to 3 times as strong as the VA metals Nb and Ta. The reason for this difference is itself a reason for attaching importance to these VIA metals.

Chromium

It does not at first seem attractive as a refractory metal in view of its relatively low melting point, lowest of all of the refractory metals in Table 17.7. In fact its boiling

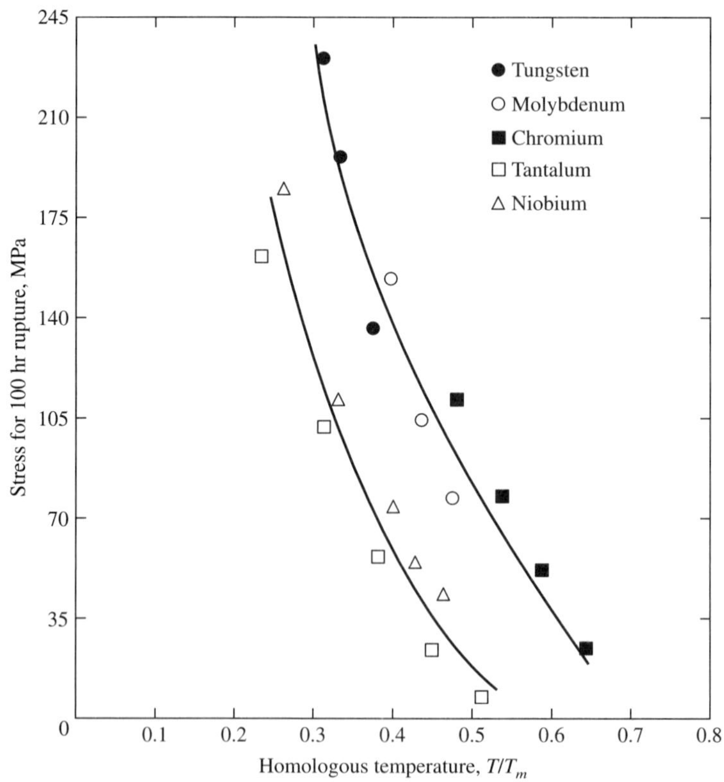

Figure 17.5 Stress for 100-h rupture life of the refractory metals in a pure state vs. homologous temperature.

point is even lower than that of iron. It is not cold-workable like the VA metals vanadium, niobium, and tantalum. The other VIA metals with very much higher melting points, molybdenum and tungsten, have already found important commercial uses. Thus most refractory metals have found a place in industry, and new uses are rapidly developing. The exception is chromium. Were it not for the fact that we live in an atmosphere of nitrogen, oxygen, and moisture, there would be little future for the element chromium. However, we do live in this corrosive atmosphere, and our structural materials must be stable in this environment. The alloys based on iron, nickel, and cobalt that have been developed to satisfy our moderately elevated temperature needs achieve stability in air largely because of their high chromium content. The annual consumption of chromium of well over 1 million tons per year, all as an alloying agent plus a small amount for surface plating, far exceeds the total consumption of all other refractory metals.

Considering the effectiveness of chromium in imparting oxidation resistance to other metals, it could be expected that the metal itself would have good resistance to oxidation. This is not the case. Even though chromium is superior to the other refractory metals in this respect, the chromium oxide film, which protects iron- and nickel-based alloys, will not prevent continuing oxidation of metallic chromium at temperatures above 925°C.

Research has shown that the rare-earth elements, e.g., yttrium, have an astonishing effect in reducing oxidation of chromium. The reason for this pronounced effect has not been established, although it is known that the oxide formed is chromic oxide containing very little yttria and that the oxidation occurs by diffusion of chromium ions through the oxide layer to the air-oxide interface.

A major problem in the case of chromium is its brittleness at room temperatures or, relatedly, its high brittle-fracture transition temperature. Interstitial impurities are of course the significant determinant of this temperature. With as little as 2 ppm of nitrogen, the transition temperature is 65°C, but an increase to only 20 ppm of nitrogen raises the transition temperature to 360°C. Therefore, to have chromium metal with some room-temperature ductility, it is necessary (1) to have a cold-worked structure, which moves dislocations away from nitrogen atoms and (2) to keep the nitrogen concentration to a level of less than 1 ppm, or (3) to add cerium which ties up nitrogen in the form of insoluble cerium nitride crystals. In addition, and as for all metals, the surface condition of chromium can also be a determining factor with regard to brittleness.

Molybdenum

It is one of the most abundant refractory metals, especially in the United States, which produces about 80% of all molybdenum consumed. This metal has been available commercially in fabricated shapes for many years, but only recently have purer metal and alloys based thereon been developed for possible large-scale uses at very high temperatures. Molybdenum not only has a very high melting point, exceeded only by tungsten and tantalum of the refractory metals in Table 17.7, but it also has the advantage of relatively light weight or a low density.

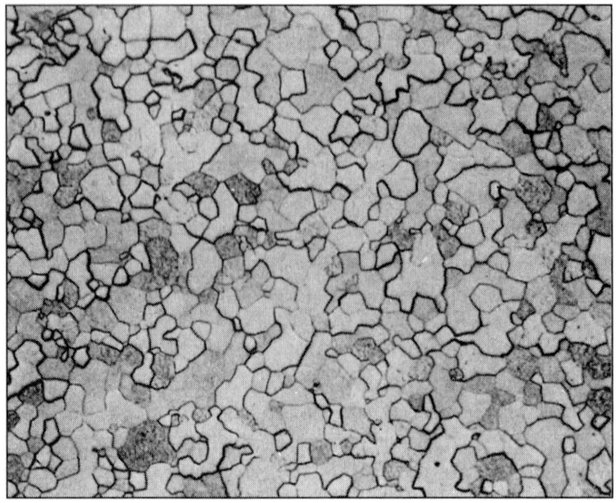

Micrograph 17.7 Structure of wall at base of slip-cast and sintered molybdenum crucible has normal pore-free polygonal crystal structure; same etch as Micro. 17.6; ×100.

Application of the newer techniques to fabrication of molybdenum has shown that electron-beam melting, while quite feasible, does not necessarily reduce the brittle transition temperature to that value needed for many applications. Consumable-arc-melted molybdenum is finer-grained and frequently more ductile at room temperatures than the purer electron-beam-melted metal. Molybdenum can also be slip cast and sintered to produce shapes (Micro. 17.7) or, more recently, fabricated by hot isostatic pressing.

The greatest problem in the high-temperature uses of molybdenum is oxidation to which this metal is more subject than the other refractory metals (see, for example, Fig. 17.6). The magnitude of this problem is indicated by the fact that the end product of oxidation, MoO_3, melts at 795°C and is very volatile at 700°C, both temperatures being well below those at which the metal is otherwise suitable for use. In flowing air at a temperature of only 1100°C, at which molybdenum may have a strength of about 275 MPa or more, a catastrophic surface recession rate for molybdenum of 11 cm/day has been observed. When the oxygen partial pressure is as low as 76 torr, that is, in a moderately good vacuum, surface oxidation of molybdenum is prevented, but sufficient oxygen is absorbed at grain boundaries to result in noticeable embrittlement.

The high-temperature properties of molybdenum are so excellent that much research has been directed at such metallurgical means as alloying for minimizing the oxidation of molybdenum and for reducing its brittle transition temperature. Great improvements in the latter may be achieved by alloying with rhenium, but substantial concentrations of 10 to 30% are needed, and rhenium is both scarce and expensive. No substantial improvement in oxidation behavior has yet been found by alloying molybdenum; the use of protective coatings is the only means showing any promise.

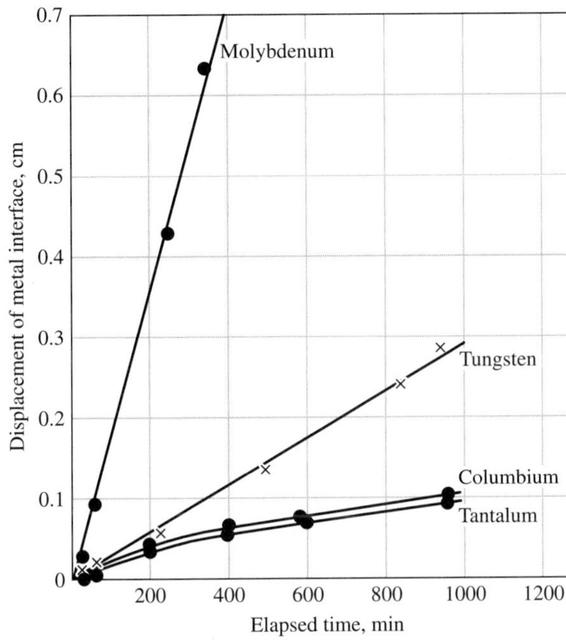

Figure 17.6 Sealing of molybdenum, tungsten, tantalum, and niobium in flowing air at 1095°C.

Molybdenum-alloy development has been under way for a longer time than for the other refractory metals. There are commercial alloys, such as Mo + 0.5% Ti or 0.5% Ti + 0.07% Zr, available in as-worked or stress-relieved rolled states, which have stress-rupture strengths at 1095°C of 235 and 360 MPa, respectively, compared with 105 MPa for pure molybdenum. This strengthening results from a dispersion of fine TiC particles. A major objective of alloying additions, including titanium, has been to raise the recrystallization temperature of molybdenum, thus retaining the strain hardening resulting from cold working to higher service temperatures. However, once the service temperature exceeds the recrystallization temperature of the alloy being used, it appears that the alloying advantage is lost and pure molybdenum is about as strong.

Tungsten can be added to molybdenum in very large amounts, since these two very similar metals form a continuous series of solid solutions. An alloy of 25% W and 0.1% Zr, with a strength of 500 MPa at 1315°C, is probably the strongest commercial molybdenum alloy thus far developed for very high-temperature service.

Tungsten

It has the highest melting point of all the metals, highest elastic modulus, and lowest vapor pressure at high temperatures. However, it is one of the densest of the metals, which, for most structural applications, is a disadvantage. Tungsten has long been used in wire form as filaments for incandescent bulbs and electronic tubes, as rods for welding electrodes, and in the carbide form for tools. The properties of tungsten metal in sheet and more massive forms required for high-temperature structural applications are sufficiently different to require new and intensive development work.

Tungsten is generally prepared by reduction of ammonium paratungstate to metallic powder which is pressed, presintered at 1200°C for better handling, and then finally sintered at about 3000°C, a temperature attained by resistance to passage of an electric current. Alternatively, both electron-beam melting and arc melting have been used to produce small tungsten ingots. Sintered or arc-cast metal can be forged, starting at 1760°C and continuously working down to 985 to 1100°C. In very high-velocity forming or extension forming, such appreciable reductions can be obtained (by rapid working, which minimizes heat losses) as to permit forward extrusion of tungsten shapes.

Tungsten sheets can be produced by cold-rolling micrometer-sized powder to a "green" strip, which subsequently is sintered and hot-rolled, starting at about 1485°C. When sheet has been reduced to about 0.5 mm, rolling can be carried on at close to room temperatures. Preferred orientation of rolled tungsten sheet is a problem, as it is, indeed, with all refractory metals. Typically, the rolled texture leads to a susceptibility to fracture in the 45° direction. Both cross rolling and intermediate annealing during rolling can reduce the degree of this texture.

Although less susceptible to catastrophic oxidation than molybdenum, tungsten oxidizes more rapidly than niobium and tantalum (Fig. 17.6). The end product of oxidation is WO_3, which melts near 1475°C and is decidedly volatile at temperatures somewhat below this. Below 1000°C, oxidation initially tends to be parabolic, but as the blue suboxide converts to porous yellow WO_3, the oxidation rate changes toward becoming linear with time. The presence of water vapor increases the rate of evaporation of WO_3, and, relatedly, oxidation data from different sources show considerable variation or scatter. There are some indications that niobium added as an alloying element can greatly reduce the oxidation of tungsten at some temperatures, but it appears likely that for service at 1650°C or above, coatings will be needed for tungsten structures.

The strength of tungsten for service at very high temperatures cannot generally be increased by solid-solution alloying. Alloys containing 10 to 25% Mo, while slightly stronger than tungsten at intermediate temperatures, are weaker than the pure metal above 1900°C. The only effective means of strengthening tungsten here appears to be dispersion strengthening, particularly with thoria, which is the most stable of all oxides. Tungsten with 1% ThO_2 dispersed in its structure shows strength of nearly 140 MPa at 2200°C, compared with 40 MPa for the pure metal (Fig. 17.7). The effect of ThO_2 depends of course on the size and distribution as well as the amount of the ThO_2 particles. Since material of this type produced to date has particles considerably larger than is believed desirable for optimum dispersion strengthening, it appears possible that alloys of this kind with even better high-temperature strength can be developed.

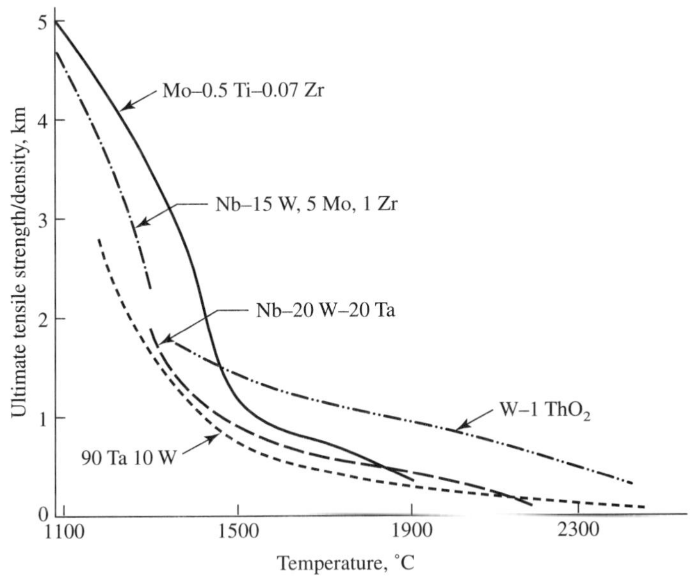

Figure 17.7 Comparison of several refractory alloys on basis of (ultimate tensile strength)/(density) ratio vs. temperature.

17.6 REFRACTORY METAL COATINGS

The relatively poor resistance of the refractory metals to high-temperature oxidation can seldom be adequately improved by alloying. Therefore, unless used in an inert atmosphere, protective coatings are required if the full high-temperature service potential of these metals is to be realized. The first requirement of a coating is that it protects the underlying metal from oxidation and, in many cases, from diffusion of gases into the underlying metal. It must be chemically compatible with the metal, meaning that it should not itself supply harmful elements for diffusion into the base metal or allow diffusion to result in formation of brittle intermetallic compounds with the base metal. The coating should be physically compatible with the base metal over a range of temperatures so that it does not crack or spall off upon thermal cycling. Since coatings are usually brittle, this means that a coating should have a slightly lower thermal coefficient of expansion than the base metal so that the coating is under compression at lower temperatures.

The favored method of applying coatings to structural parts of refractory metals is vapor plating by a process similar to the pack-cementation process used in pack carburizing of steel. For example, a silicide coating 1 mil in thickness can be formed on molybdenum or tungsten by packing the part in silicon powder having 10% NaF as an activator and heating to 1035°C for 8 h. This pack-cementation process is also used to form coatings of Cr-Ti-Si. Another method of forming a straight silicide coating is to immerse the molybdenum or tungsten part in an Al-Si eutectic bath at 1035°C. A protective $TaAl_3$ coating can be formed by hot-dipping tantalum in aluminum and then heating the coated part at a high temperature so that diffusion will form the aluminide. Complex mixtures of metals and oxides constitute *cermet* coatings, which are sometimes applied by plasma arc spraying.

The chemistry and metallurgy of coatings and the methods of applying them to obtain and maintain adhesion of a nonporous, oxidation-resistant surface constitute a complex technology indeed. Refractory-metal coating development has not yet reached the stage where fully reliable coatings can be specified for any of these metals. Even the relative evaluation of coatings is extremely variable, depending on the number of specimens tested, the number of thermal cycles, the stress simultaneously applied, and so on. Generalizations are almost impossible, since each service application has its own specific conditions and requirements.

PROBLEMS

1. Differentiate between precipitation hardening and dispersion hardening of alloys.
2. What would be the desirable properties of a second-phase precipitate in a superalloy? Would nonmetallic inclusions such as oxide particles be useful constituents in such an alloy?
3. What is the theoretical basis for describing the properties in terms of the ratio T/T_m instead of in terms of the absolute temperature? Why does creep become important in all pure metals at $T > 0.5T_m$?
4. Derive the parabolic-rate law of oxidation.
5. Give an example of a common metal that forms a protective oxide layer in which the oxide is porous.

REFERENCES

Davis, J. R. (Ed.): *ASM Specialty Handbook: Heat Resistant Materials,* ASM Int., Materials Park, OH, 1997.

Stoloff, N. S.: "Wrought and P/M Superalloys," *Properties and Selection: Irons, Steels, and High-Performance Alloys,* vol. 1, *ASM Handbook,* ASM Int., Materials Park, OH, 1990, pp. 950–980.

Betteridge, W.: *Cobalt and Its Alloys,* Ellis Horwood, Chichester, UK, 1982.

Ashby, M. F.: *Materials Selection in Mechanical Design,* Pergamon Press, New York, 1992.

Rohrbach, K. P.: "Trends in High-Temperature Alloys," *Adv. Mat. Proc.,* vol. 148 (4) pp. 37–40, October 1995.

Lai., G. Y.: *High-Temperature Corrosion of Engineering Alloys,* ASM Int., Materials Park, OH, 1990.

Cabrera, N, and N. F. Mott: "Theory and Oxidation of Metals," *Rep. Prog. Phys.,* vol. 12, pp. 163–184, 1948–1949.

Yau, T. L., and R. T. Webster: "Corrosion of Niobium and Niobium Alloys," *Corrosion,* vol. 13, *ASM Handbook,* ASM Int., Materials Park, OH, 1987.

Titran, R. H., "Niobium and Its Alloys," *Adv. Mat. Proc.,* vol. 142 (5), pp. 34–41, November 1992.

Cardonne, S. M., P. Kumar, C. A. Michaluk, and H. D. Schwartz: "Tantalum and Its Alloys," *Adv. Mat. Proc.,* vol. 142 (3), pp. 16–20, September 1992.

Smialek, J. L., C. A. Barrett, and J. C. Schaeffer: "Design for Oxidation Resistance," *Materials Selection and Design,* vol. 20, *ASM Handbook,* ASM Int., Materials Park, OH, 1997.

SECTION

4

NONMETALLIC MATERIALS AND COMPOSITES ENGINEERING

chapter

18

Engineering Polymers

In general, polymeric materials provide only modest strength, toughness, and temperature resistance. However, they can also have good corrosion resistance, excellent strength, unusually high toughness, and excellent properties at low temperatures. In addition, polymers may also provide a variety of unique characteristics that cannot be obtained from other materials. These include the surface friction characteristics associated with polytetrafluoroethylene (Teflon) and related polymers, the elasticity characteristics of polyisoprene (natural rubber) and other rubbery polymers, and the strength and toughness but lightweight characteristics of the reinforced plastics.

There is such a variety of these materials in use today that one chapter in a book devoted primarily to metallic materials can only serve as an introduction to them and their application. On the other hand, their engineering significance has grown to the point where they are increasingly used to replace metallic materials in many kinds of service. Thus some introduction appears necessary so that engineers can appreciate the range of properties they offer and the basic principles by which these properties are developed. This chapter is intended to provide such an introduction.

In the discussion of the polymeric materials, the term plastic will be avoided. This term has been commonly used to designate polymers but also has a significant meaning in terms of a type of mechanical behavior, i.e., plastic deformation. To separate the material from the behavior (not all plastics are plastic in behavior) the term polymer will be used, as appropriate, rather than plastic. Polymer here implies long-chain molecules built up from shorter molecules with the spine of the molecule usually, but not exclusively, comprising carbon atoms.

In the discussion of polymers, very little will be said about the polymerization processes used to produce these materials. This is in keeping with the general emphasis of this text, and indeed it would require a book of considerable size to cover this topic in detail alone. It should be noted here, however, that polymerization processes could be divided into two groups, condensation polymerizations and addition polymerizations. In condensation polymerization, two or more different organic molecules react to produce a chain made up of combinations of the starting molecules, a small molecule, often water, being eliminated. In the case of addition polymerization, in one (or more) organic molecules containing a shared electron bond, the double bond is split by the action of an initiator. This produces unpaired electrons on the starting molecule, which subsequently bond with surrounding molecules of the same type by addition until a long chain is produced.

The conditions required to initiate these reactions are complex, as is their control, but significant changes in regulating polymerization processes can affect polymer properties. In general, the effects of processing can be understood by study of the polymer structures that result, and thus it will be the structure of polymers that will be given the most attention in this chapter.

18.1 BONDING AND STRUCTURE IN POLYMERS

Considering the polymer as a simple chain consisting of repeating units, a number of characteristics of the chain can be defined to describe its structure. It must first be remembered that carbon has four electrons in its outermost electron shell and thus prefers four bonds in a tetrahedral configuration surrounding the carbon atom (similar to the structure seen in Fig. 1.12 on page 11). This means that the chain of atoms is not, in fact, straight, but a three-dimensional zigzag structure like that in Fig. 18.1. The structure shown here is for a simple polymer based on ethylene, C_2H_4, and thus represents the simplest of polymer structures. More complex polymers may have chains that are more convoluted or have side branches on the main one. These variations will have a significant influence on polymer properties and will be discussed later.

One characteristic of the polymer molecule of significance is the degree of polymerization (DP). This is the number of repeating units of identical structure in the chain. Each repeating unit, which is about equal chemically to the starting molecule, or monomer, is called a mer. Thus, for the structure shown in Fig. 18.1, the

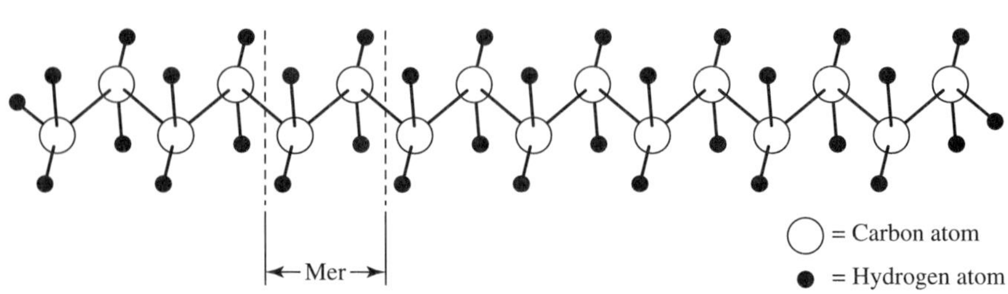

Figure 18.1 Model of the structure of linear polyethylene. There are eight mers in this chain.

monomer is ethylene, C_2H_4, while the mer, having the same chemical formula, has two unsatisfied covalent bonds. The structure seen in Fig. 18.1 has 8 mers, and so its degree of polymerization is 8. For the simplest polymers with a structure based on that of ethylene, the hydrocarbons containing 5 to 20 mers are gases and liquids at ambient temperatures. When the number of mers is over 50, the polymer is a soft solid, a wax. Higher degrees of polymerization, 200 to 20,000, produce the solid polyethylene, which is used in a number of moderate service engineering parts and widely used in household items.

The bonding in polymers, especially the simplest ones, is only by van der Waals forces. While there are strong covalent bonds between the carbon atoms in the chain and between the carbon and hydrogen (and other) atoms, this bonding is often confined to the chain. The forces that bind the chains together are therefore usually transient electrostatic dipoles, and the resulting structure is weak. When stressed, the simpler chains tend to slide over one another, and failure is by interchain separation rather than by breaking the intrachain bonds. As will be illustrated later, increasing the van der Waals bonds can enhance these interchain bonds. Some polymers are cross-linked: i.e., there are covalent bonds between the chains. Cross linking is usually developed when there are unsaturated carbon bonds, double bonds between carbon atoms that can be broken so that individual atoms or molecules can be used to link adjacent chains. Heavily cross-linked polymers may develop a rigid framework of covalent bonds. This produces higher strengths but limits the ductility of the material. Even without cross linking, most polymers have side branches on their chain and thus are not strictly linear. Side branching also alters the properties of a polymer, usually increasing strength by making interchain sliding more difficult. Linear, branched, and cross-linked structures are compared in Fig. 18.2.

Unlike metals, most polymers are predominantly in a noncrystalline or glassy state. They do not have a sharp melting point like metals. Instead, they pass into this state from a viscous liquid or melt state on cooling through a temperature known as the glass-transition temperature T_g. Below this temperature range, local molecular motion virtually ceases and there is a marked change in properties. These polymers are then hard, stiff, brittle, and usually transparent. In the glassy state, the structure of the polymer is not unlike that of an inorganic glass. These molecules have a short-range order, but no long-range crystal structure is developed.

Some polymers are partly crystalline, and their structure is characterized by a degree of crystallinity (usually stated as a percentage). This is a measure of the extent of long-range three-dimensional order within the polymers compared with a measured or calculated value for the same polymer in a highly crystalline state. For example, x-ray diffraction is used to compare the area under diffraction peaks from specific crystalline planes produced by the crystalline regions with that area under the broad amorphous (short-range-order) peak produced by the amorphous regions. The exact value of degree of crystallinity depends on the model used to predict diffraction behavior for a fully crystalline state and thus may vary depending on the methods used and model employed. The reported values for

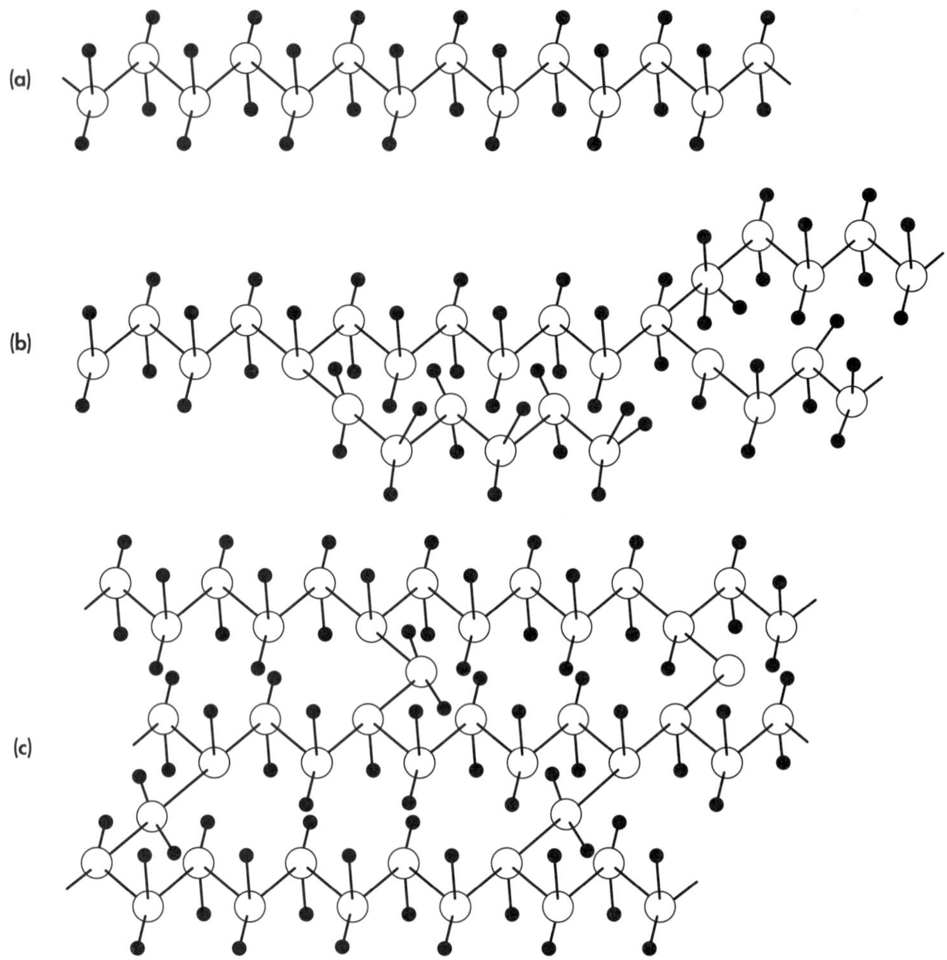

Figure 18.2 Model of (*a*) linear, (*b*) branched, and (*c*) cross-linked polymer structures.

different polymers range from nearly zero to over 95%. The simpler polymers with the least complicated chains (little or no side branching, relatively linear) crystallize more easily than others. The heavily cross-linked polymers are generally not crystalline, as cross linking occurs at high enough temperatures to ensure that on subsequent cooling to below T_m, the crystalline melting point, they will not have sufficient mobility to align in a crystalline or partially crystalline state. As a result, they will cool to below T_g and enter the glassy state without any appreciable crystallinity. Although both T_m and T_g are referred to as specific temperatures, they are, in fact, the temperature ranges over which changes occur. The T_m of a polymer can be detected by observing changes in evidence of crystallinity in x-ray diffraction on heating or cooling or more commonly by changes in specific volume using differential thermal-analysis techniques. A curve derived by such

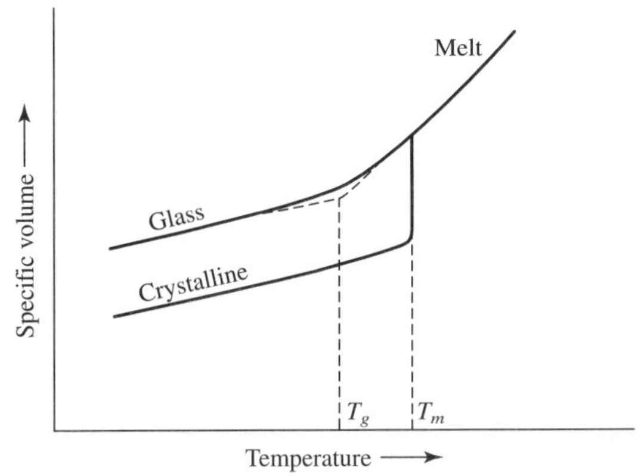

Figure 18.3 Specific-volume-vs.-temperature curve for a polymer. If the polymer has a sufficiently linear structure without extensive branching or cross linking, it may follow the crystalline structure line on cooling. If not, it will pass below T_g without crystallizing and become a glass. Many polymers have both crystalline and glassy regions in their structure.

a method is seen in Fig. 18.3. Here the specific volume is seen to decrease with decreasing temperature from the melt down to T_m. If crystallization can occur, the volume decreases fairly sharply and follows the specific-volume-vs.-temperature line for the crystalline state. This behavior is similar to that of most metals. If crystallization does not occur for structural or other reasons, the material continues to contract linearly with temperature until T_g is reached, at which temperature the slope of the curve changes slightly and continued cooling is of material in the glassy state. The T_g and T_m of most polymers are simply related, as both are influenced by the chain structure. Depending on symmetry, T_g (in kelvin) is approximately one-half to two-thirds T_m (in kelvin).

A fully satisfactory model for the structure of crystalline, or more commonly, partly crystalline, polymers has not yet been achieved. Early theories, based for the most part on x-ray diffraction experiments, conceived of the polymer being made up of small regions, on the order of several hundred angstroms in width, in which chains were folded back and forth or at least aligned. The chains also extended into other regions, where they were randomly oriented. Amorphous regions therefore surrounded the aligned regions, called crystallites, and the whole concept was called the fringed-micelle or fringed-crystallite model. Microscopic examination of crystalline or partly crystalline polymers reveals that they do have prominent structural features called spherulites, as seen in Micro. 18.1. The spherulites are apparently coincident with the crystallites, while the surrounding material is either noncrystalline or crystallized later (at a lower temperature). The spherulites do appear to have a lamellar structure, with growth accomplished by growth of the individual lamellae from some growth nucleus.

Current theory for highly crystalline polymers is that the structure should not be thought of as crystalline spherulites surrounded by amorphous regions but as

Micrograph 18.1 Spherulites in a polymer viewed between crossed polarizing filters. The Maltese cross pattern reveals the crystal orientation in the spherulites. The large spherulites apparently crystallized first, with the remaining material crystallizing later. Thin section of polymer. About ×200.

being more analogous to metals, in which the structure is crystalline but contains crystal defects. This is called the defect-crystal model. These defects give rise to noncrystalline scattering in x-ray diffraction patterns. The internal structure in the spherulites can be more readily understood using this model, but further work on polymer structures, especially those with low degrees of crystallinity, must be done to establish its general validity.

Mechanical deformation, elastic or plastic, particularly between T_m and T_g, will enhance crystallinity and also result in a preferred orientation of the molecular chains. Such deformation can be made during fabrication such as drawing of a fiber or drawing of a sheet. This will result in an increase in strength because the molecules are aligned by the process and the applied stresses are taken by the strong covalent bonds down the length of the fiber. This is not the situation in bulk polymers, where the molecular orientations are random.

The extent and type of cross linking in polymers have a significant influence on their thermal behavior. When polymers, which are held together primarily by van der Waals bonds, are heated, their intermolecular forces are easily overcome and they become soft and deform easily. In such a condition, the material is readily fabricated and thus its melt properties, particularly its melt viscosity, are important features in its economic use. These materials are called thermoplastic. The heavily cross-linked polymers do not soften readily on heating, since the thermal energy required to overcome the covalent bonding forces is much greater than for van der Waals bonds. These polymers will eventually degrade (depolymerize) on heating and thus cannot be fabricated by heating and forming after their initial setting. These materials are therefore more limited in fabricability and must usually be formed during or before polymerization and cross linking. These polymers are called thermosetting, and although they are harder to fabricate than thermoplastic ones, they are generally stronger owing to the extent of covalent bonding (cross linking) present.

In addition to the nature of the bonding between the polymer chains, structural variations within the chain may have a significant influence on properties. When we consider a polymer that has only carbon and hydrogen atoms present, like that shown in Fig. 18.1, the arrangement of hydrogen atoms is of little consequence, since they are small enough not to interfere with each other when adjacent and their positions are limited to those dictated by the bonding requirements of the carbon atoms. In the example of Fig. 18.1 all possible carbon bonds are filled, and this structure is saturated. If the polymer is slightly more complex, e.g., that based on the vinyl chloride molecule, C_2H_5Cl, a new condition is introduced. Such a structure is seen in Fig. 18.4. The chlorine atoms introduced into the structure are large compared with the hydrogen atoms, and their arrangement on the lattice is therefore of some significance. The chlorine atoms may be placed on the individual mers in a more or less random manner, without regard to any other chlorine atoms. This is called an atactic structure (Fig. 18.4*a*). If the chlorine atoms are placed on the chain in the same location in each mer, the structure is called isotactic (Fig. 18.4*b*). If the chlorine atoms are placed on the chain in a more or less regular way but alternate positions on either side of the chain rather than assuming the

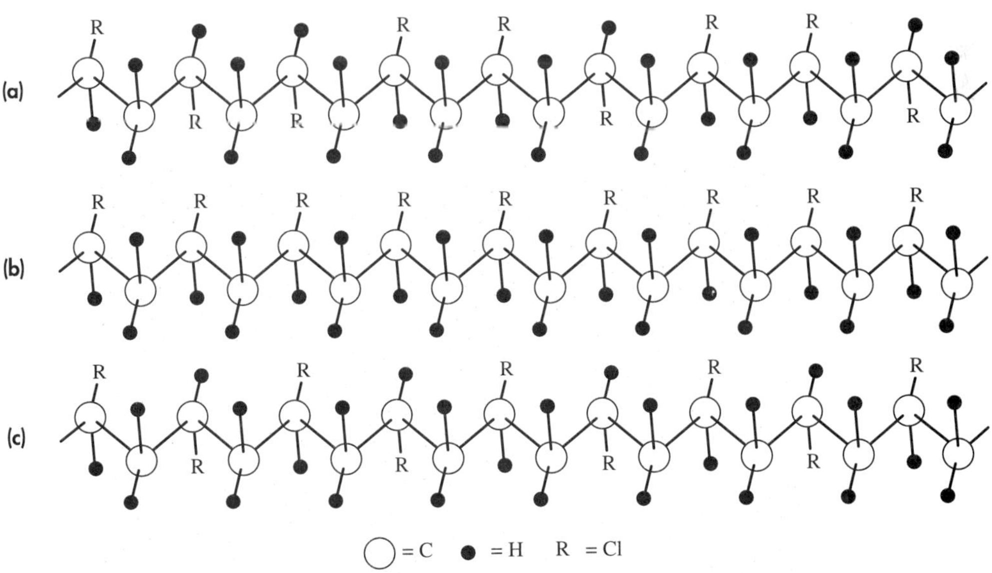

Figure 18.4 Models of (*a*) atactic, (*b*) isotactic, and (*c*) syndiotactic polymer structures. The circles represent carbon atoms, dots are hydrogen atoms, and the R's are, in this case, chlorine atoms.

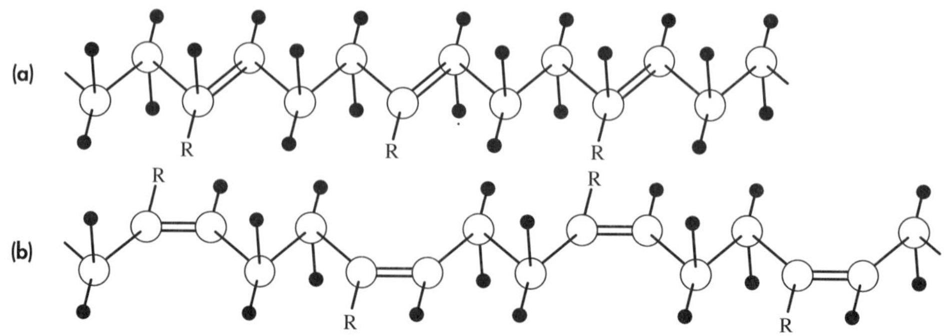

Figure 18.5 Models of (*a*) cis and (*b*) trans structures in dienes. The circles represent carbon atoms, dots are hydrogen atoms, and R's are chlorine atoms or simple organic molecules.

same position on each mer, the structure is referred to as syndiotactic (Fig. 18.4*c*). The position assumed by the chlorine (or in more complex molecules, other atoms) has an influence on properties because atactic structures decrease the regularity of the chain compared with isotactic or syndiotactic ones and thus are less easily made crystalline. Again, placing a side group (R in Fig. 18.4) in the isotactic position may result in interference and repulsion of the adjacent molecules (especially if they are large), causing the resulting molecule to be more curved than would otherwise be the case.

Curvature in the polymer chain may also be produced by the arrangement of bonds between the carbon atoms in the chain. This is illustrated for diene chains in Fig. 18.5. Here the molecule or atom, R, is placed on an unsaturated carbon chain in either the cis or trans position. In the cis position, the unsaturated bonds are on the same side of the chain, while in the trans position they are on opposite sides. The difference between these two possibilities is most important in butadiene rubbers, where the cis structure makes the molecule tend to coil rather than remain more or less linear and is believed responsible for much of the elasticity of rubber.

18.2 GENERALIZED PROPERTIES OF POLYMERS

The mechanical behavior of polymers is influenced by a variety of structural parameters, as described above. There are, however, some general principles influencing behavior that can be applied to most of the simple structures and many of the more complex ones. The strength properties of many polymers are directly related to molecular weight as well as to degree of crystallinity. The tensile related properties

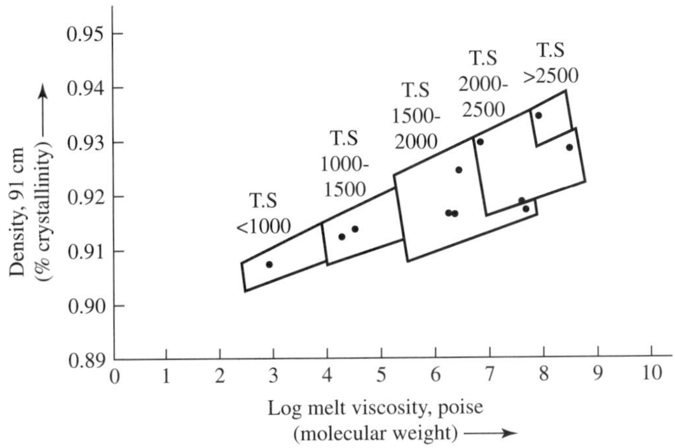

Figure 18.6 The influence of crystallinity and molecular weight on the tensile strength of polyethylene. Crystallinity is determined from density in this case and molecular weight from melt viscosity. Both molecular weight and crystallinity influence tensile strength.

are often expressed in relations with the form

$$\text{Property} = a - b/\overline{M}_n$$

where $\overline{M}_n$ is the *number average molecular weight* and a and b are constants. The number average molecular weight is a characterization of a polymer with a mixture of molecular weights

$$\overline{M}_n = \Sigma[(X_i)(\text{MW})_i]/\Sigma X_i$$

Here X_i is the number of molecules in each size fraction, and $(\text{MW})_i$ is the mean molecular weight in each size fraction.

Crystallinity may also increase tensile strength. These two relationships are seen for polyethylene in Fig. 18.6. Equally important characteristics of polymers influenced by crystallinity are stiffness and yield strength. In polymers having some crystallinity, stiffness, or modulus of elasticity, may change more than an order of magnitude with changes in crystallinity, as can be seen in Fig. 18.7. Increasing the yield point and stiffness of a polymer without a corresponding increase in tensile strength, however, usually results in an increase in brittleness. Thus a balance between yield strength and tensile strength is required in polymers, as it is in metals, to provide adequate tensile ductility for engineering applications.

Another mechanical property dependent on molecular weight and crystallinity is flex (fatigue) life. Increasing molecular weight improves flex life, while increasing crystallinity decreases it. This property is probably a complex one, with results depending on how the test is performed. Since the specimen stiffness and yield point also change with crystallinity, tests run at constant specimen deflection will give results different from those of tests run at constant specimen stress.

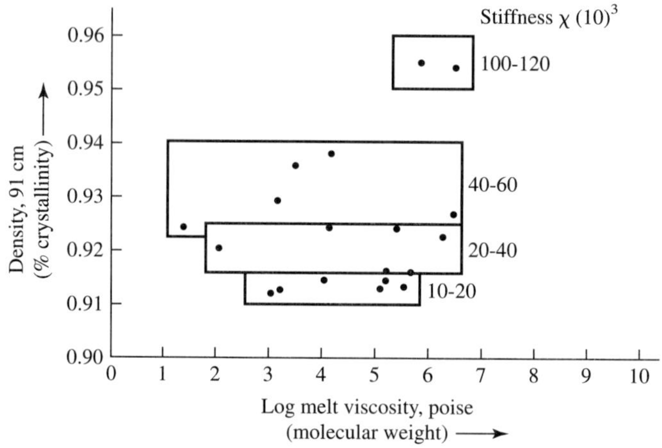

Figure 18.7 The influence of crystallinity and molecular weight on the stiffness or elastic modulus of polyethylene. Crystallinity is determined from density in this case and molecular weight from melt viscosity. Molecular weight has little effect on stiffness compared with crystallinity.

One aspect of the behavior of polymers that must also be considered is their tendency for time-dependent deformation. Actually, polymers exhibit both elastic (immediate) and viscous (time-dependent) behavior, a combination referred to as viscoelasticity. When a polymer, particularly an amorphous polymer, is placed under an applied stress, there is an immediate elastic response, as seen in Fig. 18.8. This response is followed by a time-dependent viscous flow, which decreases with

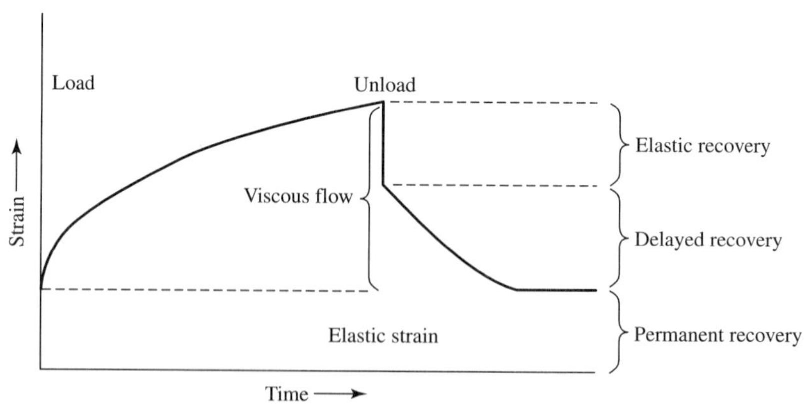

Figure 18.8 Viscoelastic behavior of a polymer. Loading produces an immediate elastic strain followed by viscous flow. Unloading produces an immediate elastic recovery, followed by additional recovery over a period of time. There is, however, a remaining permanent strain.

increasing time until a steady state is reached (similar to first- and second-stage creep in metals). If the material is unloaded, Fig. 18.8 shows that there is an instantaneous (elastic) strain recovery followed by a time-dependent one; however, a permanent deformation remains.

The time-dependent strain recovery is often referred to as plastic memory. For example, a flat disk of many polymers can be cold-drawn through a die with a mandrel into the form of a cup. Upon emerging from the tooling, the straight walls of the cup show an elastic recovery to a tapered wall, larger in diameter at the open top than at the closed bottom. With time or particularly with the application of some heat, the taper on the cup will increase; i.e., the material will tend to go back toward its initial shape of a flat disk; hence plastic memory.

If the tensile-test conditions are changed from constant stress to constant strain, the stress at that strain level will decrease with time, giving rise to the phenomenon of stress relaxation. This is significant in some types of service conditions at elevated temperature. In addition, if thermoplastic amorphous polymers are heated to temperatures in the region of the glass-transition temperature, internal stresses are dissipated by stress relaxation.

A property of amorphous polymers that is important in some engineering applications and is usually lost upon crystallization is transparency. Crystalline polymers are usually translucent or opaque, while the amorphous ones are transparent. As an amorphous polymer is cooled below T_m and its crystallinity increases, the crystalline regions are significantly denser than the noncrystalline ones. This produces a difference in refractive index between the two regions and leads to scattering of light as long as the crystallized regions are not small in size compared to the wavelength of the light. In turn, this will result in translucency or opaqueness in the partly crystalline polymers.

Copolymerization of two or more monomers will usually modify their mechanical properties, decreasing crystallinity and thus stiffness and yield point. Other effects, especially in polymers that do not crystallize, are more complex and depend on the nature of the monomers employed. The addition of plasticizers, fillers, and other modifiers may also significantly change polymer properties. Plasticizers are usually low-molecular-weight polymers (or monomers) that separate the polymer chains and reduce crystallinity. This makes the polymer more flexible and less brittle and can be used in glassy polymers to increase ductility and toughness below the T_g of the unplasticized material, i.e., in effect, to lower T_g.

Fillers are of many types. They are added to impart a variety of special properties to polymers. The fillers may be divided into particulate and fibrous types. Particulate fillers are usually materials like silica (sand, quartz), silicates (mica, talc, asbestos), glass (granules, flakes, and spheres), metal powders, inorganic compounds (limestone, alumina), cellulosics (wood flour), and other polymers (rubber). The fibrous fillers include cellulosics (cotton fibers), synthetic fibers (nylon, polyester, acrylic), glass, carbon, boron, and even single-crystal fibers of alumina, silicon carbide, and others. In general, the particulate fillers provide improved compressive strength, wear resistance, temperature resistance, impact resistance, and dimensional stability. The fibrous fillers will impart higher tensile and impact resistance.

While mechanical property tests of plastics are performed in a manner similar to those on metals and the results are expressed in the same terms, they frequently cannot be interpreted in the same way. Metals really are more plastic, in the sense of capability for permanent deformation, than polymers but can never be subjected to anywhere near as much elastic deformation. Because of the pronounced viscoelastic behavior of polymers, they are far more strain-rate-sensitive than the fully crystalline metals. As another example of behavioral differences, the slope of the plastic part of the stress-strain curve of metals denotes work hardening, an increase in strength associated with impediments to the easy movements of dislocations. Plastics engineers have called the equivalent part of a stress-strain curve for polymers work hardening, but it is a completely different phenomenon. In polymers, it is an alignment of molecular chains during flow so that the applied stress is in the direction of the strong carbon-carbon bonds, or backbone of the polymer chains.

18.3 OLEFIN, VINYL, AND RELATED POLYMERS

The largest volume production and sale of polymers is of the group known as the olefin polymers, the most common example of which is polyethylene (PE). As indicated in Figs. 18.1 and 18.2, the structure of this material is simple and its strength low, especially in the branch or low-density (LD) form. The intermediate-density types are partly crystalline, with percent crystallinity increasing as branching decreases. The high-density (HD) form is a more crystalline linear polymer with improved strength and stiffness and good chemical inertness. The low-density polyethylene is used for toys, household items, plastic bags or films, coatings, and squeeze bottles. The greatest use of the high-density form is for rigid containers, e.g., bottles for detergents, bleaches, etc. It is also used for battery parts, pipe, and structural panels. A low T_m limits the range of application of polyethylenes, but, as one of the cheapest polymers and one with a low T_g, that is, not brittle when cold, its production volume exceeds that of all others.

Closely related in properties to polyethylene is a more recently developed olefin, polypropylene (PP). As indicated in Table 18.1, it has a methane side group and can be isotactic, atactic, and syndiotactic. The isotactic form is crystalline with a T_m of 165°C and is the most useful. It has higher strength and stiffness than PE and is the lightest of the polymers, with a specific gravity of 0.905. Polypropylene undergoes several transitions below the classical one at T_g, a phenomenon typical of crystalline or nearly crystalline polymers believed related to the mobility of short segments of the chain. In the case of polypropylene, this results in a loss of toughness around 0°C. The high T_m for this material, 165°C, gives it added temperature resistance in comparison with polyethylene, useful in many applications involving boiling water or steam. Both polyethylene and polypropylene are used filled and copolymerized for some applications.

Table 18.1 Structure and transition points of olefin, vinyl, and related polymers

Polymer	Mer structure	T_m, °C	T_g, °C
Polyethylene, HD LD	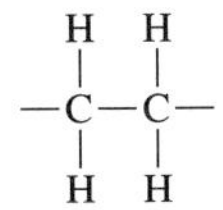	137 ~115	−120 −120
Polypropylene	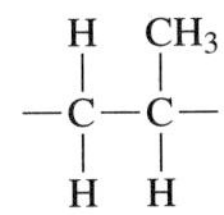	176	−18
Polystyrene	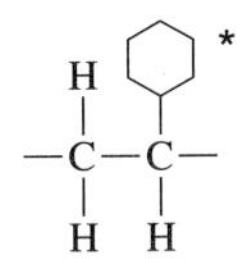	…	−50
Polyacrylonitrile	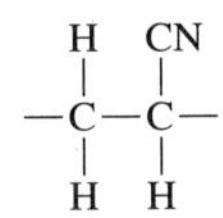	…	105
Polymethyl methacrylate	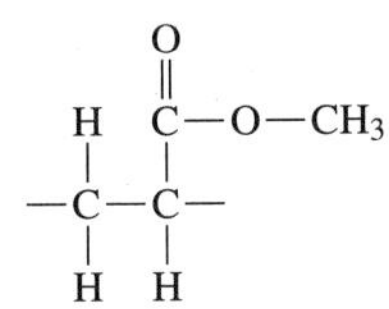	…	105
ABS		…	105
Polyvinyl chloride	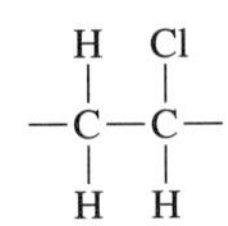	212	87
Polytetrafluoroethylene	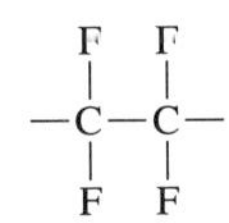	327	

*  = benzene ring = 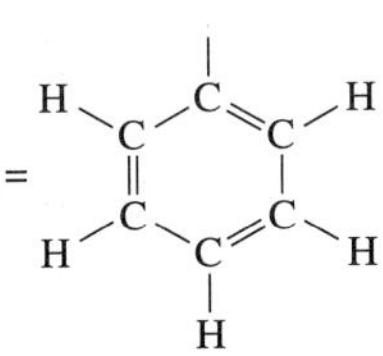

There are a large number of engineering polymers of similar structure to polypropylene known as the vinyl polymers, most common of which is polyvinyl chloride. The vinyl group is monosubstituted ethylenic polymers; the substituting molecules may be chlorine, as shown in Table 18.1, as well as benzene ring (styrene) and other groups involving carbon, hydrogen, and oxygen.

Table 18.2 Properties of olefin, vinyl, and related polymers*

Polymer	Tensile Strength, MPa	Tensile Modulus, GPa	Elongation, %†	Izod Impact Strength, J/cm of Notch
Polyethylene, HD	21–28	0.7–1.4	16–1000	0.5–2.7
LD	7–21	2.1	50–800	10.7
Polypropylene	28–35	1.1	300	0.2–1.1
Polystyrene	35–56	3.2	1–2	0.3
Impact	21–49	2.2	3–80	0.5–4.8
Polymethyl methacrylate	42–84	2.1–3.5	4–5	0.27
ABS	42–56	2.1–2.8	15–25	1.1–2.7
Polyvinyl chloride, rigid	35–63	2.1–4.2	5–100	0.5–10.7
Plasticized	7–28	0.2	350	
Polytetrafluoroethylene	14–49	0.7	350	1.6–3.2

*Properties vary extensively with processing and compositional variations.
†Total elongation prior to fracture; much of this, in contrast to metals, is elastic stretch.

Polyvinyl chloride, often referred to as vinyl or PVC, was for a time the most widely used of these polymers. As seen in Table 18.2, it has attractive strength in the unplasticized, or "rigid," forms. The mechanical properties still make this popular for bottles and containers and (when plasticized) in sheet and fiber forms for clothing and covers. It is also widely used in the copolymerized state with vinyl acetate and vinylidene chloride (a disubstituted ethylenic in which two chlorines are contained in the mer). These are more stable and somewhat stronger than PVC.

More important from the engineering standpoint are two additional materials in this group that are widely used by themselves and are combined to produce a major class of polymers, the acrylic-butadiene-styrene (ABS) resins. The first group are those materials referred to as acrylic polymers, usually polyacrylonitrile or polymethyl methacrylate. The polyacrylonitrile is used often in the form of fibers, which are a principal constituent in several types of commercial clothing, e.g., orlon, acrylon. In this form, the polymer has good strength, stiffness, toughness, and flex life. Polymethyl methacrylate is a clear, colorless, transparent plastic with good tensile properties (Table 18.2) and is used in a variety of cast solid forms, e.g., plate and rods, and also in compositions for molding and extrusions. It is a linear thermoplastic, which is for the most part syndiotactic, but because of its bulky side-group structure (Table 18.1) it is amorphous. Its good optical properties and outdoor weathering behavior make it very popular in signs, lights, and glazing applications. It is better known by one of its trade names, PMMA or Plexiglas.

The second constituent of the ABS group, another significant polymer in its own right, is polystyrene. This has been a leading low-cost thermoplastic and has been used

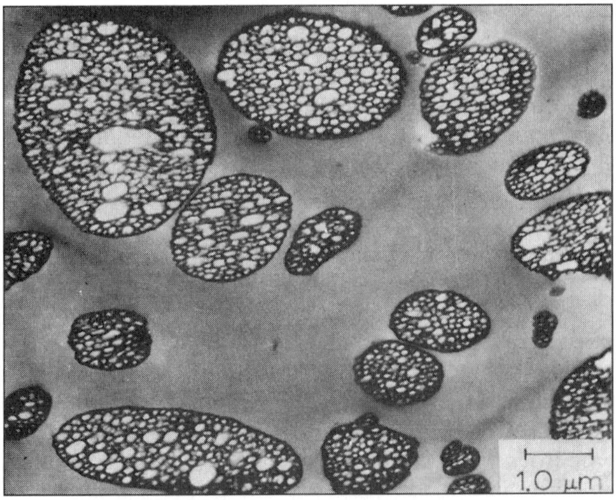

Micrograph 18.2 Rubber-toughened polystyrene. Dark-colored rubber regions contain occluded particles of polystyrene, which produce a characteristic "salami" structure.

for a host of home appliances, toys, and, more recently, construction-material applications. Like polymethyl methacrylate, it has excellent optical properties and good strength. It is atactic and amorphous and is one of the most easily fabricated of the thermoplastics. On the negative side, it tends to be brittle and has a low softening temperature (100°C) and poor solvent resistance. All these problems can be overcome by additives or copolymerization. The impact resistance of this material can be improved by incorporating rubber into the polystyrene matrix, usually 5 to 10%, which results in a product called impact polystyrene. The rubber is present in the structure as a second phase, producing the microstructure seen in Micro. 18.2. This structure is one of the most analogous to metal microstructures to be found among the polymers. The overall result is a substantial increase in toughness, as can be seen in Table 18.2.

The most attractive combination of polystyrene, as indicated above, is in the form of ABS, made of a blend of acrylonitrile, butadiene (a rubber), and styrene. This combination is also a two-phase mixture, much like the impact polystyrene, the result being rubber inclusions in a glassy matrix. The rubber is a butadiene-styrene copolymer while the matrix is an acrylonitrile-styrene copolymer. The best properties are obtained when there is a grafting (bonding) between the matrix and the inclusions. The ABS resins have higher temperature and solvent resistance than polystyrene. They also have good strength, impact resistance, and formability, and moderate cost. They are extensively used in a variety of applications such as pipe, components for the transportation and home-appliance industries, telephone equipment, baggage, and even automobile grills. They can even be cold-formed in dies much like metals. Their properties are listed in Table 18.2.

Many carbon-chain thermoplastics have not been considered here because of their limited use in engineering applications. One group with structure very similar to the materials discussed thus far has good engineering potential, but its applications have been somewhat limited due to cost. This is the group of fully fluorinated polymers, commonly known as teflons. The commonly used materials in this group are polytetrafluoroethylene (PTFE), polychlorotrifluoroethylene (PCTFE), and fluorinated ethylene propylene (FEP). These are highly crystalline, orientable polymers without much side branching and have relatively high crystalline melting points (220 to 325°C). Their outstanding characteristics are their stability over a wide range of temperature, good low-temperature toughness, good elevated-temperature resistance, low dielectric constant, extreme chemical inertness, and low surface-friction characteristics. All these characteristics have led to important uses of this material. As can be seen from Table 18.2, however, the mechanical properties of the material are intermediate, and the service characteristics indicated above, particularly temperature resistance and inertness, have made it a relatively difficult material to fabricate. Since the material does not flow well even above its crystalline melting point, techniques much like those used in powder metallurgy are employed. Typical applications for this material include chemical pipes and pumps, liners, electrical equipment, and insulating shields.

18.4 THERMOPLASTIC POLYMERS

The basic difference between the thermoplastic polymers and the olefin and vinyl polymers lies in the nature of the spine of the polymer chain. A heterochain group of materials has a more complex chain than the previously discussed groups in that oxygen, nitrogen, silicon, and ring structures containing these elements lie in the basic chain. In this category fall a number of commercial fibers of importance as well as a variety of engineering plastics. Some materials with this structure also fall into the class of elastomers, which will be discussed later.

Perhaps the most generally known of the materials in this group are the polyamides and the polyesters. The former are more familiarly known by their now generic name nylon, while the latter are known by a variety of trade names including Dacron, Kodel, and Mylar. The nylon group is the most widely used, and the structures of the various commercially used grades differ primarily in the number of CH_2 groups that separate the oxygen and nitrogen positions in the chain. They are described by the number of carbon atoms in the monomer chains, a single number being used for the nylons produced from a single amino acid and several numbers being used for those produced from diamines (the first number) and dibasic acids (the second number). Two common nylon structures are indicated in Table 18.3.

The greatest commercial use of nylons is of the 6, 66, 610, 8, 11, and 12 varieties and their copolymers. These materials are split in their use between the applications that might be considered primarily engineering, e.g., bearings, gears, cams, slides, small parts, and tire cord, and those that are primarily concerned with the textile industry. Until recently, tire cord consumed a major section of the nylon-fiber market.

Table 18.3 Structure and transition points of heterochain thermoplastics

Polymer	Mer structure	T_m, °C	T_g, °C
66 Nylon (polyamide)	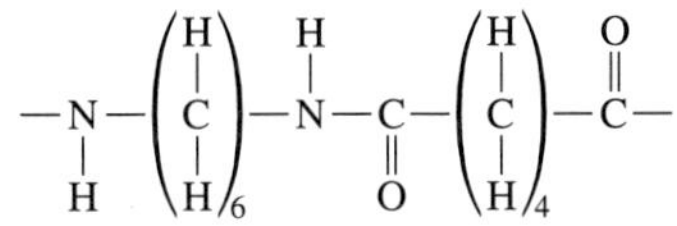	265	50
11 Nylon (polyamide)	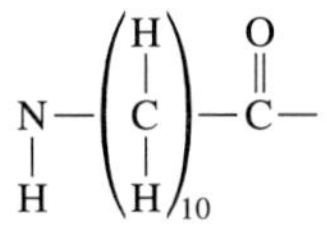	194	
Polyethylene terephthalate (polyester)	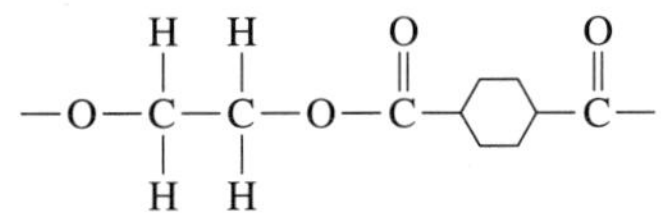	265	80
Polyoxymethylene	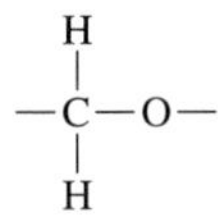	180	−50
Polycarbonate	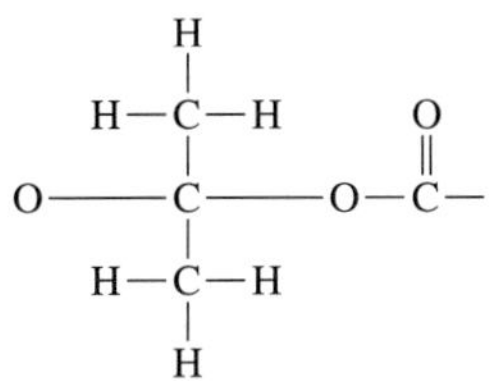	230	150
Polydimethylsiloxane (silicone)	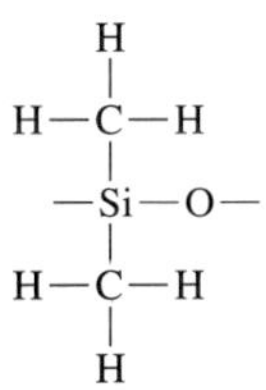	...	−123

In general, the nylons are tough, opaque solids with moderate melting points. Their outstanding characteristic is their ability to be drawn into fibers, becoming tough, transparent materials of relatively high strength. In this form they are crystalline and highly oriented. Good strength, toughness, and abrasion resistance characterize the 66 nylon, with mechanical properties being maintained up to 125°C. Low-temperature properties are maintained reasonably well. Solvent resistance is good. Moisture resistance is fair; moisture acts as a plasticizer in this material and thus its toughness depends to some extent on moisture content. While 66 nylon might be used for its good strength properties alone, it also has in its favor good fabricability, light weight, and self-lubricating bearing properties.

Table 18.4 Properties of thermoplastic polymers*

Polymer	Tensile Strength, MPa	Tensile Modulus, GPa	Elongation, %†	Izod Impact Strength, J/cm of Notch
Polyamides:				
66 Nylon	84	2.8–3.5	60–300	3.2–6.4
30% FRP‡	147–196	9.8	7	0.5–1.1
11 Nylon	56	0.7–1.4	300	1.1
Acetal:				
Polyoxymethylene	70	3.5	25–75	0.5–1.1
Polycarbonate	63–77	2.1–2.8	130	6.4–9.6

*Properties vary extensively with processing and compositional variations.
†Total elongation prior to fracture, much of which is elastic stretch.
‡Fiber-reinforced polyamide.

The mechanical properties of several nylons are shown in Table 18.4. The 6 nylon is used primarily in tire cord and general-purpose parts having lower strength requirements than 66. The 610, 8, 11, and 12 varieties are used for insulation and jacketing and, in some limited applications, as filaments.

The thermoplastic polyesters are used primarily as fibers and films, the most important of which is polyethylene terephthalate. In the as-spun form, these fibers are crystalline and relatively stiff, retaining good mechanical properties up to 150°C. It has good solvent resistance and low moisture absorption. In the form of film it is quite strong, with tensile strengths equaling 175 MPa, and with very high flex strength. The impact toughness is also high, which makes this material popular in a wide range of applications involving tape, e.g., magnetic recording tape.

Other significant engineering thermoplastics are the acetals, polyethers, and polycarbonates. The acetals, of which polyoxymethylene is the best example, are about 75% crystalline and have all the attractive properties of high-molecular-weight partially crystalline polymers. They have a relatively high melting point of 180°C, good stiffness, strength, toughness, wear, and abrasion resistance. As a result of this combination, they have been used as replacements for metals in a wide variety of applications such as automotive parts, pumps, pipe, and valves, in many places replacing cast iron, zinc, brass, and aluminum. From their structure (Table 18.3) it can be seen that the acetals are, in fact, a member of the polyethers in which the ether linkage is a methylene group. Other polyethers are used in pipes, pumps, and tubing, where they may also replace metals.

The polycarbonates are similar to the acetals in their general engineering usefulness. They are partly crystalline thermoplastics with good strength and outstanding toughness. They are readily fabricated and are transparent, an unusual optical property for polymers with a high degree of crystallinity. Typical applications include electrical parts, pump housings, structural parts, and applications utilizing their optical properties such as lamp globes. The acetals, polyethers, and polycarbonates are all polymers with excellent potential for engineering use and are replacing some of the more conventional polymers in some applications. Their price is generally higher than polypropylene, ABS, or the intermediate-priced nylon, and thus their widespread application is closely related to the economic picture.

A much smaller group of complex thermoplastics of which some mention should be made are the polyimides. These materials are relatively expensive, but because their structure is stiffened by complex chains made of rings connected at several points (heterocyclic polymers), they have very good high-temperature resistance. They retain moderate properties at 300°C for months and at 400°C for several hours.

Thermoplastic polymers include a number of natural fibers, e.g., wool, silk, and the cellulose fibers like cotton and rayon. Wool and silk are polypeptides made of amino acid complexes, while cotton and rayon are structures of linear carbon-chain linkages between complex carbon-oxygen rings. From the engineering standpoint, the cellulosic products rayon and cellulose acetate are the most important. Rayon is a regenerated cellulose product known as viscose or viscose rayon. This product is used in both the stretched and unstretched form, but the stretched form has the most applicability to engineering use because considerable orientation and crystallinity are introduced by this treatment. This fiber has seen extensive use in the past as tire cord. The same material is used in film form as cellophane, a coating material now generally replaced by other polymers. Cellulose acetate is another cellulose product used in sheet and molded forms. Its cost is low, but its strength is not high, and it has a low softening temperature and high moisture absorption as well.

A special, perhaps extreme, case of the heterochain polymers are the silicones. These materials have a relatively simple chain containing alternating oxygen and silicon atoms down the spine and thus contain no carbon. These materials are usually heavily cross-linked and are used in the low-molecular-weight form as dielectric fluids (liquids) or in higher-molecular-weight form as elastomers or polymer coatings.

18.5 THERMOSETTING POLYMERS

Thermosetting polymers, as described before, change under the influence of processing or fabrication to heavily covalent cross-linked three-dimensional networks. In some cases, heat is required to produce the cross linking, which occurs coincidentally with polymerization. In others, no heat is required, and it may even occur as a separate postpolymerization step. This is usually characteristic of the elastomers, which are to be considered separately, rather than the normal thermosetting resins.

The thermosetting resins that have been used for the longest time and are still widely used today are the phenolics. These resins are produced from phenol and formaldehyde and are outstanding in their heat resistance and dimensional stability, as well as moderate strength. They are often used extensively filled with a variety of both particulate and fibrous fillers, which serve to improve their heat resistance, electrical properties, or impact resistance. Without proper fillers, usually elastomers, their impact resistance is quite low. Their cost is low to intermediate, depending on

Table 18.5 Properties of thermosetting polymers*

Polymer	Tensile Strength, MPa	Tensile Modulus, GPa	Elongation, %	Izod Impact Strength, J/cm of Notch
Phenolic:				
Phenol-formaldehyde	35–63	5.6–9.1	1	0.3
Glass-filled	63–84	0.7–1.4	1–2	0.5–4.8
Rubber-filled	28–63	0.28–0.42		
Amino:				
Urea-formaldehyde	35–70	9.1–11.0	1	0.2
Melamine-formaldehyde	35–63	7.0–9.1		
Cellulose-filled	49–70	9.1–11.0	1	0.2
Polyester:	42–91	6.3–8.4	3	0.2
Glass-filled	35–70	10.5–17.5	1–5	0.5–8.5
Epoxy	63–105	2.8–6.3	4–5	0.5

*Properties vary extensively with processing and compositional variations.

the filler used. The properties of several filled phenolics are listed in Table 18.5. The phenolics are used not only as filled molding resins but also in alcoholic solutions for impregnating paper, wood, and other materials to produce countertops, wall coverings, and industrial laminates. They are used in bonding resins for adhesives and in varnishes. The molding resins include those designated by the trade name Bakelite.

A second important group of thermosetting polymers are the amino resins, of which the urea-formaldehyde and melamine-formaldehyde are the most important. Like the phenolics, they are produced as molding resins, laminating resins, and adhesives, as well as other more specialized products. The amino resins are clear or colorless and thus can be used in applications where light or pastel colors are required. Their strength and hardness are typically higher than the phenolics and their toughness lower. However, these materials are usually extensively filled, and properties depend heavily on the filler used. Properties of some filled amino resins are found in Table 18.5. Applications of molding resins include appliance housings, electrical parts, and housewares.

The polyesters, already discussed as thermoplastic polymers, may also fall into the thermosetting category when produced with extensive cross linking from double-bonded carbon atoms. The starting materials are a combination of a dibasic acid, a dihydric alcohol, and a monomer, usually styrene. They are produced as both laminating and molding resins. The strength of the polymer is intermediate and the toughness low in the untilled form, but it is improved by the use of fillers. Glass-fiber reinforcement is common and, as shown in Table 18.5, substantially improves both strength and toughness. Typical uses include such engineering applications as boat hulls, auto and truck body parts, structural shapes, sheet, and paneling. In these applications they are, for the most part, glass-fiber-filled.

The epoxy resins, actually related to the polyethers, are a group of tough, strong, flexible and chemical-resistant thermosets, which are used in the form of molding and laminating resins. They are moderately expensive, which limits their use, but are nonetheless popular for filled electrical and tooling parts, as surface coatings, and as adhesives.

Finally, a group of thermosetting materials used primarily as foams are the urethanes. Closely related to the elastomers (polyurethane-elastomers are also produced), these heavily cross-linked foams have been used as reinforcement in hollow structural shapes with a resulting saving of weight. Their resistance to oil and moisture makes them particularly suitable as foams for sandwich insulation and structural panels.

18.6 ELASTOMERIC POLYMERS

A significant group of polymers used in a number of engineering applications are the elastomers. These materials combine the flexibility of the thermoplastic polymers with, on a limited and controlled scale, the cross linking of the thermosets to provide a unique set of properties. The structure of the elastomers is more complex than this simple statement implies, as it also involves coiling and cross linking of the elastomer chains so that the remarkable elastic elongations characteristic of this group can be achieved. The mechanics of this elasticity are still not fully understood. Apparently the elastomers are made of cis unsaturated hydrocarbon chains, which tend to coil. The coils themselves contribute to elasticity, much as stretching of a spring under load permits extensive elongation. Vulcanization, a cross-linking process, is also necessary. Without vulcanization, the coils merely slide over one another without being extended, and soft thermoplastic results. The process of vulcanization cross-links the polymer-chain coils to the extent that stress on the bulk material forces uncoiling of some of the molecules rather than sliding. Carried too far, the cross linking binds the molecules tightly together, and extension of the molecular coils cannot occur; i.e., it becomes hard and stiff. By control of the vulcanization process, therefore, a variety of rubber products ranging from soft, weak rubber-gasket materials to hard and stiff molded products can be produced. Generally, vulcanization is accomplished using sulfur, the amount of which can be varied between as little as a few percent to as much as 30% in order to produce this effect.

Vulcanized rubbers are still not entirely satisfactory, and fillers must be used to adjust mechanical properties such as strength, wear resistance, abrasion resistance, and stiffness. The most common filler used to produce this effect is carbon black, and its role is a complex one in that it apparently also contributes to some interchain interactions to supplement the bonds produced by vulcanization. Other significant reinforcing agents are cords, fibers, and belts, often used in rubber tires, and other compounds added to improve oxidation resistance. Rubbers that have unsaturated carbon bonds tend to be attacked by oxygen, producing either softening or embrittlement (resulting from both cross linking and degradation).

Compounding is a specialized term in rubber technology that refers to additives, which are used primarily to make processing easier rather than improve properties. The most common compounding materials are oil extenders (adding oil as a plasticizer), activators, and accelerators added to hasten vulcanization. The process of vulcanization itself, called curing, must be carefully controlled not only for sulfur content and the mix used but also for duration. Apparently both cross linking and degradation occur simultaneously in vulcanization, and overcured rubbers undergo reversion; i.e., they soften and become weak much like those that have undergone oxidation attack.

The most common rubbers used today are those which are made from copolymers of styrene and butadiene, the SBR rubbers, followed by other rubbers containing butadiene and materials of more limited application, e.g., butyl rubber, nitrile, and neoprene. Polyisoprene, or natural rubber, is now only a modest fraction of the rubber market in either its naturally produced or synthetic form. The mechanical properties of these materials are found in Table 18.6.

The most widely used rubber, SBR, is very similar to natural rubber when properly reinforced and manufactured. Tensile strengths exceed 3500 lb/in^2, and wear properties are good. Weatherability is also good, but elevated-temperature resistance is not particularly attractive compared with polyisoprene. It is used for a wide variety of commercial products such as tires, molded goods, flooring, belts, hoses, and clothing. Other butadienes are used primarily in blends with other rubbers. Butyl rubber, or polyisobutylene, is a special-purpose rubber because it lacks the strength and wear resistance of SBR. It has very low gas permeability, however, and this makes it attractive for adhesives, caulking compounds, seals, and tire inner tubes. Nitrile rubbers are polymers of butadiene and acrylonitrile. At high acrylonitrile levels (40%) they are extremely oil-resistant and are used in applications requiring this property. Neoprene (polychloroprene) is noted for its oil resistance but also has otherwise good general properties and is used in a wide variety of services. It has high tensile strength without use of carbon black filler. It is somewhat higher-priced than other rubbers, which limits its applications.

Table 18.6 Properties of elastomers

Elastomer	Tensile Strength, MPa	Elongation, %	Use Temp., °C Upper	Use Temp., °C Lower	Oxidation Resistance	Gas Permeability	Solvent Resistance
SBR	26	550	100	−55	Fair	High	Poor
Butyl	21	400	120	−50	Good	Low	Poor
Polyisoprene	31	600	100	−60	Poor	High	Poor
Nitrile	17	550	140	−15	Good	Moderate	Good
Neoprene	28	800	100	−45	Good	Moderate	Good
Silicone	10	600	250	−90	Excellent	. . .	Excellent

PROBLEMS

1. Describe the polymers that might be recycled easily and explain why.
2. Describe the structural differences you would expect between LD and HD polyethylene. Why is HD polyethylene more crystalline than LD?
3. If it is economically feasible to replace inorganic glassware in hospital service with a polymer, what characteristics would be needed to compete successfully in this application? Which polymers would you recommend?
4. Arrange the following polymers (*a*) in the order of anticipated heat resistance, (*b*) in the order of increasing strength, and (*c*) in the order of increasing toughness: polyethylene, polyvinyl chloride, 66 nylon, polycarbonate, and polytetrafluoroethylene.
5. Polyethylene terephthalate is quenched as a thin film and has a density of 1.30 g/cm^3. It is annealed, and the density increases to 1.40 g/cm^3. Explain the phenomenon involved.
6. Based on its structure, which of the following polymers would have the greatest tendency to assume a crystalline state: polypropylene, polymethyl methacrylate, polytetrafluoroethylene, and polystyrene? What factors would influence this behavior?

REFERENCES

Seymour, R. B.: *Polymers for Engineering Applications,* ASM Int., Materials Park, OH, 1990.

Budinski, K. G., and M. R. Budinski: *Engineering Materials: Properties and Selection,* 6th ed., Prentice-Hall, Upper Saddle River, NJ, 1999.

Rosen, S. L.: *Fundamental Principles of Polymeric Materials,* Wiley, New York, 1982.

Strong, A. B.: *Plastics: Materials and Processing,* 2d ed., Prentice-Hall, Upper Saddle River, NJ, 2000.

Tadmor, Z., and C. G. Gogos: *Principles of Polymer Processing,* Wiley, New York, 1979.

Allcock, H. R., and F. W. Lampe: *Contemporary Polymer Chemistry,* 2d ed., Prentice-Hall, Upper Saddle River, NJ, 1990.

Rodriguez, F.: *Principles of Polymer Systems,* 2d ed., McGraw-Hill, New York, 1982.

Fried, J. R.: *Polymer Science and Technology,* Prentice-Hall, Upper Saddle River, NJ, 1995.

VanVlack, L. H.: *Elements of Materials Science and Engineering,* 6th ed., Prentice-Hall, Upper Saddle River, NJ, 1989.

Engineering Plastics, vol. 2, *Engineered Materials Handbook,* ASM Int., Materials Park, OH, 1997.

Harper, C. A.: *Handbook of Plastics, Elastomers, and Composites,* McGraw-Hill, New York, 1996.

chapter

19

Ceramics and Glasses

Ceramic materials are extensively used because of their high hardness, resistance to high temperature, water, and other chemicals, and their special optical, thermal, and electrical properties. Historically, ceramic products made of clay were among the first objects to be manufactured and used by man. Sun-dried clay brick was used in prehistoric and historic times and indeed is used today as building material in some parts of the world. Other early engineering materials, which could be considered ceramics, are the natural stone materials (marble, granite, sandstone, etc.) that have also been used for thousands of years as materials of construction. Six of the seven wonders of the ancient world were basically ceramic,[1] as are Stonehenge and the Great Wall of China. Many of the concrete structures built by the Romans 2000 years ago are still in use today as foundations for modern buildings and bridges.

In this chapter on ceramics, three groupings will be considered: the traditional ceramics such as clay products, the crystalline ceramics, and the glasses. The clay products, because of their more limited engineering use, will be considered briefly, using the example of aluminosilicates. The crystalline ceramics are used for abrasion and wear resistance as well as refractory service. Refractory brick materials, which could be considered with the clay products, will be added to the crystalline group for convenience. The glasses, in the form of pipe and containers, see extensive engineering service. The divisions or boundaries among these groups are somewhat arbitrary since there are many crossover points among the three. This fact should be recognized even though, in this chapter, they will be treated separately.

19.1 A CERAMIC PHASE DIAGRAM (Al_2O_3-SiO_2)

The popularity of the clay products comes from the ease with which they are formed. Clays consist mainly of hydrated aluminosilicates (Al_2O_3-SiO_2 compounds), alkaline-earth (Ba, Ca) oxides or compounds, alkali-metal (Na, K) compounds, and

[1] The Colossus of Rhodes is believed to have been made of bronze.

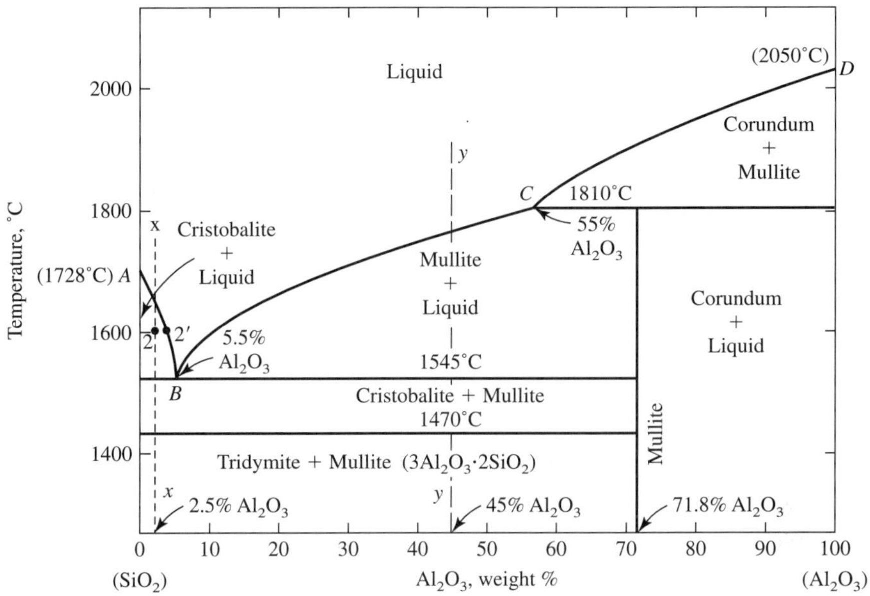

Figure 19.1 The Al_2O_3-SiO_2 phase diagram.

often some iron oxide. The properties of the clay products are sensitive to composition of the clay and also the particle size, shape, and size distribution of the starting materials. The melting points of SiO_2 and Al_2O_3 are high, 1728 and 2050°C, respectively, but the phase diagram between the two oxides (Fig. 19.1) has a low-melting-point eutectic (1545°C) at 5.5% Al_2O_3. The other constituents in most of the clays modify this diagram and allow some fusion at temperatures in the order of 900°C. Clays of various compositions permit the manufacture of the range of products referred to as brick, earthenware, stoneware, and porcelain.

Porcelain is the purest of these products, generally consisting of Al_2O_3 and SiO_2 in the composition range between SiO_2 and mullite (3 Al_2O_3-2 SiO_2), that is, about 20 to 50% Al_2O_3. Small amounts of other materials, generally alkali oxide, are added to adjust the fusion temperature.

19.2 TRADITIONAL CERAMICS: CLAY, REFRACTORIES, AND ABRASIVES

Clay

Fabrication of clay products requires the addition of water to make them plastic. The flaky clay particles can move more freely over one another when they are coated by the water molecules and tend to align parallel to each other. The water is distributed in the interstices between the particles and the surface pores of the particles themselves. Sufficient water must be provided to prevent the mixture from becoming weak and crumbly; this minimum amount of water is the plastic limit. More than the minimum water may be added, but an upper limit will be reached where more water renders the clay wet, weak, and sticky. This is called the water limit. The exact amount of water required is quite dependent on the type of clay and the surface conditions of the clay particles. A number of additives are blended into the clay to give it specific properties either in fabrication or in subsequent service.

When the clay has formed, it is dried and fired. During drying, the water in the interstices escapes, and the dried clay is stronger than before. The rate of drying is important. If dried too fast, the normal shrinkage accompanying the operation will cause cracking of the product. It is necessary, therefore, to control the rate such that the water is removed from the surface at about the same rate as it is able to diffuse to the surface through the interstices. Increased firing that follows drying removes the surface water first and then at higher temperatures, up to 600°C, the chemically bound water. As temperatures approach 900°C, oxidation reactions occur that remove organic materials in the clay. At 900°C, fusion, or vitrification, of the product begins. As the temperature is increased between 900 and 1400°C, vitrification may become complete, although this is rarely done because, while fully fused products are strong, during the process the product also becomes soft at the high temperature and collapses. More commonly, the product is only partially fused, resulting in unreacted Al_2O_3 and SiO_2 particles in a glassy matrix. When a nonporous, glazed surface is desired, a surface coating of lower melting-point ceramic composition is applied to the object after dry firing, and then on subsequent high firing, the glaze material becomes completely liquid, and upon cooling, a glass.

Earthenware and common brick are usually fired between 800 and 900°C, while face brick is fired at somewhat higher temperatures. Face brick is used for the external faces of buildings; since it requires lower porosity, higher compressive strength, and a denser structure than common brick, it is more extensively vitrified. Stoneware and porcelain are also highly vitrified, especially porcelain, which may be nearly translucent in thin sections.

Brick materials and earthenware are used in construction industries. The porcelain and stoneware products are used in chemical process industries for their chemical resistance, particularly to acids. In general, however, they are low in shock resistance, are attacked by alkalies, and have only modest tensile strength. These properties limit the service conditions under which they can be used. Like other ceramic materials, they have much better strength in compression than in tension.

Refractories

One of the most important uses of ceramic materials is as refractories. Some properties are shown in Table 19.1. Compared with metallic materials, the strong covalent and ionic bonds of the ceramics produce extremely stable compounds, and as a result very high temperatures are required to cause these materials to melt. The

Table 19.1 Properties of refractory oxides

Oxide	Melting Point, °C	Density, g/cm^3	Thermal Conductivity, W/m•°C at temperature, °C
Al_2O_3	2050	3.97	4 @ 1315
BeO	2550	3.03	29 @ 1000
MgO	2800	3.58	59 @ 1100
SiO_2	1728	2.65	
ThO_2	3300	9.69	3 @ 1000
ZrO_2	2677	5.56	3 @ 1315

most important characteristics of refractories are their temperature resistance, insulating ability, and resistance to slag (elevated temperature chemical attack).

Refractory materials may be obtained in many forms, but the most common form is as refractory brick. Fireclay bricks differ from others in that they generally are pure mixtures of Al_2O_3-SiO_2 that range from 20 to 40% Al_2O_3. The properties of the fireclay brick depend on their Al_2O_3 content and their temperature of firing, both of which vary. Examination of the Al_2O_3-SiO_2 phase diagram (Fig. 19.1) indicates the variations in constitution and related property variations, which may occur in these materials. Both fireclay and mullite ($3Al_2O_3$-$2SiO_2$) may be used as refractories. If compositions of pure mullite, about 72% Al_2O_3, are used, however, little or no glassy binding phase will be present. Instead, bonding will occur by sintering of the individual particles of Al_2O_3 and SiO_2 to form strong covalent compounds. This requires a correspondingly higher firing temperature than for materials, which contain more eutectic melting glassy-matrix phase. The glassy matrix is referred to in ceramics as the ceramic bond. Most refractories are fired at high enough temperatures to produce substantial amounts of ceramic bond, ensuring good cold strength and minimizing shrinkage and dimensional changes during elevated-temperature service. While ceramic bond improves the strength of the refractory when cold, it decreases its strength at elevated temperatures since it softens on heating to temperatures above about 1300°C. Thus fireclay brick will soften at 1300°C and above, while mullite will not. The presence of extensive ceramic bond in refractories allows for easier fabrication of the refractory but is not particularly helpful in terms of elevated-temperature service. Again, for high-temperature dimensional stability, well-fired brick with low porosity is best. However, porosity improves the insulating characteristics of the brick and thus has some advantages. For this reason, complete firing of refractories may not be desirable.

The chemical resistance of refractory brick may also be of significance. Refractories are often classified as being acidic, basic, or neutral. This terminology comes from the behavior of the oxides in aqueous solution. Its significance for elevated-temperature service of the refractory can be related to the susceptibility of the refractory to high-temperature, liquid-state chemical attack.

Abrasives

Many materials used for refractory service can also be used for their hardness and wear resistance. The material that has the highest hardness is carbon in the form of diamond, and both natural and synthetic diamonds are used for this application. Diamond dust is used as an abrasive in a slurry or, for a cutting tool or blade, in the form of fine chips mounted in a binder of lower hardness. Other hard materials include boron and metal carbides, aluminum oxide, silica, and soft oxides like MgO and Fe_2O_3. Table 19.2 lists the hardnesses of these abrasives. The characteristics of importance in an abrasive include its hardness, toughness, resistance to attrition, and friability. Since most of the materials in Table 19.2 have sufficient hardness to serve as abrasives, the other characteristics may make a significant difference in their performance. Abrasive grains deteriorate in service by loss of small chips or flakes, which dull the cutting edges and reduce their effectiveness. This is called *attrition*. The attrition is compensated for by fracture of the grains, producing new cutting surfaces. Friability may therefore be an important positive characteristic of an abrasive from the standpoint of mitigating the effects of attrition. Too much friability is undesirable, however, because it leads to low shock resistance and toughness, both properties useful in abrasives. In addition, a high melting point is useful in abrasives because local temperatures may be high and softening or rounding of a cutting tool or abrasive particles in a wheel reduces its effectiveness.

Abrasives are used in a variety of forms, and it is not possible in this brief chapter to discuss the details of this complex field. Abrasives used in grinding wheels, for example, are greatly influenced in their performance by the binder used. This may vary from a vitrified ceramic bond through a rubber binder to a baked waterglass one. Some binders tightly hold the abrasive grains in the wheel, while others are designed to wear away and allow worn abrasive particles to pull out, exposing fresh abrasive particles. The size and the speed and pressure of operation may also be significant factors in performance.

Finally, subtle surface and chemical effects influence abrasive cutting operations. Certain combinations of abrasives and metals work better than others that have similar hardness. Lubricants or coolants may also markedly change the effectiveness of some abrasive operations but have little influence on others. Current technology in abrasives has led to greater use of ceramic materials in the fused form not

Table 19.2 Hardnesses of abrasives

Abrasive	Vickers Hardness No.	Abrasive	Vickers Hardness No.
Diamond	8000	Al_2O_3	2800
BC	3700	WC	2400
SiC	3500	Quartz	1250
TIC	3200		

only in grinding applications but in machine tools such as lathe bits as well. In these applications, their inherently low toughness is overcome by improvements in the structure of the ceramics and in tool supports, and they are replacing metals in high-speed rough machining operations.

19.3 STRUCTURE AND PROPERTIES OF ENGINEERING CERAMICS

High-grade refractories are those which are made of pure or relatively pure oxides in which the ceramic bond is held to a minimum. Moreover, ceramic bond in the refractory often decreases its resistance to slags and fluxes, which further militates against its presence. Elimination of ceramic bond usually means that the fabrication of the oxide ceramic will be by slip casting or dry pressing or some method not requiring addition of other oxides or compounds. Alumina, silica, magnesia, zirconia, beryllia, and thoria are all used as pure oxide refractories. Of these, the latter three oxides are generally much more expensive than the former.

Refractories rich in MgO and CaO are basic and are attacked by slags rich in SiO_2. Conversely, SiO_2-rich refractories are acidic and attacked by basic slags. The mechanism of the attack is by the formation of low-melting-point eutectics. In the use of refractories in primary metal smelting and refining, where liquid slags are present, this may be the most important characteristic of the refractory. Moreover, in processing the metal, a specific type of slag may be required. For example, a high-oxygen-content basic (CaO-rich) slag is required in basic oxygen steelmaking because of the SiO_2 formed by oxidation of Si in the pig iron being refined. Under these circumstances a basic or neutral refractory must be used. Al_2O_3 is an amphoteric refractory, so that its action in slags depends on other constituents, lime or silica, present.

Refractories include metal carbides and carbon itself. While most carbides have high melting points, they do not resist oxidation and thus are not particularly suitable for elevated-temperature service. The most important high-temperature carbide refractories in use today are those of silicon, zirconium, and titanium. More limited use is made of those of boron. Silicon carbide has long been used in refractory service for brick, as crucibles, and as resistance-heating elements. This product is normally made with a binder of clay, silicon nitride, and silicon. It has lower softening points than the silicon carbide alone because the bond softens from 1200 to 1500°C. Self-bonded silicon carbide produced by sintering is the most refractory. Graphite and carbon can both be used as refractories. Both oxidize, however, especially above 800°C, so that for such service, protective coatings or atmospheres are required.

At ambient temperatures, carbon has the added advantage that it is inert and very easy to machine, a characteristic shared by few of the other refractories. A graphite product, pyrolytic graphite, is made by controlled deposition of carbon on a substrate from a gaseous atmosphere. This material has a highly oriented hexagonal structure with its basal plane parallel to the substrate surface. Its strength, which is high in directions parallel to the surface, is maintained to temperatures as high as 2400°C.

Coatings

The word ceramic is often used in broad terms to refer to all inorganic, nonmetallic materials. A field of expanding interest is the use of these hard, wear-resistant compounds (nitrides, carbides, oxides, etc.) as coatings, primarily for metalworking tools. Other inorganics of importance are molybdenum disulfide (a solid lubricant), gallium arsenide (a semiconductor), and cadmium sulfide (a photoelectric material), all of which are applied with techniques such as chemical vapor deposition (CVD) and physical vapor deposition (PVD).

Alumina (Al_2O_3) can be vapor-phase deposited to create a surface that is chemically inert, electrically resistant, and has a low thermal conductivity and a hardness of about HV 2100. Titanium carbide, like other carbides of the Group IV, V, and VI metals, produces an inert coating of exceptional hardness (average HV 3200) with a melting point of 3140°C. Its counterpart, titanium nitride, has a lower hardness (HV 2000) but is more resistant than the carbide to oxidation at high temperatures. Tungsten carbides (WC and W_2C) have the attraction of lower deposition temperatures but are much less widely used. Their brittleness results in a low bond strength that spalls in high-speed tool applications, although they exhibit extremely high wear resistance in low-load situations.

Plasma-assisted chemical vapor deposition (PACVD) is a process that can be used with hydrogen gas containing a small percentage of hydrocarbon at a pressure of about 0.5 Pa and an energized 13.56-MHz plasma to deposit hard carbon films. Now generally spoken of as diamondlike carbon coatings, two structures are generally distinguished as amorphous carbon (a-C) and hydrogenated amorphous carbon (a-C:H) films.

Both types of carbon films have mixed sp^3 (diamond) and sp^2 (graphite) bonds, and the ratio can be varied by changing the reactive gas composition. They have a unique set of properties: optically transparent, electrically insulating, inert to most chemicals, elastic, and hard, with a low coefficient of friction. One limitation is that they oxidize above 350°C. Another is the adhesion problem between coating and substrate, which can be alleviated by the use of graded interlayers, preferably titanium/titanium nitride/titanium carbonitride/titanium carbide/diamondlike coating.

Diamondlike coating applications are developing rapidly. Examples of use are as coatings for seals, punches, taps, low-friction applications, decorative coatings, surgical saws, and cams in Formula 1 motorcycle engines.

19.4 CHARACTERISTICS OF GLASS

Of the ceramic materials of importance to engineers, those commonly referred to as glasses represent one of the most widely used classes. As indicated in Chap. 18, glasses are actually a characteristic type of noncrystalline structure rather than a specific material, and both inorganic and organic materials can exist in the glassy state. Organic glasses were discussed in Chap. 18, and while the inorganic glasses are harder and have much higher melting points, they are similar to the organic glasses in many ways. The most commonly used general description of a glass is the one adopted by ASTM, namely, "an inorganic product of fusion, which has cooled to a rigid condition without crystallizing." This, of course, excludes the organic materials from the grouping of glasses. As indicated previously, glasses do not lack structure but lack the long-range, three-dimensional order characteristic of crystalline solids. Like the organic polymers, however, the inorganic glasses have a three-dimensional framework structure even though the order is only short-range. Like the organics also, the bonding in these inorganic glasses is covalent for the most part, producing a high-strength solid.

The difference between the crystalline and glassy structures for an inorganic glass is illustrated in Fig. 19.2. The structure is of B_2O_3, which can be more conveniently represented than SiO_2. The classic example of a silicate glass was shown in Fig. 1.14. In this comparable B_2O_3 structure made of triangular B-O building blocks rather than SiO_4 tetrahedra, a short-range or first-neighbor order exists, as shown in Fig. 19.2*a*. The boron-oxygen 2:3 ratio is maintained, but the position of the oxygens is such that no long-range order is present. Figure 19.2*b* illustrates a long-range characteristic of a crystalline glass having the same boron-oxygen ratio.

While most commercial glasses consist primarily of SiO_2, other components are almost always present. This creates no structural difficulties since short-range order places no severe requirements on the geometrical configurations within the structure. Common additions include B_2O_3, Na_2O, K_2O, CaO, Al_2O_3, and others. Not all these oxides form glassy structures, however, as only a limited number have the coordination and structural requirements necessary to form a three-dimensional covalent network.

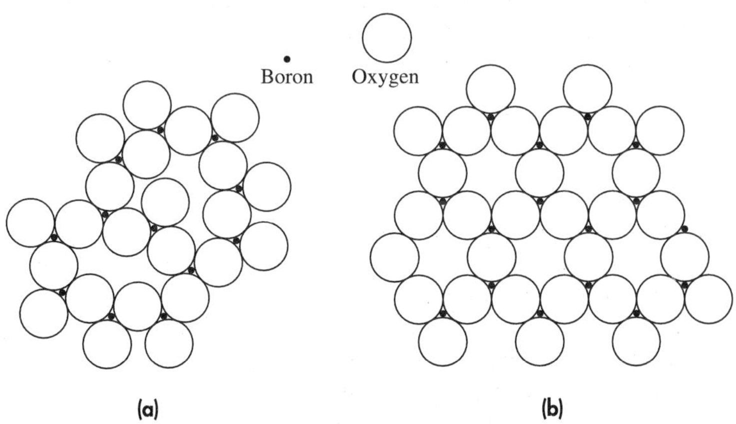

Figure 19.2 The structure of B_2O_3 in (*a*) the glassy state and (*b*) the crystalline state.

The oxides B_2O_3, GeO_2, P_2O, As_2O_5, and (of course) SiO_2 have a low enough coordination number for the oxygen and the metal ions to form the triangular or tetrahedral units that make a framework when they are bonded at three or more corners. These oxides are called *network formers*. The oxides Na_2O, K_2O, and CaO tend to polymerize the SiO_2 by introducing ionic bonds into the structure as a result of forming oxygen ion and cation pairs. The cations are not random in the structure but must occupy interstitial positions in the network of SiO_2. They are called *network modifiers*.

Some oxides such as Al_2O_3 can enter into the framework, replacing Si. Other oxides such as TiO_2, PbO, ZrO_2, CdO, and ZnO are marginal glass formers and are referred to as intermediates. The effect of the network modifiers and intermediates is significant. They may have a beneficial effect by decreasing the softening temperature of SiO_2 for easier fabrication, but they also allow crystallization of the glass.

As indicated for the organic glasses, when the inorganic glasses cool from the melt (liquid) glassmaking temperature down to the crystalline melting point, they may either become crystalline or remain in the liquid state. If they are rapidly cooled or the liquid is so viscous that the atoms cannot arrange themselves in the crystalline state, they continue to cool to a glass-transition temperature range called, in this case, the *fictive temperature* T_f. This is analogous to T_g in the organic polymers. The temperature is not a unique one but varies with cooling rate. Glasses may also undergo local densification and crystallization, called *devitrification,* on reheating or heat treatment in and above this temperature range after having been rapidly cooled through it. Most glasses resist devitrification.

Silica is an ideal glass except, as indicated before, for its high melting point, 1728°C, and the fact that it is too viscous to permit easy fabrication by molding or other glass-forming techniques. In normal low-strength commercial glass, this problem is overcome by the addition of Na_2O to SiO_2. The phase diagram of this binary system is shown in Fig. 19.3. As can be seen from this diagram, silica may exist in three crystalline forms: quartz, which is stable under equilibrium conditions to 870°C; tridymite, which is stable from 870 to 1470°C; and crystobalite, which is stable to the melting point.

Because of the sluggishness of the crystallization process, these forms can exist metastably for an almost indefinite time at low temperatures. Additions of Na_2O quickly decrease the melting temperature of the glass (and its viscosity) to the point where processing becomes feasible. The lower melting-point eutectic composition between SiO_2 and the compound Na_2O-$2SiO_2$ might be considered ideal. However, this composition, about 25% Na_2O, produces a glass that is soluble in water. This CaO must be added to the glass composition to produce water-insoluble soda-lime glass. This glass, with the approximate formula Na_2O-CaO-$6SiO_2$, is the basis for the most common glass today. Small amounts of Al_2O_3 and MgO may be added to this composition to improve resistance to chemical attack. Such glasses are low in cost and resist devitrification and thus are suitable for bottles, electric bulbs, glazing, and other general-purpose uses. In the absence of impurity oxides, soda-lime glass is colorless and transparent and is called flint glass. Iron oxide in the melt col-

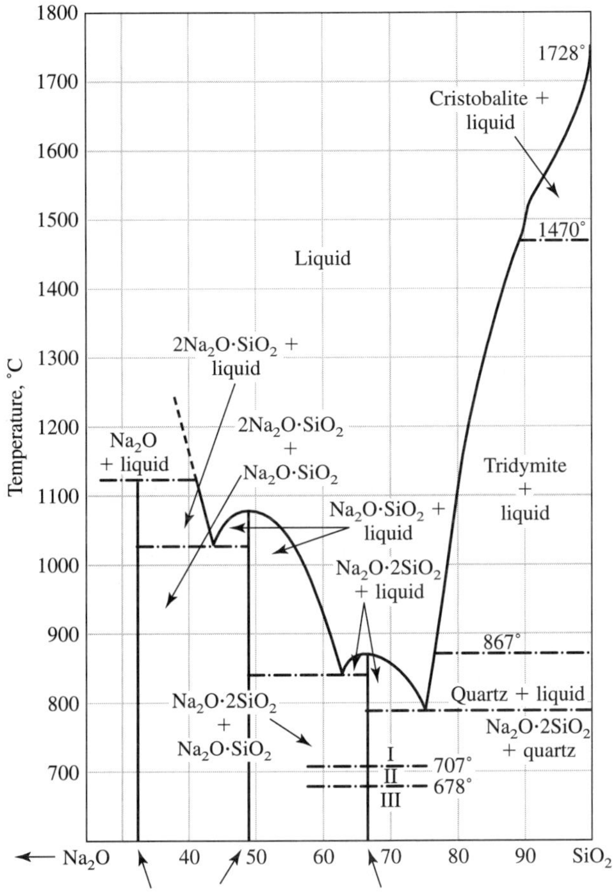

Figure 19.3 The Na_2O-SiO_2 phase diagram. Modest amounts of Na_2O substantially depress the melting point of SiO_2.

ors the glass green, and other oxides produce other colors; e.g., small amounts of cobalt oxide result in a blue glass.

Among the special-purpose glasses, the most significant from the engineering standpoint are those in which boron oxide and higher silica contents are used for resistance to chemical attack and thermal shock. These glasses, shown in Table 19.3, are the borosilicate and high-silica glasses. The borosilicate glasses have good chemical resistance and a low thermal coefficient of expansion, which together make them popular for piping, laboratory ware, and in some domestic service. This glass is better known by its trade name Pyrex. A similar product is aluminosilicate glass, which has higher heat resistance and the good shock resistance of the borosilicate glass.

High-silica glasses are produced by chemically treating the borosilicate glasses. The B_2O_3-Na_2O-SiO_2 glasses break down into immiscible liquids when heated to

Table 19.3 Compositions of commercial glasses

	Component, %							
Glass Type	**SiO_2**	**Na_2O**	**K_2O**	**CaO**	**PbO**	**B_2O_3**	**Al_2O_3**	**MgO**
Soda-lime	70–75	12–18	0–1	5–14	. . .	. . .	0.5–2.5	0–4
Lead	53–68	5–10	1–10	0–6	15–40	. . .	0–2	
Borosilicate	73–82	3–10	0.5–1	0–1	1–10	5–20	2–3	
Aluminosilicate	57	1	. . .	5.5	. . .	4	20.5	12
High-silica	96	. . .	. . .	. . .	. . .	3		

a high enough temperature but fortunately one at which the physical shape of the formed product is not destroyed. The formation of the two glasses permits a redistribution of the B_2O_3 such that a SiO_2-rich phase (96%) and a SiO_2-poor phase are produced. The low-silica phase containing B_2O_3 is removed by etching with hot acid. The remaining high-silica product is porous and translucent, but reheating to 1200°C results in a clear, highly shock-resistant glass with temperature resistance to 900°C.

The highest-silica products are, of course, fused silica, which contains over 99% SiO_2. The fusion temperature for this product is high, making it extremely difficult to produce. Because of its low temperature-expansion coefficient, it is highly resistant to thermal shock and temperature. Also it is mechanically quite strong; when it is clear, it is transparent to ultraviolet, visible, and infrared radiation. With respect to optical properties, there are two forms of silica. One, a transparent form, is known as fused quartz, while the other, a translucent form, is known as fused silica. The fused silica is chosen for thermal shock and chemical resistance. Fused quartz is used in applications requiring special optical properties. The fused quartz is more difficult to produce and therefore more expensive.

Other glasses are used for specific applications where one or two particular properties are needed. Lead glass, also listed in Table 19.3, has a high refractive index and thus a higher luster. Because of this optical property, it is used primarily for high-quality tableware. Engineering applications include radiation shielding and optical ware.

Processing of glass products requires a knowledge of the temperature ranges over which they can be formed and in which they can be annealed to remove or reduce residual stresses. Figure 19.4 illustrates these temperature ranges for the commonly used classes of glass. The fabrication of glass must take place at a high enough temperature for the material to flow, which in practical terms means the viscosity must be low enough to permit flow at reasonable stresses. This range of viscosity is considered to be between 10^4 and 10^8 P. The temperature range over which the various types of glasses described above go through these viscosities is considerable. As can be seen in Fig. 19.4, soda-lime glass passes through this range between

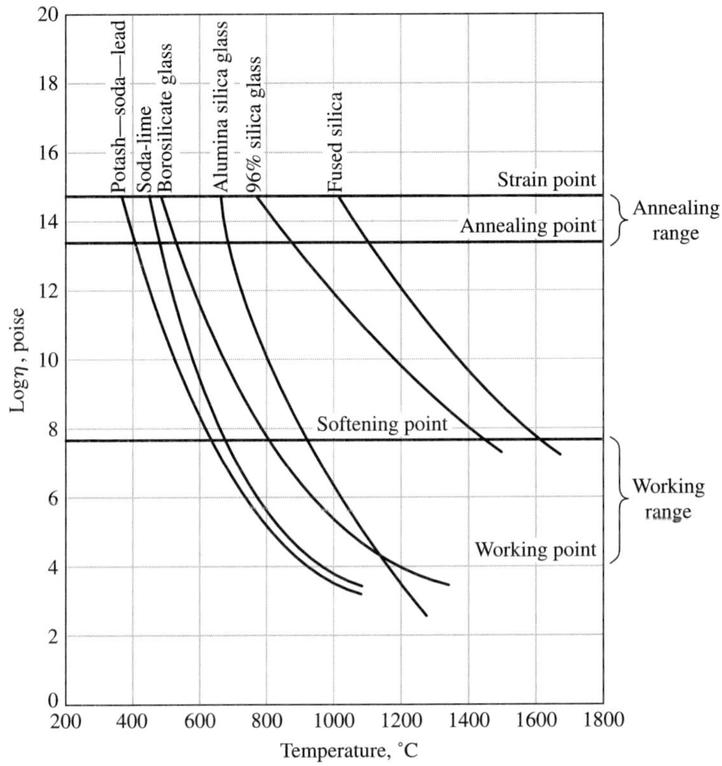

Figure 19.4 The viscosity-temperature curves for several standard glasses. The annealing and working ranges occur at substantially different temperatures for the various types of glasses.

1000 and 700°C temperatures, which allow for heating and forming well within the normal span for readily achievable technology. For borosilicate glass, these temperatures are between 1200 and 800°C. For the high-silica glass and fused silica, the corresponding temperatures are all above 1400°C, and thus processing of these glasses is difficult and expensive.

Most glass-forming operations leave some residual stresses in the glass after it cools, and the ambient temperature strength of the glass is adversely affected if they are not controlled. For products that have only modest engineering-strength requirements, sufficient control of stresses is readily achieved through subsequent low-temperature annealing or by controlled cooling rates after forming. For other products, such as optical glass, improved control is essential because stresses cause double refraction and low strength. This control is usually accomplished by a heat treatment in the range called the annealing range in Fig. 19.4. Within this range the glass has a sufficiently low viscosity to allow residual stresses to be relieved by viscous microflow, although subsequent cooling rates in and below this range must be slow to keep the product stress-free. In optical glass, the cooling rate near the fictive temperature (defined here as the temperature at which viscosity is $10^{14.6}$ P) must be around 1°C/h. Below this temperature, higher cooling rates can be employed to room temperature. The viscosity is then about 10^{20} P. For large parts or complex shapes, cooling rates may be slower. Perhaps the most unusual example of this type was the Mt. Polomar observatory telescope mirror of cast borosilicate glass (500 cm in diameter). It was cooled at a rate of 1°C/day in the temperature range between 500 and 300°C; that is, it required 200 days to cool through this range.

The residual stress pattern in glass, usually considered a disadvantage, may be an asset as well. Glass is sometimes given special cooling treatments to produce residual compressive stresses on the glass surface to increase its static and impact strength. This process is called tempering. The procedure is to heat the glass to the annealing range and then to chill the outer surfaces with an air blast. This causes the surface to contract and cool while the center is still viscous and can readjust to the stress. Later slow cooling of the center causes it to contract when the surface is now hard, producing compressive stresses in the surfaces of the glass and tensile stresses in the center of the cross section. In addition, the rapid cooling from above T_f causes the surface to have a greater final volume (lower density) than the center, which is more slowly cooled and in which more atomic rearrangement and a higher degree of long-range order can develop. The overall result of this effect is additional surface compressive stresses. Incidentally, the objective and result of such tempering are directly comparable to shot peening of metals (see Chap. 13).

The same result can be achieved in chemically strengthened glass by adding metallic ions to the glass surface, e.g., in a molten salt bath, to alter the density of the surface of the glass or change its thermal-expansion coefficient. In either case, the surface layer properties are adjusted so that after processing and cooling the surface layers are in compression. Again, a comparable process and result are the carburizing of low-carbon steel (Chap. 8).

The value of surface-compressive stresses in glass stems from the fact that essentially all fractures of glass originate from tensile stresses at Griffith-type microcracks at or near the surface. The presence of prior surface compressive stresses in two directions has two highly beneficial effects:

1. Substantially higher bending stresses and associated elastic deformation can be applied before the tensile stress at a Griffith flaw reaches the critical value.
2. When fracture occurs, it occurs in the directions of the two principal stresses, resulting in small block fragments, which are harmless, rather than long, sharp shards that are extremely dangerous.

Tempered glass is typically used for reasons of safety, e.g., automotive glass and home shower enclosures, and in these applications may be thinner than regular glass because of its higher strength. Since chemically strengthened glass is more expensive to produce, its uses are limited to shapes not adapted to the tempering process.

Finally, there has been increasing interest in, and use of, what appears to be a contradiction in terms, crystalline glasses. These materials are produced and fabricated in the amorphous form and are then rendered crystalline by thermal treatment. They contain a relatively high proportion of network modifiers, common additions to the SiO_2, which include Li_2O, Al_2O_3, MgO, and ZnO. These are usually nucleated by agents such as TiO_2, ZrO_2, CaF_2, or occasionally microscopic metal particles like Pt, Au, Ag, and Cu.

The glasses are fabricated to finished articles by normal techniques and then given a two-step heat treatment. The first step is to heat to a nucleation temperature T_n at which the nucleating agent initiates the formation of minute crystalline regions. The viscosity at T_n is between 10^{11} and 10^{12} P. The temperature is then raised at a rate of 5°C/min to the crystal growth temperature T_{cr}, which is 100°C below the crystalline T_m. The resulting product is microcrystalline with crystallite sizes from 0.01 to 1 micrometer in diameter dispersed in a glassy matrix at a concentration of from 1012 to 1015 nuclei per cubic centimeter. This glass is now opaque, much like the crystalline organic polymers, because of scattering at interfaces between the crystals and the matrix. These glasses are stronger and have greater impact resistance than the noncrystalline glasses. A commercial glass of this type is the material with the trade name Pyroceram.

PROBLEMS

1. Describe the sequence of physical events leading to the formation of compressive residual stress in the surface layers of tempered glass. How does modifying the thermal expansion coefficient of the surface layers by chemical addition (and in which direction, increase or decrease) accomplish the same purpose?

2. The ASTM definition of glass is "an inorganic product of fusion which has cooled to a rigid condition without crystallizing." How is this definition adequate or inadequate to describe glasses?

3. Figure 19.4 shows the temperature-viscosity curve for common glasses. Do the low annealing and working temperature ranges for lead glass help or hinder its manufacture and use? The strain point in this figure comes at a viscosity of about 10^{15} P. Why does it not extend below this level of viscosity?

4. When grinding a hard material, why does the hardness of the abrasive alone not control the effectiveness of the abrasive? What abrasive characteristics would be most useful in this application?

5. List in the order of greatest temperature resistance (softening temperature) from greatest to least the following: porcelain, soda-lime glass, fused quartz, mullite, and earthenware. Where would diamond fit in this listing and why?

6. Which refractory brick might be most suitable for service in a furnace designed for melting glass for blowing or molding?

REFERENCES

Mangonon, P. L.: *The Principles of Materials Science for Engineering Design,* Prentice-Hall, Upper Saddle River, NJ, 1999.

Ceramics and Glasses, vol. 4, *Engineered Materials Handbook,* ASM Int., Materials Park, OH, 1991.

Meyers, M. A., and K. K. Chawla: *Mechanical Behavior of Materials,* Prentice-Hall, Upper Saddle River, NJ, 1999.

Kingery, W. D., H. K. Bowen, and D. R. Uhlmann: *Introduction to Ceramics,* 2d ed., Wiley, New York, 1976.

Wachtman, J. B.: *Structural Ceramics,* Academic Press, New York, 1988.

Kalpakjian, S.: *Manufacturing Engineering and Technology,* 4th ed., Prentice-Hall, Upper Saddle River, NJ, 2001.

Reed, J. S.: *Introduction to the Principles of Ceramic Processing*, Wiley, New York, 1988.

Brinker, C. J., and G. W. Scherer: *Sol-Gel Science, The Physics and Chemistry of Sol-Gel Processing,* Academic Press, New York, 1990.

Jacobs, J. A., and T. F. Kilduff: *Engineering Materials Technology: Structures, Processing, Properties and Selection,* 4th ed., Prentice-Hall, Upper Saddle River, NJ, 2001.

chapter

20

Composite Materials

When two or more forms of materials are mixed to combine the best properties of each for a specific type of application, a new class of materials is formed, which we refer to as composite materials. In a strict sense, precipitation-hardened metal alloys may be thought of as intrinsic composites, since they have a metal matrix with dispersed, microscopic, intermetallic compound particles. We have maintained the forming properties of the ductile matrix while adding an additional strength factor from the brittle precipitates. The resultant material has therefore been improved, if the application requires both strength and ductility.

There are many attractive properties of composites that make them unique in the material selection process. Thermal expansion can be either very low or tailored for the design. Characteristics such as corrosion resistance, fatigue and fracture behavior, electrical conductivity and transparency, thermal insulation, wear, and flexural and torsional stiffness can often be adjusted.

By convention, the reinforcing constituent, or dispersed phase, of the composite is macroscopic, compared to the microscopic precipitates of an alloy. However, the concept of improving the final product by summing properties of each constituent (the principle of combined action) is the same. The *rule of mixtures* defines this relationship for many properties. As an example of the effect on mechanical properties, the modulus of elasticity of the composite is

$$E_C = f_M E_M + f_R E_R \qquad \text{mixture of moduli}$$

where E_C is the modulus of the composite, M and R refer to the matrix and reinforcing constituent, respectively, and f is the volume fraction. Likewise, although the bonding of constituent and matrix are major considerations, tensile strength of a composite may sometimes be approximated by

$$\sigma_C = f_M \sigma_M + f_R \sigma_R \qquad \text{mixture of tensile strengths}$$

where σ_M is the tensile strength of the matrix and σ_R is that of the dispersed phase. Physical properties often follow the rule of mixtures, provided the reinforcement is uniformly dispersed, continuous, and unidirectional. Examples include

$$\rho_C = f_M\rho_M + f_R\rho_R \qquad \text{mixture of densities}$$

$$K_C = f_M K_M + f_R K_R \qquad \text{mixture of thermal conductivities}$$

The summation properties of composites are therefore functions of the properties of both the matrix and dispersed phases and their fractional amounts. Equally important are the shape, size, distribution, and orientation of the reinforcing constituent. The most common classifications for composites are based on these factors.

20.1 FORMS AND PROPERTIES OF COMPOSITE REINFORCING MATERIALS

Composites may be classified by the description of their reinforcing phase into particle-reinforced, fiber-reinforced, or structural; the last class includes laminates and panels. The strength and stiffness of the reinforcing phase can vary significantly as shown for fibers in Figs. 20.1 and 20.2. Researchers have also found that by

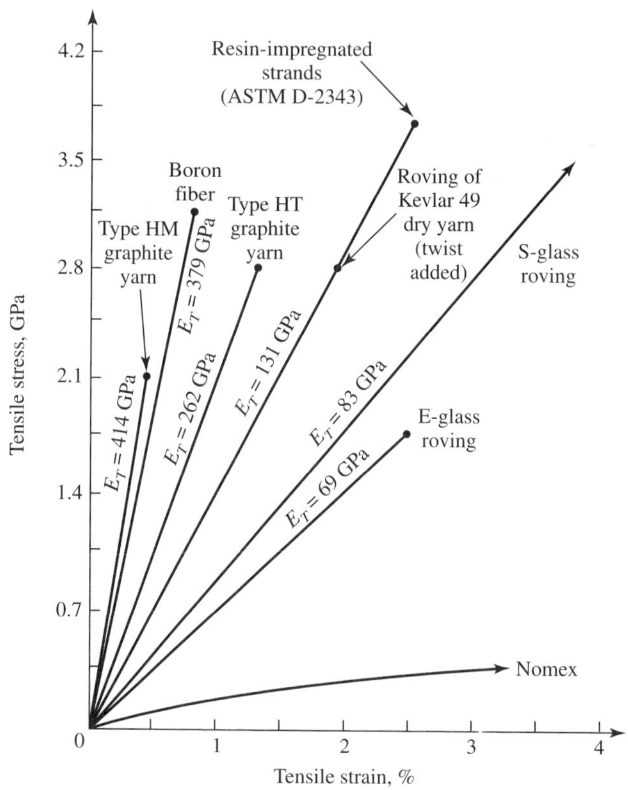

Figure 20.1 Stress-strain behavior of selected reinforcing fibers.

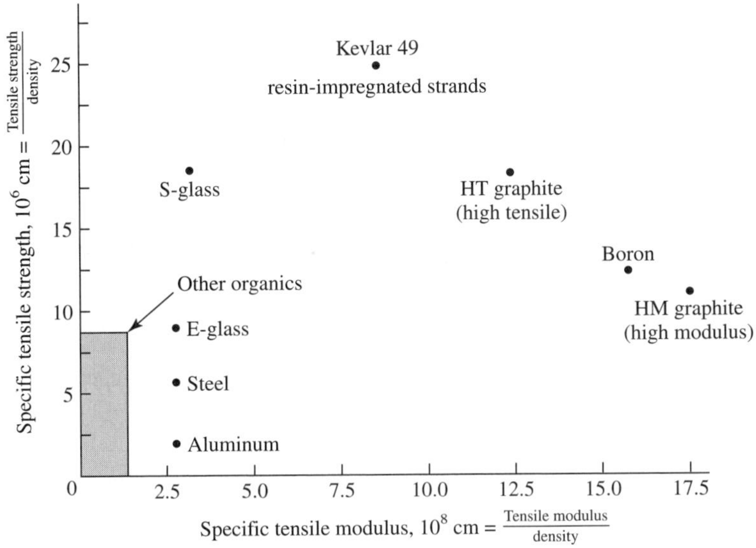

Figure 20.2 Specific tensile strength and modulus for selected reinforcing fibers.

increasing the volume fraction of fiber in a composite, its strength and stiffness increase. However, beyond a maximum volume fraction of about 80%, the fibers are no longer surrounded completely by the matrix. If the properties vary with direction, the material is termed anisotropic. There are a few specialized symmetries including orthotropic and cubic. Advanced polymeric composite structures can be severely anisotropic along different axes. Strength, stiffness, and coefficient of thermal expansion may vary by a factor of 10 in different directions.

Glass Fibers

Glass fiber is one of the most common classes of material for composite reinforcement, available in several forms. It owes its popularity to a combination of useful properties: chemical inertness, strength, flexibility, light weight, and ease of processing. Commercial, continuous glass fibers are made by the direct melt process of extruding molten glass through multiple orifices and rapidly drawing them to a fine diameter (3 to 20 μm). Single filaments are combined into strands after being "size" coated to resist abrasion.

Among the many types, E-glass is a general-purpose fiber composed of calcium aluminoborosilicate with a 2% maximum alkali content. Composition is typically 52 to 56% SiO_2, 16 to 25% CaO, 12 to 16% Al_2O_3, and 8 to 13% B_2O_3. It is selected for its strength and high electrical resistivity (see Table 20.1). S-glass attributes its higher strength to a magnesium aluminosilicate composition. S-2 uses a different coating, although it has similar composition. C-glass has the more typical

Table 20.1 Properties of selected reinforcing fibers

	Tensile Strength, MPa	Modulus of Elasticity, GPa	Coefficient of Thermal Expansion, 10^{-6}/K	Melting Temp., °C	Dielectric Constant at 20°C and 1 MHz
E-glass	3450	72	5.0	1720	6.3
A-glass	3040		8.6		6.9
S-glass	4600	87	5.6	1720	5.1
Alumina	2070	380		2015	
Silicon carbide	3930	485		2700	
Zirconia	2070	345		2677	
Boron	3450	380		2030	
Tungsten	4000	400		3410	

soda-lime borosilicate composition, known for its chemical stability, especially in acids. A-glass is the high-alkali form and ECR is a chemically resistant type of E-glass.

Tensile strength values of glass fibers are commonly given as the pristine single-filament strengths in air at room temperature. Sometimes the multifilament strand strength may be given, which could be as much as 20 to 40% lower than the pristine filament value because of surface damage during the strand-forming process. A significant loss of strength occurs in moisture, known as static fatigue. Decreases of 50 to 100% have been observed at room temperature in air with 50% relative humidity. Poisson's ratio does not vary greatly with composition or temperature in silicate glasses. It has a value of 0.22 ± 0.02 for E-glass.

Finished forms of glass fibers are typically continuous rovings, woven rovings, fiberglass mats, chopped strand, and textile yarns. Rovings are untwisted bundles of filaments. The continuous form results from bundling a collection of strands into a single large strand, then packaging it in cylinders. The process of chopping the roving and spray-up to a mold is commonly used for products such as bathtubs, fiberglass boats, and shower stalls. The woven roving is in fabric form with plain or twill weaves that can provide unique multidirectional strength. Mats contain nonwoven randomly oriented fibers that are produced in both continuous and chopped configurations for applications such as corrosion-resistant liners and surfacing. Chopped strand is used extensively as reinforcement in injection-molded products. Twisting and plying fine fibers produces strands of yarn with strength capable of being woven into fabric.

Additional forms of glass reinforcement are microspheres. They have exceptionally well controlled particle size, strength, and density. There are many types, including siliceous, ceramic, glass, polymeric, and mineral. Solid glass microspheres are fabricated from A-glass in sizes from 5 to 5000 μm, the size most often used in polymers being 30 μm. They may be specified with a coupling agent coating that enhances bonding and eliminates absorption of liquids into the separations around the spheres. There is also a hollow form of microsphere, with the primary use being reduction of weight in plastic systems.

Carbon Fibers

Carbon-reinforced composites are known for their combination of light weight, very high strength, and high modulus of elasticity. The high cost is restrictive to many industries, but its use has remained high for aerospace applications. To fabricate them, precursors such as rayon, polyacrylonitrile (PAN), or pitch, an aromatic compound, are often drawn into thin continuous fibers and oxidized. This is the first of three processing stages and is referred to as stabilization. The second stage, carbonization, pyrolizes the precursor material. This high-temperature process decomposes the organics and drives off all elements except carbon. Increasing the carbonization temperature from 1000 to 3000°C decreases the tensile strength but increases the modulus of elasticity. This third stage is termed graphitization.

There are carbon fibers and graphite fibers, although the terms are often used interchangeably. The difference is the temperature at which they are made. PAN-based carbon, produced at 1315°C (2400°F), has a composition of 93 to 95% carbon, whereas the higher-modulus graphite, produced at 3010°C (5450°F), is usually 99+% carbon. Graphite fibers are the stiffest fibers available, typically 1.5 to 2.0 times stiffer than steel. The fibers themselves are composites, since only a small part of the carbon is converted to graphite in the form of crystallite platelets. The higher the graphite content, the stiffer the fiber.

Aramid Fibers

Introduced in the 1970s as a replacement for steel in radial tires, aramid fibers have found a strong niche in the composite world. Aramid polymers are typically used in high-performance applications where resistance to fatigue, damage, and stress rupture is important. They owe their popularity to an excellent combination of strength, stiffness, and low density. Because of this unique combination of features, they have outstanding stiffness-to-weight and strength-to-weight ratios.

Chemically, aramids are aromatic polyamide polymers that receive their strength from a backbone containing benzene rings. Kevlar is the best-known example, produced by Dupont under the trade names Kevlar 29 and Kevlar 49. These fibers produce tough, impact-resistant, composite structures with half the stiffness of graphite. Both types of Kevlar have tensile strengths of about 2344 MPa and a strain to failure of 1.8%. The tensile modulus of Kevlar 49 is about twice that of Kevlar 29 and above that of titanium. One advantage is that their density is much less than that of glass and carbon fibers. They also resist flame, organic fuels, solvents, and lubricant but are attacked by acids and alkalies. In comparison to brittle glass and carbon, they are pliable and easily woven. A disadvantage of their use is that they are limited to low-temperature applications.

Metal and Ceramic Fibers

Metal fiber reinforcements have several advantages over other systems. In particular, they are relatively easy to produce, are less sensitive to surface damage, are extremely strong and resistant to high temperature, are inherently more ductile than ceramic fibers, and are easier to handle than glass. Steel wire in tires is still the major large-scale application.

The most commercially successful continuous metal fibers have been boron, silicon carbide, graphite, aluminum oxide, and tungsten. Borsic fibers used in borsic-aluminum composites are prepared by vapor deposition of boron onto thin (10-μm diameter) tungsten wires that are then coated with silicon carbide to reduce reactions with the aluminum matrix. Continuous silicon carbide fibers are produced by a two-step CVD process on a resistively heated carbon monofilament. Pyrolytic graphite (PG) 1 μm thick is first deposited on the carbon to smooth the substrate and enhance electrical conductivity. In the second step, the PG-coated substrate is exposed to silane and hydrogen gases, which decompose to form a beta-SiC continuous layer on the substrate. Graphite fibers for MMCs are prepared by the same carbonizing process as they are for polymer matrix composites, which are described above.

20.2 FORMS AND PROPERTIES OF COMPOSITE MATRIX MATERIALS

The term matrix refers to the nonfiber phase of the composite. Its function is simply to aid in the processing of the fibers and act as a transfer medium from the environment to the fibers. The matrix usually controls physical properties. Mechanically, it sustains only a small portion of the applied load. The materials cover the range from metals to polymers to ceramics. Polymer matrix materials are the most common and are characterized by low density, relatively low strengths, a nonlinear stress-strain relationship, and a high strain-to-failure. They can contain higher-volume fractions of reinforcement than the metal or ceramic matrix materials. Available as either thermosetting or thermoplastic compounds, the former has an advantage of higher temperature resistance and the latter the ability to be heated and formed repeatedly. Metal matrix composites may include aluminum, copper, nickel, magnesium, and intermetallic compounds.

We generally want the fibers and matrix to be securely bonded for successful transmission of the load from the matrix to the higher-stiffness fibers. An exception is the ceramic-reinforced ceramic matrix composite that relies on poor bonding to act as a crack-arresting system and to compensate for mismatches in coefficients of thermal expansion. To ensure a good bond, glass fibers are often sized with a silane-type agent to improve bonding and moisture resistance in fiberglass composites. Matching coefficients of expansion ensure that breaking of bonds is minimized as the fibers expand and contract at different rates than the matrix. Delamination, or debonding between layers, may be reduced with techniques such as three-dimensional weaving.

Polymer Matrix Resins

As already discussed, the reinforcing constituent contributes an improved strength and stiffness factor to the composite. The function of the matrix, on the other hand, is to add some impact resistance but mostly to serve as a medium to hold the reinforcement in position and transfer the applied load to the stronger phase. Since the matrix completely surrounds the reinforcement medium, it also protects it from abrasive damage and corrosion. For polymer matrix systems, both thermosetting and thermoplastic resins are used. Table 20.2 gives some perspective on the qualities of the thermoplastic materials.

There are five classes of thermosetting resins that serve the majority of the matrix applications for polymer matrix composites. They are the epoxies, bismaleimides, phenolics, polyesters, and polyimides. The final matrix or so-called neat resin system is typically a blend of ingredients combined for a desired set of properties. A formulation may involve several resins combined with curing agents, catalysts, fillers, and control agents, each contributing to the final matrix structure properties.

Epoxies provide a balance of properties including environmental and moisture resistance, fatigue resistance, interlaminar shear strength, and ease of processing. Their properties remain stable over a broad temperature range and their strength is superior to polyesters and phenolics. Bismaleimides are used in applications that require temperature performance between the epoxies and the polyimides. They have epoxylike processing and a similar balance of properties but tend to be more brittle. Phenolics have high heat and chemical resistance, good dielectric properties, high surface hardness, and dimensional stability. Volatile by-products emitted during curing limit their use in some composite applications. Alkyds and unsaturated polyesters are the most common polyester resins, having good arc-track resistance,

Table 20.2 Properties of selected thermoplastic resins

Types of Resin	Properties
Nylon	Good dimensional stability, high heat distortion, good tensile and flexural strength
Vinyl	Superior electrical properties, excellent moisture and chemical resistance, self-extinguishing
Fluorocarbon	Very high heat and chemical resistance, nonburning, high dimensional stability, lowest coefficient of friction
Polystyrene	Low cost, good stiffness, good impact strength, moderate heat distortion
Polyethylene	Light weight, low cost, can be welded, good flexibility, good toughness
Acrylic	Optical clarity, good gloss, weather resistance, excellent electrical properties
Polycarbonate	High mechanical properties, self-extinguishing, high dielectric strength
Polysulfone	Good electrical properties at high temperatures, good transparency, high mechanical properties, heat resistance, can be electroplated

dimensional stability, and outstanding dielectric properties. They yield moderate composite mechanical properties, have poor chemical and hydrolytic resistance, and have a higher shrink factor than other thermosetting resins. The incentives to their use are the ability to cure quickly and a lower cost. The commercial success of polyimides can be attributed to a good balance of thermooxidative stability and a high glass transition temperature. Like the phenolics, their use is limited because of the high levels of volatiles released during cure. They are being used in applications requiring good performance at temperatures (110 to 190°C) above the capabilities of the bismaleimides.

Prepregs

Prepreg is the term used by the composite industry for a system of continuous fiber reinforcement in a partially polymerized matrix. The most common reinforcements are carbon, glass, and aramid fibers. Both thermoplastic and thermosetting resins may be used as the matrix. Prepregs are typically supplied in one of four forms: unidirectional fiber tapes, woven fabrics, roving, or chopped mat.

Woven fabrics provide more uniform composite properties than unidirectional fiber tapes at a cost of lower strength. However, they are used extensively because of the ease of fabrication and two-dimensional strength. Chopped mat provides lower-cost structures with uniform strength characteristics. Composite strength properties are generally lower than those of woven fabrics.

Once the prepreg system is selected for a given selection, the user begins with either a hand or mechanically assisted layup of the prepreg onto a shaped surface. Curing with simultaneous heat and pressure completes the process. Typical thicknesses of the prepreg sheets range between 0.1 and 0.25 mm; widths vary from 2.5 to 152.5 cm. Since the curing process continues at room temperature, the prepreg must be stored below 0°C.

20.3 METAL MATRIX COMPOSITES

The obvious advantage of a metallic material over a polymer or ceramic is the combination of high temperature resistance, moderately high strength, and good ductility. Additional benefits when the material is reinforced to become a composite include (1) increased creep strength, which is important in applications such as turbine blades, (2) a decrease in the coefficient of thermal expansion, which reduces thermal strains, (3) increased strength and modulus per unit density, thereby reducing structural weight, and (4) increased wear resistance. Some mechanical properties of metal matrix composites (MMCs) are listed in Table 20.3.

In addition to aerospace applications, metal matrix composites have found a wide field of applications in areas such as the nuclear, biotechnology, heavy machinery, sporting goods, and automotive industries. For example, the usefulness of carbides and nitrides in pure form for wear resistance is limited because of the brittle nature at ordinary temperatures. However, when they are dispersed in a metallic

Table 20.3 Tensile properties of selected metal matrix composites

Material	Fabrication Method/Form	Tensile Strength, MPa	Modulus of Elasticity, GPa	Elongation, %
Al-Cu	Squeeze cast	261	70.5	14.0
Al-Cu + Al_2O_3	Squeeze cast	374	95.4	2.2
Al-Cu-Mg (2124-T6)	Powder-rolled plate	474	73.1	8.0
Al-Cu-Mg (2124-T6) + SiC ($V_f = 0.17$, 3 μm)	Powder-rolled plate	590	99.6	4.0
Al-Cu-Mg (2124-T6) + SiC ($V_f = 0.10$, 10 μm)	Spray-rolled sheet	343	91.9	3.8
Al-Li-Cu-Mg (8090-T6)	Spray-formed plate	505	79.5	6.5
Al-Li-Cu-Mg (8090-T6) + SiC ($V_f = 0.17$, 3 μm)	Spray-formed plate	550	104.5	2.0

matrix, they significantly increase the wear resistance of the component. About 40% of the metal cutting tools used in Japan are TiC-based cermets of the type TiC-Mo_2C-Ni/Mo (or Ti, Mo) C-Ni/Mo. They display longer life, lower wear, and increased cutting speed. Electronic connectors, storage battery plates, and superconducting wires use MMCs extensively. Superior damping properties can be realized using silicon and TiC particulate reinforced aluminum composites. Light bicycles, sailing masts, and golf club heads use systems such as SiC-reinforced aluminum.

Functional gradient materials (FGM) are composites designed with a compositional gradient to withstand thermomechanical loading by having a heat-resistant antioxidation surface layer with a mechanically tough material on the inside. A transition in the material properties such as modulus, coefficient of expansion, and thermal conductivity is therefore designed through the thickness. An additional advantage is that thermal relaxation occurs throughout the material.

Continuous Fiber-Reinforced MMCs

Boron was the first high-strength, high-modulus reinforcing fiber to be used in metal matrix composites. Moderate success was achieved with a boron-reinforced titanium matrix composite. However, without surface coatings and diffusion barriers, its strength and stiffness degrades in the severe processing environments. Today the most useful commercial systems of continuous fiber-reinforced MMCs have been the borsic-aluminum, SiC-aluminum, and graphite-aluminum composites. Some advanced composites include titanium alloys reinforced with silicon carbide fibers as shown in Micro. 20.1. All of these provide exceptionally high strength and modulus levels. A boron-reinforced MMC offers useful mechanical properties up to 510°C, whereas an equivalent boron-epoxy system is limited to 190°C.

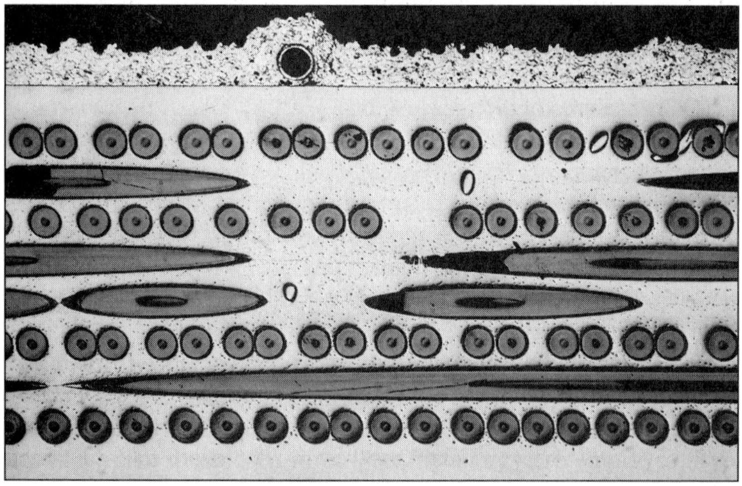

Micrograph 20.1 Metal matrix composite of Ti-6Al-4V with SiC reinforcing fibers.

Discontinuous Fiber-Reinforced MMCs

The majority of today's metal matrix composites are discontinuously reinforced with particulate or whisker forms of Al_2O_3, SiC, TiC, or B_4C in aluminum, magnesium, and other matrix alloys. Discontinuous fibers are often described by their aspect ratio of length to diameter; the strength improves when the ratio is high. Typical fiber diameters are 10 to 150 micrometers. An example of the effectiveness of this type of reinforcement is the significant increase in the modulus (50%) of aluminum that can be achieved by the addition of a high-volume fraction of carbides. The costs of discontinuous fiber-reinforcement materials, processing, and machining are considerably less than for continuous fiber composites.

The most common method for fabricating metal-ceramic and metal-metal composites is powder metallurgy. A prealloyed powder is blended with the reinforcement particulate or whiskers, cold isostatically pressed, thoroughly outgassed, then forged or extruded. Most work is carried out below the matrix solidus temperature, but it is sometimes necessary to maintain the temperature above the solidus to minimize stresses and avoid damage to the reinforcing medium.

20.4 POLYMER MATRIX COMPOSITES

The most widely used composite materials have always been the polymer matrix systems because of their relative ease of processing, low density, chemical resistance, and good mechanical and dielectric properties. Today, more than 80% of all reinforced PMCs have a thermoset matrix. High-temperature thermoset matrix composites use the polyimides and bismaleimides as described above. Major applications include aerospace and electrical industries in helicopter firewalls, printed wiring boards, and aerospace wing skin ribs. These composites may generally be used for long durations up to 315°C or for short durations up to 480°C.

Phenolic resins are considered for medium-temperature use, with properties remaining stable to 200°C. The two main types of resins are termed resole and novolac, the latter being a two-stage general-purpose molding compound. For high-strength requirements at medium temperatures, epoxies are an option with a usable range of 230 to 260°C. Phenolic-chopped fiber systems have found their niche mostly in the automotive, electrical, and appliance market. Glass-fiber-reinforced PMCs are provided in prepregs for layups and tape wrapping.

The composite material system most often specified is a polyester thermosetting resin matrix reinforced with E-glass. It is the most versatile, most economical composite, with good electrical properties and exceptional resistance to acids. Its usefulness is limited to low-temperature applications below 125°C. More technically demanding applications specify thermoplastic matrices such as polycarbonate resins. Their use, however, as a composite matrix material is limited, but there is still a high interest. Table 20.4 clearly indicates the influence of the reinforcement on mechanical properties.

Table 20.4 Properties of selected thermoplastic polymer matrix composites

Resin Type	Tensile Strength, MPa	Flexural Modulus, GPa	Impact Strength, J/cm (Notched/ Unnotched)	Heat-Deflection Temperature, °C
Amorphous resins:				
Nylon				
Base resin	66	2.6	>21/no break	125
30% glass fiber	148	7.9	0.64/3.7	140
30% carbon fiber	207	15.2	0.64/4.3	145
Polycarbonate				
Base resin	62	2.3	1.40/>21	130
30% glass fiber	128	8.3	2.00/9.34	150
30% carbon fiber	165	13.1	0.96/5.34	150
Crystalline resins:				
Nylon 66				
Base resin	80	2.8	0.48/>21	77
30% glass fiber	179	9.0	1.5/11	255
30% carbon fiber	241	20.0	0.80/6.4	257
Polybutylene terephthalate (PBT)				
Base resin	59	2.3	0.48/>21	85
30% glass fiber	134	9.7	1.4/9.1	210
30% carbon fiber	152	15.9	0.64/3.5	210

20.5 CERAMIC MATRIX COMPOSITES

Desirable properties of ceramic materials include their high melting points, high compressive strength, good strength retention at high temperatures, and excellent resistance to oxidation. Early progress in ceramic matrix composite development was hampered by the inherent brittleness of the materials, shrinkage associated with sintering, reactivity, and the generation of internal stresses due to thermal mismatch. However, there has been a maintained interest in their success, since their exceptionally high temperature resistance and strength capabilities make them the only economically possible material selection for structural applications at 1650°C and higher.

It is convenient to classify ceramic matrix composites by their type of reinforcement into three classes: metal fibers and wires, carbon (including graphite) fibers, and ceramic fibers. Suitable metals are limited by requirements of chemical compatibility, coefficients of thermal expansion, and selected mechanical properties. In addition, they must be able to retain their integrity throughout the high-temperature fabrication process. In contrast to polymer matrix composites that commonly have 70 vol % or more reinforcement, ceramic matrix composites (CMCs) are limited to less than 50 vol % before properties begin to degrade.

Most metal-reinforced ceramics use refractory elements including molybdenum, tungsten, tantalum, and niobium. These choices have two inherent problems: susceptibility to oxidation and very high density. Therefore, the use of metal-reinforced ceramic matrix composites (CMCs) has been restricted to special short-term applications in the range of 500 to 800°C. Likewise, carbon fibers have the same disadvantage of rapid degradation when exposed to air at temperatures in excess of 400°C. Unlike the refractory metals, carbon fibers have low density, chemical inertness, high strength and moduli, along with better availability. Their combination of high-temperature resistance, high strength, and low density makes them attractive for applications in the absence of air. Attention has understandably been focused on oxidation-protective coatings for carbon and on ceramic fibers. Increasing availability of ceramic fibers such as silicon carbide and alumina has improved the possibilities of producing high-temperature ceramic-fiber-reinforced ceramic matrix composite systems. Although there has been limited work with true ceramic matrix materials, some success has been obtained with no loss of strength to temperatures above 900°C.

20.6 CARBON AND GRAPHITE COMPOSITES

For applications involving temperatures up to 2200°C, carbon and graphite are indeed superior individual materials. However, combining carbon or graphite fibers in either a carbon or graphite matrix results in a class of materials even more advanced than the constituents. They are known as carbon-carbon composites (CCCs) and they have unusual property characteristics that cannot be realized in other material systems. Although not very impressive at ambient temperatures, they can retain their usable properties and remain dimensionally stable at temperatures from room temperature to an impressive 2760°C. The combination of high-temperature resistance, light weight, and stiffness makes them excellent candidates for solid rocket motors, missiles, and space vehicles.

Carbon-carbon composites are fabricated by multiple impregnation of porous "all-carbon" configurations with a liquid carbonizable precursor such as pitch. The system is subsequently pyrolized in the same manner as carbon fibers are made. A second method applies the pyrolized carbon through a chemical vapor infiltration (CVI) process, using gaseous hydrocarbons such as methane, propane/propylene, and benzene. A major problem of CVI is uniform deposition through thick sections, which may often be overcome by applying a pressure gradient across the section.

In comparison to polymer matrix composites, the strain-to-failure of the matrix in a CCC is usually lower than the fiber reinforcement and is frequently microcracked from the processing stresses. Therefore, the carbon matrix fails first. One would expect the fatigue of carbon-carbon composites to be a problem because of this extensive matrix cracking. However, these materials have been shown to be as proportionally good as resin composites. Another variation from polymer matrix materials is that, since the moduli of the fiber and matrix may be similar, the load is more evenly shared between the two constituents than in PMCs.

Continuous fiber reinforcement may be unidirectional or multidirectional. Both are biocompatible and can be tailored to be structurally compatible with bone for applications such as fixation of fractures and hip joint replacement. There are also special oxidation-resistant and structurally reinforced grades for friction materials in aircraft and vehicular brake systems, hot pressing molds, thermal protective panels, and hot structural parts in space systems and advanced gas-turbine engines.

PROBLEMS

1. What design considerations might have to be considered in using a metal matrix composite in a temperature range where either the matrix or the reinforcement undergoes a phase transformation such as age hardening?
2. What types of corrosion mechanisms do you think would be most likely to occur or most detrimental in a metal-fiber-reinforced polymer matrix composite? In what environments should this class of composite be avoided?
3. Based on your general knowledge of materials, prepare a list of several composite material systems (*a*) in order of increasing heat resistance, (*b*) in order of increasing strength, and (*c*) in order of increasing relative cost. Consider several possible constituent material combinations.
4. Compare and contrast the fracture modes of the matrix and reinforcing media for both polymer matrix composites and carbon-carbon composites.
5. List two attractive characteristics and also two undesirable features when using discontinuous, random-oriented fiber reinforcement. Do the same for continuous, oriented fiber reinforcement.

REFERENCES

Schwartz, M. M.: *Composite Materials,* vol. 1, *Properties, Non-Destructive Evaluation and Repair,* Prentice-Hall, Upper Saddle River, NJ, 1997.

Engineered Materials Handbook, vol. 1, *Composites,* ASM Int., Materials Park, OH, 1987.

Herakovich, C. T.: *Mechanics of Fibrous Composites,* Wiley, New York, 1998.

Warner, S. B.: *Fiber Science,* Prentice-Hall, Upper Saddle River, NJ, 1995.

Callister, W. D. Jr.: *Materials Science and Engineering: An Introduction,* 5th ed., Wiley, New York, 2000.

Harper, C. A.: *Handbook of Plastics, Elastomers, and Composites,* McGraw-Hill, New York, 1996.

Suresh, S., A. Mortensen, and A. Needleman: *Fundamentals of Metal Matrix Composites,* Butterworth-Heinemann, Stoneham, MA, 1993.

Weeton, J. W., D. M. Peters, and K. L. Thomas: *Engineer's Guide to Composite Materials,* ASM Int., Materials Park, OH, 1987.

INDEX

Note: Page numbers followed by "*n*" indicate footnotes.

D

Q

R

S

T

U